D0862858

BEHAVIOURAL ECOLOGY
An Evolutionary Approach

BEHAVIOURAL ECOLOGY
An Evolutionary Approach

SECOND EDITION

EDITED BY

J. R. KREBS
Lecturer in Zoology at the
Edward Grey Institute of Field Ornithology,
Department of Zoology,
University of Oxford; and
E. P. Abraham Fellow of Pembroke College

AND

N. B. DAVIES
Demonstrator in Zoology at the
University of Cambridge; and
Fellow of Pembroke College

BLACKWELL SCIENTIFIC PUBLICATIONS

OXFORD LONDON EDINBURGH
BOSTON PALO ALTO MELBOURNE

© 1978, 1984 by
Blackwell Scientific Publications
Editorial offices:
Osney Mead, Oxford OX2 0EL
8 John Street, London WC1N 2ES
23 Ainslie Place, Edinburgh EH3 6AJ
52 Beacon Street, Boston
    Massachusetts 02108, USA
667 Lytton Avenue, Palo Alto
    California 94301, USA
107 Barry Street, Carlton
    Victoria 3053, Australia

First published 1978
Second edition 1984
Reprinted 1986

Set by Santype International Ltd,
Salisbury and printed and bound by
R. J. Acford Ltd, Chichester, Sussex

Distributed in the USA and Canada
by Sinauer Associates Inc,
Publishers, Sunderland,
Massachusetts 01375

British Library
Cataloguing in Publication Data

Behavioural ecology.—2nd ed.
    1. Animals, Habits and behavior of
    2. Animal ecology   3. Evolution
    I. Krebs, J. R.   II. Davies, N. B.
    591.51   QL751
ISBN 0-632-00987-X
ISBN 0-632-00998-5 Pbk

Contents

List of Authors

JACK W. BRADBURY *Department of Biology, University of California, San Diego, La Jolla, California 92093, U.S.A.*

H. JANE BROCKMANN *Department of Zoology, University of Florida, Gainesville, Florida 32611, U.S.A.*

THOMAS CARACO *Department of Biology, University of Rochester, Rochester, New York 14627, U.S.A.*

ERIC L. CHARNOV *Department of Biology, University of Utah, Salt Lake City, Utah, U.S.A.*

TIMOTHY H. CLUTTON-BROCK *Large Animal Research Group, Department of Zoology, University of Cambridge, Downing Street, Cambridge CB2 3EJ, U.K.*

NICHOLAS B. DAVIES *Department of Zoology, University of Cambridge, Downing Street, Cambridge CB2 3EJ, U.K.*

RICHARD DAWKINS *Animal Behaviour Research Group, Department of Zoology, University of Oxford, South Parks Road, Oxford OX1 3PS, U.K.*

STEPHEN T. EMLEN *Section of Neurobiology and Behavior, Division of Biological Sciences, Cornell University, Ithaca, New York 14853, U.S.A.*

ALAN GRAFEN *Animal Behaviour Research Group, Department of Zoology, University of Oxford, South Parks Road, Oxford OX1 3PS, U.K.*

TIMOTHY R. HALLIDAY *Department of Biology, The Open University, Milton Keynes MK7 6AA, U.K.*

PAUL H. HARVEY *School of Biological Sciences, University of Sussex, Falmer, Brighton BN1 9QG, U.K.*

HENRY S. HORN *Department of Biology, Princeton University, Princeton, New Jersey 08540, U.S.A.*

ALASDAIR I. HOUSTON *Department of Zoology, University of Cambridge, Downing Street, Cambridge CB2 3EJ, U.K.*

JOHN R. KREBS *Edward Grey Institute, Department of Zoology, University of Oxford, South Parks Road, Oxford OX1 3PS, U.K.*

JOHN MAYNARD SMITH *School of Biological Sciences, University of Sussex, Falmer, Brighton BN1 9QG, U.K.*

ROBIN H. McCLEERY *Edward Grey Institute, Department of Zoology, University of Oxford, South Parks Road, Oxford OX1 3PS, U.K.*

GEOFFREY A. PARKER *Department of Zoology, University of Liverpool, Liverpool L69 3BX, U.K.*

LINDA PARTRIDGE *Department of Zoology, University of Edinburgh, West Mains Road, Edinburgh EH9 3JT, U.K.*

H. RONALD PULLIAM *Department of Biological Sciences, State University of New York, Albany, New York 12222, U.S.A*

DANIEL I. RUBENSTEIN *Department of Biology, Princeton University, Princeton, New Jersey 08540, U.S.A.*

SARA J. SHETTLEWORTH *Department of Psychology, University of Toronto, Toronto, Ontario M5S 1A1, Canada*

SANDRA L. VEHRENCAMP *Department of Biology, University of California, San Diego, La Jolla, California 92093, U.S.A.*

Preface

Although based on our previous edited book *Behavioural Ecology: an evolutionary approach*, this volume contains a completely new collection of chapters. The overall theme is still that of how behaviour is influenced by natural selection in relation to ecological conditions. As before, the emphasis is on quantitative and experimental methods and on integrating theoretical ideas with field and laboratory evidence. Some chapters cover fields not dealt with in the earlier book (for example learning, Chapter 7) and chapters which consider the same topics as previously have all been completely rewritten or extensively revised.

In the five years since *Behavioural Ecology* appeared the subject has changed in several ways. *Behavioural Ecology* was the first text book to make extensive use of techniques such as game theory (to analyse evolutionarily stable strategies) and optimization models, ideas which are now in everyday use by behavioural ecologists. Indeed, in retrospect some of the early models and interpretations already seem naive, although others have stood up well to the test of time. Behavioural ecologists have also become more critical of both theory and evidence. What has been termed the 'romantic era', the explosion of new ideas, is over, and the long, hard grind of sifting ideas and collecting evidence has begun. This critical vein is illustrated, for example, by Grafen's assessment of kin selection and related theory in Chapter 3, Bradbury and Vehrencamp's appraisal of the well-known 'polygyny threshold' model in Chapter 10, and Clutton-Brock and Harvey's careful exposé of the pitfalls of the comparative method in Chapter 1.

The organization of the book follows a similar pattern to *Behavioural Ecology*, although the first section, on methods and ideas, is totally new. We brought this in because it has become apparent that the central theoretical ideas in behavioural ecology are often misunderstood. The three chapters consider the comparative method for studying adaptation, the concept of evolutionarily stable strategies, and levels of selection. All of these are important themes underlying the arguments in the rest of the book.

The following sections, as in our previous book, consider the problems that face animals in surviving and reproducing, and how they might solve these problems so as to achieve the evolutionary goal of maximizing gene survival. In the second section, Predators and Prey, foraging, learning about food, defending resources and living in groups are considered. The next section looks at reproductive strategies, the problems of sexual versus asexual reproduction, mate choice, investment in offspring and the trade-off between survival and reproduction that constitutes an overall life history strategy.

The final section looks at conflict and cooperation in animal societies. The 'environment' of an individual includes not only its physical milieu but also other individuals such as predators and competitors. We consider how these selective forces have influenced the evolution of complex societies in both vertebrates and insects, and then show how similar ideas can be applied to plants, unlikely as this may seem at first sight. This section on societies is completed with a discussion of animal communication, the process which underlies all forms of social interaction.

Behavioural ecology is a beguilingly simple subject because much of it is based on natural history which is familiar to everyone. However, we hope that two important messages from this book are, first, that the theories have many subtleties which are easily misunderstood from superficial reading, and secondly, that testing the ideas rigorously is by no means straightforward. At the same time the book shows how much scope there is for the imaginative student to make major contributions.

Oxford, March 1983

Nicholas Davies
John Krebs

Acknowledgements

We thank Bob Campbell and Simon Rallison of Blackwell Scientific Publications for their enthusiasm and help in the preparation of this book. Authors of various chapters would like to acknowledge the following for their help. Chapter 2: Jane Brockmann, Steve Frank, Jeff Lucas, Meg McVey, and Mark Ridley. Chapter 4: Alex Kacelnik and Alasdair Houston for comments and Steve Lima, Ron Ydenberg and Dave Stephens for access to unpublished work. Chapter 5: Jerram Brown, Steve Emlen and Graham Pyke for comments and Mary Bayham for typing. Chapter 6: Anthony Arak, David Gibbons, David Harper and Arne Lundberg for comments. Chapter 12: Sandy Vehrencamp (parts of Chapter 12 are based on the ideas in a joint paper, Emlen & Vehrencamp 1983). Chapter 13: Alan Grafen, Charles Michener, Joan Strassman, Robert Matthews, Christopher Starr, Jon Seger, Barbara Thorne, Christine Nalepa, Yael Lubin, Mary Jane West-Eberhard, Bert Hölldobler and Jeffrey Lucas for comments and discussion. Chapter 15: Robert Hinde and Jim Markl for comments and discussion.

PART 1
METHODS AND
IDEAS

Introduction

There is an examination question for zoology undergraduates which goes something like this: 'Progress in biological research is determined by the development of new techniques. Discuss.'. If the answer were a straight 'yes', one would have to report that the frontiers of behavioural ecology had moved little. In fact the techniques used in the 1980s are hardly different from those used by ethologists 50 years ago: careful observation with notebook and stop watch, simple experiments, individual marking of animals, and so on. There are, it is true, technical innovations borrowed from other fields, microprocessors, portable tape recorders, radiotransmitters, electrophoretic analysis, to name but a few, but the studies described in probably 90% of references at the back of this book could technically have been done in the 1930s. Does this mean that there is nothing by way of progress to write about? Fortunately, the answer to the exam question is not a simple 'yes'. While it is true that one can point to fields of biological research which opened up as a result of new techniques, electron microscopy for example, it is equally easy to point to fields which opened up because of a new *idea* rather than a technique. The study of stereochemistry of biological molecules, for example, arose out of a more or less chance thought by one individual (see Ogston A. L. (1976) *Nature (London)*, **276**, 676). So even with no new techniques, behavioural ecology could be making progress with its ideas.

Furthermore an idea ('notion, thought, any product of intellectual action') can serve as a technique ('a method of performance, manipulation'). A new way of thinking about a problem can become a technique just as a new device for measuring, viewing or recording. The first part of this book is about some of the techniques and ways of thinking which have become part of the behavioural ecologist's technical armoury for studying evolutionary questions.

Questions about adaptation are implicitly, if not explicitly, about *differences* between entities: individuals, species, or higher taxa. In Chapter 1, Clutton-Brock and Harvey give an up-to-date account of one of the classical methods of studying adaptive differences— interspecific comparison. In earlier times comparisons between species were qualitative in nature, for example pointing out behavioural or anatomical differences between species associated with habitat differences, as in the classical studies of ground- and cliff-nesting gulls. More recently, quantitative analyses using multivariate statistics to partition possible causes of variance have been employed, as illustrated by Clutton-Brock and Harvey's own work on mammalian social organization, ecology and morphology. As well as illustrating the power of this approach, Chapter 1 points to some of the pitfalls, problems such as confounding variables, independence of taxonomic units and intraspecific variation.

While Clutton-Brock and Harvey focus on differences between species, Parker in Chapter 2 introduces one of the methods used to study individual variation within a species from an adaptive viewpoint. Game theory, and in particular Maynard Smith's concept of evolutionarily stable strategies (ESSs), has encouraged behavioural ecologists to look for the possibility that two or more modes of behaviour might coexist in stable equilibrium because of frequency-dependent benefits. Although many behavioural ecologists were already using 'ESS thinking' ('in a population of individuals of type A would a rare mutant B invade?'), Maynard Smith's mathematical formulation revealed the full potential of game theory for evolutionary biology. Parker's taxonomy of game-theoretic models distinguishes between *pure* and *mixed* ESSs. The former may be *conditional* or *fixed* strategies, and the latter may be *polymorphisms* (in which case ESS stands for 'evolutionarily stable state') or *stochastic alternatives* within an individual. Games may be played between two individuals (contests) or between many (scrambles). Well-known contests include the war of attrition and the hawk–dove game, while scrambles include parent–offspring conflict and sexual competition.

Many of the specific models developed from game theory are not the kind that one goes out to test—it would be hard to see how one could test the hawk–dove model for example—but they are useful heuristic devices for thinking about evolutionary problems. The outcomes they suggest depend, of course, on their assumptions. The oft-quoted ESS for settling contests by an arbitrary asymmetry such as 'resident wins', for example, turns out to depend critically on discontinuous risk (i.e. contestants can play either low or high risk but cannot finely adjust risk to gains expected).

Game-theoretic models, together with their close cousins optimality models, are purely phenotypic. Strategies are not modelled with an underlying genetic mechanism in mind, yet genetic variation lies at the heart of evolutionary adaptation. This dilemma is discussed by Grafen at the start of Chapter 3. The question is whether genetic constraints limit the feasible set of optimal or evolutionarily stable strategies. To take a simple example of where they would, consider a single locus for which the optimal genotype is a heterozygote. Because of the laws of Mendelian inheritance, the population can never evolve to a pure culture of optimal individuals. Grafen goes on to discuss the ways in which behavioural ecologists might be able to justify their use of non-genetic models.

One of the most widely used theoretical concepts in behavioural ecology is inclusive fitness. As Grafen shows, it is also one of the most widely misunderstood (by, among others, the editors of this volume— Krebs & Davies 1981!). The common misconception is to define inclusive fitness as 'the offspring reared by Ego, plus the offspring of Ego's relatives devalued by their coefficient of relatedness. In fact only the *extra* young of relatives resulting from Ego's behaviour and only Ego's young which are *not* the result of others' help should be counted.

Less susceptible to confusion, because it focuses on the *differences* in fitness resulting from a particular behaviour, is Hamilton's rule for the evolution of altruism, $rb - c > 0$.

Finally, Grafen deals with the 'new breed' of group selection models that have emerged in the last few years. He shows first that there is a blurring of the distinction between group and kin selection in the new models and secondly that whether or not one calls them models of group selection is largely a matter of definition.

Chapter 1
Comparative Approaches to Investigating Adaptation

T. H. CLUTTON-BROCK and

PAUL H. HARVEY

1.1 USES OF INTERSPECIFIC COMPARISONS

The origins of most questions about adaptation lie in comparisons between species. At the most general level, it is the intricate fit between organisms and their environments that prompts us to ask adaptive questions at all (Williams 1966; Lewontin 1978). More specifically, the existence of well-defined contrasts in behaviour or morphology between closely related species provokes investigation. For example, studies of the natural history of gulls showed that while the ground-nesting black-headed gull (*Larus ridibundus*) removes the eggshells from its nest shortly after the chicks hatch, dropping them some distance from the nest, the closely related kittiwake (*Rissa tridactyla*) does not. This led Niko Tinbergen to suggest that eggshell removal is an adaptation to avoiding predation and, by a series of elegant experiments, to demonstrate that the presence of eggshells beside nests significantly increases the chances of predation (Tinbergen *et al.* 1962; Tinbergen *et al.* 1963; Tinbergen 1963). Associated research on kittiwakes by Cullen (1957) confirmed that rates of nest predation were low, due to the inaccessibility of nesting sites, and showed that there were many other behavioural differences between the two species that were related to their contrasting nesting sites (see Fig. 1.1).

Comparisons also provide a way of ordering information for many different species which generates hypotheses concerning the functional significance of interspecific differences. For example, Jarman's (1974) demonstration that interspecific differences in body size, feeding ecology and social behaviour in African antelope were closely related to each other, provided a basis for a diversity of functional hypotheses concerning interspecific differences in body size and grouping behaviour. In some cases, the contrasting behaviour of related species poses questions that are later answered when new facts give a clue to the adaptive significance of the difference. It was known for some years that female black-and-white colobus monkeys (*Colobus guereza*) passed newly born infants for care and attention to other females belonging to the same troop, while red colobus (*C. badius*) did not (Clutton-Brock 1974; Struhsaker 1975). The adaptive significance of the difference was unknown until long-term studies in

	Nesting behaviour of black-headed gull	Nesting behaviour of kittiwake
Nest site:	On ground	On cliffs
Nest construction:	Sparse and cryptic	Deep and strong
Nest dispersion:	Nests widely spaced within colony	Nests close together
Parental behaviour:	Adults seldom defaecate or allopreen at nest Eggshells removed from nest site by parent after hatching Adults recognize their chicks	Adults defaecate and allopreen at nest Eggshells commonly left on nest Adults do not recognize their chicks
Chick behaviour:	Chicks camouflaged and crouch when disturbed Chicks hide from an early age	Chicks strikingly marked and usually ignore disturbance Better locomotion until close to fledging
Chick mortality:	Predators common: 10% of eggs laid fledge Death from falling rare	Predators rare: 60% of eggs laid fledge Death from falling comparatively common

Fig. 1.1 Nesting behaviour of black-headed gulls and kittiwakes. (From Tinbergen *et al.* 1962, 1963; Cullen 1957.)

Uganda showed that while female black-and-white colobus usually remained in their natal group, female red colobus usually moved to another troop and females belonging to the same troop were seldom related to each other (Struhsaker & Leland 1979). Similarly, the reasons why male primates of certain species regularly carry and care for dependent offspring only became clear when enough primate species had been studied to show that extensive male care only occurs in monogamous species (Clutton-Brock & Harvey 1977). Other questions posed by interspecific comparisons are still partly or fully unanswered. In some primate species, females develop large swellings in the perineal region around the time of ovulation. With few exceptions, these swellings are only found in species living in large, multi-male troops (Clutton-Brock & Harvey 1976)—but the functional significance of this association is not yet clear. While interspecific differences in the relative frequency of dispersal among males and females provide many important insights into the significance of social relationships between members of the same sex (see above), the reasons why in some species females disperse and males remain, whereas in others males disperse and females are the sedentary sex, are still not clear. Although several hypotheses have been suggested (Greenwood 1980; Wrangham 1981a), none yet provides a satisfactory explanation for the distribution of male versus female biases in dispersal.

These examples all concern the distribution of discrete traits, but comparisons can also be important in posing questions about the form of quantitative relationships between specific traits and ecological variables. In a wide variety of organisms, sexual dimorphism in body size and the development of weaponry increases in larger species (Gould 1975; Clutton-Brock *et al.* 1977). Why should this be? Similarly, the well-established trend for the size of neonates relative to maternal size to decline in bigger mammals still requires a satisfactory explanation. So too does the puzzling tendency for the slope of log population density on log body weight to approximate to -0.75 across both herbivore and carnivores species (Damuth 1981; Gittleman 1983).

As well as generating questions and ordering information, interspecific comparisons can be used to test specific hypotheses. For example, if eggshell removal by gulls has evolved because it reduces the chance that predators will detect the nest, it should be common among other ground-nesting species apart from black-headed gulls and absent in other cliff-nesting species apart from kittiwakes. Comparative studies of other gulls confirm that this is the case (Cullen 1960; Cullen & Ashmole 1963; Hailman 1965).

In the previous example, qualitative data (the presence or absence of eggshell removal) were used to test a qualitative hypothesis (that ground-nesting gulls remove eggshells, while cliff-nesters do not). However, it would have been possible to treat the data in a more formal quantitative fashion by counting the number of ground- and cliff-nesting gulls that do and do not remove the eggshells after hatching, and casting the data in a 2×2 contingency table to test the prediction that eggshell removal is significantly more common among ground-nesting gulls than among cliff-nesters. Though this has, to our knowledge, never been done for gulls, quantitative comparisons have often been used to test other qualitative hypotheses. For example, on the grounds that competition for mates is more intense among males in polygynous species than in monogamous ones, and that large size improves an individual's fighting ability, a relationship might be expected between sexual dimorphism and degree of polygyny. Studies of reptiles and amphibians (Shine 1978, 1979; Berry & Shine 1980), pinnipeds, ungulates (Alexander *et al.* 1979), carnivores (Gittleman 1983) and primates (Clutton-Brock *et al.* 1977) have all shown that sexual dimorphism in size is consistently greater in species where polygyny is well developed than in monogamous species or those where males can only monopolize a small number of females (see Fig. 1.2a). If this explanation for the existence of size dimorphism is correct, we might also expect the development of weaponry relative to male body size to show a similar distribution. Are weapons used in intraspecific combat, like the canines of primates or the antlers of deer, largest *relative to male body size* in strongly polygynous species and smallest in monogamous ones? Quantitative

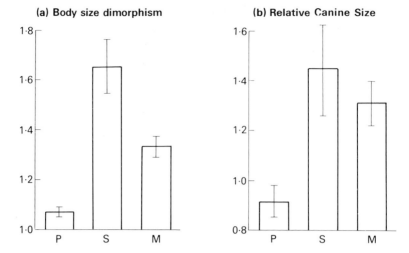

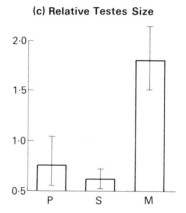

P—pair-living or monogamous
S—single-male
M—multi-male

Fig. 1.2 **(a)** Body size dimorphism (adult male divided by adult female weight), **(b)** Relative Canine Size (a measure of canine size dimorphism: Harvey *et al.* 1978) and **(c)** Relative Testes Size (a measure of testes size after body size effects have been removed: Harcourt *et al.* 1981) for primate genera with different breeding systems. Sample size for P, S and M respectively: body size dimorphism 12, 9, 14; Relative Canine Size 4, 7, 9; Relative Testes Size 4, 7, 8. Bars indicate one standard error in each direction from the mean.

comparisons have confirmed that this is the case (see Fig. 1.2b; Harvey *et al.* 1978; Clutton-Brock *et al.* 1980), supporting the explanation of size dimorphism suggested above.

These results lead to further predictions. Competition between males is not confined to the pre-copulatory stages of mating, and sperm competition is likely to occur in species where females mate with several males during a single oestrus (Parker 1970b). Since the volume of ejaculate is related to testis size, it is reasonable to predict that in species which typically form multi-male breeding groups, males should have relatively larger testes than in monogamous or harem-forming species (Short 1979). This prediction is particularly interesting for, unlike the previous two, it suggests that harem-forming species should resemble monogamous ones rather than those living in multi-male groups. Systematic studies both of primates (Harcourt *et al.* 1981; Harvey & Harcourt 1982; see Fig. 1.2c) and of cervids (Clutton-Brock *et al.* 1982) have shown that this is, in fact, the case.

Although these comparisons involve quantitative data, all the hypotheses that we have described so far have been qualitative ones, concerning the distribution of a particular trait or the direction of a relationship. A more powerful use of interspecific comparisons is to test quantitative hypotheses concerning the *form* of particular inter-specific relationships. Unfortunately, it is only in exceptional cases that our understanding of the costs and benefits of a particular trait is sufficient to allow us to predict the way in which it should be related to a specific environmental variable. For example, we do not know enough of the costs and benefits of large bodies to males and females in different breeding systems to be able to predict the form of the relationship between sexual dimorphism and the degree of polygyny. When the ratio of male to female weight is plotted against a measure of the degree of polygyny, should it increase rapidly or slowly? Should the relationship be linear or exponential? We currently have no firm basis for prediction.

In a few cases, however, adaptive hypotheses are sufficiently simple to allow us to predict the form of particular relationships, and data are available which can be used to test them. In several verte-brate groups, including lizards (Turner *et al.* 1969), birds (Schoener 1968; Mace & Harvey 1983), small mammals (Mace *et al.* 1983), carni-vores (Gittleman & Harvey 1982) and primates (Milton & May 1976; Clutton-Brock & Harvey 1977), average home-range or feeding-territory size for each species increases with body weight. An obvious explanation of these trends is that they occur because larger animals need more food than smaller ones (McNab 1963). This argument pro-vides a basis for a number of predictions about both the slope and the elevation of the relationship. First, since the standing crop of foliage is generally greater than that of fruit, while fruit is usually a more abun-dant food source than insects or vertebrates (see Ricklefs 1973), the home-ranges of herbivores of a given body size should be smaller than those of frugivores, insectivores or carnivores. The data appear to bear out this prediction for all the vertebrate groups mentioned above (for example see Fig. 1.3).

Secondly, within dietetic groups, home-range sizes should be directly proportional to the metabolic needs of the animals con-sidered, and should increase as approximately body weight$^{0.75}$ (Kleiber 1961). The data tend not to support this prediction since plots

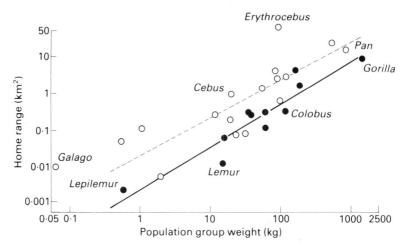

Fig. 1.3   Mean home-range size plotted against weight of the group that inhabits that home-range for different primate genera. ○ insectivores and frugivores, ● folivores. The upper line shows the least-squares regression for insectivores and frugivores, the lower one that for folivores. (From Clutton-Brock & Harvey 1977.)

of log home-range size against log estimated metabolic needs tend to have slopes appreciably greater than one for birds, small mammals, primates and carnivores (Mace *et al.* 1983; Harvey & Clutton-Brock 1981; Gittleman & Harvey 1982; Mace & Harvey 1983).

Thirdly, on the assumption that as the body size of a carnivore increases so the size of acceptable prey diminishes, whereas larger herbivores generally accept a wider range of food than smaller ones (Bell 1969), it has been predicted that the slope of average home-range size on body weight should be steeper for carnivores than for herbivores and frugivores. The results are ambiguous: some studies support this prediction (Schoener 1968; Harestad & Bunnell 1979) but analyses of data from an extensive range of primates (Clutton-Brock & Harvey 1977; Harvey & Clutton-Brock 1981) and carnivores (Gittleman & Harvey 1982) show that the slopes of home-range size on estimated metabolic needs are very similar in the two orders.

Although these examples concern behavioural traits, the comparative approach can, of course, be used to investigate the functional significance of any characteristic that varies between species. Studies of adaptation relying on the comparative approach are legion and involve traits as diverse as genome size (Grime & Mowforth 1982), chromosome number (Sherman 1979; Seger 1982), body size (Jarman 1974; Eisenberg 1981), brain size (Eisenberg & Wilson 1978; Harvey *et al.* 1980; Martin 1981a), life history variables from gestation or incubation length to lifespan (Lack 1968; Western 1979; Millar 1977, 1981; Eisenberg 1981), mating systems (Orians 1961; Crook 1965; Verner & Willson 1966; Eisenberg *et al.* 1972; Emlen & Oring 1977), feeding time, dietetic diversity and prey size (Hespenheide 1971;

Rosenzweig 1968; Clutton-Brock & Harvey 1977), sex ratio (Clutton-Brock & Albon 1982), population density (Damuth 1981) and biomass (Clutton-Brock & Harvey 1979b). In all these cases interspecific comparisons have provided an important way of making sense of the diversity of animal life.

1.2 PROBLEMS OF COMPARISON

While a considerable number of books have been devoted solely to the problems of devising adequate experiments to test biological questions, only recently has any significant attention been paid to the methodological problems of comparison. Our intention in reviewing the pitfalls of the comparative method is not to discourage the use of comparisons: many important questions can only be approached in this way. Moreover, it is worth remembering that while ideal comparisons are uncommon, ideal experiments which control for all possible confounding variables are also rare.

In the following sections, we discuss seven questions that all interspecific comparisons need to consider, although many of the problems involved are, in fact, common to both comparative and experimental approaches to understanding adaptation.

1.2.1 What kind of comparison is most useful?

Although we did not distinguish between different types of comparison in section 1.1, in practice interspecific comparisons mostly fall into one of three categories. First are comparisons between a small number, usually a pair, of closely related species that differ in some important aspect of their ecology (see, for example, Orians 1961, 1980). The principal advantage of comparison at this level is that it is possible to compare more detailed variables than in comparisons involving many species. For example, while the frequency of nest predation can be compared between kittiwakes and black-headed gulls, the data would not be available to allow similar comparisons to be made across a large number of gull species. In addition, it is often possible to support an adaptive argument concerning a particular trait by reference to other associated characteristics (see, for example, Fig. 1.1).

The main limitation of paired comparisons is the frequent impossibility of deciding which of an array of ecological differences between two species is responsible for the evolution of a particular behavioural or morphological difference. While this is reasonably obvious in the case of the kittiwake and the black-headed gull, the ecological reasons for particular interspecific differences are often unclear and it can be difficult to determine which of several ecological differences is the likely cause of a particular morphological or behavioural contrast. In these circumstances, it is all too easy to construct plausible arguments to account for any difference (Lewontin 1978). In addition, even when the ecological causes of a particular contrast between two species are obvious, as in the case of eggshell removal, paired comparisons provide no guarantee of the generality of the relationship.

The second kind of comparison involves considerable numbers of species, which are allocated to different ecological categories. For example, Crook (1964, 1965) compared the social organization of ploceine weaver birds living in either forest or savannah habitats. His well-known paper demonstrated that forest-dwelling species, which are primarily insectivorous, mostly live in monogamous pairs which defend large territories, while the savannah-dwelling species, which are principally granivorous (seed eating), nest in large colonies or grouped territories, feed in flocks and are mostly polygynous (Table 1.1).

Table 1.1. Distribution of species showing different types of breeding behaviour in relation to habitat and food type in species of weaver birds from the subfamily Ploceinae (from Crook 1964; Lack 1968). Sociality categories are: solitary (S), group territories (GT) and colonial (C).

		Pair bond		Sociality		
Habitat	Main food	Monogamous	Polygynous	S	GT	C
Forest	Insects	17	0	15	0	1
Savannah	Insects	5	0	4	0	2
Forest	Insects + seeds	3	0	2	0	0
Savannah	Insects + seeds	1	4	1	0	4
Grassland	Insects + seeds	1	?1	1	0	0
Savannah	Seeds	2	10	0	1	16
Grassland	Seeds	0	15	0	13	3

Comparisons of this kind are most useful when ecological distinctions are simple and well-defined and when comparisons involve phylogenetically similar species. When the same approach was tried with primate societies (Crook & Gartlan 1966), it proved less satisfactory because differences between species allocated to the same habitat category were in many cases as pronounced as differences between categories (Clutton-Brock 1974; Martin 1981b)—partly because important ecological parameters, such as diet type, seasonality of food supplies and liability to predation, differ widely among primates occupying the same kind of habitat. Another reason why the primate comparison was less satisfactory was that it included all species within an order (instead of within a subfamily), thus increasing the effects of phylogenetic influences and the probability that phylogenetically diverse species which had solved the same ecological problems in different ways might be included in the same category.

An additional limitation of attempts to categorize species is that it is often difficult for anyone who does not have a detailed knowledge of the distribution of the trait in question to assess the validity of any association between behavioural and ecological traits—selected data combined with eloquent argument can too easily appear to establish a relationship where none exists. For this reason, the third category of interspecies comparison, which involves the use of formal statistics to test associations between interspecific differences in behaviour and ecology, is usually preferable. Where both variables being considered are discontinuous, standard categorical tests can be used, and bivariate or multivariate tests where they are continuously distributed (Harvey & Mace 1982).

However, investigating quantitative relationships between particular traits has the important limitation that it restricts analyses to variables that have been measured along an ordinal scale on a reasonable sample of species, and there are consequently many questions that it is impossible to examine because the data are not available. In addition, this approach makes it necessary to take a number of decisions concerning the precise variables to be used, how they are to be measured and the techniques of analyses that are suitable (see below). The validity of any analysis will depend upon these assumptions.

1.2.2 Are the variables biologically relevant?

The first question that any attempt to make quantitative comparisons needs to ask is whether the variables are biologically relevant, for not all quantitative measures are meaningful. In particular, it is important that ecological variables reflect the animal's experience of its environment rather than that of the observer. For example, an observer can measure the brightness of a male bird's plumage and compare such measures across species (see Baker & Parker 1979; Hamilton & Zuk 1982), but unless the animals themselves (or their predators) judge brightness on a similar scale, the rating may be meaningless. Similarly, many observers record the mean group size in the population they are working on. But it is often the case that the mean group size is substantially smaller than that in which an average individual finds itself (see Jarman 1982 and Table 1.2).

Where variables of greater complexity are used, it is even more important to consider the biological significance of the calculation. On the assumption that the home-range size of primate groups should increase with their nutritional requirements, Clutton-Brock and Harvey (1977) examined the intergeneric correlation between the weight of groups and the average home-range size (see Fig. 1.3). However, metabolic requirements increase as body weight$^{0.75}$ and, as Martin (1981a) has pointed out, it would have been more sensible to examine the relationship between the metabolic requirements of

Table 1.2. A population consists of 10 groups, which contain varying numbers of territorial (T) and bachelor (B) males, adult females (F) and unsexed juveniles (J). Observer estimates can differ markedly from the experience of individual animals. (After an example in Jarman 1982.)

Composition of the population				
Structure of individual groups				Number of groups with that structure
T	B	F	J	
1	0	12	8	1
1	0	8	5	1
1	0	14	10	1
0	0	1	1	1
1	0	0	0	4
0	6	0	0	1
0	8	0	0	1

Observer's estimate	*Animal's experience*
Mean group size = 8	Average size of group in which animal lives = 17.1
Mean number of females found together = 8.75	Average female is one of 11.6 in her group
Adult sex ratio in population 0.6 male : 1 female	Average female is in contact with 0.97 adult males
Mean number of juveniles found together = 6	Average juvenile is one of 7.9 in its group
Average number of bachelors per group = 1.4	Average bachelor is in a group with 7.1 other bachelors, while other animals are in groups that contain no bachelors.

groups and their range size, though re-analysis does not, in fact, change the overall conclusions (Harvey & Clutton-Brock 1981).

Broad interspecific comparisons are generally forced to rely on data that vary widely in quality among species and, in any substantial data set, some errors are likely to be present as a result of transcription alone (see Sadleir, in press). For example, it was thought for some years that Bornean orang utans (*Pongo pygmaeus*) weighed rather more than twice as much as conspecifics from Sumatra, despite the fact that their skulls seemed to be of similar size. However, this difference eventually proved to be the consequence of a confusion between pounds and kilograms (see Erkhardt 1975).

A more important kind of error arises from inadequate sampling. Where behavioural parameters vary widely but only one population—perhaps only a single group—has been sampled, the value may not be representative of the modal value for the species. For example, although home-range size of mammals can vary between populations of the same species by as much as two orders of magnitude (see Gittleman & Harvey 1982), studies of comparative home-range size (e.g. Milton & May 1976; Clutton-Brock & Harvey 1977) usually combine estimates for species where different groups have been studied in several different areas with estimates based on the range size of a single troop, without weighting some estimates more than others since there is seldom any logical basis on which to assess the reliability of a particular estimate. Consequently, individual points may often be unrepresentative.

To put this problem in perspective, it is worth remembering that many physiological parameters vary widely with time within individuals—heart rate is a good example. As a result, in comparisons between two sets of animals, the reading for each individual may not represent that individual's modal state. This is seldom perceived as an important problem, for most experimenters are concerned with the average values for the different samples of animals rather than with values for individuals, and variation of this kind is unlikely to create a relationship when none exists, though it may obscure a genuine relationship. For precisely the same reasons, random variation is not an insurmountable problem in interspecific comparisons—though it is important that little reliance should be placed on values for individual species.

In practice, if the range of values that a variable can take is large and the data are well distributed throughout the range, random errors often have surprisingly little effect on the estimated correlation coefficient or form of a relationship. However, there are several circumstances under which random error can exert an important influence on both the correlation coefficient and the exponent. First, if the range of values is small, the effects of random error can be considerable. For example, in a recent analysis of life history variables across mammals, Millar (1977) concluded that time to weaning is not closely related to adult body weight. His analysis was based on estimates for 98 mammal species and gave a correlation coefficient of 0.17 between weaning age and body weight, with an exponent of 0.05. Weaning age is notoriously difficult to measure since the frequency of suckling declines gradually. Nearly 70% of values lay between 14 and 28 days and nearly half of the values used were multiples of seven (e.g. 14, 21 or 28 days), indicating that they had originally been measured in weeks (see Fig. 1.4). Therefore it is reasonable to assume that the error in relation to the range of values was large and the low correlation coefficient could be a consequence of this rather than of the absence of any close relationship between body size and weaning age.

Secondly, if the values used are distributed very unequally throughout the range, extreme values can exert a strong effect on both the correlation coefficient and the exponent and, if inaccurate, can produce misleading results. The data set used by Millar included 81 rodent species and only two mammals over 4 kg (the racoon, *Procyon lotor* (5.5 kg), and the black-tailed deer, *Odocoileus hemionus* (57 kg))—see Fig. 1.4. The analysis used an incorrect estimate of 21 days for the weaning age of black-tailed deer, slightly lower than that

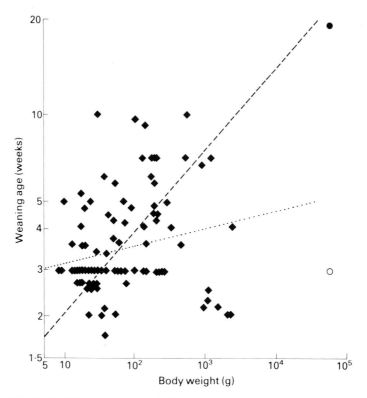

Fig. 1.4 Weaning age plotted against body weight for Millar's (1977) set of 98 species. The incorrect value for the black-tailed deer (*Odocoileus hemionus*) is shown by the open circle and a more accurate estimate (see text) by the closed circle. The two lines are best-fit estimates through all points except the black-tailed deer. The shallower, dotted line is the least squares regression and the steeper, dashed line is the reduced major axis (see p. 21). In this example the slope of the reduced major axis is more in accord with other studies and with the corrected datum point than is the regression estimate.

of the majority of rodents sampled and some 120 days shorter than the real weaning age (see Sadleir 1980a, b). Under precisely these circumstances individual errors can have a considerable effect: other studies, using more carefully selected values for less heterogeneous assemblages of species, indicate that weaning age and body weight are more highly correlated and that weaning age increases with body weight as approximately body weight$^{0.25}$ (Blaxter 1971; Mace 1979; Reiss 1982; Russell 1982).

A third case where random error can be important concerns multiple regression. This technique assumes that all independent variables are measured without error, but in practice different variables will be subject to varying amounts of error and this may lead to problems of interpretation. Sacher (1959) examined the relationships between life span of primate species and their brain size and body weight. Brain volume and body weight were closely correlated but, when the partial regression coefficients were examined (that is the effect of either brain or body size on life span, with the other size variable held constant), brain size appeared to be a more important influence than body size. However, estimates of body weight are likely to be subjected to greater error than estimates of brain weight, since they can be strongly influenced by body condition and reproductive state. Consequently, it is possible that body size exerts a stronger influence on life span than brain size, even though the latter may predict life span more accurately.

1.2.3 How broad a taxonomic array of species should be included?

The problems of random error can be reduced if large numbers of species are included in the analysis. However, although samples including very wide taxonomic arrays of species—such as all mammals (the famous 'mouse to elephant' curves), all birds (Lack 1968) or even all living organisms (Bonner 1965; Southwood 1981)— have been used in the past, there are several disadvantages in using comparisons based on very broad taxonomic groups. First, these comparisons seldom sample different subtaxa with similar thoroughness: for example, Millar's analyses of life history variables in mammals, which included 82 rodents, eight lagomorphs, four insectivores, four bats, one carnivore and an ungulate, is an extreme case. Since relationships commonly differ between different phylogenetic groups, results may not reflect the form of a relationship at the higher taxonomic level: clearly, Millar's analyses tell us more about rodents than about mammals as a whole.

A second problem is that the form of relationships often changes with the taxonomic level at which the analysis is carried out. For example, many anatomical and physiological variables are closely related to body size and it is common practice to plot them against size across different species belonging to the same class or order (e.g. Millar 1977; Western 1979; May & Rubenstein 1982). However, if the same relationships are examined at lower taxonomic levels, such as the family or genus, exponents are often lower. Figure 1.5 shows an actual example involving an analysis of brain size in primates (Harvey, in preparation), although the biological significance of this trend is still unknown.

It is also the case that when many variables are plotted against body size, plots for different taxonomic groups differ in the intercept on the ordinate (Clutton-Brock & Harvey 1979a). Figure 1.6 shows the relationship between a hypothetical variable and body size in three different families belonging to the same order. Taxonomic effects of this kind have been found in analyses of the relationship between body weight and brain size (Gould 1975; Harvey *et al.* 1980), home-range size (Clutton-Brock & Harvey 1979b), gestation length, neonatal weight, litter and clutch size, breeding age and lifespan (see Lack

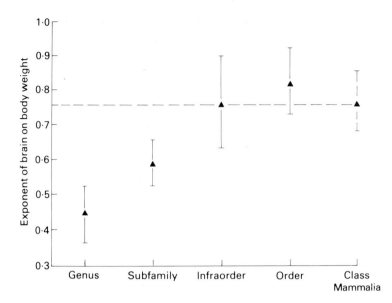

Fig. 1.5 The exponent relating brain weight to body weight at different taxonomic levels among the primates. The exponent is calculated as the slope of the major axis line on the logarithmically transformed data. Bars indicate 95% confidence limits around the mean. The generic slope is the common slope estimated among species within genera, the subfamily slope is among genera within subfamilies and so on. The dashed line is the slope among 15 orders in the class Mammalia. The primate data are from 114 species.

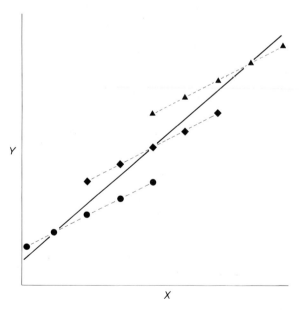

Fig. 1.6 Idealized relation between a life history variable (Y) and body size (X) both plotted on a logarithmic scale. Generic points are shown, and those belonging to the same family are given the same symbol. Though differences in elevation and slope have been exaggerated, the figure illustrates a common phenomenon: the continuous line which represents a best fit through all generic points has a steeper slope than the intra-family lines.

1968; Mace 1979). In fact relationships of this sort may prove to be the rule rather than the exception.

The relevance of this to attempts to make broad comparisons is that if an overall line is fitted to the points arranged as in Fig. 1.6, its slope will be steeper than slopes calculated within individual families and, as a result, larger species within each family will appear to deviate negatively from the line while smaller species will deviate positively. There is often no way of saying which is the most biologically meaningful slope, so that explanations of the biological meaning of slopes must be treated cautiously (see Harvey & Mace 1982).

1.2.4 What type of analysis should be used?

In bivariate studies, such as the investigation of the relationship between home-range size and body size, the usual way of estimating the relationship between one variable and another is to plot the least-squares linear regression on the logarithmically transformed data, since when the relationship is of the form

$$y = a \cdot x^b$$

where, for example, y is home-range area, x is body size and a and b are constants, it follows that

$$\log y = \log a + b \cdot \log x$$

and b is the slope of the regression of log y on log x.

However, the use of least-squares regression is sometimes inappropriate for comparative studies because the technique assumes that values for the independent variable are assessed without error and that causation is in a particular direction. With many size-related variables, like weaning age, measurement error may be considerable (see above) and it is perfectly possible that the pathway of causation is complex: each variable may influence variation in the other, or both may be influenced by a third variable. The net result of these effects is that regression analysis will give too shallow a slope when the correlation coefficient is not one (Harvey & Mace 1982; Martin 1981b). As the correlation coefficient is reduced through increased error in measurement of the independent variable, then we should be increasingly wary of interpreting regression slopes.

Other lines of best fit assume different distributions of error and will usually give steeper slopes than regression analysis (Harvey & Mace 1982). For example, the reduced major axis (Kermack & Haldane 1950) assumes that error variance is the same proportion of the total variance on each axis, which may often be closer to reality than the assumption that one variable is measured without error. When Millar's (1977) data for weaning age on body size (with O. *hemionus* excluded) are analysed by reduced major axis the slope is 0.28, far nearer the usual 0.25 estimate found in other studies (Blaxter 1971;

Rudder 1979; Gittleman 1983; Russell 1982; see Fig. 1.4). However, since we seldom have any estimate of error on the two axes, it is rarely possible to place any confidence in estimates of slope when correlation coefficients are below $+0.95$ or above -0.95.

Whether least-squares regression or some other line of best fit is used, it is often necessary to transform data for analysis. There are two common reasons for this. First, it may be necessary in order to make them more normally distributed. A useful way of testing for an improved fit to normality is by calculating the skewness statistic (Fisher 1954; Sokal & Rohlf 1969)—normal distributions are not skewed. The second reason is to produce a linear relationship between variables—for example, an exponential relationship becomes linear after both variables are logarithmically transformed (see above). An increased correlation coefficient is usually a suitable test of a better fit to linearity of bivariate distributions (Meyer 1970). Recently, Smith (1980) has cautioned against the uncritical use of logarithmic transformations in comparative studies, arguing that these seldom improve the proportion of variance accounted for. However, using Smith's compilation of 60 data sets, Harvey (1982) showed that for intraspecific studies logarithmic transformation generally makes little difference, while in cross-taxonomic comparisons the logarithmic transformation usually reduces skewness and relationships become more linear.

1.2.5 How should the effects of confounding variables be removed?

In order to investigate the relationships between interspecific differences in behaviour, physiology or morphology and particular ecological variables, it is often necessary to remove the effects of confounding variables (Clutton-Brock & Harvey 1979a). In some cases, this can be achieved by partitioning the sample of species and examining relationships within particular subsets of species which share the same habitat type or feeding niche. Mace, Harvey and Clutton-Brock (1981) examined the distribution of comparative brain size across 128 genera of small mammals, to test whether folivores have smaller brains than a combined category of frugivores, insectivores and granivores. This was indeed the case, but comparative brain size was also related to habitat (forest-dwelling genera tend to have larger brains than grassland forms), stratification (arboreal species have larger brains than terrestrial or fossorial ones) and activity timing (nocturnal genera tend to have larger brains than diurnal ones). Since folivorous small mammals are rarely forest-dwelling, arboreal or nocturnal, it was possible that the initial difference between the two dietetic groups was a consequence of the association between diet and habitat type, stratification or activity timing. By comparing genera belonging to the same habitat and timing categories but differing in diet (by two-way analyses of variance), the correlation between relative brain size and diet type was shown not to be a result of an association of this kind.

An important disadvantage of partitioning data sets to remove the effects of particular ecological variables is that this inevitably reduces the effective sample size. An alternative is to use some form of analysis that attempts to control for the effects of the confounding variable. There are three principal ways of achieving this. First, where only one of the independent variables has been measured on an ordinal scale and the other is nominal (i.e. not ordered, such as diet type), values of the dependent variable can be plotted against the independent variable and comparisons of slope or elevation made between nominal categories. The home-range analyses described above are of this type. Secondly, where two independent variables have been measured on continuous scales, either partial correlation or multiple regression is generally used. It is important to remember that such techniques should not be used for variables which cannot be rank-ordered; that if several independent variables are included in the analysis, it is often difficult to interpret the results; and that if enough independent variables are included in the analysis, it will eventually be possible to account for a considerable proportion of the variance even if none of them has a significant effect on the dependent variable. Thirdly, multivariate analyses such as factor or principal-component analysis can be used (e.g. Jorde & Spuhler 1974; Spuhler & Jorde 1975), but the results differ enormously depending upon which of numerous analytical procedures is applied (see Gould 1981). In comparative studies, assumptions are usually insufficiently formulated to allow us to choose among the different procedures.

Because its effects on behaviour, physiology and anatomy are wide-reaching, the commonest confounding variable of all is body size. The great majority of behavioural, physiological and morphological variables are size-related (Clutton-Brock & Harvey 1979a), and to compare values between animals of different sizes it is first of all necessary to allow for the effects of size. Analysis of the ecological correlates of brain size provides one of the best examples of the difficulties of removing size effects, because the relationship has been widely investigated (Gould 1975), although very similar problems apply to comparisons involving many other size-dependent variables, including most life history measures (Clutton-Brock & Harvey 1979a).

A number of studies have predicted that animals with different life styles might require different-sized brains (see Jerison 1973). However, a comparison of absolute brain size between two ecologically distinct groups of species, such as frugivorous versus folivorous primates, would not answer this question, for brain size and body size are closely related to each other and folivorous species tend to be bigger than frugivorous ones (Clutton-Brock & Harvey 1977). Instead, we need to compare the brain sizes of different species *relative to their body sizes*.

Initial attempts to compare the relative brain sizes of different species used ratios of brain weight to body weight as a measure of relative brain size. However, since brain size (like many other variables) does not change proportionately with body weight, but increases across mammals as a whole as body weight$^{0.75}$ (Eisenberg 1981; Martin 1981a), these ratios are misleading because they systematically underestimate the relative brain size of larger species. Instead, we need to determine the overall relationship between brain weight and body weight and examine the extent to which frugivorous versus folivorous species deviate from this line. However, as we have already described (see Fig. 1.5), relationships between brain weight and body weight differ between taxonomic groups, and slopes calculated within families are shallower than those calculated across all species within larger taxonomic groupings. In these circumstances, relative values for size-related variables (i.e. deviations from the overall slope) will tend to be positive for the smaller species within any family and negative for the larger ones (Fig. 1.6). Since differences in body size within families are usually related to many other ecological and behavioural variables (Eisenberg 1981), there is a very real danger that this effect can create the appearance of relationships where none exist.

One way of reducing this problem is to examine the deviations of different species from slopes of brain weight on body weight *set separately for each family* included in the analysis. However, the same problem occurs at this lower level, since slopes of brain weight on body weight are generally shallower among species belonging to the same *genus* than among all those belonging to the same family. As we have argued elsewhere (Clutton-Brock & Harvey 1979a), the most satisfactory way out of this dilemma is to restrict comparisons to the taxonomic level immediately below that at which the line of best fit is set. For example, if lines are fitted separately for each family, comparisons should be restricted to deviations for generic points, while if values for species are to be compared, size effects should be removed at the generic level.

In the analysis of relative brain size described above, we calculated average values for different primate genera and examined these relative to slopes of brain weight on body weight calculated for each family (Clutton-Brock & Harvey 1979a; Harvey *et al.* 1980), thus distinguishing between variation in brain size at a high taxonomic level (between families) and variation at a lower level (between genera), and concentrating our attention on the latter. Generic deviations from family slopes (which we called 'comparative brain size') were closely related to variation in diet and home-range size, thus supporting previous suggestions that species living on complex and widely dispersed food supplies may need larger brains for their body size than those living on more concentrated and predictable food supplies. In addition, our analysis showed that differences in brain/body size relationships among families followed the same trend: families whose component species tended to have relatively large home-ranges also tended to have large brains for their body weights. This distinction between high- and low-level variation in particular traits is, we believe, a useful one and may help to avoid many of the obvious problems of broad interspecific analyses.

A different approach to removing size effects is suggested by Smith (1980), who argues that it may be more helpful to restrict comparisons to species with similar body weight, thus avoiding the problems of calculating relative values. For example, since humans weigh about 60 kg, Smith favours comparing human brain weights with those of other mammals weighing between 50 and 70 kg, arguing that size effects are minimal within this range. However, even if a suitable range could be devised on statistical grounds, this approach has the inevitable limitation of small sample size if comparisons are to be restricted to related species. To increase sample sizes sufficiently for statistical analysis, the taxonomic net must be cast wide: Smith suggests comparing humans with a diverse group of species including warthogs and alligators. As a result many differences between species are likely to be a result of variation at a high taxonomic level and are consequently difficult to interpret. There are probably several reasons why humans have comparatively larger brains than alligators.

1.2.6 Can different species be regarded as independent?

If, as we have suggested, relationships across species are subjected to statistical testing in the same way as correlations across individuals of the same species, it is important to be reasonably certain that values for different species can be treated as independent points. For example, congeneric species often show very similar forms of behaviour. This may be because they are independently adapted to similar environments (convergence), in which case it is reasonable to treat values for different species as statistically independent of each other. Alternatively, they may be similar because both their behaviour and their ecology is constrained by their common phylogenetic inheritance (homology), and in this case points for congeneric species should not be regarded as independent of each other.

Effects of this kind can have an important influence on the credibility of analyses of interspecific data. In Crook's comparison of grassland versus forest-dwelling weaver birds (see Table 1.1), 14 of the 16 grassland species belonged to the single genus *Euplectes* and the association between grasslands and polygyny is considerably less impressive if genera are treated as single points. Similarly, Sherman (1979) demonstrated an association between eusociality and chromosome number in the Hymenoptera. Three of the 20 families considered were primarily eusocial and, using species as independent points, Sherman made a statistical comparison of chromosome number between each of the three eusocial families and each of the other 17 families. But (as Sherman himself noted) there may be phylogenetic constraints on chromosome number, in which case the statistical analyses would be invalid because the assumption of independence of data points (here species within families) would be violated.

Using statistics alone it is impossible to tell whether similarities within taxa are based on homology or convergence. However, it is possible to investigate the distribution of variation among taxonomic levels using nested analyses of variance (see Sokal & Rohlf 1969). For example, analysis of the taxonomic distribution of variances of home-range size, oestrous cycle length and neonatal weight among primates shows that a substantial proportion of variation in home-range size occurs at low taxonomic levels (Table 1.3) (as we might expect on the grounds that this appears to be a particularly plastic variable), while most of the variation in cycle length and neonatal weight occurs between families or subfamilies (see Table 1.3). In the first case it is reasonable to treat values for different species as independent, but in the latter two, phylogenetic effects may be important and when examining how these variables are related to ecological differences, it would be advisable to repeat analyses using average values for different genera or subfamilies to see if they give the same results as analyses at the species level.

This raises an additional difficulty if the sample includes genera whose component species fall into different ecological categories. For example, when Harcourt *et al.* (1981) were comparing testis size among primates whose breeding systems differed, they decided to perform their analyses at the generic level but found that different baboon (*Papio*) species have different breeding systems. How should such a genus be classified? There is no simple answer to this question, but we have found that one useful approach is to calculate averages for all congeneric species belonging to the same ecological category (see Clutton-Brock & Harvey 1977).

Another way of identifying the taxonomic level at which it is reasonable to regard values as independent is by using the cladistic

Table 1.3. The percentage of variation in home-range size, oestrous cycle length and neonatal weight (all variables logarithmically transformed) among primate species that is attributable to different taxonomic levels. The figures are based on the results of a three-level Model II Nested Analysis of Variance with unequal sample sizes.

Variance component	—Among Families —Within Order	Subfamilies Families	Genera Subfamilies	Species Genera
Home-range size	32	21	24	23
Oestrous cycle length	20	64	12	4
Neonatal weight	74	17	4	5

technique of outgroup comparison to estimate the minimum number of times a particular trait has independently evolved (Ridley 1983). This is achieved by first identifying the lowest taxonomic level at which variation occurs. If some species in a genus are arboreal while others are terrestrial, then other genera within the same family would comprise the outgroup, and if they all contained only arboreal species we could apply the law of parsimony to conclude that arboreality is primitive to the family. The terrestrial habit would be presumed to have evolved only once since the appearance of the genus with both arboreal and terrestrial species. The number of times a trait has evolved, measured in this way, can be far smaller than the number of species or genera in which the trait is found. Unfortunately, this technique is not well suited to the analysis of continuously distributed variables such as home-range size or the majority of life history variables, and its application may be limited.

1.3 PROBLEMS OF INTERPRETATION

Although the previous section has been concerned with problems of statistical inference, we have so far avoided the more fundamental difficulties of interpreting the results of comparisons. Undoubtedly the most important of these is the danger that even the closest association may not indicate any causal relationship (see Krebs & Davies 1978). For instance, apart from differing in their susceptibility to predators, the nest sites of kittiwakes and black-headed gulls vary in their proximity to the adults' feeding sites as well as in temperature and shelter. How can we be sure that it is the difference in susceptibility to predators rather than some other ecological difference that is responsible for the contrasting behaviour of kittiwakes and black-headed gulls (see Fig. 1.1)?

Since the closer the association between a behavioural variable and an ecological one, the more likely it is that the relationship is causal, one way is to examine whether the two associations hold within different groups of animals. The association between group biomass and home-range size (see Fig. 1.3) is found within each of the four main taxonomic groups of primates (prosimians, New World monkeys, Old World monkeys and apes; see Clutton-Brock & Harvey 1979b) as well as within several species (see Struhsaker 1967; Waser 1976).

Even if an association between some environmental parameter and a behavioural or physiological trait is a very close one, the explanation of the association may be wrong. For example, large groups may evolve as a response to heavy pressures from predators—but this may be because the proximity of other individuals helps in predator detection (Kenward 1978), because it reduces the probability that the individual will be selected by the predator (Hamilton 1971), because individuals cooperate to fight off the predator (Wilson 1975) or for other reasons (see Harvey & Greene 1981; Chapter 5). In practice, functional explanations concerning species differences are no different from those concerning other functional hypotheses and are ideally tested by manipulative field experiments (see Krebs & Davies 1978). However, where this is impossible, many different kinds of circumstantial evidence are useful. In particular, it is often revealing to examine whether other morphological or behavioural traits likely to be influenced by the same selective pressures also differ between the species concerned (see Tinbergen 1963; Curio 1973). The contrasting defaecation behaviour of black-headed gulls and kittiwakes (Fig. 1.1), as well as the presence of cryptic coloration in black-headed gull chicks and its absence in young kittiwakes, supports the suggestion that eggshell removal in black-headed gulls may represent a response to predation pressure. Secondly, it is often possible to use the explanation to make independent predictions concerning the distribution of the trait in question. For example, evidence that home-range size or feeding-territory size is related negatively to the proportion of foliage in the diet (Clutton-Brock & Harvey 1979b) and positively to the number of sympatric competitors (Yeaton & Cody 1974) supports the suggestion that larger species have bigger territories because they require more food.

Lastly, even where it is reasonably certain that a causal relationship exists between two variables, it may be impossible to tell in which direction it runs. For instance, does body size influence home-range size or does home-range size affect body size? When faced with these questions, it is important to be clear about the time-scale to which they refer; within contemporary populations body size probably controls home-range size (since manipulating home-range size does not have a marked effect on body size), but on an evolutionary time-scale, it is possible that the occupation of feeding niches requiring large home-range size affects selection pressures for body size (see Pennycuick 1979). A common assumption in these cases is that the ecological niche a species occupies ultimately controls its behaviour and morphology, and in many cases this may well be so: it would seem substantially more likely that eggshell removal is an adaptation to ground nesting than that ground nesting is an adaptation to behaviour. However, particular morphologies or behaviours can pre-adapt animals to invade certain ecological niches. For example, the kea (*Nestor notabilis*) is a New Zealand parrot which uses its beak to rip through the skin of sheep and feed on the fat beneath. The beak is like that of normal seed- and fruit-eating parrots but its form presumably pre-adapted the kea to invade the sheep-eating niche (see Futuyma 1979; section 4.6).

Questions about the direction of causation on an evolutionary time-scale can be answered if the fossil record is sufficiently complete to record the order in which related traits evolved. However, in practice this is seldom the case and these problems are generally inaccessible to the comparative method—or for that matter to an experimental approach. This does not invalidate the use of comparisons to investigate the functional significance of the trait, for we can expect the products of evolution to be clusters of functionally related traits, although it does mark a boundary which our understanding of adaptation is unlikely to reach in the foreseeable future.

1.4 CONCLUSIONS

Interspecific comparisons provide the stimulus for posing many adaptive questions, as well as an important way of testing hypotheses about function. Paired comparisons of closely related species are often the only way of investigating the functional significance of detailed aspects of behaviour, but they provide no indication of the generality of relationships between behavioural and ecological traits. Quantitative comparisons involving data from many different species can be useful in establishing general trends as well as in testing specific hypotheses about the form of relationships. However, it is important that the variables used are biologically meaningful, that estimates extracted from the literature are sufficiently accurate for the purposes of the study, and that particular taxonomic groups are not grossly over-represented in the sample. Where closely related species show very similar values, it may be necessary to avoid the assumption that they are statistically independent of each other, or to repeat analyses at different taxonomic levels. Where comparisons involve animals of different sizes, it is important to be aware of the fact that body size is correlated with most behavioural and ecological variables, and that size effects frequently differ between taxonomic groups.

Interpreting interspecific comparisons also requires care. Associations between ecological parameters are commonly responsible for correlations which have no causal basis, while it is often the case that several different functional hypotheses predict the same correlations between behavioural and ecological traits. Lastly, even where it is reasonably certain that two variables are causally related to each other it may not be possible to determine the direction of causality, although this may depend on the time scale in question.

Chapter 2
Evolutionarily Stable Strategies

GEOFFREY A. PARKER

2.1 THE ESS APPROACH—WHAT IS IT, AND WHY DO WE NEED IT?

My aim in this chapter is to introduce to fellow behavioural ecologists/sociobiologists the evolutionarily stable strategy or ESS (Maynard Smith 1972) approach, in the hope that they may come to share my enthusiasm for it.

The chapter is not intended for the converted, nor for ESS theorists; I have striven to reduce mathematical detail to a minimum. An excellent, more detailed survey is now available, written by Maynard Smith (1982a).

2.1.1 Strategies and models

In biology, a strategy is simply one of a series of alternative courses of action. A strategy might, say, be a particular way of allocating parental effort between producing sons and producing daughters. To understand why evolution often results in systems with equal numbers of each sex, we need to ask what will happen in a population of individuals that play all possible alternative strategies of sex ratio (see Chapter 8). Whether all the strategies can ever exist in a heritable form in real populations is an important question, but should not deter us from considering what would happen if they could. What we wish to do with our strategy set (the series of alternative strategies) is to derive some *a priori* expectation of evolution, which we can then compare with observations from nature.

In defining a strategy set, there are two things to consider: (i) what are the (biologically) plausible strategies, and (ii) how much do we wish to simplify the list of alternatives in order to make the analysis easier? Both must relate to the question that we are attempting to answer. Suppose the question is: 'Why is the sex ratio often 1 : 1, given that there can be only two sexes?', then the obvious strategy set to consider is the continuous range between the extremes of producing only sons and producing only daughters. If the question is: 'Why do animals damage each other in contests rather rarely?', it is probably satisfactory to consider just two alternatives such as 'hawk' and 'dove' strategies (p. 39).

Strategy sets can incorporate either continuous or discrete arrangements of strategies. The distinction is obvious; a continuous strategy set could be 'play any value of x between 0 and x_{max}', and a discontinuous set could be 'play either x_A, x_B or x_C'. In general, continuous strategies are more realistic biologically, but discrete strategies are sometimes more useful. It depends upon the question we wish to ask.

2.1.2 ESSs

A strategy is an ESS if, when adopted by most members of a population, it cannot be invaded by the spread of any rare alternative strategy (Maynard Smith 1972). In looking for an ESS, we are simply seeking a strategy that is robust against mutants playing alternative strategies.

Why do we need to think in terms of the ESS to understand animal behaviour? The answer is straightforward. The payoff (= value in terms of Darwinian fitness) to a strategy must often depend strongly on the strategies being played by other individuals. This is notoriously true of behaviour, especially social behaviour, but the effect can apply to any aspect of biology. We need the ESS approach to cope with this problem of frequency-dependent selection, i.e. when payoffs to actions depend critically on the types and frequencies of the other strategies adopted by members of the population.

An ESS or 'competitive optimum' will differ from what we may term a 'simple optimum', in which fitnesses are not frequency dependent (see Dawkins 1980; section 5.4.3). A strategy that is a simple optimum is the best strategy against some fixed condition, such as an environmental feature like the onset of winter. For instance, there could be a trade-off between expenditures on reproduction and on provisioning for hibernation. If the genetic system will allow it, evolution can produce an optimum allocation between the two expenditures, at which point fitness will be maximized. An ESS is a competitive optimum, in the sense that competitors do not have a fixed effect like the onset of winter. Thus as evolution proceeds, the best strategy to adopt may change because other individuals will be behaving differently. Further, there may be more than one stable equilibrium, and the one that is reached will depend on the frequencies of strategies that were present initially in the population. Finally, at an ESS, fitness is unlikely to be at a maximum value; other (unstable) population states may yield higher average rates of reproduction.

These differences can cause some confusion. At a simple optimum, fitness is maximized both at the individual and population levels. This is equivalent to what a population geneticist means if he says that $\bar{w}$ (mean fitness) has a maximum value. But if a sociobiologist talks of fitness maximization, he generally means that selection will

$\bar{w}$ = mean fitness

favour playing the strategy that yields highest reproductive success in relation to competing strategies. This is what he means by using the shorthand 'individuals maximize their fitness by . . .'; he does not usually wish to imply that selection will proceed to a maximum value for $\bar{w}$. This difference in terminology has led to misunderstandings, and O'Donald (1982) is probably right in saying that sociobiologists should be less sloppy in their prose.

How does the ESS approach differ from classical work of population geneticists (e.g. Clarke & O'Donald 1964) on frequency-dependent selection? The difference probably has most to do with emphasis. Population geneticists have been mainly concerned with dynamics of gene frequencies, and with the properties of equilibria. The aim has often been to investigate effects of different types of selection on the alternative genotypes at a locus. Strategy sets are simplified drastically (often to just two strategies related to two alternative alleles) in the interests of analytical tractability. ESS theorists, on the other hand, have not been concerned with questions about the genetic system, and usually assume that strategies replicate asexually or that the organism is haploid. This sacrifice of genetic rigour allows greater freedom to consider wider, more complex strategy sets, and more elaborate fitness interactions between strategies. The different interests of the two disciplines have led both to make unrealistic simplifications, but in opposite directions.

2.1.3 History

Before Maynard Smith's first papers on ESS (Maynard Smith 1972; Maynard Smith & Price 1973; Maynard Smith 1974a), several biologists had applied what we would now call 'ESS reasoning' to specific problems. Hints of this are clearly evident in Darwin's (1859) discussion of sex ratio, a problem that was later solved by Fisher (1930) using a cryptic verbal ESS argument. Population geneticists interested in frequency-dependent selection (see above) had also thought in ESS terms. Perhaps the most notable example of ESS thinking prior to 1972 is Hamilton's (1967) investigation into the 'evolutionarily stable sex ratio' with local competition for mates.

Some field biologists had also used ESS reasoning. In work on dung flies, Parker (1970a) argued that variations in male behaviour would be held in stable equilibrium because of frequency-dependent gains. Fretwell & Lucas (1970) independently applied the same philosophy in a general model of habitat use. Parker (1970b, c) also considered the fitness of mutant strategies in populations fixed at different states, in order to derive what would be the stable state.

Gadgil (1972) had interpreted male dimorphism in terms of equal fitness of the two morphs. The problem of the evolution of the two sexes is also strongly frequency dependent, and was initially approached by computer simulations to find the stable state (Parker *et al.* 1972). There are doubtless many other examples of 'ESS before ESS'.

33

*Evolutionarily
Stable
Strategies*

However, credit for the development of the ESS concept quite rightly goes to J. Maynard Smith. Not only did he suggest the term ESS and appreciate its general significance for biology, but he and G. Price also made the first crucial step towards formalizing the ESS in terms of the mathematics of game theory. This made the study of evolutionary games accessible to game theorists. My own view is that Maynard Smith's ESS concept is the most important recent development in evolutionary theory.

During the past few years, the study of ESSs has grown exponentially. At the present state, it is convenient to divide ESS theory arbitrarily into 'pure' and 'applied' components. Pure ESS theory concerns the fundamental properties of an ESS itself. Applied ESS theory concerns the analysis of specific games in biology; to date most ESS theory has been of this type, though the amount of pure ESS theory grows steadily.

Although some introduction to the formal rules of ESS is necessary, this chapter mainly concerns how we can apply ESS theory to specific biological circumstances.

2.2 TYPES OF ESS—SOME TERMINOLOGY

There are two basic types of ESS:

(i) pure ESS—for example, in condition c, play the unique strategy A;
(ii) mixed ESS—for example, in condition c, play pure strategy A with probability p_A, B with probability p_B, C with p_C, etc.; the values of p_A, p_B, p_C, etc., are prescribed by the ESS.

There are essentially two sources of adaptive variation in behaviour (Parker 1978a; Maynard Smith 1979). First, conditions may vary as c_1, c_2, c_3, etc. The ESS may be a different pure strategy, A_1, A_2, A_3, etc., in each condition. There is no reason to expect the fitness payoff for individuals playing A_1 to be equal to that of individuals playing A_2 or A_3. For this reason, such ESS arrangements have been termed 'best of a bad job' strategies (Maynard Smith 1982a). They are a form of conditional ESS (Dawkins 1980), i.e. a system in which the circumstances an individual finds itself in prescribe the strategy (pure or mixed) that it should play.

Secondly, adaptive variation in behaviour may result from a mixed ESS. In some instances, conditions may be identical for all individuals, but no pure strategy can be stable on its own, so that the solution is a mix of strategies. At equilibrium, each pure strategy component of the mixed ESS must, on average, do equally well in terms of fitness payoff. No one pure strategy can do better than any other—payoffs must be equal.

A mixed ESS could be achieved in two ways (Maynard Smith & Parker 1976). The population could be genetically polymorphic, with the ESS strategy frequencies corresponding to the appropriate geno-type frequencies. A strategy is best defined as what an individual does, not what a population does. For polymorphic equilibria, the term 'ESS' is best seen as an abbreviation of 'evolutionarily stable state' (Dawkins 1980), since each individual would consistently play only one pure strategy, determined by its genotype. There are major difficulties to be overcome in explaining how complex distributions of strategies could be maintained from generation to generation with sexually reproducing diploidy; at present we have no adequate ana-lytical techniques for coping with a complex polymorphic ESS. Alter-natively, a mixed ESS could be achieved by genetic monomorphism, with individuals playing strategies randomly within the ESS probabil-ities. Each individual behaves like a random number generator. This notion does not always appeal to behaviourists, despite the fact that we obtain so much variation in behaviour under constant conditions.

A conditional ESS is really a set of solutions to a set of conditions. It may sometimes be convenient biologically to divide conditional strategies into:

(i) environmentally determined strategies—individuals play strategies in relation to external cues;

(ii) phenotype-limited strategies (Parker 1982)—individuals play stra-tegies in relation to their own phenotype, or their phenotype relative to that of a contestant.

2.3 BIOLOGICAL GAMES—WHAT TYPES ARE THERE?

In games theory, a distinction is often made between two-person games and *n*-person games. In biology, this distinction can also be useful. The following classification is a modification of that given by Parker (1982); it is intended more for biological than for mathematical convenience.

Contests

Individuals meet in pairs, so that the game concerns the interaction between pairs of strategies. Maynard Smith (1982a) has termed these 'pairwise interactions'. Obvious examples of contests are fights between pairs of individuals over resources such as territories, females, nest sites, or food items.

Scrambles

Many, probably most, competitive games do not involve pairwise interactions, they involve *n* players—and *n* may sometimes be the number of individuals in the population. Maynard Smith (1982a) calls these sorts of interactions 'playing the field'. In the same way that contests can be subdivided into continuous-strategy and discrete-strategy versions, so can scrambles. But because they have come to be rather distinct biologically (unlike the two sorts of contest), I feel it is worth emphasizing the two sorts of scramble by using the following terms.

(a)   One-option scrambles *Continuous*

A one-option scramble is an n-player game in which a given individual can gain more of some fitness-related commodity by increasing its expenditure beyond that of its competitors. An ESS concerns a level of expenditure which can be varied continuously. One-option scrambles are a form of biological treadmill that is somewhat equivalent to advertising in economics. The main benefit of extra advertising concerns the increased gains made during the transient stage before other competitors are forced to match this expenditure. Similarly, in the biological version, gains to self are related to self's expenditure compared to the expenditure of others. Models of one-option scrambles have been constructed for parent–offspring conflict (Parker & MacNair 1979), territoriality (Knowlton & Parker 1979; Parker & Knowlton 1980), sexual advertisement (Andersson 1982a; Parker 1982), and arms races (Maynard Smith 1982a; Parker 1983a).

(b)   Alternative-option scrambles *discrete*

These are also n-person games, but instead of there being basically just one way to obtain benefits, there are two or more alternative options. I formerly called such games 'alternative strategy competitions' (Parker 1982) because of the case (much discussed in the literature) of alternative mating strategies.

Alternative-option scrambles are in a sense the discrete-strategy version of one-option scrambles, but not quite. In alternative-option scrambles, the gains from each option are usually inversely proportional to the number of competitors adopting that option, and there may even be a continuous set of options available to exploit. The most simple case is the 'ideal free distribution' of Fretwell and Lucas (1970). At equilibrium, we should expect the distribution of individuals amongst the options, for example habitats, to be such that all competitors achieve equal gains; good habitats will contain more individuals than poor habitats. Evidence for ideal free distributions have been found both in the mate searching of dung flies (Parker 1970a, 1974a, 1978b) and toads (Davies & Halliday 1979) and in fish and ducks foraging for food (Milinski 1979b; Harper 1982; section 5.4.3).

Other examples of alternative-option scrambles include the typical producer–scrounger systems such as food stealing (see Brockmann &

Barnard 1979), food hoarding (Andersson & Krebs 1978), 'digging and entering' in *Sphex* wasps (Brockmann *et al.* 1979), and alternative mating strategies (Alcock *et al.* 1977; Thornhill 1979a; see reviews by Rubenstein 1980; Dunbar 1982).

2.4 THE LOGIC OF ESS ANALYSIS—SOME NOTATION

To derive the conditions under which a strategy can be an ESS, we make the assumption that mutations occur as rare events, and hence occur singly. Consider a strategy I that is fixed in a population. Strategy J is any rare alternative strategy to I; J is really every other strategy in the strategy set, examined separately, in competition with I.

First, consider a contest. $E(J, I)$ is the expected payoff to an opponent that plays J against an opponent that plays I; $E(I, I)$ is the expected payoff of I played against itself, and so on. Payoffs are *changes* in fitness that result from contests, *not* the overall fitness experienced by individuals during a lifetime. Let p = the frequency of strategy I; q = the frequency of strategy J. The 'fitnesses' of strategies I and J will be:

$$\text{fitness of I} = W(I) = pE(I, I) + qE(I, J) + W_0$$

$$\text{fitness of J} = W(J) = pE(J, I) + qE(J, J) + W_0$$

where W_0 is a constant which ensures that fitnesses cannot be negative; it is fitness *excluding* the effect of contests.

This leads us to what are now known as 'Maynard Smith's first and second ESS conditions'. Strategy I is an ESS if $W(I) > W(J)$ for each alternative J when the mutant strategy J is rare. If J is rare, $0 < q \ll 1$, so that we can write that I is an ESS if:

ESS condition (1) $\qquad\qquad E(I, I) > E(J, I)$

In terms of game theory, I must be a *best reply to itself*. What if $E(I, I) = E(J, I)$—in other words, some strategies J are *alternative best replies* to I? We then need:

ESS condition (2) $\qquad\qquad$ If $E(I, I) = E(J, I)$

$$E(I, J) > E(J, J)$$

Stated verbally, condition (2) requires that if J is an alternative best reply to I, then I must be a better reply to J than J is to itself.

Although the first condition is usually all that is needed when the ESS is a pure strategy, the second condition must usually be met if the ESS is a mixed strategy. This is because all the component pure strategies of the mixed ESS I (these are called the *supports* of I) must do equally well against I. Hence by definition, if mutant strategy J consists of some support(s) of I, J must be an alternative best reply to I, and so we require condition (2).

How do we find a mixed ESS? We first seek for probabilities p_A,

$p_B, p_C \ldots p_N$, or in the case of a continuous strategy, for a probability density function, such that each support has an equal expectation against the mixed strategy. When we have done this, we have found an equilibrium strategy, but still need to show that it is a stable equilibrium. To do this, we must use ESS condition (2).

What about the ESS for scrambles? An alternative pair of conditions has recently been proposed by Hammerstein (in preparation). Let the fitness of a single J strategist in a population of I strategists be written as $W(J, I)$. I will be an ESS if:

ESS condition (1) $W(I, I) > W(J, I)$

When $W(I, I) = W(I, J)$, then we need an equivalent of condition (2). But this will now be rather different; we need to show that the fitness of I is greater than the fitness of J in a population of I strategists that contains a small proportion q of J strategists. Let $W(J, P_{q, J, I})$ be the fitness of a J strategist in a population that contains $qJ + (1 - q)I$ individuals. Then:

ESS condition (2)   If $W(I, I) = W(J, I)$

then for small values of q

$$W(I, P_{q, J, I}) > W(J, P_{q, J, I})$$

Hammerstein's modifications are more general than the original ESS conditions; they can also be used for contests, in which case they reduce to the conditions given originally by Maynard Smith.

2.5   ANIMAL CONTESTS

Animal contests probably offer the best demonstration of the need for the ESS approach. Though students of the psychology of 'aggression' ponder endlessly over its causes, as behavioural ecologists let us simplify the world by assuming that animals fight each other to gain some fitness-related commodity. Thus if two individuals meet by an item of food or a territory, either would gain in terms of fitness if it were able to repel the other individual and exploit the resource alone.

What should an animal do to win? This clearly depends on how his opponent is likely to respond. If the response is likely to be 'fight to the death', it will probably pay to withdraw and search for an alternative, unoccupied resource. In contrast, if an opponent is likely to flee after receiving a few threats, it will clearly pay to offer some demonstration of belligerence.

There have been essentially three types of approach to the modelling of animal contests, depending on the sort of decision rule for winning.

1. *War of attrition models* (Maynard Smith & Price 1973; Maynard Smith 1974a). Victory is decided by persistence—the winner is the individual that chooses to continue longer than his opponent. The choice of persistence time is made before the contest.

2. *Hawks–doves models* (Maynard Smith 1972; Maynard Smith & Parker 1976). If there is escalated fighting, the winner is the individual who injures his opponent.

3. *Information acquired during a contest* (Maynard Smith & Parker 1976). See section 2.5.3.

The theory of animal disputes began with the assumption that contests are *symmetric*, i.e. opponents are equal (Maynard Smith 1972; Maynard Smith & Price 1973). There arose a growing need to include inequalities between opponents in the models (Parker 1974b), and the theory of *asymmetric* contests then began (Maynard Smith 1974a; Maynard Smith & Parker 1976). We shall look first at symmetric contests, since these provide a useful introduction to the parallel asymmetric forms and to the simple application of ESS techniques.

2.5.1 Symmetric contests

Model 1: War of attrition

Question: What is the ESS when winning a contest depends on choosing a longer persistence time than your opponent?

Strategies: A strategy is a choice of persistence time t, which can vary continuously between zero and infinity.

Analysis: Let V = the value of winning

Let c = the rate at which costs are expended during the contest; at time t cumulative costs are therefore ct

(i) *There is no pure ESS*
Imagine a population that plays the pure strategy T

$$\text{The mean payoff,} \quad E(T, T) = \frac{V}{2} - cT$$

It is easy to see that if $cT < V$, any mutant playing $t > T$ will spread.

$$\text{If} \quad cT \geqslant \frac{V}{2}, \quad \text{any mutant playing } t = 0 \text{ will spread.}$$

(ii) *A mixed ESS*
Suppose that there is a mixed ESS—call it strategy I—that plays a probability distribution of times t, $p(t)$.

Now if I is an equilibrium strategy, each component pure strategy time t' played against I must have the same payoff. So we can write:

$$E(t', I) = \underbrace{\int_0^{t'} (V - ct)p(t)dt}_{t' \text{ wins}} - \underbrace{\int_{t'}^{\infty} ct'p(t)dt}_{t' \text{ loses}} = \text{constant A}$$

We can now differentiate:

$$\frac{dE(t', I)}{dt'} = \frac{dA}{dt'} = 0$$

Remembering that p(t) is a probability density so that

$$\int_0^\infty p(t)dt = 1$$

we can eventually show that

$$p(t) = \frac{c}{V} \exp\left(-\frac{ct}{V}\right) \qquad (2.1)$$

We know that strategy I that plays this distribution is an equilibrium strategy—but is it a stable equilibrium, an ESS?

To prove this we must rely on satisfying ESS condition (2), since we know that any mutant strategy J is an alternative best reply to I. Those interested in seeing how this is done should consult Maynard Smith (1974a).

Comments: Thus the ESS is a mixed strategy 'select a persistence time t according to the negative exponential distribution given in equation 2.1'. This is, of course, the distribution of bids, not the distribution of contest lengths, which can be calculated as:

$$p_0(t) = \frac{2c}{V} \exp\left(-\frac{2ct}{V}\right) \qquad (2.2)$$

(Parker & Thompson 1980).

The mean contest cost in the war of attrition is always exactly $V/2$, so that the expected payoff to each opponent is in fact zero (Maynard Smith 1974a). The result of fighting to obtain a resource is eventually equivalent to withdrawing immediately, without fighting. This is to be expected, since playing $t = 0$ is one of the component pure strategies in p(t).

The war of attrition has been further developed theoretically by Norman, Taylor and Robertson (1977) and by Bishop and Cannings (1978). As yet, there is no good evidence that any animal contest obeys the rules for a symmetric war of attrition.

Model 2: Hawks–doves game

Question: What is the ESS when two levels of fighting are possible—escalated, damaging fighting versus amicable, conventional settlement?

Strategies: Hawk (H)—fight at escalated level, retreat only
if injured.
Dove (D)—retreat immediately if opponent escalates,
otherwise settle quickly without escalation.

Analysis: Let V = the value of winning
                Let C = the cost of injury

This is a discrete strategy game, and we can construct the payoff matrix:

against:

	Hawk H	Dove D
Hawk H	$\dfrac{V - C}{2}$	V
Dove D	0	$\dfrac{V}{2}$

Payoff to: (rows)

Thus doves lose against hawks, but do equally against other doves (gain V/2). Hawks win against doves, but get equal chances of victory or defeat against other hawks (gain (V − C)/2).

First of all, pure dove can never be an ESS, since it can always be invaded by hawk.

(i) *Injury costs less than the resource value*

If C < V, we can easily show that pure hawk is an ESS. By concordance with ESS condition (1), H is an ESS if:

$$E(H, H) > E(D, H)$$

$$\therefore \quad \frac{V - C}{2} > 0$$

$$\therefore \quad V > C$$

(ii) *Injury costs more than resource value*

If C > V, it is clear from (i) that pure hawk cannot be an ESS. If the ESS is a mixed strategy 'play hawk with probability p', then at equilibrium the expected payoff to a randomly selected hawk must equal that of a randomly selected dove:

$$\underbrace{p\left(\frac{V - C}{2}\right) + (1 - p)V}_{\text{expected payoff to H}} = \underbrace{p \cdot 0 + (1 - p)\frac{V}{2}}_{\text{expected payoff to D}}$$

$$\therefore \quad p = \frac{V}{C} \qquad (2.3)$$

This equilibrium strategy can be shown to be stable (Maynard Smith & Parker 1976)

Comments: An immediate conclusion from equation 2.3 is that the tendency for dangerous escalation should decrease as the cost of injury increases. Whether this is so remains to be established, though sometimes the evidence is persuasive. For instance, in intraspecific combats, poisonous snakes rarely bite each other. However, caution is necessary: the reason that severe damage is relatively rare in animal combat may be more to do with skills at avoiding damage than with inhibitions about inflicting it.

A second conclusion of equation 2.3 is that the more valuable the resource, the greater is the probability that an animal will take risks to defend it. This conclusion would seem fairly trite to most ethologists, and we need not belabour the evidence.

The value of the hawks–doves game is mainly that it allows us to think about levels of fighting that differ qualitatively; its use is mainly heuristic. However, there are one or two biological examples that appear seductively 'hawk–dovish'. In African elephants (*Loxodonta africana*), a male may enter temporarily the physiological and behavioural state of 'must'; such males are characteristically frenzied and hawkish, and are prone to make attacks spontaneously and with enthusiasm.

Finally, two general comments should be made about the war of attrition and hawks–doves models. First, it is important to note that the resource value V is not a direct measure, such as the calorific value of a contested food item. It is really an opportunity cost, related to the search cost of finding an alternative, unguarded resource of equivalent value. Secondly, there has grown a misconception that the war of attrition is appropriate only for energetically expensive displays, but not if injury occurs, and vice versa for hawks–doves. This is not, in fact, the distinction between the models. War of attrition is entirely appropriate when injuries occur in contests, if:
(i) injury does not affect subsequent performance;
(ii) the average contest costs increase as a continuous function of time spent fighting;
(iii) animals can therefore regulate mean contest costs continuously by a continuous choice of persistence times.
The (continuous) war of attrition is a model in which animals can regulate contest costs continuously. The hawks–doves game is a model in which there is an abrupt change in contest costs if an animal alters its fighting strategy; it is about different levels of fighting.

2.5.2 Asymmetric contests and assessment

There are essentially three sorts of asymmetry in contests (Maynard Smith & Parker 1976).

(i) *Asymmetries in resource value.* One opponent may have more to gain from winning than the other opponent.

(ii) *Asymmetries in 'resource-holding potential' (RHP).* Opponents may differ in fighting ability, or in their ability to defend a resource, due to such features as positional effects.

(iii) *Uncorrelated asymmetries.* These are arbitrary asymmetries that are uncorrelated with any effect on payoff. A favoured candidate here is the 'owner–interloper' asymmetry—in most contests, one individual will be present at the resource before the other, and there is no reason *a priori* why this asymmetry should correlate with any asymmetry in (i) or (ii). There are, however, many biological reasons why there could be a correlation between RHP and/or resource value and being an owner or an interloper (Parker 1974b).

Hammerstein (1981) has termed the first two types of asymmetry 'payoff-relevant' asymmetries, in contrast to 'payoff-irrelevant' (uncorrelated) asymmetries.

ESS for discrete-strategy contests

Model 3: Asymmetric hawks–doves game

Question: What is the ESS for the hawks–doves game when there is an uncorrelated asymmetry that both opponents can perceive accurately?

Strategies: Since it is an asymmetric contest, let one opponent occupy role A (he might be 'owner') and the other, role B (e.g. 'interloper'). Each opponent has not only the choice of playing hawk or dove (or some mix of the two); he can now choose his strategies conditionally on being in role A or B. A strategy now consists of what to do in role A, and what to do in role B.

Analysis: For reasons that parallel the symmetric hawks–doves game, we find that if $C < V$, the ESS is 'always play hawk whatever role'. If escalated fighting is costly, so that $C > V$, there are now two *pure* ESSs:

(i) 'play hawk in A, play dove in B',

(ii) 'play hawk in B, play dove in A'.

We can prove that these are both ESSs from ESS condition (1). Roles A and B occur half the time. Suppose that 'play hawk in A, dove in B' has reached fixation in the population. Call it strategy I. Then

$$E(I, I) = \frac{V}{2}.$$

Let mutant strategy J be any alternative strategy to I; it plays hawk in role A with probability p_A, and plays hawk in role B with probability p_B. Then

$$E(J, I) = \frac{1}{2}\left[1{-}p_A\, \frac{V}{2} + p_A V \right] \quad \text{when J is in role A}$$

$$+ \frac{1}{2}\left[(1 - p_B)0 + p_B\left(\frac{V - C}{2}\right) \right] \quad \text{when J is in role B}$$

$$= \frac{V}{4}(1 + p_A) + p_B\left(\frac{V - C}{4}\right)$$

Now $p_B(V - C)/4$ is negative since $C > V$, and $V(1 + p_A)/4$ is smaller than $E(I, I) = V/2$. Hence 'play hawk in A, dove in B' is an ESS. By similar reasoning 'play hawk in B, dove in A' will also be an ESS. We simply change A for B throughout the analysis.

Comments: This model outlines the important point that a particular game may have more than one ESS. Bistable or multistable equilibria have long been discussed by population geneticists; behavioural ecologists sometimes find the concept rather surprising.

Note that if the population is fixed at one of the two ESSs, no escalated fighting (hawk) will be seen. This can be called a 'conventional settlement' (Maynard Smith & Parker 1976).

Much debate centres on 'bourgeois' (Maynard Smith 1976a), the strategy 'play hawk when owner, dove when interloper'. Does it exist as an uncorrelated asymmetry in nature? There can be no doubt that prior residence is often used to settle disputes, but it is very difficult to prove that prior residence does not correlate with payoff-relevant asymmetries. A few examples of 'antibourgeois' (play dove when owner, hawk when interloper) also appear to occur (see Dawkins & Krebs 1978). Perhaps the important point to note is that both bourgeois and antibourgeois could exist as uncorrelated ESSs, but only (as we shall see later) when animals cannot regulate their contest costs continuously.

Now consider a payoff-relevant asymmetry. We shall apply exactly the same framework of analysis, but roles A and B may now be bigger : smaller; or more to win : less to win. As before, there are two ESSs. Following Maynard Smith and Parker (1976), we can call these the 'commonsense' solution (bigger individual, or individual with more to win, plays hawk) and the 'paradoxical' solution (smaller individual, or individual with less to win, plays hawk). Although either solution can be an ESS when the asymmetry is relatively trivial, only the commonsense solution can be an ESS when the asymmetry is a relatively strong one (Maynard Smith & Parker 1976). Thus where the resource is of equal value to both opponents, we should not expect to see a very large animal back down in favour of a very small one, though we might see a slightly bigger opponent submit.

The most complete analysis of asymmetric contests that obey hawks–doves rules and where there is perfect information about the asymmetry is that by Hammerstein (1981), who has also examined the case where payoff-relevant and payoff-irrelevant asymmetries occur simultaneously. In nature, it rather looks as if animals often respect size when there is a moderate to large size difference, and ownership when there is not (e.g. Kummer *et al.* 1974; Hazlett *et al.* 1975; Riechert 1978), an observation which fits the results of Hammerstein's analysis rather well.

Note that all the ESSs for these asymmetric, discrete-strategy games are pure, conditional strategies. Indeed, Selten (1980) has shown that if animals always occupy different roles (A ≠ B), the ESS must be a pure strategy not a mixed one. Selten's theorem does not apply when animals can make mistakes about their role, so that A = B in at least some contests.

ESS for continuous strategy contests

The asymmetric war of attrition has been considered by Parker and Rubenstein (1981), and more rigorously by Hammerstein and Parker (1982). The discrete war of attrition—provided it has big enough gaps in the choices of persistence times that an animal can play—behaves rather like the asymmetric hawks–doves game: both commonsense and paradoxical solutions can exist. This is to be expected, since in both models animals cannot regulate contest costs continuously. The continuous asymmetric war of attrition, however, appears to have just one ESS—it is a commonsense ESS.

First of all, there is no ESS for the asymmetric war of attrition if opponents have perfect information about the asymmetry. Candidates for ESSs were described by Maynard Smith and Parker (1976), they are (i) 'in role A choose a large persistence time T that inflicts on the opponent costs which exceed the opponent's resource value; in role B retreat immediately', and (ii) the parallel strategy, with A and B reversed. However, these cannot now be ESSs, for the following reason. If the population is fixed for one or other strategy, escalation (in this case, playing T) is never shown. Thus there is no selection to maintain the value of T, which will therefore alter because of drift. If T becomes too low, then the strategy becomes unstable and can be invaded by a convention-breaking strategy.

There is a way out of this problem if we consider mistakes in role assessment (Parker & Rubenstein 1981; Hammerstein & Parker 1982). When animals make mistakes, it is possible for them both to occupy the same role; they may both assess themselves to be on the same side of the asymmetry. This has the effect of allowing selection to mould the entire conditional ESS, so that no part of it is subject to drift. The ESS is: 'choose a persistence time from probability distribution $p_A(t)$ when in role A, and a persistence time from $p_B(t)$ when in role B'.

These two distributions are non-overlapping, but there is no gap between them (Fig. 2.1). The distributions are separated at point S, and the more likely an animal is to make mistakes about roles A and B, the higher the value of S (Fig. 2.1). We can call role A the 'winning' role and B the 'losing' role, since the t values in $p_B(t)$ are all smaller than those in $p_A(t)$.

This ESS is a mixed, conditional ESS. Mostly, $p_A(t)$ will play (and win) against $p_B(t)$. When mistakes are very rare indeed, $p_B(t)$ collapses towards a unit probability at $t \to 0$, and

$$p_A(t) \to \frac{\bar{c}}{\bar{V}} \exp\left(-\frac{\bar{c}t}{\bar{V}}\right)$$

which is the same negative exponential distribution as in equation 2.1 for the symmetric war of attrition; $\bar{c}$ and $\bar{V}$ are now mean values for the two opponents.

What determines roles A and B, and hence choice of $p_A(t)$ or $p_B(t)$? This is decided by what could be termed the 'assessor rule'. Suppose that one opponent will get V_a if it wins, and that it will expend fitness at rate c_a if it fights. The corresponding values for the other opponent are V_b and c_b. The winning role A is associated with having the higher value for V/c, and role B with the lower value for V/c; i.e. opponent a should attempt to choose $p_A(t)$ when:

$$\frac{V_a}{c_a} > \frac{V_b}{c_b} \tag{2.4}$$

Sometimes he will make mistakes, and choose $p_B(t)$.

What does the assessor rule mean biologically? It is obviously a commonsense ESS. If opponents have the same fighting ability or resource-holding potential $(c_a = c_b)$ then the winning role is associated with having the higher resource value $(V_a > V_b)$; if resource values are equal $(V_a = V_b)$, the winning role is associated with having greater fighting ability $(c_a < c_b)$. But the rule also shows how the two payoff-relevant asymmetries interact when they are *contradictory* (the opponent with the higher resource value has lower RHP). Since V is a number of units of fitness, and c is a rate at which fitness units are expended, then V/c defines a critical time beyond which a given opponent must get a negative payoff from the contest, even if it wins. Thus the assessor rule tells us that a contestant should attempt to choose the winning role A if he would reach this critical time later than his opponent, were both to fight indefinitely (Fig. 2.1c).

If mistakes are rare, the assessor rule approximates to 'retreat if you would be first to spend more in a contest than the resource is worth to you, otherwise persist'. Ironically (and more by good luck than by deep insight), this sort of rule was proposed some time ago (Parker 1974b; see also Popp & Devore 1979); its mathematical vindication proved rather complex (Hammerstein & Parker 1982).

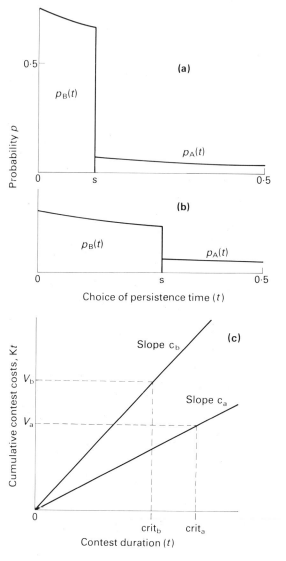

Fig. 2.1 (a, b) ESS choices of persistence times for the continuous asymmetric war of attrition. An individual should choose from distribution $p_A(t)$ if he estimates himself to be in role A, and from $p_B(t)$ if he estimates himself to be in role B. In the examples shown in diagrams a and b, the rate c at which fitness is expended is equal for both opponents ($c_a = c_b = 1$) but the resource is twice as valuable for one opponent as for the other ($V_a = 2$, $V_b = 1$). In diagram a, mistakes in role assessment are rare (90% of the time individuals are choosing from the correct bid distribution). In diagram b, mistakes are much more common (60% correct). As mistakes increase, the separation point s increases and more escalation will be observed. **(c)** The rule for choosing role A or B. An individual should choose role A if $V_a/c_a > V_b/c_b$, i.e. if $crit_a > crit_b$. In the case shown, there is a contradiction between the two asymmetries; the individual who should choose role A has better fighting ability ($c_a < c_b$) but a lower resource value ($V_a < V_b$). The cumulative contest costs exceed the resource value first for the opponent who has V_b and c_b; he should therefore select role B. See also Popp & Devore 1979, Hammerstein & Parker 1982.

Animals usually respect size and weaponry differences in a commonsense fashion; just one or two examples must suffice here. Geist (1971) was one of the first ethologists to adopt an individual rather than group selection approach to the evolution of fighting behaviour; he argued that horn size was an important determinant of dominance status in bighorn sheep (*Ovis canadensis*). The outcome of dungfly (*Scatophaga stercoraria*) struggles is strongly influenced by the relative sizes of the opponents (Sigurjonsdottir & Parker 1981). Sometimes, the cue assessed may not be size or strength directly. Red deer (*Cervus elaphus*) have roaring contests, and victory may go to the individual able to maintain the higher rate of roaring (Clutton-Brock & Albon 1979). Rather similar is the elegant work of Davies and Halliday (1978) on common toads (*Bufo bufo*). Decisions to attack a male that is guarding a female are dependent on visual cues and on the pitch of the croak emitted by the guarding male; deep croaks tend to repel attacks. It is important that when an indirect cue is used to estimate size and strength, the cue cannot be faked, and gives a reliable measure of RHP (see also Chapter 15).

Evidence for assessment of resource value asymmetries is more scanty. It may be difficult to estimate differences in resource value of, say, a food item. However, contest strategy may change with hunger. In the pygmy sunfish, small fish tend to be subordinate to large fish, but if starved, a subordinate will escalate and often win (Rubenstein 1977; see also Hazlett *et al.* 1975 for similar experiments with crayfish). Sometimes, direct assessment of relative resource values may be simple to achieve. In hamadryas baboons (*Papio hamadryas*), Bachmann and Kummer (1980) found that respect of ownership of a female tends to be positively correlated (for middle- and low-ranking males) with a preference of the female for the owner. They interpret this in terms of a resource-value asymmetry, since females that have a strong preference for another male may be costly in terms of herding effort and the risk of subsequently losing her.

2.5.3 Information, and the lack of it

Comparison of symmetric and asymmetric contest models shows clearly that if animals can perceive the asymmetries present in contests, this will greatly alter the ESS. Information will usually be imperfect. I briefly summarize some of the sorts of problem that relate to information (see also section 15.4).

Signals of intent

An animal should obviously not behave in such a way as to assist his opponent, if this will in turn cause a loss to the signaller. Thus in a war of attrition, evolution could hardly favour a signal that relates to an opponent's choice of persistence time. The only rational strategy, whatever the real bid, would be to signal a choice of infinite persistence, and the only rational response to ignore it. Caryl (1979) analysed data on fighting and found that information about attack is very poorly encoded in displays, as we would expect. However, there is some evidence for signals of intent, and this poses some fascinating problems (Maynard Smith 1982a; Rohwer 1982; see section 15.4).

Rather different is the question of signals of RHP; as we have discussed, it will often pay to assess relative RHP and to respond accordingly. Cues of RHP that are not constrained by costs are likely to become exaggerated into bluffs (see Maynard Smith & Parker 1976; section 15.4).

Information about resource value V

An animal may have accurate information about the value of a resource to itself, but may not know the value to its opponent. Bishop *et al.* (1978) have analysed this situation for the war of attrition, assuming that contests are asymmetric only in resource value (RHP is symmetric). If the distribution of resource values in the population is continuous (a large population with accurate monitoring of self's V), the ESS is to choose a time t that increases as self's evaluation of V increases. Thus if $V_a > V_b$, it always holds that $t_a > t_b$. In contrast, if there is a set of discrete states of V (e.g. $V_1 < V_2 < V_3 < \ldots V_n$) the ESS is a set of non-overlapping but adjacent bid distributions $p_1(t)$, $p_2(t)$, $p_3(t) \ldots p_n(t)$, in which the persistence times in $p_1(t)$ are all less than those in $p_2(t)$, and so on.

Information acquired in a contest

Perhaps the most realistic model of animal combat would be one in which opponents have some information about RHP before the start of a contest, but gain further information as the battle continues. Maynard Smith and Parker (1976) considered a game in which victory is decided by persistence, but the decision when to withdraw is made during the contest itself, not beforehand as in the war of attrition. The contest can be seen as a series of rounds; at the end of each round, and in the light of the current information, each opponent makes the decision whether to continue or to withdraw. If the probability of winning one round is only a poor predictor of who will win the next, the result is a mixed ESS for when to give up after losing a round. If the probability of winning a given round is a very good predictor of what will happen subsequently, the ESS becomes a pure ESS for giving up at an early stage after losing (see also Parker & Rubenstein 1981).

I would guess that models of this type, based on the much wider strategic possibilities of being able to give up at any point in a contest, will eventually replace the simpler hawks–doves and war of attrition models. Such models require the techniques of 'games in extensive form', and are currently being developed (Selten & Hammerstein, in preparation).

2.6 ONE-OPTION SCRAMBLES

2.6.1 Differentiating the fitness function

When an individual can increase its relative benefit from contests against other, similar members of the population by increasing some expenditure continuously, we can find the ESS by the technique of differentiation of the fitness function. I demonstrate this using a simple example.

Model 4: The sexual advertisement scramble

Question: What is the ESS for male investment in sexual advertisement to females, when a male's share of the available females increases with his commitment to advertisement relative to that of other males in the same group?

Strategies: a strategy is a continuous choice of advertisement level x

Analysis: Assumptions are as follows:
(i) F females are attracted to each lek (group of n males)
(ii) A male i gains a proportion $x_i/\bar{x}$ of the available females (i.e. equal to his advertisement level divided by the mean advertisement level).
(iii) The cost of advertisement is measured by the survival probability $S(x)$; S is a monotonic decreasing function of x.

If there is a pure strategy x_* that is an ESS against any alternative strategy x, we can see from ESS condition (2) that

$$W(x_*, x_*) > W(x, x_*)$$

As we alter x from one end of its range to the other, for x_* to be an ESS requires that

$$W(x_*, x_*) - W(x, x_*) > 0$$

except where $x = x_*$, where the fitnesses are equal. This point $(x = x_*)$ will be a maximum. Hence we can find the ESS value by differentiating:

$$\left[\delta \frac{[W(x_*, x_*) - W(x, x_*)]}{\delta x} \right]_{x = x_*} = 0$$

Normally, $W(x_*, x_*)$ will not contain any term including x (but see Knowlton & Parker 1979), as we assume that we are dealing with a large population. Hence we need simply that:

$$\delta\left[\frac{W(x, x_*)}{\delta x}\right]_{x=x_*} = 0 \qquad (2.5)$$

and

$$\delta^2\left[\frac{W(x, x_*)}{\delta x^2}\right]_{x=x_*} < 0 \qquad (2.6)$$

Equation 2.6 establishes that x_* is a maximum, not a minimum. This technique is known as 'differentiation of the fitness function'. It is very valuable for finding the ESS for continuous-strategy scrambles, but may only give an ESS that is stable against invasion by nearby strategies ('local stability' rather than 'global stability').

Now, for the sexual advertisement model, the fitness of a mutant male playing x in a population playing x_* is

$$W(x, x_*) = \frac{F \cdot x \cdot S(x)}{\bar{x} \cdot n} = \frac{F \cdot x \cdot S(x)}{[(n-1)x_* - x]/n}$$

Differentiating as in equation 2.5 gives the solution (Parker & Knowlton 1980):

$$-S'(x_*) = \frac{S(x_*)}{x_*} \cdot \frac{(n-1)}{n} \qquad (2.7)$$

The negative sign counteracts the fact that the gradient $S'(x_*)$ is negative. Thus at the ESS, the gradient of $S(x)$ should equal $S(x)$ multiplied by a constant and divided by x.

Comments: To get a feel for what equation 2.7 means, let $S(x) = \exp(-kx)$, where k is a constant that scales how fast survival decreases as x increases. Then

$$x_* = \frac{(n-1)}{kn}$$

In other words, the ESS expenditure will be close to zero when there are few males in a group, as we would expect intuitively. Expenditure x_* will rise to an asymptotic value of $1/k$ at large numbers of competitors. That $x_* = (n-1)/kn$ is a maximum can be established by using equation 2.6.

2.6.2 Two continuous strategies: parent–offspring conflict

I dealt with the previous example at some length to demonstrate how one-option scrambles can be analysed. The next example uses essentially the same technique, but is a little more complex. It is a model of parent–offspring conflict (Trivers 1974). At first sight, it looks like a contest, but detailed investigation reveals the model to be closer to a scramble. I use it to demonstrate (i) how we can use the technique of differentiation of the fitness function for games involving two continuous-strategy sets, and (ii) how we must sometimes include genetics in ESS models. It is the simplest example from a series of models derived by Parker and MacNair (1979).

Assume that a mother has M units of parental investment (PI; Trivers 1972) to expend in her lifetime; she dies when M is expended. She produces each offspring singly, and can put any amount, m, of PI into each offspring. The species is a sexually reproducing diploid. All the offspring of a female have the same father (they are full sibs); the father cannot remate after the female dies. Let the fitness or viability (f) of an offspring be an increasing function of the amount of PI it obtains (see Fig. 2.2a). The mother's optimal investment (m_f) occurs when:

$$f'(m_f) = f(m_f)/m_f \qquad (2.8)$$

This is a simple optimum, it does not depend on what other mothers are doing.

What is the ESS amount of PI (m_o) for an offspring to take, assuming that the mother willingly supplies any amount that the offspring demands? At the ESS (Parker & MacNair 1978):

$$f'(m_o) = f(m_o)/2m_o \qquad (2.9)$$

which is clearly different from the parental optimum in equation 2.8. The ESS for the offspring is for a greater amount of PI (see Fig. 2.2a).

Model 5: Parent–offspring conflict

Question: How will this conflict be resolved, given that a parent may reduce its sensitivity to the demands of its offspring?

Strategies: (i) The parent can select any value of m that it will, on average, invest in each offspring. It is, however, sensitive to the offspring's solicitation, so that if an offspring deviates in its solicitation level, m is allocated in proportion to the ratio: deviant level/mean level.

(ii) The offspring can choose any level of solicitation, x. We shall assume that solicitation carries costs; survival decreases monotonically with x as function $S(x)$, see Fig. 2.2b, similar to the sexual advertisement game in section 2.6.1.

Analysis: The game contains two continuous functions, $f(m)$ and $S(x)$; parents select an investment level m, and offspring select a solicitation level x.

To find an ESS, we use an extension of the technique in equation 2.5; I follow the notation devised by Maynard Smith (1982a). Suppose that the ESS consists of the pair of pure strategies, m_*, x_*. This ESS will be stable against any unilateral deviation in one of the strategy pair. Call $W(mx_*$, $m_* x_*)$ the fitness of an mx_*' mutant in an $m_* x_*$

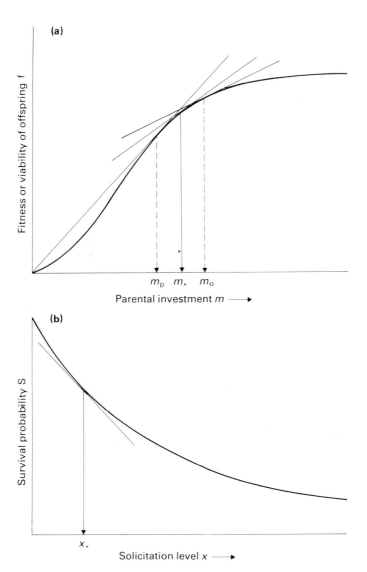

Fig. 2.2 (a) $f(m)$ the relationship between the viability f of an offspring and the amount m of parental investment it receives. The parent's optimum is given by the tangent, at which point the gradient $f'(m_p) = f(m_p)/m_p$. The ideal ESS for the offspring (in the case discussed in the text) has $2f'(m_o) = f(m_o)/m_o$; m_o is always greater than m_p. The resolution of this conflict is where $\frac{4}{3}f'(m_*) = f(m_*)/m_*$; m_* is a compromise between m_p and m_o if costs of solicitation increase continuously as in **(b)**, which shows how an offspring's survivorship S may decline with increasing levels of solicitation. At the ESS $-4S'(x_*) = S(x_*)/x_*$, and the offspring will therefore usually show demands for more PI than the m_* it is receiving from its parent.

population; $W(m_* x, m_* x_*)$ is the fitness of an $m_* x$ mutant. Following equation 2.5, stability requires that

$$\left[\delta \, \frac{W(mx_*, \, m_* x_*)}{\delta m} \right]_{m = m_*} = 0 \qquad (2.10)$$

$$\left[\delta \, \frac{W(m_* x, \, m_* x_*)}{\delta x} \right]_{x = x_*} = 0 \qquad (2.11)$$

and, following equation 2.6, to show that these are maxima, we
require that their second derivatives should be negative.

What is $W(mx_*, \, m_* x_*)$, the fitness of a mutation that causes a
parent to allocate PI at a deviant level? Now, when the parent plays
m_*, the offspring solicits at level x_*, so if the parent alters to
$m \neq m_*$, we assume that the offspring will alter their solicitation
from x_* to $x_* m/m_*$. Thus if the mutant parent pays out more than
m_*, the expenditure on solicitation is reduced; if the mutant parent
pays out less than m_*, expenditure on solicitation increases. The
number of surviving offspring of a mutant parent is therefore

$$W(mx_*, \, m_* x_*) = \frac{M \cdot f(m) \cdot S(x_* m/m_*)}{m}$$

What is $W(m_* x, \, m_* x_*)$, the fitness of a mutant with a deviant
solicitation level? We will assume that if an offspring shifts its
solicitation expenditure from x_* to x, it alters the amount of PI that it
gets, from m_* to $m_* x/x_*$. Thus it can get more PI by soliciting more,
though this will cost it more.

Assume that x is a dominant mutation (the same solution is
obtained if it is recessive) so that half the sibship will be x phenotypes and half will be x_*. The numbers of surviving replicas of the x
allele will be

$$W(m_* x, \, m_* x_*) = \frac{\frac{1}{2} M \cdot f(m_* x/x_*) \cdot S(x)}{\frac{1}{2}(m_* x/x_* + m_*)}$$

in which the denominator $\frac{1}{2}(m_* x/x_* + m_*)$ is the mean PI per offspring, $\bar{m}$.

We now find m_* and x_* by differentiating the two fitness functions, using the rules of equations 2.10 and 2.11. By substituting
between the two equations obtained after the differentiation, we can
get the solution:

when parent, play $\qquad\qquad f'(m_*) = \dfrac{\alpha f(m_*)}{m_*} \Bigg\}$

$\qquad\qquad\qquad\qquad\qquad\qquad\qquad\qquad\qquad\qquad\qquad\qquad (2.12)$

when offspring, play $\qquad -S'(x_*) = \dfrac{\beta S(x_*)}{x_*} \Bigg\}$

in which α and β are the constants 0.75 and 0.25. These are the ESS
pair of strategies; we can solve for m_* and x_* if we know the explicit
forms of $f(m)$ and $S(x)$. We also then test that these are maxima. Note

that the second part of equation 2.12 is equivalent to the ESS for the sexual advertisement scramble, equation 2.7; see Fig. 2.2b.

Comments: The assumptions leading to the calculation of the mutant fitnesses require some biological justification. The notion is that it will be of selective advantage for a mother to have some sensitivity to the changing demands of her offspring, but she cannot tell whether a super-demanding offspring is really in need of extra PI or is simply a 'cheating' mutant. Similarly, a given offspring's solicitations can be to some extent reduced (reducing the solicitation costs) if it is supplied with a higher than average amount of PI.

Several points emerge about the ESS in equation 2.12.

(i) The investment level, m_*, will usually be a compromise between the ideal interest of the mother (m_f) and that of the offspring (m_o) (see Fig. 2.2a).

(ii) The compromise depends on the mating system and the nature of sib-competition. The case analysed gives the lowest level of conflict as measured by x_*; the highest level occurs when offspring compete against other brood members (rather than displace future sibs) and where the offspring are half-sibs. However, altering these sorts of conditions does not fundamentally affect the ESS in equation 2.12, it merely alters the values of the constants α and β, which always have the property that $\alpha + \beta = 1$.

(iii) For any given case, the solutions for x_* and m_* are dependent only on the shape of the function they characterize. For instance, m_* is dependent only on the shape of $f(m)$ and constant α, not on $S(x)$. Hence altering the solicitation costs should not (in the long term) affect PI allocation, though it will change solicitation level.

(iv) At the ESS, we should actually *observe* conflict, because offspring will be soliciting in a costly way for more PI than they are receiving. The very fact that we do observe some conflict over such things as weaning (Trivers 1974) lends some support to the models.

Why is this a scramble rather than a contest? There is no doubt that we could construct a contest model of parent–offspring conflict, if we again include some consideration that the contest concerns relatives (see Grafen 1979). But the present model is a scramble, for the following reason. Parents are in a scramble against other parents, and offspring scramble against offspring. (Remember that to find m_* we examine the fate of a rare m played against the m_* population, and similarly for x_*.) It is a one-option scramble because parents have only one option, and offspring only one option concerning the type of strategy they play. The two scrambles are interdependent because each forms a sort of homeostatic constraint for the other.

2.7.1 Ideal free searching

Ideal free searching represents a case of a mixed ESS (Parker 1978a, b; Milinski 1979b; Maynard Smith 1979, 1982a). The animal has a series of alternative options (different parts of the environment) to search for some fitness-related commodity. The value of each option generally declines as the number of competitors exploiting it increases. An ESS is found by setting payoffs equal for all individuals, and then solving for the number of competitors that must then have chosen each option, under the condition that no individual can do better by a change. There is good evidence that such ESSs occur in nature (see section 3.3).

Because they have been the subject of a previous review (Parker 1978b) and are also discussed by Pulliam and Caraco in Chapter 5, I shall attempt here only to distinguish briefly between two extreme examples. In both instances, payoffs are measured in terms of the rate at which resource items (prey or mates) are captured within a patch. For simplicity, it will also be assumed that movement between patches costs nothing, and that competitors have perfect information about the patch profitabilities.

Type I. No resource renewal; resources deplete with exploitation

A series of equal-sized patches are ordered 1 to n in terms of decreasing profitability. Profitability is a simple function of prey density—the best patch (No. 1) has the highest prey density and therefore offers the highest rate of capture of prey. Provided that there is no 'interference' (reduction in search efficiency due to interactions between the competitors), the instantaneous rate of prey capture is a function *only* of the prey density and is not dependent on the number of competitors in a patch. All predators should exploit patch 1 first, until its prey density is depleted to that of patch 2. The predators should then divide into two halves between patches 1 and 2 until both are depleted simultaneously to the prey density in 3, whereupon a third of the predators should enter patch 3, and so on. An ESS consists of maintaining the instantaneous rates of gain equal at all times for all predators by gradually expanding the options exploited as the existing options become depleted. This is essentially the model proposed by Fretwell and Lucas (1970).

Type 2. Continuous resource renewal

Resource items arrive at the various patches at specified rates; the best patch has the highest input rate. Resource items (often females) are captured instantaneously on entering the patch. The ESS consists of instantaneous division of the competitors between patches in accordance with the input matching rule (Parker 1978b); the number of competitors in patch i must equal the resource input rate into patch i.

Most empirical tests of ideal free theory appear to have involved systems of type 2 (Parker 1974a, 1978b; Davies & Halliday 1979; Milinski 1979b; Harper 1982). In nature, most systems probably contain elements of both type 1 and type 2.

2.7.2 Producers and scroungers

Producer–scrounger type games (Barnard & Sibly 1981) include a wide variety of biological phenomena, ranging from alternative mating strategies (sneaks/guarders; callers/satellites) to food piracy (see section 2.3). The unifying features of these scrambles are as follows.

(i) Certain individuals (producers) invest time and energy in guarding or creating some resource, which other individuals (scroungers) can then parasitize. They are, in fact, analogous to host–parasite systems, or to the male–female phenomenon in which males can be seen as parasites upon female investment.

(ii) The fitness of scroungers is strongly frequency dependent; for stable coexistence of the two strategies the fitness of scroungers (W_s) is higher than that of producers (W_p) when scroungers are rare, but lower when scroungers are common. This is, in fact, the scramble equivalent of the hawks–doves contest model.

A diagrammatic representation of the system is given in Fig. 2.3 (see also Barnard & Sibly 1981). At the ESS frequency of scroungers, p_*, the fitness of both strategies is equal. Although in most cases the fitness of both strategies declines as the frequency of scroungers increases, in others the fitness of only one strategy may be decreasing; that of the other strategy may actually be increasing. This could apply to certain alternative mate-searching systems such as that in the bee *Centris pallida* (Alcock *et al.* 1977; see section 2.8). The same sort of model should apply for both intra- and interspecific cases; at the equilibrium p_* the fitnesses are equal in the sense that the proportions of producers and scroungers will stay constant.

2.8 PHENOTYPE-LIMITED ESS, AND LIFE HISTORY SWITCHES

Mixed ESSs commonly collapse into conditional ESS if we allow that there will be phenotypic differences between competitors. For instance, in contests a mixed ESS (for the symmetric contest) can be changed by RHP asymmetries into a conditional (sometimes pure) ESS. The same is true for alternative-option scrambles (Parker 1982).

Suppose that a population consists of a series of phenotypes, A, B, C ... N, which may even form a continuous distribution. To find

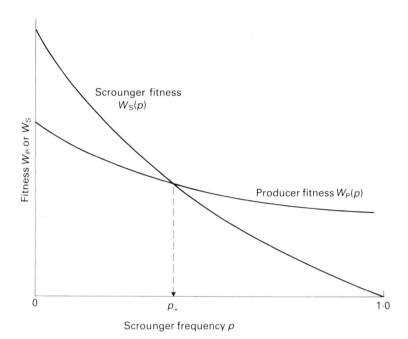

Fig. 2.3 The producer–scrounger model: an alternative-option scramble. Scroungers exploit the investment made by producers, so that the fitness of a scrounger, W_S, decreases with the frequency, p, of scroungers in the population. The fitness of producers, W_P, is less strongly frequency-dependent. The equilibrium p_* is where $W_S(p) = W_P(p)$. (After Barnard & Sibly 1981.)

a 'phenotype-limited' ESS (a set of strategies specified by each phenotype) we use the following rule (Bishop *et al.* 1978; Parker & Knowlton 1980; Parker 1982). For a phenotype-limited strategy to be an ESS, no alternative strategy for a given phenotype must be able to invade when played by that phenotype. Call the full phenotype-limited set of specifications strategy I, composed of I_A played by A, I_B played by B, etc. For I to be an ESS against a deviation J_A in the strategy of A, using Hammerstein's ESS conditions (section 3.4) we require that

$$W(I_A, I) > W(J_A, I)$$

or, if
$$W(I_A, I) = W(J_A, I) \qquad (2.13)$$

then for small q, $\quad W(I_A, P_{q, J_A, I}) > W(J_A, P_{q, J_A, I})$

Equivalent conditions must be satisfied for each element of I (I_A, I_B, I_C ... I_N); thus at the ESS no individual can profit by deviating. The fitnesses of all phenotypes will not be equal; in a sense each phenotype is playing a 'best of a bad job' strategy.

The following example appears to be a common one in nature; it is a phenotype-limited ESS version of the producer–scrounger model. Thus there are two alternative options, P and S. The population consists of a continuous set of phenotypes with frequency distribution

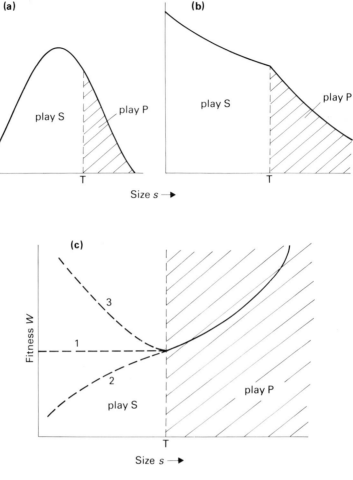

Fig. 2.4 (a, b) p(s), the distribution of size in two hypothetical populations in which there is a competitive advantage to large size in strategy P, which is one of two alternative options, P and S. The ESS is to play P if s > T, play S if s < T. The distribution in **a** is for some species (e.g. an insect) in which size does not increase with age; the distribution in **b** is for a species (e.g. a mammal) in which growth continues after the onset of maturity. Diagram **c** shows possible relationships between fitness W and size s at the ESS. In curve 1, size is unimportant in strategy S; in curve 2, large individuals do best out of those competitors in S; in curve 3, small individuals do best out of those in S. (a and b after Parker 1982.)

p(s); for convenience, we shall imagine that *s* is size (or age if size correlates with age). Two plausible forms for p(s) are shown in Fig. 2.4a and b.

We shall assume that in P inherent competitive ability increases with size. Thus whatever the mix of phenotypes exploiting P, payoffs always increase with size. In *both P and S*, payoffs are a function of the frequency of the two strategies, following Fig. 2.3. The rules we use to find the ESS are these: *(i) the phenotype at the switch point T (see Fig. 2.4a, b) must do equally well in either strategy, P or S; (ii) all phenotypes with s < T must do better in S than P; (iii) all phenotypes* *with s > T must do better in P than S.* First assume that in S all phenotypes do equally well. Then a phenotype at T must be able to achieve the same fitness in P as all individuals in S (curve 1, Fig. 2.4c). Alternatively, different competitive principles may apply in S (curves 2 and 3, Fig. 2.4c), but the switch point will still be solved in the same way.

The importance of this result is that when we include continuously varying phenotypic differences that have some relevance to payoff, we end up with the conditional pure strategy: 'when smaller (or younger) than T, play S; when larger (or older) than T, play P'. Further, there is evidence that a population which lacks phenotypic limitation (and plays the mixed ESS in Fig. 2.3) will evolve to the conditional ESS (Parker 1982).

In a natural P–S system, this conditional ESS will look at the population level much like the mixed ESS in which individuals simply play P with probability p_* (Fig. 2.3). P and S will coexist at stable equilibrium, and even the mean fitnesses of each strategy could be similar if size affects competitive abilities in opposite ways in P and S (Fig. 2.4, curve 3). But in terms of an individual's strategy, there will be a vast difference.

Several biological examples appear to fit the phenotype-limited ESS (see also Chapter 9). In the bee *Centris pallida*, larger males patrol emergence sites and fight over emerging females, whereas smaller males tend to be 'hoverers', searching other areas to pick up females that have escaped the patrollers (Alcock *et al.* 1977). Male scorpion-flies *Hylobittacus apicalis* need a prey item to attract a female—they may either 'hunt' for a prey or 'steal' one by mimicking a female (Thornhill 1979a). The payoff to hunting appears to be lower than that to stealing. In another scorpionfly, *Panorpa*, there are three alternative ways for males to attract a female (supply dead arthropods; use a salivary mass; force copulation) and these strategies are related to male size (Thornhill 1983).

In insects, adults do not grow and so a size-phenotype may well play a single strategy throughout life. In vertebrates, size often increases with age, and the life history switch often operates (for example in red deer (Clutton-Brock *et al.* 1979); elephant seal (Le Boeuf 1974); impala (Jarman 1979)). In the bluegill sunfish (*Lepomis macrochirus*), however, Gross and Charnov (1980) found two alternative life history pathways ('cuckolder' and 'parental care'), which diverge at an early age. The mechanism of the split is not yet understood, but the fitnesses of the two pathways appear to be equal. Within cuckolders, an age/size switch brings about a change from sneaking (darting into nests to obtain fertilizations) to satelliting (mimicking females to gain entry to nests).

A mixed ESS may exist in the digger wasp *Sphex ichneumoneus* (Brockmann *et al.* 1979). Female wasps dig burrows which they provision with food (paralysed katydids) for their offspring. Sometimes, however, the female may lay her eggs in the burrow of another female. The two strategies, 'digging' and 'entering', do not obviously depend on phenotype; individuals switch from one strategy to the other (Brockmann & Dawkins 1979). One population studied showed approximately equal fitnesses for the two strategies, as would be required for the mixed ESS; another population, alas, did not.

2.9 SOME DIFFICULTIES

The ESS approach is still in its infancy. It does not entirely lack problems, both for the theorist and for the field biologist. I conclude by listing a few areas (not all of them problems!) that may repay further study, or may require some caution.

2.9.1 Theoretical issues

(i) *Multiple ESS, and historic constraints.* Where a game has more than one ESS, the only way to deduce which ESS is likely to achieve fixation is from a knowledge of the frequencies of strategies at the start of the game (i.e. it depends on the historic succession of ESS to previous, less complex games). Ideally, we need to examine each step in the possible sequences of changes in games, mutation by mutation, until the present state of strategic complexity is reached.

(ii) *Dynamics and oscillations.* Associated with problem (i) is the difficulty that ESS theory is mainly about stability, and much less about the dynamics of evolving to the ESS. Sometimes there is no ESS to a game, and strategies cycle endlessly (e.g. Parker & MacNair 1979).

(iii) *Genetics.* 'Will a sexual population evolve to an ESS?' (Maynard Smith 1981). This is a complex problem beyond our present scope, but the evidence (reviewed in Maynard Smith 1982a) appears to be reasonably optimistic. Obviously, an ESS cannot be achieved if there is no genetic mechanism that prescribes it, though the system will often proceed towards the ESS as closely as the genetic mechanism will allow.

2.9.2 Methodological issues

There are sometimes difficulties in applying ESS techniques to the study of real animals. For further discussion of some of the items listed below, see Dawkins (1980) and Davies (1982).

(i) *Strategy sets and payoffs.* It is sometimes difficult to identify what strategies are biologically plausible. If a population is at an ESS, this is the only strategy that is likely to be observed in nature; we can hardly be expected to sit and await mutations! It is difficult to ascribe real payoffs to interactions between strategies we seldom if ever see.

(ii) *Continuous strategies, and fitness functions.* Point (i) above is mainly a problem when alternatives are discrete (discontinuous) strategies. For continuous-strategy sets there is usually little difficulty, except perhaps in defining the correct bounds to the strategy range. But how do we build up continuous fitness functions? Ideally, we would use a field experiment in which the appropriate variables were manipulated and fitnesses measured. This is by no means always possible. An alternative technique has been to use the natural variance observed in the field to construct some picture of the fitness functions. However, there is a real problem here that the natural variance might itself be adaptive, representing sets of local optima ('best of bad job' strategies).

(iii) *Mixed ESS*. It will often be extremely difficult to prove that a real strategy is a mixed ESS. As mentioned in section 2.8, the fitnesses of two alternative strategies may indeed approach equality, yet the ESS may be a conditional pure ESS. Further, it will be difficult to prove that animals are behaving probabilistically rather than behaving conditionally in relation to some variable that the observer cannot detect. Mixed ESS are probably most likely to occur in contests where animals lack good information about asymmetries, and in scrambles where phenotypes are discontinuous, or where an animal cannot 'know' its phenotype accurately. This poses the very real difficulty of measuring precisely how much information animals have.

Chapter 3
Natural Selection, Kin Selection and Group Selection

ALAN GRAFEN

3.1 INTRODUCTION

'Animals maximize their inclusive fitness.' 'Animals do not sacrifice their own fitness for the good of their group.' Such statements can be read in many of the chapters of this book, as well as in many recent interesting papers in behavioural ecology and animal behaviour. But they are not obviously true—there are counterexamples in theory to both of them—and my main purposes in this chapter are to trace the logic underlying them, to identify the justification that has been found for them, and to display the connections between these usable principles of behavioural ecology and the more fundamental principle of natural selection.

This chapter will mainly justify what many people already believe: but this is an important exercise. The central concept of inclusive fitness is routinely misdefined and occasionally misused. Some confusion still surrounds the problem of group selection. Only by concentrating on the logical grounds of our orthodoxy can we clarify and defend it.

The rest of this chapter is divided into three parts. The first describes how population genetics is fundamental to behavioural ecology, but also how we may be able to avoid dealing with its complications in everyday practice. The second traces the logic underlying inclusive fitness. The last discusses what natural selection tells us about the evolution of animals in groups.

3.2 POPULATION GENETICS UNDERLIES BEHAVIOURAL ECOLOGY

The starting point for much behavioural ecology is that animals are maximizers of one sort or another—efficient predators or foragers, or elusive prey. The only ground for believing this is that natural selection made them so. If not now then at some time in the past (Dawkins 1982a, pp. 20–24), there existed heritable variation in hunting and foraging techniques and in ploys to escape predators. Changes in allele frequencies have made animals good at what they do (see also section 4.1).

The behavioural ecologist, though, does not usually know the genetics underlying the character he studies. While he would be interested to know this genetic system, it is not of primary importance to him. His main aim is to uncover the selective forces that shape the character. The behavioural ecologist has to hope in his ignorance that his method will work almost regardless of which particular genetic system underlies the character (Lloyd 1977). This hope raises two questions. First, is it justified? Secondly, is the assumption so powerful and plausible that a whole research strategy should be based on it?

3.2.1 The phenotypic gambit

Let us start with a brief caricature, with examples, of an important method in behavioural ecology. It has two elements.

(i) *A strategy set*. This is a list or set of (perhaps all) possible states of the character of interest. Here are three examples of strategy sets. McGregor, Krebs and Perrins (1981) studied the song of male great tits, and in particular their repertory size. The strategy set they used was simply every different repertory size they observed: integers from one to five. Brockmann, Grafen and Dawkins (1979) studied the nesting of great golden digger wasps. These wasps sometimes acquire their nest by digging one, and sometimes by entering an already existing one. Brockmann and co-workers were interested in the relative frequency of these two ways of acquiring a nest, and so the strategy set was simply all possible proportions of digging rather than entering: numbers from zero to one. In the hawk–dove game devised by Maynard Smith and Price (1973), the strategy set consists of the two strategies, hawk and dove.

(ii) *A rule for determining the success of a strategy*. The success of a strategy is the number of offspring left by an animal adopting it, or alternatively its inclusive fitness (see section 3.3.1). The rule for determining success may involve the frequency with which strategies are adopted in the population. We may observe the operation of the rule, as McGregor, Krebs and Perrins did. They counted how many offspring every male fathered in his lifetime, and averaged within all males sharing the same repertory size. If it is necessary to know how the successes of strategies change when their frequencies change, then we may model the rule. Brockmann, Grafen and Dawkins did this, and used data to estimate parameters in the rule. When the purpose is to investigate theoretically the consequences of a particular form of frequency dependence, a rule exhibiting this form is simply assumed: in the hawk–dove game the rule is represented in the payoff matrix.

The phenotypic gambit is to examine the evolutionary basis of a character as if the very simplest genetic system controlled it: as if there were a haploid locus at which each distinct strategy was represented by a distinct allele, as if the payoff rule gave the number of offspring for each allele, and as if enough mutation occured to allow each strategy the chance to invade.

The gambit implies that all strategies that occur in the population are equally successful, and that they are at least as successful as any non-occurring strategy would be if it arose in small numbers. The application of the gambit to a given strategy set and payoff rule is a powerful test of the joint hypothesis that the strategy set and payoff rule have been correctly identified, and that the gambit is true.

In their first model, Brockmann, Grafen and Dawkins rejected this joint hypothesis when two existing strategies turned out not to be equally successful. They adopted a new strategy set in their second model. See Dawkins (1980, 1982a) for a full discussion of what conditions an act must satisfy to be a 'strategy'.

The joint hypothesis might be false because the genetic system underlying the character does not produce the same phenotypic effects as the very simplest genetic system, the one assumed in the gambit. The mere fact that the prediction of equal success is rejected does not reveal which element in the joint hypothesis is false. The research strategy implied by the phenotypic gambit is to treat such rejection as evidence that the payoff function or strategy set is wrong, and not that the gambit is wrong. Maynard Smith (1978a) discusses this more fully.

3.2.2 Is the gambit true?

Taken literally, the gambit is usually false: few species studied by behavioural ecologists are haploid. But will the genetic system that does underlie the character produce the same phenotypic effects as the genetic system the gambit assumes?

Two points are important here. First, an example is known in which the gambit would be extremely misleading. In some human populations affected by malaria, there are three distinct phenotypes corresponding to the three possible genotypes at a diploid locus with two alleles (Allison 1954). One type almost invariably dies from sickle-cell anaemia before reproducing. The other two types differ in their resistance to malaria. The coexistence of these three phenotypes with markedly different fitnesses would be very puzzling to a behavioural ecologist applying the phenotypic gambit. The mechanics of Mendelian segregation prevent the whole population from sharing the optimal phenotype, because it is produced by the heterozygous genotype. Here, as undoubtedly elsewhere, it is essential to know the underlying genetics in order to understand the distribution of phenotypes observed in the population.

The second point is that such cases are probably rare. Only certain features of genetic systems, such as overdominance in the sickle-cell case, can sustain dramatic differences in fitness, and these features are not known to be common. Maynard Smith (1982a) has analysed how well different genetic systems support the simplification represented by the gambit, and he concludes that by and large they do so very well. The sorts of character studied by behavioural ecologists are likely to be controlled by many loci, and this reduces the scope for the maintenance of large fitness differences.

Genetic systems are themselves subject to evolution. In its simplest form, this is the creation of a new allele by mutation, but more substantial changes could occur. In the sickle-cell case, a (functional) gene duplication of the locus would allow one locus to fixate for each allele. Every individual in the population could then have the 'intermediate' genotype that confers malarial protection without sickle-cell anaemia. The existence of fitness differences between genotypes creates selection for evolution of the genetic system itself.

The behavioural ecologist hopes that most genetic systems support the gambit, and that those that do not are rare or transient. If the discrepancies produced by different genetic systems are smaller than the accuracy of data, then field workers can safely ignore them. We know that this might not be so, and we should be anxious to find out whether this hope is justified. The dependence of behavioural ecology on population genetics is such that the soundness of our methods depends on arguments concerning population genetics, but our method is designed to avoid doing genetics.

We have seen that the gambit cannot certainly be made with safety. It is a leap of faith. But should we then refuse to use it in our research? To answer this, suppose that we did refuse. What would behavioural ecology be like? It would be very different. Detailed studies in which the precise nature of a character is examined as an adaptation would have to be accompanied by a study in which the genetic mechanism underlying the character was uncovered so precisely that an explicit genetic model could be constructed. The motto would be: no decimal points without genetics. The range of characters that could be studied would be drastically reduced. Genetically simple and well studied characters are rarely of evolutionary interest. They are usually straightforwardly disadvantageous mutants maintained by judicious artificial selection in strains which have spent tens of generations in the laboratory. If we had to work out the genetics of every character chosen for its evolutionary interest, the size of the study would become very large. In some cases it would be impossible to complete the study within the lifetime of a scientist.

Another serious point is that if the gambit is generally true, then the genetics discovered would be almost an irrelevant complication in understanding the evolutionary significance of the character. The gambit makes truly phenotypic explanations possible, and the effort expended in discovering the genetics would be wasted. Better to allocate that effort to studying in an evolutionary way characters of evolutionary interest, and in a genetic way characters of genetic interest.

These are the reasons why the gambit is so attractive—they

should not be mistaken for reasons why the gambit is true. Neverthe-less, these advantages seem to me to justify continuing to employ the gambit, always provided we remember that we may be wrong. We should also recognize the urgency of the need to provide a proper justification for employing this convenient simplifying assumption.

3.3 INCLUSIVE FITNESS AND HAMILTON'S RULE

Textbooks on behavioural ecology or animal behaviour usually have a section on kin selection or inclusive fitness in which the reader is advised that what animals really maximize is inclusive fitness. They then either fail to define inclusive fitness, or define it wrongly (Grafen 1982); this section is intended to set out what inclusive fitness is. Even in their eulogies the textbooks are not usually right, so this section will also explain why inclusive fitness and Hamilton's rule are extremely useful additions to our theoretical armoury, although by no means a replacement for 'number of offspring' as a measure of repro-ductive success. It also deals briefly with what is currently known about the scope of their applicability.

3.3.1   What is inclusive fitness exactly?

Inclusive fitness (Hamilton 1964) is a device that simplifies the calcu-lation of conditions for the spread of certain alleles. These alleles have an effect, through their bearer's phenotype, on how many offspring other animals in the population produce. We can see the simplifica-tion by comparing the analyses of a very simple model of sib altruism by standard population genetics and by inclusive fitness. Maynard Smith (1982b) carries out a similar exercise.

Suppose males in a species disperse little, so that every breeding male has exactly one brother of the same age breeding nearby. A single locus controls how a male behaves towards his sib, and the population is at fixation for an allele a at that locus. We consider an allele A that alters the behaviour of its (homozygous and heterozygous) bearers so that each bearer has c fewer offspring, and the bearer's sib has b more offspring.

Will A spread when rare? When A is rare, the homozygote AA is extremely rare and can be neglected. The number of offspring a mated pair produces depends only on the male's genotype and that of the male's brother. How many offspring will the pair have on average if the male is aa? If his brother is also aa, then the pair produces the standard one offspring. If his brother is Aa then on average the pair produces $(1 + b)$ offspring. The overall average for an aa male there-fore depends on the chance that his brother is Aa. Let the overall proportion of Aa be p. Then an aa male has an Aa brother with chance $p/2$. The average number of offspring of an aa male is then $(1 + bp/2)$.

The chance that an *Aa* male has an *Aa* brother is $(1 + p)/2$, and the *Aa* male loses c offspring through the effect of the *A* allele, so the average number of offspring of an *Aa* male is $1 - c + (1 + p)b/2$.

I digress to explain how the chances $p/2$ and $(1 + p)/2$ are calculated. In the absence of any information, an animal would calculate the chance that his sib was *Aa* to be p. If the animal is itself *aa*, then it knows that half of the available four parental alleles are not *A*; and so the chance of his sib containing *A* is halved, i.e. it is $p/2$. If the animal itself is *Aa* then a more complex calculation is required. If the chance that an animal is *Aa* is p, then the chance that any allele is *A* must be $p/2$. One of the parents of an *Aa* individual is *A?* and the other is *a?*, where *?* means '*A* with chance $p/2$ and *a* with chance $1 - p/2$'. The sib therefore has a chance $1/2 + p/4$ of receiving *A* from the *A?* parent, and a chance $p/4$ of receiving *A* from the *a?* parent. The total chance of receiving *A* is therefore $(1 + p)/2$. Charnov (1977) uses this method of calculation. It is approximate because it depends on p being small, and it assumes Hardy–Weinberg equilibrium. Now let us return to the calculation of the condition for Λ to spread.

The fraction of *Aa* males in the next generation goes up if *Aa* males have more offspring than *aa* males. What must be true of our variables p, c and b for this to be so? Well,

$$1 - c + b(1 + p)/2 > 1 + bp/2$$

reduces to:

$$b/2 - c > 0.$$

We have just derived the condition for *A* to spread by calculating simply the number of offspring produced by *Aa* and *aa* males. How is the advantage of altruism shown using this approach? Through the extra probability that the brother of an *Aa* male is *Aa* and so increases the male's own number of offspring.

Inclusive fitness arises from a different accounting procedure (Abugov & Michod 1981), in which instead of counting the effect of everybody's actions on one individual's number of offspring, we calculate the effect of one individual's actions on everybody's numbers of offspring. The count is weighted by the relatedness (for a precise definition see section 3.3.4 below). Inclusive fitness was invented and defined (mathematically) by Hamilton (1964). His paper is devoted to proving that the alternative accounting procedure that underlies inclusive fitness gives the same answer as the standard and logically prior procedure, an example of which we have just worked through. Hamilton described inclusive fitness as:

> 'the animal's production of adult offspring . . . stripped of all components . . . due to the individual's social environment, leaving the fitness he would express if not exposed to any of the harms or benefits of that environment, . . . and augmented by certain fractions of the quantities of the harm and benefit the indi-

Table 3.1. Illustration of the two different accounting procedures implied by two measures of reproductive success: number of offspring and inclusive fitness. The 'advantage to *Aa*' is the same in both systems. The genotype that has more offspring will also have higher inclusive fitness. Number of offspring counts the effects of everybody's acts on an individual; inclusive fitness counts the (weighted) effects of the acts of one individual on everybody.

Measure of RS	Number of offspring		Inclusive fitness	
Genotype	*aa*	*Aa*	*aa*	*Aa*
Basic nonsocial fitness	1	1	1	1
Cost of act		c		c
Benefit of act	$pb/2$	$(1 + p)b/2$		$b/2$
Total	$1 + pb/2$	$1 - c + (1 + p)b/2$	1	$1 - c + b/2$
Advantage to *Aa*	$b/2 - c$		$b/2 - c$	

vidual himself causes to the fitnesses of his neighbours. The fractions in question are simply the coefficients of relationship . . .' (Hamilton 1964).

Applying this to our example, the *aa* males have an inclusive fitness of one, because the extra b they sometimes receive (i.e. when their brother is *Aa*) is disregarded as a 'help from the social environment'. The *Aa* males have $(1 + b/2 - c)$ because their relatedness to their brother is $1/2$, which we use to devalue b. The condition for *Aa* males to have a higher inclusive fitness than *aa* males is then that

$$1 + b/2 - c > 1, \quad \text{or} \quad b/2 - c > 0,$$

which of course is the same answer as before. We have just demonstrated a (simple and) special case of Hamilton's 1964 result. Table 3.1 illustrates the two accounting procedures. The two methods give the same answer by different means.

3.3.2 How not to measure inclusive fitness

> Them things that you're liable
> To read in the Bible
> They ain't necessarily so.
>
> *Porgy and Bess*

In the example of the last section we saw what inclusive fitness is. Many textbooks give one of two erroneous definitions, and studies have calculated inclusive fitness from data using one of these definitions. It is instructive to examine these errors.

Erroneous Definition 1 (from Barash 1980, p. 212): 'the sum of individual fitness (reproductive success) and the reproductive success of an individual's relatives, with each devalued in proportion as it is more distantly related.'

Erroneous Definition 2 (from Wilson 1975, p. 586): 'The sum

of an animal's own fitness plus all its influence on fitness in its relatives . . .'. I assume it is intended to weight the influences by relatedness.

The original definition of inclusive fitness, as given in the quotation from Hamilton (1964) in section 3.3.1 above, has a component from 'self' and a component from others. ED1 counts all relatives' offspring, whereas inclusive fitness counts only those the relatives had because of the actions of 'self'. ED1 and ED2 count all the offspring of 'self', whereas inclusive fitness does not include those offspring gained through the actions of others (the 'harms or benefits of that environment' in the above quotation from Hamilton).

The reason such erroneous definitions persist is that in most cases they are not applied. In a general discussion the definition itself is never called on with any precision and so the error is in a sense silent, even unimportant. But as soon as data are used to calculate inclusive fitness, the precise definition obviously does matter.

Measures of reproductive success must have one essential property. If bearers of one allele have a higher (lower) reproductive success than non-bearers, then the allele must increase (decrease) in frequency. ED1 and ED2 lack this property. Number of offspring and inclusive fitness do have this property. See Grafen (1982) for a fuller discussion.

In the face of the obvious difficulties of calculating the differences in number of offspring that helping causes, an alternative to inclusive fitness is to use number of offspring as a measure of reproductive success. If these data are available, it will be much simpler to calculate.

When it is desirable to use the inclusive fitness approach in analysing field data, it is better to aim at using Hamilton's rule than to calculate inclusive fitness itself. The next section discusses how to do that. The inclusive fitness approach allows us to separate the success of an allele into components of 'own offspring' and 'relatives' offspring'. Also, it may sometimes be the only approach we can use with certain data. See the discussion of Noonan's example in the next section.

3.3.3 Hamilton's rule and how to use it

Hamilton's rule is that animals are selected to perform actions for which $rb - c > 0$, where r stands for relatedness. Inclusive fitness in Hamilton's 1964 paper was just a tool used in the construction of the rule, and the only reason we have dealt with it at length is that it is surrounded by so much confusion. Hamilton's rule is more important and more illuminating than inclusive fitness; it is also easier to apply data to it because the form of the rule encourages us to use the correct logic of differences. The rule has been derived recently by Charlesworth (1980), using a simple population genetics approach, and by Hamilton (1975) and Seger (1981) using the selection mathematics of Price (1970, 1972).

The first application of Hamilton's rule to data complete with decimal points was by Brown (1975). We will come later to his example of helping at the nest in birds.

The very first step in applying Hamilton's rule is to choose the decision we are interested in—being as explicit as possible about the alternative course of action. To calculate b and c we must think through all the consequences on lifetime number of offspring that follow from doing one thing rather than another. The simple difference in number of offspring will also include the extra b's contributed by those relatives, and therefore does not give a proper estimate of c.

A difference of c in the animal's lifetime number of offspring results from choosing to do Y rather than X. Any consequences that would follow from doing Y not X should be taken into account—decreased longevity, retribution and so on. It may seem at first sight that a simple way to estimate this from data is to take the difference in lifetime number of offspring between animals that do X and animals that do Y. However, this seemingly reasonable procedure may give the wrong answer. The reason is that animals that do X will have relatives who tend to do X, and animals that do Y will have relatives that tend to do Y. The principle is that an animal that helps n times, and is helped m times, should have $mb - nc$ offspring as a result.

The same caution applies to measuring b.

The value of r has been assessed in a number of ways. Bertram (1976) modelled the structure of his lion prides to arrive at relatednesses; Brown (1975) used simple ancestry; and Metcalf and Whitt (1977b) used electrophoresis. See section 3.3.4 below for a further discussion of r.

Finally, before proceeding to examples, Hamilton's rule in the form $rb - c > 0$ has definite advantages over the more popular form $b/c > 1/r$ for the purposes of statistical testing. For one thing the first form is correct whatever the signs of r, b and c. Also, if r is known from *a priori* grounds, then the mean and variance of the difference $rb - c$ are calculable simply from the means and variances of b and c. The ratio b/c on the other hand has mean and variance that depend in a more complicated way on the distributions of b and c.

Example 1: Brown's analysis of helping at the nest in the superb blue wren

Juveniles in many bird species sometimes stay behind at their parents' nest and help rear their siblings instead of leaving to try to raise offspring of their own. Using data on the superb blue wren (*Malurus cyaneus*) from Rowley (1965), Brown (1975) knew how many young a nest produced in a year according to whether or not there was a helper present that year. He estimated the benefit to the parents as the difference that the helper made, and estimated the cost to the juvenile as the average number of offspring produced by an unhelped pair.

We now examine the assumptions made in assessing b and c in this way. It is possible that parents survive better if they are helped by their young, and so produce more offspring themselves in later years. It is also possible that helpers do not help, but that able parents have many offspring each year, a fraction of whom stay behind. This would produce a correlation between number of 'helpers' and number of young raised, but b would be correctly assessed as zero. The general assumption lying behind the measure of b used is that the number of young produced in a year depends only on the number of helpers in that year. It is possible that experience of breeding in the first year is better or worse preparation for breeding in the second year than helping, or that there is differential mortality in the two groups. It is probably true that helping rather than breeding alone affects the chance of being helped in the second year. These would upset the measure of b, as it is the lifetime number of offspring that matters. Here the general assumption is that the rest of the juvenile's life is unaffected by its decision in the first year.

The point to notice about these assumptions is that although crucial they are interesting and biological and perfectly amenable to investigation, as is borne out by many later studies by Brown and his associates. They discover (among other things) with increasingly powerful methods whether or not helpers in the gray-crowned babbler really help (Brown *et al.* 1978; Brown *et al.* 1982).

In the example of the superb blue wren, the basic data were that pairs with helpers produced 2.83 offspring on average and those without produced only 1.50. The benefit of staying was therefore 1.33 to the parents. For females, the cost of staying was 1.50, as it was for males who could find mates; while for males who could not find mates, the cost was zero. The value of r was one, because the choice is between creating siblings and creating offspring, which are equally related. (Alternatively, we can say that the helper helps both his parents increase their number of offspring, and the sum of his relatedness to his parents is one.) The conclusions were that females should not help because

$$rb - c = 1.33 - 1.50 = -0.17 < 0,$$

and that males who could find mates should not help by the same calculation. Males who could not find a mate should help because

$$rb - c = 1.33 - 0 = 1.33 > 0.$$

Brown (1975) also discusses the case of the Florida scrub jay, using data from Woolfenden (1975), in which females should help. Emlen (1978) discusses at more length the application of Hamilton's rule to helping in birds.

Example 2: Noonan's study of joint nesting in Polistes fuscatus

Noonan (1981) studied the founding of nests by females of the social brown paper wasp *Polistes fuscatus*. In this species, nests may be founded by from one to about ten females; one of these females becomes the queen, who does most of the egg-laying. These females are almost always closely related. One of the questions Noonan asked was whether a female who joined a group of relatives as a worker did better than a female who decided to become a solitary queen. Specifically, did the fact that she was helping close relatives swing the balance in favour of social cooperation? We will not follow Noonan's own analysis.

To apply Hamilton's rule, it is vital to be precise about the wasp's decision. Suppose a female is confronted with a nest of $N - 1$ relatives and she knows that she is the last to decide whether or not to join it. If she does not join it we will assume, in two separate applications of the rule, first that she leaves and becomes a single foundress and second that she leaves and dies.

To estimate the benefit to her relatives, we must calculate how many young her relatives would produce as a nest of $N - 1$ females without her, and then in a nest of N females with her as an extra worker. The difference is the benefit she confers on her relatives. The cost to herself is the difference between the eggs of her own she lays as a worker in a nest of N females, and the young she rears in her alternative role. This is the number of young reared by a workerless queen in the first application, and the zero young reared by dying in the second. Noonan's paper contains all the necessary information to carry out these calculations, and the results are shown in Table 3.2.

The main conclusions are as follows.
1. In terms of her own number of offspring, a female is much better off as a solitary queen than as a worker, and much better off as a worker than dead.
2. In terms of her contribution to her nest-mates' number of offspring, her effect depends strongly on how many workers there already are at the nest. If she would be the only worker, then her effect is strongly positive; if she would not, then she has a negative effect on their reproduction. She lays more eggs as a worker than the extra she provides for the nest as a whole.
3. If she would be the only worker, then her relatedness to the queen may well favour her joining. The condition is that $r > 0.48$.
4. If she would not be the only worker, then her relatedness to the members of the nest will act against her joining. She would do better to join strangers and parasitize them. The value of r can even be high enough to swing the balance in favour of dying rather than joining close relatives in a group.

The implications of these conclusions do not matter to us here, but

Table 3.2. A kin selection analysis of the decision of a female *Polistes fuscatus* whether to join a nest of relatives and become a worker in a nest with a total of N females. Two alternatives are considered: becoming a solitary queen, and dying. Data are from Table 2-2 in Noonan 1981. The conditions for joining are calculated using '$rb - c > 0$ means join'. We assume that r is never negative. Nests of sizes 1, 2, 3, and 4 were in fact common (Noonan 1981).

	N		
	2	3	4
Number of eggs of a solitary queen	18.25	18.25	18.25
Number of eggs of a worker in nest of size N	4.6	4.8	3.6
Cost of joining instead of becoming a solitary queen	13.65	13.45	14.65
Cost of joining rather than dying	−4.6	−4.8	−3.6
Number of eggs by the rest of the nest if joined (making N)	46.4	40.5	39.1
Number of eggs by the nest if not joined (making $N - 1$)	18.25	51.0	45.5
Benefit to rest of nest of being joined	28.15	−10.5	−6.4
When should a female join rather than become a solitary queen?	$r > 0.48$	Never	Never
When should a female join rather than die?	Always	$r < 0.46$	$r < 0.56$

one general point does. Why do we apply Hamilton's rule when Noonan's data are good enough to allow us simply to calculate number of offspring? In the spirit of section 3.3.1, should not the two methods give us the same answer and would not number of offspring be easier? One reason to apply Hamilton's rule is to see whether a trait is advantageous through an individual's own reproduction alone, or whether the effect on relatives' reproduction swings the balance. Another reason is that in counting number of offspring we must average over all animals who would have decided (i.e. had the genes for deciding) to join and those who would have decided to go it alone. But we do not know which all these animals are. The problem arises because not all animals are called on to make the decision, and so not all the animals' strategies are laid bare. We do not know who would have done what. Using the inclusive fitness approach, we can legitimately concentrate on only those animals who faced the decision. (The 'work' of deciding who would have done what is in effect done for us by the calculation of relatedness.) This is a crucial advantage to the inclusive fitness/Hamilton's rule approach when not all animals are faced with the decision of interest.

3.3.4   The validity of Hamilton's rule

Hamilton's rule holds good only under certain assumptions. There are different definitions of r, and the scope of the rule depends on the definition of r employed (Michod & Hamilton 1980; Seger 1981). Here we are concerned mainly with applications, and so restrict ourselves to forms of r that can be estimated from data. Charlesworth's derivation of Hamilton's rule (Charlesworth 1980) makes the roles of the assumptions clearest, and we follow his treatment in what follows. The latest, but perhaps not the last, word on the validity of inclusive fitness and Hamilton's rule is a review by Michod (1982).

Assumption 1: Additivity of costs and benefits

An animal that is helped m times, and helps n times, should experience a change of $mb - nc$ in its number of offspring as a result. This assumes that effects add. Addition will not always be the most plausible way for costs and benefits to combine. If the trait affects survival, then multiplication may be more appropriate (Charlesworth 1978). To see this, consider an animal that is exposed on two separate occasions to a 75% chance of dying. Each occasion quarters its fitness, and the overall effect is to reduce its fitness to one-sixteenth of what it would have been. The assumption of additivity is untroubling if the b and c are small, because then additivity will hold at least approximately. 'Small' is relative to average lifetime reproductive output.

In fact the assumption of additivity is really two assumptions in disguise. Two things go wrong if additivity is broken. First, an animal that pays one cost and receives one benefit does not have a net gain of $b - c$. This is partly just a measurement problem—if we define b and c as the average effects measured in number of offspring, then we may avoid it. Secondly, and more seriously, the cost and benefit of an action will depend on the genotype of the actor and recipient respectively. If two helps are not twice as good as one, then altruists will tend to lose out; they receive more than their share of helping and so receive more of the substandard second helps. This would invalidate the theory on which inclusive fitness is based at a quite fundamental level (Seger 1981).

Assumption 2: The gene frequency among potential donors and receivers is the same

A potential donor is an animal that is faced with the decision being investigated. A possible though unlikely exception is considered by Charlesworth (1978). Suppose that a dominant altruistic mutant gene caused all of its bearers to commit suicide for the benefit of their sibs. The sibs that benefited could certainly not be altruists! The altruistic allele would be inevitably selected against, indeed would have a

fitness of zero, no matter what the values of b, c and r. In a sense this
assumption says that r has to mean what we think it means—the
extent of genetic similarity at the locus of interest.

Assumption 3: Weak selection

This assumption hinges on r. The essential property of r in Charlesworth's derivation of Hamilton's rule is as follows (Charlesworth
1980). Let the set of animals S be the possible recipients of an act by
an animal I, and let T be the set of all animals in the population. Let
$p(A|Z)$ be the probability that an allele selected at random from a
given locus in entity Z is allele A. (For example, $p(A|T)$ is the popu-
lation proportion of A.) Then the r that is relevant for the decision is
implicitly defined by:

$$p(A|S) = r \cdot p(A|I) + (1 - r) \cdot p(A|T)$$

In words, the possible recipients of the act are partly like I and
partly like T; and r measures how like they are to I. The condition
that $r = 0$ means that the possible recipients are genetically representative of the population as a whole; $r = 1$ means that they are genetically of the same constitution as the potential donor.

A remarkable thing about this definition is that when there is no
selection going on, r is the same for all alleles at all loci with the same
inheritance pattern (e.g. autosomal, X-linked, Y-linked). Furthermore,
it can be calculated from family trees. However, when selection is
occurring, r will not be the same at all loci and will only be calculable
approximately from family trees. Consider an animal and a class of
its relatives at a period in the life cycle when selection is occurring. Selection is a systematic change in the relative proportions of
animals with different genotypes; and so the genetic similarity of
the animal to surviving members of the class of relatives must change
as selection occurs. The assumption of weak selection is necessary
to make ancestry an accurate enough guide to the 'true' r as defined
above.

Weak selection should rarely be a problem in practice. If $rb - c$ is
large in magnitude when calculated with an r derived from ancestry,
then the deviation from the true value of r will not affect the sign of
$rb - c$. Alternatively, if $rb - c$ is small in magnitude, then selection is
weak and ancestry is a reasonable guide to the true r. Thus the direction of selection will be preserved, although it is true that the
strength of selection may be misjudged.

It is interesting that these problems do not disappear if r is measured by electrophoresis. The deviation of the ancestral from the true
r occurs at loci undergoing selection, and there is no reason to believe
that the loci sampled electrophoretically are undergoing the same selection as the locus affecting the decision of interest.

3.4 GROUP SELECTION

Group selection has a bad reputation among many behavioural ecologists and evolutionists that obfuscates debates on some recent models that go under the name of group selection. In section 3.4.1 I discuss that kind of group selection which quite rightly elicits disapproval, and which is certainly still with us today. Section 3.4.2 deals briefly with the first generation of models of group selection, and section 3.4.3 deals more extensively with a quite different class of models of the natural selection of animals in groups.

3.4.1 The bad and dangerous

The reputation of group selection comes not from mathematical models, nor from deliberate discussion of group selection, but from a certain naivety practised by laymen and many biologists alike in their day-to-day thinking about the adaptedness of animals to their environment. An adaptation is 'for something'—and that certain naivety is to propose an adaptive explanation without stopping to think what that 'something' is. It is this lack of thought, not any deliberate and considered choice of the group as that 'something', which is the target of most accusations of 'group selectionism'.

This lack of thought is a target deserving of attack—and as sustained and effective an attack as we can muster. However, 'group selection' is a poor name for this lack of thought, because the 'good of the species' and the 'good of the ecosystem' are as prominent as the 'good of the group' in its effects (to be found in journalistic and academic publications).

The fundamental case for careful thought is that adaptations arise only by natural selection, and that natural selection does not normally promote adaptations for the good of any unit larger than the organism. Two excellent 'self-help' guides in careful thought are books by G. C. Williams (1966) and R. Dawkins (1976). (The organismal approach suggested here is not in conflict with the 'gene selectionism' of Dawkins (1982a, b). In his language, we are saying that the individual is usually a well-adapted vehicle for gene replication, while groups usually are not.) In the next two sections we will see possible exceptions to the general rule, but we must not allow them to distract us from a most important lesson about adaptations: a very convincing case is needed to explain why an adaptation should be for anything other than the organism. Lack of thought is the basis for that group selection which is mad, bad and dangerous to know, and the exceptions that follow are no licence for that laziness.

3.4.2 The old

The first generation of mathematical models for the exploration of group selection can be understood by reference to Fig. 3.1, which is adapted from Maynard Smith (1976b), who gives references to examples of these models. A composite of those models works as follows. There are a large number of discrete locations each of which is capable of supporting one group. Migration between locations is restricted. There are two alleles, A_0 and A_1, at a given locus. Animals with A_1 are more cooperative than those without, but at a cost to personal fitness. The consequences are that within any group A_0 quickly displaces A_1, but that groups consisting of A_1 are better off than groups of A_0. 'Better off' means either less likely to go extinct in a year, or able to produce more migrants to leave and try to colonize empty locations.

There are three kinds of location (neglecting the transient mixtures of A_0 and A_1): E (for empty), A_0 and A_1. The traffic between these kinds occurs for a number of reasons. Extinction sends A_0 and A_1 to E. Migration to empty locations sends E to A_0 or A_1. Migration to occupied locations sends A_1 to A_0.

The question for any given model is exactly how the cooperation of A_1 animals affects extinction and number of migrants produced by the group, and whether migrants are allowed to join occupied locations or only empty ones. The factors promoting the spread of A_1 are (1) cooperation reducing extinction, (2) cooperation increasing number of migrants produced by the group, (3) migrants allowed to join only empty locations, and (4) small number of founders in a group. The last is important because the more founders there are, the more likely it is that there is at least one A_0 among them.

The final consensus on these models was that the conditions for A_1 to be successful were too stringent to be realistic. Wynne-Edwards, whose book *Animal Dispersion in Relation to Social Behaviour* (1962) sparked the whole controversy, wrote in 1978:

'but in the last 15 years many theoreticians have wrestled with it, and in particular with the specific problem of the evolution of altruism. The general consensus of theoretical biologists at present

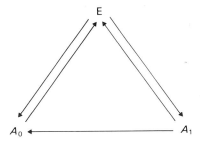

Fig. 3.1. This shows the possible states of sites in an 'old' group selection model. The arrows represent possible transitions. E means empty; A_0 and A_1 refer to groups with only that allele at the locus of interest. Mixtures of A_0 and A_1 are considered too transient to matter. The figure is adapted from Maynard Smith 1976b.

is that credible models cannot be devised by which the slow march of group selection could overtake the much faster spread of selfish genes that bring gains in individual fitness. I therefore accept their opinion.' (Wynne-Edwards 1978).

Even before this was written, a new sort of group selection had been discussed that did not require unreasonable assumptions. The new group selection of Price (1970, 1972) and Hamilton (1975) is the topic of the next section.

3.4.3   The new

Many animals live in groups, and these groups may be grouped by natural features such as rivers or mountains, and even these super-groups may be grouped by, for example, the ocean. The focus of attention of the new group selection is on this hierarchy of grouping and the effect it might have on natural selection. As a preliminary caution, we must not allow the rhetoric of hierarchies and groups within groups to intoxicate us. A sober appraisal is required, for as we shall see it is possible to have hierarchies and groups within groups that have no effect whatsoever on the workings of natural selection. The next three parts deal in turn with altruism, the sex ratio and, briefly, a different approach to explaining the workings of the following models.

Altruism

The real hero of this section is Hamilton's rule, and indeed we follow Hamilton's treatment closely here (Hamilton 1975). To understand why grouping might have an effect, we first examine the definition of r given in section 3.3.2 above. It was that for an animal I, a set of other animals S, and the total population T, r from the animal to the set satisfies:

$$p(A \mid S) = r \cdot p(A \mid I) + (1 - r) \cdot p(A \mid T).$$

In words, the genetic composition of S is a weighted average of the genetic compositions of the animal I and the population T; r measures the weight given to the animal I's genotype in that average.

Taking S to be the rest of the group to which an animal belongs, what factors might cause r to be non-zero? There are only two (Maynard Smith 1976b). One is common ancestry, and the other is preferential assortment. Taking common ancestry first, the grouping of the population may make it difficult for us to ascertain properly all the relevant kin links. As an example, Hamilton points out that in a virtually closed group the genetic similarity builds up eventually to $1/(2M + 1)$, where M is the absolute number of migrants each generation, independent of the group's size. The ties in a large group 'make up in multiplicity what they lack in close degree' (Hamilton 1975).

So unsuspected kin links may increase r above initial expectations. An example of how to estimate relatedness in a grouped population is provided by Bertram (1976). By modelling the way lion prides are formed, he calculated the genetic similarity between the different kinds of animals in a pride—the females, the males and the young. Just using the observed kin relations in a pride would have underestimated the genetic similarities. Bertram's work is a good example of the new group selection applied to data, although it is expressed entirely in terms of kin selection. An important message from it is that animals in groups may not all be equally related, and the distinctions are very interesting.

The other way for r to be raised in groups besides common ancestry is preferential assortment. If altruists share a preference for certain habitats or microhabitats, then altruists will tend to be in groups with other altruists and that is all that is required to make r positive according to the definition above. There are a number of reasons why preferential assortment is not a plausible source of genetic similarity in nature. They are based on the fact that while ancestry provides relatedness that is the same for all loci, preferential assortment only causes relatedness at the loci that cause it, and at linked loci. It is unlikely that the locus for altruism is closely linked to the loci for habitat preference. Even if it were, there would be selection at unlinked loci to suppress the altruism; for while the r at the altruism locus may be positive, the r at unlinked loci is zero and Hamilton's rule applies equally to both sorts of loci. Finally, there would be selection for a 'free-rider' allele (if one arose) at a locus unlinked to the altruism locus. It would have the effect of creating the same habitat preference as that of altruists, whether or not its bearer was an altruist.

For all these reasons the only plausible cause of genetic similarity among group members is common ancestry. Furthermore, the only moving force in the new group selection is genetic similarity.

Hamilton (1975) discusses a model of towns, with low migration, in which relatedness and altruism build up together; he suggests this as a possible genetic basis for xenophobia. The model would apply equally well to non-human groups. It is the most considered attempt so far to show how the new group selection might be important in nature. I wish now to discuss two important caveats in interpreting this model. The model itself works because, as noted above, the relatedness in a community that receives M migrants each generation builds up to $1/(2M + 1)$.

The first caveat is that the migration rates must be low to achieve a noticeable relatedness. One migrant every two generations produces sibling level relatedness; eight per generation produces second cousin level relatedness. In animal groups in which all of one sex disperses we expect very little effect.

The second caveat is more fundamental. When we say in the ordi-

nary, ungrouped, model of kin selection, that an act has effects b and
c, do we really mean that there are b − c extra offspring in the population as a whole surviving to maturity and breeding? Perhaps, but
probably not. Normally we would expect that a winter bottle-neck or
some other factor would limit total population size in a way unaffected by the act of altruism. We probably mean to say that the
donor loses c, the recipient gains b, and then because the total population number remains the same the population as a whole loses
b − c. If we let d represent the general decrement associated with the
act (and d will equal b − c if the act does not increase carrying
capacity), and let r_e be the average relatedness of the potential donor
to those suffering the general decrement, then we can write Hamilton's rule more fully as:

$$rb - c - r_e d > 0$$

Formally this is just expanding b to include all the effects of the act.
The subscript 'e' means economic, describing the forces that
regulate population size. Hamilton (1964) discussed this near conservation of fitness in the population as a whole.

If the population as a whole suffers the general decrement, then
the expansion of Hamilton's rule is unnecessary. By the definition of *r*
above, the relatedness to the population as a whole is zero and so the
extra term falls out. However, in the case of grouped populations the
extra term will often be very important. The most important factor in
creating intra-group relatedness, isolation, will also cause the general
decrement of an act to be felt by those with whom an animal is isolated. This means that r_e will tend to be proportional to the intra-
group relatedness. In the extreme case where all the general
decrement falls on the group, and the act does not increase carrying
capacity, there will be no selection for intra-group altruism; for what
the animal gives in bulk to one member, he takes away in dribs and
drabs from all the others.

There are then two parallel factors at work in the new group selection. One is the pattern of relatedness among groups, and the other
is the pattern of joint dependence of offspring of group members on
the same resources. The resolution of these two, through the expansion of Hamilton's rule above, determines how grouping affects the
workings of natural selection on altruism.

There remains one loose end to tie. It is obvious from this discussion that with random grouping there is no selection for altruism.
In this case r = 0, and the spread of an allele is determined by the
selfish criterion c < 0. Matessi and Jayakar (1976), Cohen and Eshel
(1976) and Wilson (1975, 1980 and references therein) have claimed
that there is selection for altruism in this case. This arises because
they all defined altruism not in terms of absolute fitness as Hamilton
(1964, 1972) did before them, but in terms of relative fitness within
the group. This is discussed further at the end of this section.

In Chapter 8, Maynard Smith explains the basic Fisherian model (Fisher 1958) and the idea of local mate competition. The first model showing an effect of grouping on sex ratio was constructed by Hamilton (1967). In this model, n inseminated females arrive at a site and lay their eggs, which develop and then mate among themselves. The males die, the inseminated females disperse and the next generation begins. The unbeatable sex ratio (for a diploid species) is that a fraction $(1/2)((n-1)/n)$ of resources should be allocated to producing sons.

This female bias can be seen as the result of two factors acting against the Fisherian force towards equality of investment. The first is diminishing returns to producing sons—that is, each additional son creates fewer and fewer additional grandoffspring for his mother. The second is that making a daughter increases the number of grand-offspring males produce for their mother, through increasing the available number of mates; creating sons does not affect daughters' success. This asymmetry arises because it is assumed that males contribute nothing to the care of offspring. We can add a third factor by supposing the n females to be related. Making a daughter would then have the additional benefit of increasing the mating success of sons of relatives. The sex ratio would therefore be even more female biased if the n females were related to each other.

These three factors are at work in a recent model of the sex ratio in grouped populations, by Bulmer and Taylor (1980). It is designed to account for female-biased sex ratios in the wood lemming. In the model, n foundresses arrive at a site, and g generations take place within the site before population-wide dispersal occurs and again groups of n foundresses form. Bulmer and Taylor consider separately the case where females of the dispersal generation mate within their own group before dispersal, and where they mate at random in the population. They sought that sex ratio which, once an allele producing it was common in the population, would not allow a dominant allele coding for any other sex ratio to invade. This stable sex ratio depended on the parameters n and g of the model. It was always female biased. As the number of foundresses n increased, the bias diminished, corresponding to a relaxation in the diminishing returns of producing sons. The bias increased as g, the number of generations between dispersals, increased. How can this be explained?

During the g generations between dispersals, the number of animals in each site grows exponentially at a rate determined by the sex ratio produced by the females. This creates increasing returns to scale for producing daughters when measured in number of eventual dispersers. It also causes relatedness to increase as the g generations proceed. If each of n unrelated foundresses has two daughters and one son, then each member of the next generation has as sibs a fraction

$2/(3n - 1)$ of the rest of the population. Now the exponential growth of the group means that producing daughters benefits all members of the group, and of course a female only cares about the effect on the reproduction of the others to the extent that she is related to them.

This view of the model suggests two predictions about the model's behaviour. First, the sex ratio bias should be more extreme if variation in reproductive success among males or females exists, since this also increases the relatedness between group members. (In an extreme case where all offspring have the same parent of either sex, group members are at least half-siblings.) Secondly, the sex ratio bias should increase as the g generations proceed because the relatedness between group members increases. Bulmer and Taylor do not allow the females to choose their sex ratio according to the generation to which they belong, nor do they vary variance in reproductive success.

In a similar model, however, Wilson and Colwell (1981) do vary variance in reproductive success. Their model is haplodiploid, and their equilibrial sex ratio is a genetic polymorphism between an allele producing a $1 : 1$ sex ratio, and an allele producing a sex ratio of varying degrees of female bias. They confirm the results of Bulmer and Taylor, and also show that the result of increasing the variance in male reproductive success is indeed to increase the female bias of the sex ratio.

So we see that the female bias in sex ratios that can occur in grouped populations can be attributed to diminishing returns or increasing returns to one sex, producing females being a source of reproductive success for sons, and producing females being a source of reproductive success for the sons of relatives.

Didactics

The most vocal proponents of the new group selection have been Wilson and Colwell. (Wilson 1980, and references therein; Wilson & Colwell 1981; Colwell 1981.) They have scandalized many by speaking positively of 'group selection'. The purpose of this section is to reconcile their chosen way of explaining why grouping has the effect it does with that of the previous sections. They concentrate on dividing selection into two parts: within-group selection, which is the local change in gene frequency within each group, and between-group selection which is the result of differential fecundity of groups. This difference of approach is purely didactic—there is no disagreement about matters of substance.

The different approach has, however, led to misunderstanding. Wilson and Colwell identify within-group selection with 'individual selection', and between-group selection with 'group selection'. Now individual selection already has a meaning which is quite different, and one I think is very valuable. An act is said to be favoured by individual selection when it spreads through its effect on the actor's

number of offspring alone. In terms of Hamilton's rule, this occurs
when there is a negative cost, i.e. a simple gain, in terms of number of
offspring to the actor. As r and b also enter into Hamilton's rule, this
definition does not make individual selection synonymous with
'natural selection'. See the discussion of Noonan's study in section
3.3.3 above as an example.

Another source of misunderstanding arises from the use of the
word 'altruism'. As we noted earlier, altruism will not evolve in
simple, one-generation groups that are formed at random from the
population. Matessi and Jayakar (1976) and Cohen and Eshel (1976),
as well as Wilson (1975, 1980 and references therein), redefined
altruism to refer to relative success within the group rather than
absolute success. Relative success is the individual's number of offspring divided by the average number of offspring of members of his
group. Absolute success is number of children (or number of children
relative to the whole population). Under the 'relative' definition,
'altruism' can spread. Wilson calls the acts that are altruistic under
the 'relative' definition, but not under the 'absolute' definition,
'weakly altruistic'. An alternative I prefer is 'a self-interested refusal
to be spiteful'.

Wilson and Colwell's 'between-group variance' is very closely
connected to relatedness to other group members, and the same points
can be made using both concepts. In a haploid model with constant
group size, they are connected by the following formulae:

$$ v = v_b(1 + (n-1)r) \qquad r = \frac{1}{n-1}\left[\frac{v}{v_b} - 1 \right] $$

where v_b is the variance that would arise from a binomial distribution
of the same overall proportion of genotypes. Several differences may
be noted between r and v. The expected degree of altruism depends
on r by Hamilton's rule, and not on n. Knowing that $r = 0.22$ gives
many biologists an understanding of the genetic closeness described;
the knowledge that $n = 10$ and $v/v_b = 2.98$ is (at least for the present)
less illuminating. When there is no effect of grouping, that is the
grouping is random, $r = 0$; and when spite is expected (Hamilton
1975), r is negative. The value of v itself gives little indication of the
effect of grouping; $v/v_b = 1$ for random grouping and <1 when spite
is expected. I think it will be admitted that r is the more useful and
familiar measure of genetic similarity. Familiarity is important for
clarifying the connections between the new group selection and what
is already known about kin selection.

The connections are certainly there. Bertram (1976) correctly
described his study as one in kin selection, and yet I used it as an
example of the new group selection applied to data. Once the basis of
the new group selection is understood (namely, genetic similarity due
to kinship but where groups are clearly in evidence) most kin selectionists should realize they have been new group selectionists all their

lives. It is vital to remember, of course, that when the population is grouped there may be unsuspected kin links; and that in groups that last for a number of generations, relatedness builds up as the generations proceed.

3.5 CONCLUSIONS

1. The methods of behavioural ecology depend for their correctness on various genetic assumptions.

2. Number of offspring and inclusive fitness are two different valid measures of reproductive success.

3. Inclusive fitness has often been misdefined. It includes relatives' offspring only if the animal's help is responsible for their existence. It excludes those among the animal's own offspring that exist because of help received from others.

4. Hamilton's rule in the form '$rb - c > 0$' should be used in applications of the inclusive fitness approach to data. As a general rule, inclusive fitness is applied wrongly to data but Hamilton's rule is applied correctly. This is because Hamilton's rule encourages use of the correct logic of differences.

5. New group selection models are most readily understood using Hamilton's rule. Genetic similarity as expressed in relatedness is the driving force towards altruism. The forces affecting sex ratio are increasing or decreasing returns to scale for one sex or the other, and the increase in male relatives' mating success that follows from producing a daughter.

PART 2
PREDATORS AND
PREY

Introduction

A group of starlings probing for leatherjackets on a cricket field in the spring is a commonplace sight, at least in England. Although they are industrious and energetic, running around like clockwork toys, the starlings do not seem at first to find it hard to meet their daily needs. But consider their problem. They are looking for prey buried underground in a vast, uniform field. There may be minute surface cues to help them but on the whole they have to find the prey by repeated probing until they encounter one. There are other starlings doing the same thing and it might pay to keep an eye on them (as well as looking out for predator attacks) in case they come across a good area. On the other hand, if one bird finds a good patch others will probably hurry over and jostle it out of the way. Put like this, the starling's job does not seem quite so easy and it is perhaps not surprising that students of foraging behaviour have been attracted to the idea that natural selection has, over thousands of generations, favoured animals which maximize their foraging efficiency.

Optimal foraging models are a way of investigating this notion more closely, by postulating particular maximization criteria and specifying constraints on performance. In Chapter 4, Krebs and McCleery start with what they call the 'classical' energy-maximization models of prey and patch use. From these basic ideas developed in the late 1960s and early 1970s, have emerged more sophisticated models which include nutrient constraints, time budget trade-offs, dynamic changes in internal state (e.g. hunger), and stochastic variation in the environment. Although much of the empirical literature on optimal foraging is not very incisive in testing the classical models, they have stood up quite well to the empiricist's scrutiny. Simple stochastic models of risk and information have also stimulated some neat experimental tests and proved reasonably successful. More complicated, and in principle more complete, dynamic optimization models have so far generated qualitative predictions which leave plenty of room for future research.

Pulliam and Caraco in Chapter 5 argue that the evolution of group size in animals is more suited to analysis in terms of game theory than by simple optimization models. Their conclusion, that the evolutionarily stable group size may not be one in which average fitness within the group is maximized, recalls the distinction made by Parker in Chapter 2 between 'competitive optima' and 'simple optima'. Group size is a product of individual decisions to leave or join, and what each individual does will depend on the alternatives available. If, for example, average fitness in a group is maximal at a size of five animals, but a sixth one is on its own, this solitary individual may do better by trying to join the group, even though the average fitness of the group is lowered. Conversely if the optimal group size in one habitat is five, but a solitary individual could do better if it moved to a better habitat, the group in which average fitness is maximal may not be stable. These analyses, based on Fretwell's (1972) model of the ideal free distribution (and its twin sister Orians' (1969) polygyny threshold model) suggest that the key to understanding group size from a functional viewpoint is to study individual decisions rather than traditional measures of average benefit such as peck rate or scanning rate.

These ideas are echoed in Chapter 6. Davies and Houston arrive at a similar conclusion from, so to speak, the other end. They work from an economic model of territorial defence of food to situations in which resource sharing is advantageous. Again, the emphasis is on individual decision rules and one conclusion is that the observed group size on a shared territory may not be the one that maximizes mean fitness or fitness of all individuals. The analysis of resource sharing is based on winter feeding territories of pied wagtails. The following are the main points. First, territorial individuals stay on their territories even when intake rate would be higher elsewhere (i.e. territory owners do not alter their behaviour to maximize short-term gain), while 'satellites'—birds which sometimes join a territory holder—leave when the pickings are better elsewhere. Secondly, one benefit of owning a territory is efficient exploitation of a renewing resource (insects washed up on the river bank). Thirdly, when renewal rate is high it pays a territory owner to share with a satellite, since time lost through feeding competition is more than compensated for by time saved in sharing territory defence. When the territory is shared, the resources are divided up in a way that maximizes feeding rate for both birds. However, there are some environmental conditions under which territory owners would benefit from sharing but satellites would do better elsewhere (in a flock). Thus for the owner of the territory the optimal group size is two, but the achieved group size may in fact be only one.

Although they have called it by a different name, psychologists studying animal learning have been studying for many years the same kinds of question as behavioural ecologists working on foraging. As Chapter 7 emphasizes, the recognition of this link, which would have seemed unlikely only 10 years ago, is important for both fields. Psychologists bring to behavioural ecology an array of sophisticated techniques and data on mechanisms of learning, while behavioural ecologists may be able to suggest what some kinds of learning are for, and therefore why they have the properties they do.

One area of joint interest is adaptive specializations of learning. This became familiar to psychologists through work in the 1960s on learned taste aversion in rats. It seems as though the qualitatively special properties of this kind of learning (e.g. the long delay between stimulus and consequences) suit it well to its job of teaching rats what is not safe to eat. In Chapter 7, Shettleworth suggests that there may be more examples of specialized learning or memory abilities waiting to be studied, for example the spatial memories of food-storing birds. But it is the link between learning studies and foraging theory in which the most fruitful cross-fertilization has taken place. Shettleworth's account of this complements the discussion in Chapter 4.

Psychologists have spent a great deal of effort in studying how animals in a Skinner box respond to choices between two schedules of reinforcement. The mechanisms of choice have been described in terms of both their overall effect (molar analysis) and moment-to-moment rules (molecular analysis). The general molar rule is known as the *'matching law'* and this probably results from a molecular rule of *'melioration'*, switching always to the alternative offering the higher instantaneous reward rate. For behavioural ecologists it is interesting to know whether these rules would account for performance in optimal foraging experiments (in some cases they do) and whether the cases in which they do not maximize overall reward rate are cases in which the reinforcement schedules are very unnatural. There is also an interesting analogy between the matching law, which states that responses are allocated between alternatives in proportion to the reward rates experienced in them, and Fretwell's 'ideal free distribution', which would predict that animals should allocate themselves between alternatives in relation to expected rewards.

Behavioural ecologists have begun to model learning from a purely functional standpoint, asking how animals *ought* to sample and learn about an unfamiliar environment. These models, introducing concepts such as time horizon and the value of information, may eventually shed new light on mechanisms of learning, or on well-known phenomena in the learning literature such as the 'partial reinforcement extinction effect', 'failed to show self control' and 'behavioural contrast'.

In more general terms, the interplay between proximate and ultimate accounts of behaviour implicit in this link between behavioural ecology and psychology is a healthy redress of the balance of behavioural ecology which shifted too far towards purely functional accounts in the 1970s.

Chapter 4
Optimization in Behavioural Ecology

JOHN R. KREBS and ROBIN H. McCLEERY

4.1 INTRODUCTION

Optimization models are widely used in analysis of design in biological systems. In this chapter we will discuss optimization models of behaviour, but we could equally well apply the same principles to anatomy, physiology, cell structure and cell metabolism (Alexander 1982; Baldwin & Krebs 1981). Before proceeding to the main part of the chapter, we present some brief comments about the logical basis for using optimality arguments in biology; more extensive discussions are provided by Oster and Wilson (1978), Maynard Smith (1978a), and Alexander (1982).

4.1.1 Four questions about optimization models in biology

(1) What is meant by 'design'?

The *design criteria* of a biological system are the criteria upon which selection acts; that is to say, variation between individuals in design affects their fitness (see Chapter 2 for a discussion of fitness). Suppose, for example, that one hypothesized that limb bones are designed to maximize *strength* (resistance to breaking under a bending moment) *per unit mass* (Alexander 1982). In effect this is to hypothesize that variations between individuals in bone mass and strength affect fitness, while variations in other characters such as colour or smoothness of the bone do not, except in so far as they are correlated with strength and mass. (Note that this example is given for illustrative purposes only, bone design may involve factors other than strength/mass.)

(2) Why should design be optimized?

The belief that design is optimized stems from the axiom that natural selection tends to maximize fitness (section 4.1.2). Since fitness depends on the design features, it follows that they too should be optimized. The word optimize in the technical sense does not imply 'the best conceivable', but only 'the maximum or minimum subject to specified constraints'. It is these all-important constraints that pre-clude a naive Utopian view of design in nature (Dennett 1983). We will return to this point later in our discussions of 'rules of thumb' and 'satisfication' (section 4.5).

Are there aspects of animal structure, physiology and behaviour which could not conceivably be analysed within an optimization framework, which are not products of the honing action of selection? Gould and Lewontin (1979; Lewontin 1979), and Oster and Alberch (1982) would answer 'yes' to this question: phylogenetic history and/or the mechanics of genetics and embryology are determinants of present-day structure which are not directly to do with present-day design. This point is a serious one for those seeking to explain major morphological trends in evolution, such as the evolution and radiation of the pentadactyl limb, but is less of a problem for the behavioural ecologist, because the student of behaviour has a more modest aim in using optimization theory than does the evolutionary morphologist. He takes as given the morphological and physiological traits which the morphologist tries to explain. In an optimization model of feeding behaviour, for example, the morphology of the feeding apparatus is taken as given, and its result is that the animal takes a certain time to handle each food item: this is treated as a constraint in the foraging model. The behavioural ecologist does not ask why the animal has a particular jaw size or why it requires nutrient x, but is concerned with the way in which behaviour is organized within these con-straints.

In making a model, one has to decide which factors are to be treated as constraints and which as variables. This decision depends on the modeller's biological intuition, and it sometimes becomes apparent that what was originally considered to be a constraint is in fact a variable under the animal's control (e.g. handling time, Krebs 1980). By relaxing constraints, the domain within which optimization can work is expanded, but there must be ultimate constraints, the laws of physics, which can never be relaxed.

(3) Why use quantitative models?

The hypothesis about bone design mentioned earlier could have been phrased as 'bones are designed to be light and strong'. The problem with this is that many variations of bone structure could reasonably be described as 'light and strong', so the hypothesis is in danger of being able to accommodate everything and therefore accomplish nothing. In contrast, the hypothesis that bones are designed to *maxi-mize* strength per unit mass, subject to precise constraints, makes spe-cific predictions (as well as explicit assumptions) about design. One can therefore find out if it is right or wrong. One of the oft-quoted difficulties with accounts of the adaptive value of behaviour or other traits is that they are often no more than plausible stories made

up to fit the facts (Gould & Lewontin 1979). By making testable predictions, optimization models may help to circumvent this criticism,
and may even allow one to do away with the term 'adaptation'
altogether.

(4) Why not measure variations in fitness?

Since the logical underpinning of optimization models is that variations in design reflect variations in fitness, one might well ask why
behavioural ecologists do not measure fitness in relation to design. A
practical answer can soon be seen by thinking back to the example of
bone design. A fitness-measurer's analysis of this would involve, say,
a group of leopards with varying qualities of bone strength and mass,
so that correlations between bone design and lifetime reproductive
success could be measured. Some leopards would have bones varying
in colour or smoothness but not in strength or mass. After many years
the outcome might be to show that leopards with the greatest strength
per unit mass had the highest lifetime reproductive success, but that
the other variations contributed little or nothing to variations in
fitness. The optimization approach reaches a similar conclusion by
postulating that if bones are designed to maximize strength/mass they
should have properties x, y and z. If these properties are found to
occur and *if they cannot be accounted for in other terms*, one can conclude that one has understood something about the way selection has
designed bones.

Optimization modelling, therefore, offers a potentially quicker
way of investigating design than fitness-measuring, but the extent to
which optimization models reveal how selection acts depends on the
ceteris paribus clause in the last paragraph. Studies in which variations in lifetime reproductive success are measured and correlated
with variations in individual attributes (e.g. Clutton Brock *et al.* 1982)
show directly the action of selection, but the link between selection
and design features can only be inferred from correlations which
do not indicate cause and effect. In general, the two approaches
should not be seen as alternatives, since they often complement one
another.

4.1.2 Normative and descriptive utility

Optimization models are used not only by biologists but also by
economists, psychologists (e.g. Rachlin *et al.* 1981) and decision theorists (e.g. Edwards & Tversky 1967). The way in which these various
disciplines use optimality arguments raises a distinction between the
normative and descriptive approaches. In decision theory a rational
decision maker is one that consistently makes the same choice given
the same set of options. Performing in this way implies two things:
that the options can be ordered with respect to each other, and that

the choices are made according to a maximization principle, i.e. that the rule is always to choose the first option in the ranking. The scale along which options are ranked is often called a scale of 'utility', which leads to the assertion that a rational decision maker always maximizes utility.

It is important to realize that 'utility' used in this sense is simply a descriptive term which allows the prediction of preferences to be built up from observations of past choices; to say that rational decision makers maximize utility in this sense is true by definition. In contrast, utility is also used (especially by biologists) *normatively*, claiming to predict what choices an animal *ought* to make to obtain the most advantageous option according to some functional criterion. In a biological context, normative utility is taken to represent evolutionary fitness in some way. This does not mean that the animal knows what is good for its evolutionary success in any cognitive way, simply that sets of rules which maximize fitness will tend to be selected in evolutionary time. Nor does it necessarily mean that the decision rules are directly linked to the optimization criterion at hand (see section 4.4), although some psychologists (e.g. Rachlin *et al.* 1981) assume that they are.

There is an analogy here with two of the uses of 'fitness' in the evolutionary theory. Descriptive utility corresponds to fitness in the population genetic sense, an operationally defined quantity expressing the number of offspring that an individual carrying one allele can expect to raise to reproductive age relative to an individual with another allele at the same locus. Evolution maximizes fitness in this sense by definition. The normative sense of utility on the other hand corresponds with the rather more nebulous but nonetheless important usage of the term 'fitness' to describe those capacities which render some animals better at survival and reproduction than others.

If normative and descriptive utility functions could be expressed in equivalent terms then it might be possible to derive a quantitative measure of adaptedness from the similarities and differences between what the animal ought to be doing and what it actually does. In practice this *'inverse optimality'* approach suffers from the major difficulty that it is impossible to partition the lack of fit between what the animal ought to do (normative) and what it does do (descriptive) into that due to inaccurate observations or incorrect specification of costs and benefits, and that due to a genuine failure of the animal to be fully adapted to its conditions of life. The problem is discussed more fully by McFarland and Houston (1981; see also Freeman & McFarland 1982). Nevertheless there is an element of the 'inverse' approach about much functional analysis in biology. In the studies described in this chapter for example, the measures of utility assumed to be maximized by animals are generally derived in part from *a priori* guesswork about what ought to influence fitness (normative) and in part from observation of the animal's actual behaviour (descriptive).

Optimization models have been used most extensively in behavioural ecology to analyse foraging behaviour (Pyke *et al.* 1977; Krebs 1978). This is for two main reasons: first, the components of cost and benefit in simple foraging models are relatively easy to measure using standard techniques, and secondly, closely related problems have been studied by operant psychologists, providing an additional extensive data base and armoury of techniques (see Chapter 7). The fact that animals spend much of their lives feeding rather than mating, fighting or being attacked by predators also helps to make foraging relatively easy to study.

In the following sections we first describe what have come to be known as the classical foraging models (Krebs *et al.* 1983); then we consider the implications of stochasticity in the environment for foraging animals; and finally we sketch some of the implications of foraging theory for ecological questions (many of the same points are discussed in section 7.4 from a different perspective).

4.2.1   The classical foraging models

In the first paper on optimal foraging theory, MacArthur and Pianka (1966) distinguish between selecting clumps or *patches* of food and selecting *prey* items from those available within a patch. This distinction has proved useful, although there are many cases where it is hard to apply rigorously. A patch is a place where animals feed, and in models of patch selection it is assumed that animals spend their foraging time either travelling between or foraging within patches. It is relatively easy to think of patches when describing the foraging of animals such as thrushes feeding on berries on a bush, but harder for cases such as a grazing antelope moving across a more or less continuous sward on the plains of Africa. Another example of the difficulties that can arise in distinguishing between prey and patch choices is illustrated by Tinbergen's (1981) study of starlings (*Sturnus vulgaris*). Two of the major prey were *Tipula* larvae which occurred in grazed meadows and the caterpillar *Cerapteryx* which occurred in salt-marshes: thus choice of prey implied choice of patch and vice versa.

In spite of such problems, the literature on foraging is largely organized around prey and patch models, so in the following paragraphs we retain the distinction.

The classical model of prey choice

For the sake of simplicity consider a predator hunting for just two kinds of prey which are encountered at rates λ_1 and λ_2 prey per second during T_s seconds of searching. The two prey types yield E_1 and E_2 units of net reward (e.g. calories) and take h_1 and h_2 seconds to handle: their *profitabilities* are defined as E_1/h_1 and E_2/h_2.

If the predator forages unselectively, in T_s seconds it will obtain the following amount of food (E):

$$E = T_s(\lambda_1 E_1 + \lambda_2 E_2)$$

and this will take the following total time (T):

$$T = T_s + T_s(\lambda_1 h_1 + \lambda_2 h_2),$$

in other words searching time plus handling time.

The overall rate of intake of the predator (E/T) is therefore:

$$\frac{E}{T} = \frac{\lambda_1 E_1 + \lambda_2 E_2}{1 + \lambda_1 h_1 + \lambda_2 h_2}.$$

Note that T_s has cancelled out.

Now suppose that prey type 1 is more profitable than type 2 ($E_1/h_1 > E_2/h_2$). In order to maximize E/T the predator should eat only prey type 1 if

$$\frac{\lambda_1 E_1}{1 + \lambda_1 h_1} > \frac{\lambda_1 E_1 + \lambda_2 E_2}{1 + \lambda_1 h_1 + \lambda_2 h_2},$$

i.e. the rate of energy gain from prey type 1 alone is greater than that from both types. This equation can be rearranged as follows:

$$\lambda_1 < \frac{E_1}{E_2} . h_2 - h_1 \tag{4.1}$$

to give a threshold for specializing on type 1 in terms of the time taken to search for the next item ($1/\lambda_1$).

In formulating a simple example like this we can see the essential components of an optimization model: the maximization criterion (intake rate in this case), the constraints (handling time), and the alternative courses of action for the predator (eat one or both types of prey). Furthermore, our assumptions in building the model are laid bare: they, together with the predictions are summarized in Table 4.1.

Tests of the model

Many studies that purport to confirm the diet model's predictions or assumptions report observations that, while qualitatively consistent with the model, do not constitute a rigorous test (Krebs *et al.* 1983). For example, it might be observed that a particular species eats mainly prey of high profitability. While this could be construed as supporting the model, other explanations, such as the profitable prey being more conspicuous, might equally well account for the data. In the same way, observations appearing to contradict the model (for example predators that are totally unselective) cannot be evaluated without a quantitative analysis of λ, E, and h. For instance, if the inequality of equation 4.1 is reversed, non-selective predation is predicted.

Measurements of the components of equation 4.1 are often not easy to make in the field, and therefore most of the detailed tests of the model have been done in the laboratory or in simple field situations. This is not to say that it only applies in these instances, although more complicated models may well be necessary to account for much of foraging in the wild (see later). An example of a laboratory test is the experiment of Krebs *et al.* (1977) on great tits (*Parus major*), small insectivorous birds. Two different-sized pieces of mealworm were the prey types, and the exact values of λ_1 and λ_2 were controlled by presenting the prey in sequence on a moving belt at the front of the bird's cage. The bird could see the pieces of worm for about 0.5 s, one at a time, as they moved past a small gap in a cover over the belt. In this brief moment the bird chose whether to pick up or leave the item. Picking up an item to handle it meant that time was lost from 'searching' (waiting at the belt) for the next prey, the essential trade-off in the diet model. The experiment consisted of measuring E_1, E_2, h_1, and h_2, and then varying λ_1 in such a way as to cross the threshold of equation 4.1. On one side of the threshold (low values of λ_1) the birds should have taken prey off the belt unselectively, at higher values of λ_1 they should have been more selective, taking only type 1. Figure 4.1a shows that selectivity in four

Table 4.1. Assumptions and predictions of the classical optimal diet model.

(a) Assumptions

1. Prey value is measurable net energy or some other comparable single dimension
2. Handling time is a fixed constraint
3. Handling and searching cannot be done at the same time. If this were not true there would be no need for the animal to forage selectively
4. Prey are recognized instantaneously and with no errors
5. Prey are encountered sequentially and randomly: the expected time to find the next prey of type i is always $1/\lambda_i$
6. Energetic costs per second of handling are similar for different prey
7. Predators are designed to maximize rate of energy (or other measure of value) intake

(b) Predictions

1. The highest ranking prey (in terms of profitability) should never be ignored
2. Low-ranking prey should be ignored according to the rule specified in equation 4.1
3. The exclusion of low-ranking prey should be all-or-nothing, depending on the direction of the inequality in equation 4.1
4. The exclusion of low-ranking prey does not depend on their own values of λ. Note that λ_2 does not appear in equation 4.1. This point can be grasped intuitively by realizing that when type 1 are sufficiently abundant 'time out' from search to handle type 2 is not profitable. In other words E_2/h_2 is less than what could be obtained by searching for and consuming the next type 1, so the abundance of type 2 is irrelevant

---

animals did increase in roughly the predicted way, but the change was not the step function predicted by equation 4.1. Figure 4.1b shows that a study of bluegill sunfish (*Lepomis macrochirus*) reached an essentially similar conclusion.

Partial preferences

Figure 4.1 demonstrates that the classical diet model fails by predicting a step change where in fact something closer to a sigmoid change is observed; that is, the animals show 'partial preferences' for high-ranking prey and do not totally exclude lower ranks. Why is this? At least six different explanations have been proposed, all of which may apply to one case or another.

(1) *Discrimination errors*. Rechten *et al.* (1983) showed that the failure of great tits in experiments similar to the one illustrated in Fig. 4.1a to

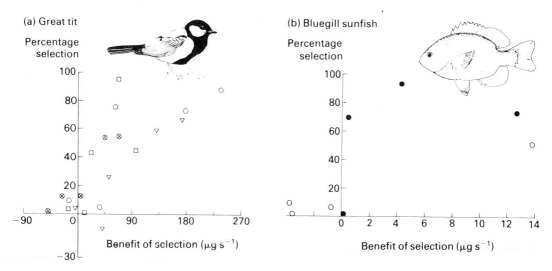

Fig. 4.1 **(a)** A test of the classical 'optimal diet model' with great tits. The graph shows on the *y*-axis the degree of selectivity for the larger (more profitable) of the two prey sizes, and on the *x*-axis the difference in intake rate obtained by selective and non-selective predation. To the right of the *y*-axis, 100% selection is predicted by the model. The four symbols are for four different animals, each tested at several values of λ_1 and λ_2. The birds become more selective to the right, as predicted, but they do not show a step function with all-or-nothing selection. The graph indicates that they take small prey when according to the model they should not. Another kind of possible error, rejecting large prey, is not shown on the graph. **(b)** Data from Werner & Hall 1974 for bluegill sunfish eating *Daphnia* plotted in a similar way. The data are for experiments with 20, 50, 200 and 300 large prey per tank, and corresponding variations in the density of medium and small prey. The closed circles refer to selection for large and medium prey (i.e. rejecting small ones) and the open circles refer to the selection of large only (rejecting medium and small). Data were taken from Tables 2 and 6 and Figure 4 of Werner & Hall 1974. As with the great tits, sunfish become more selective as they cross the threshold of equation 4.1, but they show partial preferences.

select exclusively large pieces of worm is partly because they confuse large and small prey. If the predator cannot discriminate perfectly between the various types of prey, optimal choice will inevitably produce partial preferences.

(2) *Long-term learning.* Snyderman (1983) found that pigeons came very close indeed to a step change if they were given many days of experience of the same values of E, h and λ. Thus some of the failures to find step changes in other studies may have arisen because the animals were not trained for long enough.

(3) *Inherent variation in the animal.* Gibbon and Church (1981), in studies of time measurement in animals, found sigmoid curves where step functions were predicted. They explained them in terms of stochastic variation in the animals' internal clock: a similar explanation could be applied in foraging studies (section 7.4.3).

(4) *Runs of bad luck.* In a random sequence of encounters an animal may have a run of bad luck, during which it does not encounter profitable prey. If encounter rates are estimated by the animal using a short-term rule, it may eat low-ranking prey after such a run of bad luck, as observed in shore crabs (*Carcinus maenas*) (Elner & Hughes 1978; see also Fig. 7.3).

(5) *Simultaneous encounters.* If more than one prey item can be detected at the same time, the optimal diet model predicts partial preferences (Waddington & Holden 1979; Waddington 1982). The reason for this is as follows. Suppose the predator at one moment sees two items, a large, distant prey (t_1 seconds away) and a small, near one (t_s seconds away). The expected reward rate from going for the large item ($E_1/(h_1 + t_1)$) may be less than that from the small one ($E_s/(h_s + t_s)$) even if the profitability, defined as E/h, of the large is greater.

(6) *Averaging across individuals.* Even if none of the previous arguments apply, but if data are averaged across animals, partial preferences might be expected. Each individual may have a step function in a slightly different place (due to variation, for example, in h), so that the population response is a sigmoid curve.

Exploitation of patches

A central concept in models of patch use is that of *resource depression* (Charnov et al. 1976), the idea that prey availability within a patch decreases as a result of the predator's foraging activity. This could come about by direct depletion of the prey, or because prey take evasive action, such as the dungflies (*Scatophaga stercoraria*) leaving a cowpat for the surrounding grass shortly after a yellow wagtail (*Motacilla flava*) starts to hunt for them (Davies 1977). The consequence of resource depression is that a predator arriving in a patch at first acquires the resource quickly, but the rate of acquisition diminishes with time, giving rise to a cumulative gain curve like the one

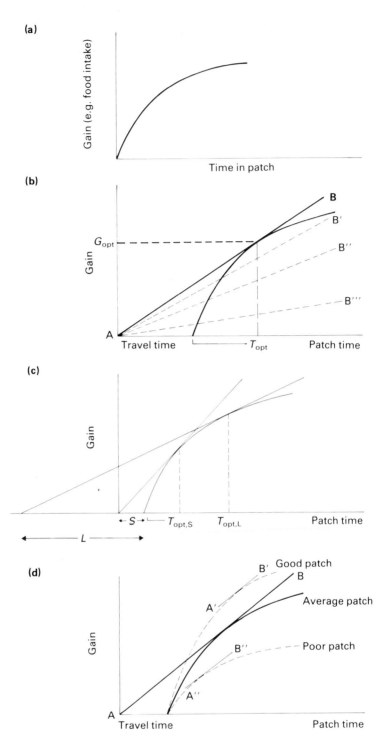

Fig. 4.2 The marginal value model. **(a)** An example of a gain curve arising from resource depression. **(b)** The 'tangent' method of finding the optimal patch resident time T_{opt}. The rays (AB) from the origin have slopes E/T and the steepest feasible line is the solid one. **(c)** As travel time increases, so does T_{opt}. For a short time (S), $T_{opt, S}$ is less than $T_{opt, L}$ for a long travel time (L). **(d)** When there are several different patch types, each should be exploited until the gain drops to the average for the environment (line AB).

shown in Fig. 4.2a. Although the resource we are talking about is food, the same principle has been applied to mate searching and copulation (Parker 1978b), collecting water (Kasuya 1982) and investing in eggs or offspring (Smith & Fretwell 1974; McNair & Parker 1979). A curve similar to the cumulative gain curve, called a *loading curve*, may arise if a predator loads up with food items in its mouth to carry back to a nest or store, and as a result of loading up gets steadily less efficient at capturing prey (Orians & Pearson 1979).

To maximize rate of gain of resource from patches with resource depression a predator should follow the rule illustrated in Fig. 4.2b, the so-called *marginal value theorem* (Charnov 1976). Imagine a predator spends T_t seconds travelling to a patch and then consumes food according to a gain curve like the one in Fig. 4.2b. Where on this curve of diminishing returns should the predator leave the patch and travel to the next one? In order to maximize energy (E) per unit time (T) the predator should leave at the point (T_{opt}) which gives the steepest slope of the line AB. This is the solid line in the figure; it has both to touch the gain curve and to intersect the x-axis at the travel time, since these are assumed to be fixed by the environment.

If travel time increases (patches are further apart) the optimal time to stay in a patch also increases (Fig. 4.2c): the longer it takes to travel, the lower the value of moving, so the further it is worth going along the curve of diminishing returns in the current patch. If patches vary in quality, the model predicts that each patch should be exploited until the gain rate within the patch drops to the average for the environment (Fig. 4.2d).

As with the diet model, the marginal value theorem makes a number of simplifying assumptions and some testable predictions; these are listed in Table 4.2.

Table 4.2. Assumptions and predictions of the marginal value model described in Fig. 4.2.

(a) Assumptions

1. Each patch type is recognized instantaneously
2. The travel time between patches is known by the predator
3. The gain curve is smooth, continuous, and decelerating. An analogous model can be produced for discontinuous gain (e.g. Kacelnik *et al.* 1981)
4. Travel time between and searching within a patch have equal energy costs. If costs differ, corrections have to be made (e.g. Cowie 1977)

(b) Predictions

1. If travel time and the gain curve are known, then T_{opt} (Fig. 4.2b) can be predicted
2. If there is more than one patch type in an environment, all should be reduced to the same marginal gain rate
3. If the gain curve is known, the relationship between T_{opt} and T_t can be specified

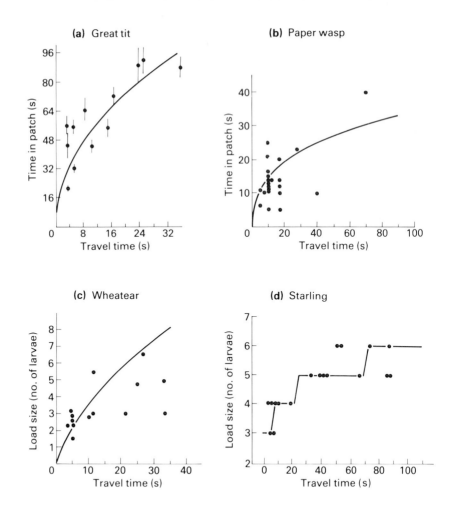

Fig. 4.3 **(a)** Cowie's (1977) test of the marginal value model with great tits. The predicted line takes into account the differential cost of travelling and searching. The twelve dots are X ± s.e. of six birds, each in two experimental treatments. **(b)–(d)** Three other tests of the m.v.t. in which travel time was manipulated and patch time or resource extracted measured as the dependent variable. Predicted relationships are lines, data points are dots. **(b)** Paper wasps sucking up water to carry to the nest (Kasuya 1982). **(c)** Wheatears (*Oenanthe oenanthe*) carrying insect larvae from feeding trays to the nest (Carlson & Moreno 1982) (data reconstructed using an average loading curve $y = 4.06x^{1.42}$). **(d)** Parent starlings collecting mealworms from patches where they are delivered on a progressive interval schedule. The predicted line in this case is stepped because the prey were discrete items delivered on a deterministic schedule (Kacelnik, in press). In b and c, corrections for energy costs were not made in calculating the prediction which therefore refers to gross intake. In d, the prediction is based on maximizing net energy returns to the parent plus its young.

The most persuasive tests of the marginal value theorem are ones in which travel time was manipulated and T_{opt} or G_{opt} (Fig. 4.2b) predicted. For example, Cowie (1977) studied captive great tits foraging in a large aviary for small pieces of mealworm hidden in sawdust-filled plastic pots (patches) attached to the branches of five artificial trees. Cowie tested each of six birds in two treatments, long and short travel times. Travel time between patches was manipulated by putting lids on the pots that were either easy or hard to remove; thus a long 'travel time' was spent by a bird prising off a difficult lid. All patches had the same number of prey and therefore the same gain curve. By measuring the gain curve it was possible to predict the relationship between T_t and T_{opt}; the fit to the predicted line was quite good (Fig. 4.3a), and there is no difference between observed and expected points when the two are compared statistically. The same model has been tested with a variety of other animals: water-boatmen sucking the contents of mosquito larvae (Cook & Cockrell 1978), wasps collecting water to take to their nest, chipmunks loading up with seeds to store in their burrows (Giraldeau & Kramer 1982), dungfly males copulating with females (Parker 1978b), and so on. The results of three of these studies are shown in Figs 4.3b–d. In each case, patch residence time or gain from the patch is predicted and travel time is varied. All four studies in Fig. 4.3 show qualitatively the predicted trend, but only two sets of results, for great tits and starlings (4.3a and 4.3d), show a good quantitative fit. We will return in our concluding comments to the question of how convincing these tests are.

To summarize, the classical models, in spite of their daunting list of simplifying assumptions, seem to do quite well in predicting behaviour for quite a range of animals: time and energy are apparently major factors in the design of foraging animals. However, there are many facets of foraging which are not encompassed by the models discussed so far, and in the following sections we consider some of these.

4.2.2 Prey quality

The simplest way to modify the optimal diet model to take into account the fact that a predator may require a particular nutrient, perhaps not available in all prey, is to consider the nutrient requirement as a fixed constraint (e.g. sodium for moose, Belovsky 1978). In Belovsky's moose model it was implicitly assumed that once the animal has met its minimum requirement of sodium, further sodium intake has no advantage. The moose were found to maximize their daily energy intake subject to the sodium constraint, implying that fitness increases with increasing energy intake. But suppose that for both sodium and energy intake the fitness consequences were similar, for example a negatively accelerating rise to a plateau? This situation can be analysed with an extension of the kind of model used for moose, but without knowing the exact form of the fitness functions it is not possible to specify the precise mixture of nutrients that should be eaten (e.g. Rapport 1980; McFarland & Houston 1981).

How important are nutrient constraints in determining prey choice in the wild? For herbivores it is generally recognized that plant qualities other than energy are important in diet selection (Owen Smith & Novellie 1982; Freeland & Janzen 1974; Milton 1979; Rapport 1980). Here the simple energy-maximizing model is unlikely to be adequate, although it may be possible to substitute another single dimension of food quality, such as ratio of protein to fibre (Milton 1979), reducing the problem to that of the classical model. Three studies of birds with diets of animal food have also indicated that simple considerations of energy maximization cannot account for prey preferences (Goss-Custard 1977; Tinbergen 1981; Royama 1970) and it is possible, but not certain, that nutrient constraints may have been important in prey choices in these cases.

4.2.3 Stochastic foraging models

The diet and patch models considered so far assume that the animal 'knows' the quality of each patch or prey item and the travel times between encounters. ('Knows' is put in quotes because it might be that the animal may use something as simple as the concentration of odour to assess patch or prey density and may therefore behave as if it knows the environment.) But suppose that travel times, patch quality, etc., vary in an unpredictable way, as must often happen in Nature? This is the problem of stochasticity. We will discuss it in relation to patches as an example, but similar arguments could be considered for prey choice.

There are two consequences of stochasticity (Stephens & Charnov 1982): (a) the animal might respond to the *risk* of doing badly and doing well in a particular patch or patch type and (b) the animal might have a strategy for assessing patch quality, acquiring *information* by sampling.

Risk sensitivity

Risk can be most readily explained by referring to the experiment of Caraco *et al.* (1980a; see also Caraco 1981a). They offered seven yellow-eyed juncos (*Junco phaeonotus*), a small granivorous bird, a choice between two feeding sites in an aviary. In a given experiment one site always offered a constant reward (say two seeds), while the other offered an unpredictable reward (say no seeds on half the occasions, few on the other half) with the same or similar mean value to the constant site. With the same mean at both sites, animals maximizing energy gain should be indifferent to the options offered, but this is not what Caraco *et al.* found. Instead they found that after one hour of starvation the juncos preferred the predictable site when tested, but after four hours of starvation they chose the unpredictable one (actual testing was preceded by 20 learning sessions). In technical terms, the bird changed from *risk-averse* to *risk-prone* behaviour as deprivation increased.

The likeliest interpretation of this is as follows. After one hour of deprivation the constant site provides enough food to meet the animal's energy requirements, while the variable site has a 50% chance of providing more than enough and a 50% chance of not providing enough. Choosing the constant site therefore maximizes daily survival. After four hours' deprivation the reverse argument applies: the constant site has no chance of providing enough food, while the variable site has a 50% chance of doing so. In this case risk-prone behaviour maximizes daily survival. This account of risk sensitivity based on maximizing daily survival is sometimes called the 'expected energy budget rule': 'be risk prone if the daily energy budget is negative, be risk averse if it is positive' (Stephens 1981).

There are various refinements of this simple rule. Stephens and Charnov (1982), for example, point out that if the mean is correlated with the variance of a patch type, conditions for risk-proneness may differ from the fixed-mean case studied by Caraco *et al.*. Houston and McNamara (1982) consider why 50% of risk-prone animals do not die, as might be expected from the argument above. Their point is that animals may face a series of choices through a single day. If they start off risk-prone, at each choice a proportion of animals will get enough to ensure that they are risk-averse at the next decision, so by the end of the day far fewer than the original proportion of risk-prone animals will still be in negative expected energy budget.

How does this apply to the marginal value model? Stephens and Charnov (1982) show that if travel time between patches is an exponentially distributed random variable, variance in rate of food intake is maximized at patch time slightly shorter than T_{opt} (Fig. 4.2b). In other words a risk-averse forager should stay slightly longer in a patch than a risk-prone one, but the difference from the patch time maximizing intake is not great in either case. Even if animals are sensitive to risk, the marginal value model may give a fairly accurate prediction of patch time.

Information

If patches vary unpredictably in quality and cannot be recognized in advance, rewards obtained at the start of a patch visit could be used to estimate patch quality, and thus how long to stay (Oaten 1977; Green 1980; McNamara 1982; Iwasa *et al.* 1981). The main conclusion of theoretical analyses of this problem is that a forager acquiring information about a patch while foraging in it should on average stay longer than predicted by the marginal value theorem (McNamara 1982). This is because at any moment a short extra stay in the patch might yield information that the patch is in fact better than the current estimate. It is worth sacrificing a little from the maximum intake rate of the deterministic model to gain this extra information.

More quantitative predictions about sampling have so far only come from studies of simple foraging problems. Lima (1983) examined how predators in the field sample patches with no resource depression (ones which have a linear gain function in a graph like Fig. 4.2a). Downy woodpeckers (*Picoides pubescens*) were faced with 60 patches—37 cm pieces of branch with 24 holes each—of two types. The types contained either no seeds or, according to experimental conditions, 24, 12, or 6 hidden seeds. The question of interest is how the predator should use its experience of the first few holes inspected in a patch to decide whether or not the patch is empty. If the two patch types contain 0 and 24 seeds, for example, and the bird has learned this, a single unsuccessful look in a patch is enough to tell that it is going to be empty. If the patches with seeds are only half or a quarter full, more empty holes have to be inspected to decide whether or not the patch is totally empty. The predicted number of holes inspected before leaving an empty patch was calculated from a version of Green's (1980) model. For the three conditions the predicted values were 1, 3 and 6, while the corresponding observed mean values were 1.7, 4.5, and 6.3 holes per empty patch. These observed values, which were reached after the birds had experienced several days in succession of the same mixture of patch types, show rather good correspondence to the predicted values. In another respect the birds' performance was worse than predicted: in the seed-containing patches they visited almost all the holes (22/24) in all three treatments. If they could have kept track of the number of seeds eaten they should have left earlier in the 6- and 12-seed treatments than in the 24-seed experiment.

In Lima's model it was assumed that the birds knew the proportion of totally empty branches and the number of seeds per branch. Krebs *et al.* (1978) discuss how animals might acquire this sort of information. They consider the problem of a foraging animal confronted with two patches of different but unknown reward rates. They generate a model to predict how much sampling of the two sites the animal should do before 'deciding' which is the better one. The model is a simplified version of the 'two-armed bandit' model of mathematicians. The simplification includes the assumption that the animal samples each site alternately before deciding to exploit one exclusively. After each sample the model updates its estimate of patch quality and compares the expected number of rewards from continuing to sample and from deciding to exploit the currently better site.

As with other models of information the advantage of sampling once more is that it might yield information that the good site is better than current estimates suggest. The model was tested by offering nine great tits a choice of two identical feeding patches in an aviary. The feeding sites were operant devices in which the birds hopped on a perch to earn food rewards at probabilities set by the experimenters. A single hop was considered to be equivalent to a 'sample' in the model. At the start of the experiment the birds alternated between sites (sampling) but later on they spent most of their time at the more profitable site. Two predictions about duration of the period of sampling were confirmed (Fig. 4.4; see also Chapter 7).

Summary of sampling models

It is tempting to think of sampling models as being 'more general' than the marginal value model, or even that they in some way 'show the marginal value theorem is wrong'. However, the two kinds of model in fact consider different situations. The marginal value theorem predicts *when* an animal should leave a patch given that it knows patch quality and travel time, while sampling models such as Oaten's, Green's and McNamara's analyse *how the animal ought to estimate* the quality of a patch. In reality most foragers are probably neither completely ignorant nor omniscient about the quality of patches, so elements of both models may be applicable. A second point to bear in mind is that sampling models make their own simplifying assumptions, for example that the predator knows the frequency distribution of prey per patch, travel time and the distribution of intercapture intervals. Finally, as Stephens (1982) points out, the value of sampling to the foraging animal must be assessed in terms of the potential increase in rate of intake as a result of sampling. Patch sampling is therefore most valuable when variance in patch quality is high, but these are the very conditions under which it is likely to be easy for the predator to recognize differences in patch quality by direct sensory cues with little or no sampling. In short, animals with good sensory mechanisms may often approximate to the marginal value theorem in patch use.

Learning rules

Related to the problems of sampling and estimating environmental changes are a number of 'optimal learning rules' which have been proposed for foraging animals (Killeen 1981; Ollason 1980; Harley 1981; see also section 7.4.4). Such rules involve two components: (a) a way of combining past and present experience (usually with an exponential forgetting rule) to assess payoff and (b) a way of allocating foraging effort between alternatives in relation to expected payoff. Such rules allow a foraging animal to respond to changes in patch or

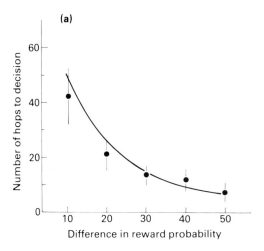

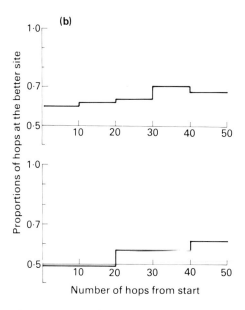

Fig. 4.4 Test of a model of sampling. **(a)** The amount of sampling of two unknown feeding sites is related to the degree of difference in reward probability at the two sites. The solid line is a prediction from a simplified optimization model, and the dots are means ±95% confidence intervals of nine birds. The y-axis is the number of sampling events prior to a 'decision' by the bird to concentrate on the more profitable site. The criterion for a decision was arbitrarily defined (Krebs *et al.* 1978). **(b)** Trajectory towards a decision to concentrate on the better side is steeper when the animals are trained to expect short experimental sessions (upper) than when sessions are long (lower). Difference in reward probability is held constant in this experiment. The model (correctly) predicts that the animal should decide more rapidly when the sessions are short, since the value of information in the future is less, so sampling is less worthwhile (Houston *et al.* 1982).

prey quality (i.e. to learn about their environment), but they are not strictly sampling rules since they do not specify the optimal strategy for acquiring information about unknown alternatives.

4.2.4 Optimal foraging and ecology

Optimal foraging theory started out as an ecological theory to account for variation in resource utilization within and between species, but most of the work we have described so far has focused on the behavioural rules of foraging animals. However, a number of more strictly ecological questions can be viewed in a new light as a result of considerations of foraging theory. An example is the functional response of predators to prey density.

Holling (1959, 1965) was the first to emphasize that changes in the number of prey eaten by an individual predator in response to changes in prey density (the functional response) are influenced by handling time. In particular, at high prey densities the predator's attack rate is limited by the time required to handle each item captured. This trade-off between handling and searching is also central to the optimal diet model, and to this extent the diet model and Holling's model are in agreement. However, in other respects Holling's characterization of the functional response may be too simple. Consider, for example, the effect of presenting the predator with a mixture of two prey types. According to Holling's model the response of the predator to each of the two prey would be similar in form to that observed when the prey are presented alone. This would not, however, be expected from the diet model. For the lower ranking type the attack rate of the predator would be expected to be zero and independent of prey density when the higher ranking type is commoner than the threshold of equation 4.1. Thus the functional response to a prey type in a mixture may depend largely on the abundance of other types, while the response to the same species on its own will depend largely on its own density.

This is just an illustrative example of how foraging theory might be employed to predict the impact of predators on their prey. More extensive discussions can be found in Abrams (1982), Krebs *et al.* (1983) and Hassell (1980).

4.3 TIME BUDGETS AND CONFLICTING DEMANDS

One of the limitations of optimal foraging models described in the previous section is that they do not take into account conflicts in the predator's time budget between feeding and other activities such as vigilance, mating and territorial defence. These activities may be interspersed with foraging (for example, foraging birds often 'look up' to scan for predators) and animals may therefore not feed as quickly as might be predicted from considerations of encounter rate and handling time alone. One way to incorporate competing activities into a foraging model is to treat them as time constraints, just as handling time is a constraint in the classical diet model. This approach is discussed in more detail with respect to territorial behaviour in Chapter 6, and one example will serve to illustrate the point here. Martindale (1982) recorded the duration of foraging trips and size of loads of food brought back for the young by Gila woodpeckers (*Melanerpes uropygialis*) in Arizona. He compared these measures during control periods and experimental periods in which he placed a stuffed woodpecker near the nest. He found that the parents made shorter trips and brought smaller loads after seeing the rival near the nest. The implication is that the parents sacrifice a certain amount of foraging efficiency in order to inspect the nest at regular intervals. This could have been modelled as a time constraint on the maximum absence from the nest, although Martindale did not do so.

In other cases *energetic* costs might act as a constraint. For example, hummingbirds do not fill their crop to capacity in each bout of foraging. This may be related to the energetic cost of carrying a crop loaded with nectar in flights between foraging bouts (De Benedictis *et al.* 1978).

Other experimental studies of time budget conflicts between feeding and other activities include those by Caraco *et al.* (1980b) on feeding, aggression and vigilance; Kacelnik *et al.* (1981) on feeding and territoriality; Lendrem (1983) on feeding and vigilance; Sih (1980) on feeding and predation risk.

4.4 DYNAMIC OPTIMIZATION AND BEHAVIOUR

The models considered so far are incomplete in that they do not consider the *state* of the animal. We may consider the state to be the set of the values of factors relevant to the animal's fitness which are altered by its behaviour. Hence the value of an activity to an animal depends on its state. For example, the relative value of time spent feeding is likely to be greater if the animal is very short of food than if it is near satiety, while the value of patrolling a territorial boundary is likely to depend on how long it is since the territory was last patrolled. The performance of an activity alters the animal's state; for example, feeding reduces food deficit and hence the value of feeding relative to territorial behaviour. The whole arrangement is therefore a feedback system (Fig. 4.5a) and it is just as logical to think of the animal controlling its internal state by means of behaviour as it is to adopt the more conventional view of behaviour being controlled by internal state.

In the following sections we try to understand from a functional point of view how all an animal's activities are organized into a sequence. Consider the problem facing an animal which is dangerously short of two commodities needed to ensure survival, say food and water, which can only be obtained one at a time. If it chooses to feed then this has beneficial results but also precludes drinking, so the effects of water deficit continue or are exacerbated. It is likely that some sequences of the two activities will be more beneficial than others. One way to think of the possible solutions to the problem is to plot changes in deficit which the animal can achieve against each other, as in Fig. 4.5b. The different possible policies then appear as trajectories in this state space (so called because the axes are state variables). In principle, choosing the best trajectory in a case like this, where the state of the system depends on the choice of behaviour,

(a)

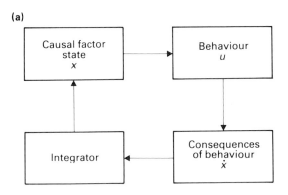

(b)

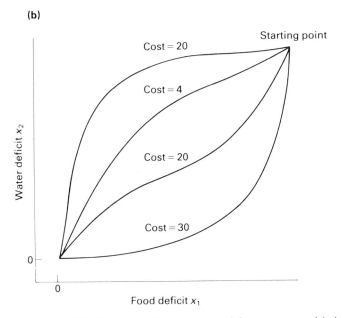

Fig. 4.5 (a) Relationship between causal factor state and behaviour. The consequence of behaviour is that the state x is changed at rate $\dot{x}$. (b) Possible paths in a state space representing food and water deficit. The overall cost of each trajectory (hypothetical) is indicated.

simply involves comparing all possible trajectories, but if there are more than a very few intermediate states between the starting point and the goal the optimum trajectory becomes impossible to find. The solution of this problem requires a branch of control theory called *dynamic optimization* theory, in contrast to the optimal foraging models described earlier, which are examples of *static optimization*. The distinguishing characteristic of dynamic optimization problems is that a solution can only be found by considering the whole trajectory to the goal point. A general form of these techniques is 'dynamic programming', which depends on the following principle: 'An optimal policy has the property that, whatever the initial state and initial decisions are, the remaining decisions must constitute an optimal policy with regard to the state resulting from the first decision' (Bellman 1957). Another way of expressing this is to say that as long as we have a complete specification of the decision problem, the way in which the current state was reached does not affect the optimal policy given this state. A familiar example is the parental investment problem: the decision whether to wean one set of offspring and start producing the next set depends only on the future costs and benefits, not on the resources already given to the current brood (see Chapter 8).

The practical consequence of this principle is that by working backwards from the final state of the system the optimal trajectory for the system can be found with far less effort than would be entailed in calculating the costs and benefits of all possible trajectories. Even so, it can involve a good deal of computation and may become impracticable for large numbers of state variables. An analytic method which is equivalent in many ways but more elegant mathematically is the Pontryagin Maximum Principle (PMP), which has been used in a number of recent ethological and ecological studies. Though the method may have been overrated (Houston 1980), it provides an heuristically powerful method of looking at optimal control in some biological cases.

4.4.1 Feeding and satiation curves

Sibly and McFarland (1976) derived an illustration of the application of PMP to the control of feeding in food-deprived Barbary doves. They proposed that the function relating costs to internal state was likely to be quadratic because the further from equilibrium the animal's state, the more likely it is to cross a lethal limit. Similarly, they proposed that the cost of acting to alleviate physiological deficits would increase quadratically as the animal was forced to pay more attention to feeding and hence would become more vulnerable to predation.

If there is no upper limit to the rate of behaviour the problem can be solved using PMP (or indeed dynamic programming) to yield the optimum solution that the rate of behaviour should be proportional to the state (Table 4.3). In other words, the optimum trajectory is for the animal to approach satiety (the goal point) exponentially. Intuitively this seems reasonable; if the animal is in an expensive state it is worthwhile to take risks in order to move the state rapidly to a safer value, but such risks are not worth taking when far from the lethal limit. Some evidence exists that free-feeding animals in a laboratory setting do indeed have satiation curves that are exponential in form (McCleery 1977).

Sibly and McFarland (1976) extended their analysis to deal with decisions taken between two activities, feeding and drinking, over a short period. With similar quadratic cost functions and dynamics for both food and water deficit, the solution turns out to be what is picturesquely described as 'bang-bang' control, meaning that in a graph like Fig. 4.5b there will be a 'switching line' passing through the origin: when the state lies above the line the animal will spend all its time drinking, while below it will exclusively feed.

Interestingly, this bang-bang aspect of control also emerges from another classic study, by Macevicz and Oster (1976), using PMP as a tool for dynamic optimization. Using a model for the economics of production of workers and reproductives in eusocial insects with an annual colony cycle, they showed that in many cases the optimum

Table 4.3. PMP and the solution of dynamic optimization problems. PMP states that minimizing costs over a trajectory is equivalent to maximizing the Hamiltonian or Pontryagin function H, which is given by H = system dynamics − cost functions.

Definitions	Example	Means
System dynamics ('plant equations'): the change in state caused by behaviour at rate u	$\dot{x} = -ru$	Change in state (food deficit) x proportional to rate of behaviour (feeding rate)
Cost function: instantaneous cost of state x, or rate of behaviour u	$C = a_1 x^2 + a_2 u^2$	Cost is proportional to the square of food deficit and the square of feeding rate
The optimum trajectory is	$\dot{x}(t) = x(o)e^{-rat}$	The food deficit should be reduced at an exponentially decelerating rate.
Generated by the control law	$u(t) = ax(t)$	Rate of behaviour should be proportional to deficit.

solution was to invest everything in workers until close to the end of the season, when all resources should be switched to making reproductives (see Oster & Wilson 1978 for a more detailed discussion). Several species of wasp do conform to this rule.

If the act of switching is itself costly (e.g. because the animal has to move somewhere else to carry out the alternative behaviour) the problem becomes too difficult to solve using PMP, since the choice of activity depends not only on the state of the system but also on which activity is currently in progress. Larkin (1981) used dynamic programming to solve this problem and showed that when it was made more difficult for Barbary doves in a Skinner box to switch from feeding to drinking, the delay in switching between activities was quantitatively as predicted.

The dynamic optimization studies of short episodes of behaviour described so far are not very realistic. It is hard to imagine that it really matters to a Barbary dove in the longer term whether it chooses exactly the right trajectory during a feeding and drinking session in a Skinner box. More importantly, the relationship between cost and state and rate of behaviour is really just a guess; in fact the studies mentioned so far can be considered examples of the descriptive approach (see section 4.1.2.), since what has been done is to guess *a priori* the form of the cost function and then to justify it by showing that the animal's behaviour fits it. Support for the quadratic form of the cost functions is actually less strong than it appears to be, for other descriptions of the satiation curve which follow from different forms of the cost functions fit the data nearly as well as the exponential form predicted by quadratic cost functions (McCleery 1977).

4.4.2 Feeding and vigilance: cost and rate

One reason for suggesting that cost is related to rate of behaviour in a quadratic or similar in form arises from the notion that increasing the rate of an activity also increases the amount of attention that must be paid to that activity, diverting attention from other matters. This might, for example, make the animal more vulnerable to predators. An attempt to evaluate this notion systematically has been made using sticklebacks (*Gasterosteus aculeatus*) by Milinski and Heller (1978). They observed that hungry sticklebacks, when offered a choice between swarms of *Daphnia* of different densities presented in perspex cells, tended to attack at first the swarm of highest density, but as the session proceeded they attacked at a lower rate and preferred lower density swarms. Milinski and Heller equated this waning response with satiation, perhaps not entirely convincingly (see McFarland & Houston 1981); they suggested that foraging at a high rate was only possible in swarms of high density and it was therefore costly because the fish had to pay more attention to the prey in order to overcome the 'confusion effect' caused by having many targets (Neill

& Cullen 1974). They suggested that this was dangerous for the fish because they would be unable to keep an adequate lookout for predators. A complication arises because although high capture rates are only possible in dense swarms, the fish can choose both the swarm density to attack and the attack rate to use at that density. Milinski and Heller used a quantitative model similar to that of Sibly and McFarland (1976) in which cost increases as the square of the departure of the internal state from its satiation point, and as the square of the rate of behaviour. The only state variable explicitly considered is food deficit, which changes linearly with rate of behaviour. They used PMP to find the optimal choice of prey density to attack and rate of capture. The main predictions of this model are as follows. (a) Capture rate and prey density chosen should both decline exponentially as 'satiation' proceeds; this seems to be borne out by the results, though only the density chosen for the first few attacks was determined (Heller & Milinski 1983). (b) Increasing the animals' assessment of the risks entailed in foraging, for example by presenting a potential predator before the trial, should lower the initial density chosen and the initial capture rate. The initial density chosen can be reduced in this way (Milinski & Heller 1978); evidence that the capture rate changes as predicted is indirect. (c) More surprisingly, the cost of a given capture rate is lower at low densities than at high densities (because there is less confusion at low densities), so the optimum rate of capture will be higher at low prey densities. This appears to be the case, at least qualitatively (Heller & Milinski 1979). Another attempt to balance the costs of feeding against those of another activity (in this case territory defence) in a dynamic optimization model has been made by Ydenberg (1982; see Fig. 4.6).

4.4.3 Long-term foraging optimization: cost and state

The idea that the function relating cost to state is quadratic arises from the intuitive notion that death is likely to result if physiological state variables are allowed to go outside certain boundary conditions. However, this ignores the fact that the consequences of death in fitness terms depend on the reproductive value of the individual at the time of death. Thus whether or not a semelparous salmon avoids death after it has spawned has little effect on its reproductive success, but whether or not it avoids capture as it enters the river has a very profound effect. The short-term dynamic models described so far make the assumption that the end point (usually satiation) is the best state for the animal to be in at the moment it is reached; there is no carry-forward effect into the next period in the animal's life. In practice this is very unlikely to be true, though it turns out that if one ignores either reproductive rate (e.g. by considering a non-breeding period) or adult mortality (e.g. by considering only a short breeding period) the formulations described are approximately correct

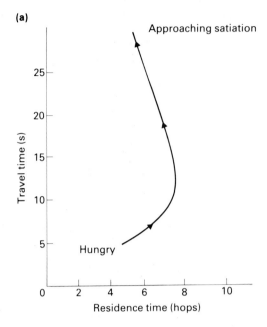

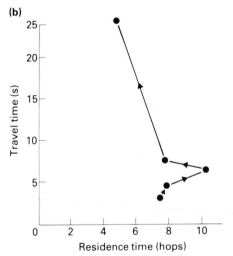

Fig. 4.6 (a) A predicted trajectory for a hungry male great tit switching between depleting food patches in an aviary. During travel between patches an intrusion site can be inspected. The bird starts at the bottom of the curve and ends at the top. The intuitive reason for the bow is this: as the hungry bird increases its travel time (to monitor for intruders) it increases its residence time to keep its rate of intake high (see Fig. 4.2c). Closer towards satiation, the value of monitoring relative to feeding increases, so the animal no longer matches residence time to increasing travel time, hence the curve bends back towards the *y*-axis. (b) An example of an observed trajectory. The dots refer to successive quintiles of the experiment.

(Freeman & McFarland 1982). It follows that one should ideally try to predict the optimum over the animal's whole life span; indeed it could be argued that we should proceed straight to the study of optimum life histories (Chapter 11). So far it has not been shown practicable to look at short-term behaviour and life history phenomena within the same framework, but at the very least perhaps one should look at longer periods than a single feeding session. One way to deal with the problem is to look at a whole breeding cycle, as was done by Katz (1974), who reasoned that state variables such as weight would return to the same values as they had had in the previous season: in Katz's model the optimality criterion was to minimize time spent foraging, subject to the constraint that starvation be avoided. In a similar study of the logger head shrike (*Lanius ludoricianus aambeli*), Craig *et al.* (1979) proposed that the bird should minimize the time spent handling prey, so as to have more time left for other (unspecified) activities. They modelled the seasonal food availability, capture rates and handling times for 10 successive months using extensive field data to set the parameters of the model. The model predicted the general course of changes in body weight in response to seasonal changes in food availability. It also predicted that the animal's weight should remain at the lower lethal limit (not surprising in view of the optimality criterion of minimizing foraging time), whereas real shrikes keep their weight above this level. Presumably this discrepancy arises because the model ignores the long-term risks of being close to the lethal boundary.

Since the seasonal model devised by Craig *et al.* runs into difficulties over long-term effects, it might be concluded that it is virtually impossible to model true cost functions on the basis of short-term behavioural studies. The problem is compounded by the fact that the actual lifetime fitness of an individual animal often depends on random events. However, this may be an unduly pessimistic view, as has been shown by McNamara and Houston (1982). They used a simulation model of the energy budget of small birds in winter to demonstrate that when there is some uncertainty about the exact timing of events relevant to an animal's state (e.g. the capture of individual food items), and when one considers only processes with a relatively short time course (e.g. food reserves in an animal which can store energy for a day or two at most) then if the animal is still alive at a point in time t_2 its state at t_2 is scarcely dependent on its state at a time t_1, provided the time interval between t_1 and t_2 is about five times as long as the time course of the processes considered. Of course the likelihood that the bird is still alive at t_2 *is* dependent on its state at t_1, but the implication of this result is that if one is considering the fitness consequences of processes which have a rapid time course, such as the effects of food reserves in small birds, then a relatively short period of observation may allow one to calculate the consequences of activity over a much longer period. This will not, however, be true of processes with a long time course.

4.5   RULES OF THUMB

Since optimization models are concerned with design features rather than specific mechanisms, they have great generality. The marginal value and diet models have been applied to various orders of insect, to birds, rodents, man and crustaceans. Although these animals behave approximately as if they have perfect knowledge of their environments and can calculate instantaneous gain rates, no one believes that foraging animals actually do the calculus of the theoretician, any more than a bird solves aerodynamic equations in order to fly. There is, in other words, an important distinction between what the animal is designed to achieve (a question about survival value) and how it does it (a question about mechanisms). It is generally assumed that foraging animals use simple *rules of thumb* to solve their foraging problems, and that these rules may approximate to the solutions predicted by optimization models.

An example to illustrate this point is the behaviour of the parasitoid *Nemeritis canescens* searching for hosts (*Ephestia cantella*) in a patchy environment. When *Nemeritis* are observed searching in an area containing concealed patches with varying number of hosts, they allocate their time to different patches in quite close accordance with a model maximizing encounter rate with unparasitized hosts (Cook & Hubbard 1977; Hubbard & Cook 1978). Waage (1979b) has shown that the behavioural mechanisms underlying this are simple. On entering a patch of hosts (detected by scent) the parasitoid walks more slowly, probing with its ovipositor, and it turns back if it reaches the edge of the patch. The edge response gradually wanes with time as responsiveness to host scent declines, so the parasitoid eventually leaves the patch. However, if a host is encountered and successfully parasitized, olfactory sensitivity increases, each oviposition having an independent incremental effect. Thus the time that a *Nemeritis* spends in a patch is a result of interaction of waning and incremental effects on sensitivity to host scents: these constitute its rule of thumb.

Four general points can be made about the relationship between rules of thumb and optimization models. (a) A rule of thumb model is likely to be more realistic with respect to a particular species, since it refers to the actual mechanisms used by the species in question (Janetos & Cole 1981), but at the same time it is bound to be less general. (b) The fact that animals use rules of thumb which only approximate 'optimal solutions' does not necessarily mean that they are suboptimal or imperfectly optimal (Janetos & Cole 1981). We pointed out in section 4.1.1 that any optimum must be defined with respect to constraints, and what appears to be suboptimal within one set of constraints may be optimal when more are added. Consider, for example, a predator which recognizes prey only by size and has a preference ranking based on size. If size and profitability are not perfectly correlated the predator will not follow the predictions of equation 4.1, but it may still maximize its intake rate subject to the constraint of discriminating prey by size alone. Thus, although at the start of this section we distinguished between survival value and proximate accounts of behaviour, the two should ideally go hand in hand in making a complete model. (c) It is possible to compare the performance of various rules of thumb in terms of the intake rate they yield. McNair (1982), for example, demonstrates that under some conditions a 'giving-up time' rule of patch exploitation (leave after x seconds of unsuccessful search) does better than staying for a fixed time or for a fixed number of captures, but that under other conditions the time and number rules do best. It would be of interest to know whether animals use the rules most suited to the conditions they frequently encounter. (d) Following on from the last point, the rules of thumb used by a predator might be expected to work well in its normal foraging environment, but the rule may work badly in unnatural situations. Waddington and Heinrich (1979) show that bumblebees (*Bombus* spp.) exploit vertical inflorescences ('spikes') by starting at the bottom and working upwards, usually departing before they reach the top. Normally this rule makes sense, since larger flowers with more nectar tend to occur at the bottom of a spike. When the normal pattern of rewards is reversed, however, the bees still tend to start at the bottom and move up.

The danger of setting the animal an unnatural task and 'fooling' its rule of thumb is particularly relevant to experiments of operant psychologists. Mazur (1981) found that pigeons failed to choose the option that maximized total reward rate but instead followed the predictions of the 'matching rule' (Herrnstein 1970), as pigeons do in most operant experiments. However, the pigeons were expected to peck vigorously at one site to obtain rewards somewhere else, a task which does not capture the essence of any natural foraging problem. In operant experiments which more closely mimic natural schedules of reinforcement, the pigeon's rule seems to approximate quite closely to maximization of overall reward rate (Staddon 1980; Houston 1983; but see Baum 1981; Heyman & Luce 1979).

4.5.1 Satisfication

There is a parallel between the notion that animals approximate to optimal solutions with rules of thumb and the concept of *satisfication* in decision theory. Simon (1956), who introduced the concept of satisfication, showed that an hypothetical organism balancing various options against one another could survive perfectly well under some conditions even though its behaviour could not be described by an optimality principle. This is akin to the point that an animal's rules of thumb may do well enough to ensure survival even though they do not actually maximize net benefit. There is, however, an important distinction to be made here. While a satisficer may not be using rules which optimize, the animal using rules of thumb is not necessarily a satisficer. The latter may be, as we have already emphasized, an optimizer within a wider set of constraints than that initially hypothesised. It remains, therefore, to be established whether or not there is any biological justification for the theoretical framework of satisfication. The crucial question is whether or not fitness is always closely associated with behavioural performance. If, for example, fitness was purely dependent on daily survival, and if survival probability is 1 when the animal accumulates more than a critical quantity of reserves, and 0 below this quantity, fitness would be independent of performance (foraging rate) over most of the range of possible performance values (McNamara & Houston 1982). In this case it might appear that the animal only needs to acquire enough food to survive, and gains no benefit by maximizing food intake. However, this no longer holds true if one considers variance in rate of food intake: even an individual with a mean rate of intake above the survival threshold still has a chance of dying, but over the long term the chance of dying is lower the further above the threshold the animal's reserves lie: organisms which minimize their probability of death (i.e. follow an optimization criterion) will eventually increase at the expense of their satisficing cousins.

Another way to think about the problem of satisfication in an unpredictable environment is to consider how animals with more or less sophisticated behavioural rules would survive in different environments: the satisficer may be able to survive only in a limited range of conditions. The point can be illustrated with reference to the control of food intake (Bolles 1980). A blowfly will eat until it bursts if the nerves carrying information to the brain about the distension of the crop are cut. Such an animal would only survive in a world where it encountered food often enough to avoid starvation but not so often that it over-filled its gut. If we allow a negative feedback loop to stop feeding when the gut is full, the range of possible environments is greatly extended, but there is still a risk of starvation if the encounter rate is too low. Adding a regulated store extends the animal's range still further, but such a system now needs a set of rules to tell it how the store should be controlled.

4.6 CONCLUSION

We introduced optimization models as a rigorous method for studying how organisms are good at doing what they do—the phenomenon of adaptation. The concept of adaptation is beset with difficulties, for reasons which include the following (Gould & Vrba 1982; Hinde 1975; Mayr 1983; Williams 1966). (i) The problem of current utility versus historical origin: traits which are used at present for one purpose (feathers for flight) may have arisen for another (feathers for heat insulation). Alternatively, traits may have originated as lucky by-products. The unfused sutures of the human neonate's cranium may appear to be an adaptation for squeezing through the birth canal, but they cannot have arisen for this purpose since they occur in birds and reptiles. (ii) The semantic issue of whether or not adaptation implies teleology, a point which was resolved by Pittendrigh (1958) when he coined the term 'teleonomy' for the study of purpose in evolution, but which still causes problems for some (Harper 1982). (iii) The fact that adaptive explanations can be found for any trait (including those that have no function) given enough ingenuity (Gould & Lewontin 1979). Optimization models, by concentrating on specific hypotheses about design criteria, help to circumvent all these difficulties.

But how successful have these models been in behavioural ecology? They can be viewed as having two roles: as 'tools for thought' (to borrow Waddington's (1977) phrase) or as generators of quantitative predictions. In the former role they have provided appropriate economic and engineering analogies for organizing data and designing experiments. They have also led to the recognition of similarities in the decision processes involved in a wide variety of activities such as mating, feeding, and habitat choice. The success of quantitative predictions, while not negligible, is more open to debate. First, many papers using optimization models have not collected sufficient or accurate enough data to make quantitative tests (Krebs *et al.* 1983). Secondly, in the relatively few studies that have collected the necessary data, and this is restricted to the optimal foraging models in section 4.2, the fit to predictions has ranged from excellent to moderate. In Fig. 4.3b and c for example, the positive correlation between patch time and travel time would probably fit a linear relationship as closely as it does the predicted exponential one. There is a need for more data and more careful analyses to assess fully the power of these models.

It is nevertheless remarkable that the very simple models tested in Figs 4.1 and 4.3 perform as well as they do, given that they cannot, for example, account for individual variation in behaviour. As we discussed in section 4.5, the accuracy of a model's predictions is also related to assumptions about the constraints, which in turn depend on knowledge of the species-specific mechanisms of behaviour. Thus the traditional aim of ethologists, to bring together causal and functional accounts of behaviour, is especially appropriate for optimization studies. Not only does specification of constraints require information on causal mechanisms, but also, as discussed by McCleery (1983), the study of motivational mechanisms cannot be complete without taking functional considerations into account.

Chapter 5
Living in Groups: Is There an Optimal Group Size?

H. RONALD PULLIAM and THOMAS CARACO

5.1 INTRODUCTION

Many, if not most, animals spend part or all of their lives in groups. For our purposes, a group is 'any set of organisms, belonging to the same species, that remain together for a period of time interacting with one another to a distinctly greater degree than with other conspecifics' (Wilson 1975, p. 585). Groups vary enormously in both size and complexity. The range encompasses everything from many fish and amphibian species that occur in groups only when temporary spawning aggregations form, to some birds, mammals and insects that live their entire lives in large, highly structured societies.

A number of authors have recently reviewed various aspects of the complexity of social organization (e.g. Crook 1970; Kleiman & Eisenberg 1973; Alexander 1974; Brown 1975). Wilson (1975), for example, lists ten characteristics of social organizations which can be used to compare the complexity of different social systems. Our paper deals almost exclusively with the determinants and consequences of *group size*, one of the ten characteristics Wilson lists. We deal briefly with group dominance structure, but only in as much as dominance influences group size. Another recent review of group size is presented by Bertram (1978).

In our paper we (1) describe selected patterns of variation in group size, (2) attempt to explain some of these patterns in terms of the advantages and disadvantages of group living, and (3) discuss how the size of a group is determined by individual decisions. With regard to the last point, we view group size as the outcome of a game in which each individual always attempts to maximize its Darwinian fitness.

5.1.1 Temporal variation in group size

Periodical changes in group size

Group size varies through time, and such fluctuations may or may not track environmental periodicities. Seasonal variation in group size occurs in many species. For example, many birds are territorial in the summer but migrate and/or feed in flocks during the non-reproductive season. Seasonal variation in rainfall and, consequently, in plant productivity induces large temporal variations in the group sizes of some African ungulates (Estes 1974; Jarman 1974) and certain primates. For example the average size of wildebeest herds increases by nearly two orders of magnitude during annual migrations to areas where food is plentiful.

Average group size fluctuates daily, or even hourly, in certain social systems. For example, in hamadryas baboons, average group size varies predictably during the course of the day (Kummer 1968). Sleeping herds are large; early-morning bands are intermediate in size; and foraging groups are relatively small. Yellow-eyed juncos roost singly at night, but forage in relatively large groups in the morning and in smaller groups in the mid-afternoon (Moore 1972). Furthermore, they feed in larger groups on colder winter days (Caraco 1979a; Caraco & Pulliam 1980). Finally, some crustaceans, including the hermit crab, exhibit endogenous circadian rhythms in their tendency to join groups or disperse.

Individuals of some species spend most of their lifetimes in the same social group (e.g. Brown & Brown 1981a; Selander 1970), whereas in other species, individuals move readily between groups which quickly form and soon dissolve (e.g. Barnard 1980; Krebs 1974; Caraco 1980). At either extreme, as well as at intermediate levels, it is sometimes useful to describe the dynamics of group size as a BIDE process (for Birth–Immigration–Death–Emigration). The BIDE family of stochastic models has been widely applied in population biology. Boswell *et al.* (1979) lucidly introduce some simple versions; Kelly (1979) summarizes some advanced models.

Cohen (1969, 1971a, 1972) attempted to interpret observed primate group sizes as realizations of BIDE processes. He pointed out that these models, carefully defined, may be applied both to truly demographic groups and to far more temporary congregations, such as sleeping groups of vervet monkeys. For groups with slowly changing membership, such as communally breeding groups of Mexican jays (Brown 1974), lion prides (Schaller 1972), hyena clans (Kruuk 1972), and many primate species (Altmann & Altmann 1970; Dittus 1977), the BIDE parameters are true demographic variables. Group size increases through birth and immigration from outside the group. Deaths and emigration reduce the size of the group. When group size and membership vary considerably over short periods, a BIDE model's parameters may concern rates of arrival at and departure from groups of different sizes, rather than demographic rates *per se*.

BIDE models can help illuminate the dynamics of group size; however, if estimates of the parameters of a BIDE model are to provide useful insights, the dynamics should be explained in terms of ecological variables and behavioural interactions (Cohen 1971a; Altmann 1974; Caraco 1980).

In this regard, BIDE models should be viewed as descriptive

models of the *proximate* control of group size variation, in contrast to the models we discuss later dealing with the *ultimate* advantages and disadvantages of group living. In our discussion of group size as a consequence of individual decisions (section 5.4), we attempt to show how the proximate and ultimate controls of group size are interrelated.

5.1.2   Dominance and group size

Several theories of social organization focus on group composition as well as group size (Crook 1972; Eisenberg *et al.* 1972; Jarman 1974; Wrangham 1981a). One might choose to classify individuals by sex, age class, reproductive condition, etc. Such distinctions acquire particular importance when, for example, the different classes of group members vary in either their response to a detected predator (Altmann 1974) or their foraging requirements (Clutton-Brock & Harvey 1977; Post *et al.* 1980).

A significant attribute in which group members may differ is their dominance status. One view of dominance supposes that the subordinate individual achieves its best possible state of affairs, given that the dominant is larger, older, or a more capable fighter. This perspective might apply particularly to a subordinate juvenile, who may attain higher status with maturity. The subordinate accepts its role despite the many advantages gained by the dominant. Dominants obtain more food (e.g. Dittus 1977) and may enjoy a more protected position within the group (Moore 1972). Dominant males sometimes acquire more mates (Le Boeuf 1974), mate more frequently (e.g. McClintock *et al.* 1982), or mate nearer a female's time of ovulation (e.g. Hausfater 1975). Dominant females may begin reproduction at a lower age (Sade *et al.* 1977), produce more offspring (Dunbar & Dunbar 1977; Wilson *et al.* 1978), or nurture their young more effectively (Drickamer 1974; Silk *et al.* 1981; McCann 1982).

Another view of dominance relationships supposes that alternative strategies provide the same net benefits. Different levels of aggressiveness or even morphological weaponry might be associated with different, but equal, 'adaptive peaks' (Fretwell 1972; Howard 1978; Hamilton 1979; Rohwer & Rohwer 1978). Furthermore, the net benefits of different levels of aggressiveness may depend on the frequency with which individuals employ the different strategies. Accordingly, the evolutionary equilibrium may be a mixture of strategies wherein all individuals expect the same net benefits (Maynard Smith 1976a; Maynard Smith & Parker 1976).

Dominance and group size are interrelated because aggressive dominants may attempt, with varying success, to control the size of a group (e.g. Pulliam 1976; Caraco 1979b), and because the costs and benefits of group membership can differ for dominant and subordinate individuals. Additionally, group size can affect the roles adopted by group members. For example, an evolutionarily stable ratio of active food 'producers' to 'scroungers' in a feeding flock may depend on group size (Barnard & Sibly 1981).

In general, since dominants and subordinates experience different costs and benefits, even when in the same group, our economic analysis of group size will frequently require us to distinguish differences in rank or role among group members.

5.2   ENVIRONMENTAL CORRELATES OF GROUP SIZE

Comparative analyses of the environmental correlates of group size variation have played an important role in the development of behavioural ecology. For example, comparative analyses of primate 'sociotypes' (Crook 1970, 1972; Crook & Gartlan 1966; Eisenberg *et al.* 1972) and Clutton-Brock and Harvey's (1977) more recent multivariate analyses of primate sociality indicate that group size in primates might be reliably predicted from a consideration of species ecology (see also Chapter 1).

Correlations between group size and environmental variables often suggest testable hypotheses but, as Altmann (1974) cautions, usually do not allow conclusions about the economics of sociality. Furthermore, differences in group size might represent different solutions to the same evolutionary problem or might be byproducts of other adaptations (see Lewontin 1979). Accordingly, we review here only a few correlations between group size and environmental variables that seem particularly reasonable to us because of their simplicity. A few other, somewhat more complex correlations also are mentioned because they suggest specific falsifiable hypotheses relating group size to resource characteristics.

5.2.1   Resource characteristics and consumer group size

Ecologists seldom question the notion that spatial and temporal patterns of essential resources influence social organization (e.g. Brown & Orians 1970; Crook 1972; Jarman 1974; Wiens 1976), and some anthropologists share this perspective (Harpending & Davis 1977; Dyson-Hudson & Smith 1978). Of course, consumers' responses to resources often are conditioned by predation pressure, constraints imposed by the mating systems, or other factors (Pulliam 1973a; Alexander 1974; Hoogland & Sherman 1976; Bertram 1978; Rubenstein 1978; Caraco 1979a).

The spatial dispersion of consumers can be limited to the proximity of a critical resource (Altmann 1974). When the number of resource locations is small, groups will be relatively large. For example, sleeping cliffs provide hamadryas baboons (*Papio hamadryas*) with protection from nocturnal predators. Since the cliffs are rare in space, the baboons sleep in large herds (Kummer 1968).

Similarly, Lee (1969) notes that over an annual cycle, the size of !Kung Bushmen settlements varies inversely with the number of water sources.

Though a single attribute of a resource sometimes exerts such a strong influence that it explains most of the variation in the size of particular groups, ordinarily we must consider several resource characteristics if we hope to elucidate relationships between social organization and resources. Resource quality and spatial variance interact to influence group size in a diversity of consumer species. When baboons exploit low quality resources, group size correlates inversely with spatial resource variance. Small groups exploit highly aggregated resources (e.g. Crook & Gartlan 1966; Crook 1972; Kummer 1968), while larger groups are associated with homogeneously distributed food (Altmann 1974). However, if we consider organisms utilizing high-quality food resources, the pattern may be just the reverse. When a high quality resource has a very patchy spatial distribution, large groups may form. Consumer aggregations at points of resource abundance (e.g. Zahavi 1971) provide ample demonstration. As the high quality resource becomes more homogeneously distributed, local concentrations may be economically defensible (Brown 1964, 1969; Gill & Wolf 1975; Davies 1980), inducing territoriality.

Though correlations between resource dispersion and group size are clear, the underlying causal relationships are not always readily apparent. For example, hamadryas baboons exploit low quality, patchily distributed food sources, and forage in small groups. These small groups most often consist of an adult male and his harem. The low quality patches may support only a few consumers, so that intraspecific competition for food keeps individuals in small groups. Alternatively, a male's defence of his mating potential may limit group size, or food competition and sexual selection may act in concert. Cost/benefit models and field experimentation are needed to differentiate properly between these hypotheses.

Resource predictability, like resource quality and dispersion, may influence group size. We view predictability as an attribute of resource abundance through time (see Colwell 1974). The autocorrelation coefficient (ρ; $-1 \leqslant \rho \leqslant 1$; see Ord 1979) is a reasonable measure of predictability. The value of $\rho(\tau)$ gives the correlation between resource abundance at times t and $t + \tau$. When ρ is near 1.0 for small and biologically germane values of τ, resource availability, whether high or low, does not fluctuate rapidly through time. Strong negative autocorrelation suggests rapid resource fluctuations, but the fluctuations possess a predictable periodicity which consumers might somehow track (Drent & Sweirstra 1977; Drent & van Eerden 1980). Zero autocorrelation implies minimal predictability, since resource abundances τ time units apart are statistically independent.

Returning to resource patterns correlated with territorial behaviour, an economically defensible resource should exhibit high quality, a relatively homogeneous spatial dispersion and high, positive auto-correlation. High quality and predictability indicate possible acquisition of substantial, sustained benefits; these conditions may be necessary for successful breeding territories. Intruder pressure and defence costs should vary inversely with the spatial variance of the resource. As previously indicated, extreme patchiness attracts very large groups, rendering defence costs uneconomical. If spatial homogeneity disperses potential intruders, defence costs may be manageable. The most thorough cost/benefit analyses have concerned individual feeding territories (e.g. Gill & Wolf 1975, 1977; Pyke 1979), and Brown (1982) suggests how such analysis can be extended to group territorial species.

While high predictability may favour resource defence by individuals or small groups, low predictability can render resources indefensible, and large groups may form. Furthermore, low temporal predictability induces spatial uncertainty, and group foraging may reduce the time each individual spends searching for the resource. Consequently, predictability often should be correlated inversely with group size (Ward 1965; Horn 1968; Clutton-Brock 1975; Krebs 1974). Our discussion of the fitness consequences of variation in group size will review a number of selective mechanisms which link low resource predictability and large group size in the context of the individual economics of sociality (Krebs *et al.* 1972; Baker *et al.* 1981; Caraco 1981b; Pulliam & Millikan 1982).

5.2.2   Other biotic correlates of group size

Birds on islands sometimes face fewer predators than do their mainland counterparts. Hawaiian birds seldom flock, and predators are scarce (Willis 1972). Pulliam (1973b) compares the sociality of a number of avian families on the Costa Rican mainland and on Jamaica. Flocking is far less common on the island, which lacks resident bird-eating raptors. Hence, we might expect that the size of some groups increases as the frequency of encounters with predators increases (e.g. Caraco 1979b).

Group size may be correlated with the probability of success in intraspecific competition under certain circumstances. For example, groups may overcome a despot's aggressive defence of resources. Flocks of chickadees can gain access to an individual's feeding territory, while solitary intruders are quickly chased away (Barash 1974). Coalitions of subordinate males sometimes collectively displace the dominant member(s) of a group (e.g. Bertram 1975). Furthermore, group size often determines the outcome of intraspecific competition between groups (e.g. Kruuk 1972; Bartz & Hölldobler 1982). The same rule may often apply to interspecific competition. Groups of lions defend carcasses against scavengers, particularly against hyena clans, better than solitary lions do (Schaller 1972). An individual of one species may be subordinate to an individual of a second species, but grouping among the former species can alleviate the impact of resource competition (Orians & Collier 1963; Robertson *et al.* 1976; Roell 1978).

5.3 THE FITNESS CONSEQUENCES OF GROUP SIZE

Hypotheses suggested by correlations between group size and ecological variables are best tested by investigating the benefits and costs to individual group members. The benefit or cost of any behavioural attribute is, respectively, the increase or decrease in inclusive fitness due to that particular attribute. The net benefit or cost of group living, then, results from the composite of all benefits and costs of all behavioural attributes of group life. In this section we discuss studies that have attempted to measure the major benefits and costs of group living. In the next section of our chapter we attempt to show how the net benefits and costs determine equilibrium group size. When asking if optimally or equilibrially sized groups are plausible, the analysis must hinge on per capita benefits and costs (Wilson 1975; Caraco & Wolf 1975; Pulliam 1976; Wittenberger 1980; Davies & Houston 1981).

5.3.1 Foraging and group size

Group size can influence the rate at which patches of food are discovered in temporally and spatially uncertain environments. Laboratory studies with minnows and goldfish (Pitcher *et al.* 1982), and with foraging birds (Krebs *et al.* 1972), have demonstrated that the time required to discover a patch of food decreases significantly as group size increases. Since the location of food by one individual alerts other group members, per capita search time can be lower in larger groups. This might be a particularly important benefit of sociality if patches are large relative to individual requirements, so that only one or a few patches need be discovered, or if the individual seeks to minimize the time until it acquires any food at all (Caraco 1981b).

The benefit of reduced search time will ordinarily reach an asymptote fairly quickly as group size increases. In larger groups, individuals will not be able to search independently, because of interference or overlap in the areas searched by group members. As the distance between patches (scaled to the searching velocity of group members) increases, per capita benefits of reduced search time per patch should reach an asymptote at larger group sizes.

Local enhancement (Crook 1965) is a term often used to describe processes by which an individual's attention is directed to the location of food by other group members. Generally, a member of a group may learn what to eat or where to eat by observing other foragers (Krebs *et al.* 1972).

The 'information centre' hypothesis proposes a similar advantage for sociality. However, in this case individuals do not immediately learn of another's discovery of food, but instead acquire the information by observing the successful forager at nesting colonies (Horn 1968; Krebs 1974) or communal roosts (Ward & Zahavi 1973; DeGroot 1980). Since the area exploited by some communally roosting birds can be quite large (e.g. Ward 1965), the per capita reduction in search time could lead to the formation of large aggregations. Evans (1982) presents evidence that black-headed gulls attract followers when departing a nesting colony, but interprets this as being due to direct benefits of flock foraging, rather than to information sharing *per se*. Furthermore, Bayer (1982) argues against information sharing as a causative factor in the evolution of colonial nesting and suggests that the evidence to date favouring the information centre hypothesis is incomplete and unconvincing.

If larger groups flush prey from hiding places, effective food density increases, and an individual's waiting time per item can decrease. Some insectivorous birds in flocks may enjoy this benefit (Morse 1970).

If we suppose that large groups serve to reduce per capita search time, the advantage of sociality should increase as resources become more ephemeral. For example, Pulliam and Millikan (1982) argue that the fraction of a patch harvested by an individual becomes less sensitive to the reduction in per capita food availability (caused by increases in group size) as patches become more ephemeral. They show that if patches of food last long enough to be completely consumed, the presence of additional group members significantly decreases per capita food consumption. However, if patches disappear in much less time than would otherwise be required for food in the patch to be completely consumed, the presence of additional group members hardly effects per capita consumption. Of course, quickly disappearing patches also can increase the importance of reductions in search time.

Under certain conditions, increasingly large foraging groups increase per capita search time. Large groups of redshank searching a given area of the intertidal can disturb their prey (shrimps), which quickly disappear under the sand (Goss-Custard 1976). This effect could very well accelerate as group size increases. Despite this counterexample, reductions in search time may represent one of the most important benefits of group living in patchy, unpredictable environments.

Group size can influence the availability of different food resources and the efficiency with which they are captured. Groups of carnivores can take larger prey than individuals can (Estes & Goddard 1967; Schaller 1972; Kruuk 1972; Mech 1970). This effect quickly reaches an asymptote, since the benefit is constrained by the size of the largest prey in the environment. For a given prey type, capture efficiency per attempt sometimes increases with group size, as does the probability of multiple kills on a single effort. These effects appear to reach an asymptote at fairly small group sizes in lions (Caraco & Wolf 1975), but seem to reach a maximum more slowly in smaller carnivores (e.g. Nudds 1978).

As mentioned above, carnivore groups defend carcasses against scavengers more effectively than do individuals. When lions defend a carcass from scavenging hyenas, the likelihood of a successful defence no longer increases once three or four lions attend the kill (Schaller 1972).

While larger groups may have access to larger prey and can capture them more efficiently, these benefits usually decelerate with increases in group size. Therefore, per capita food availability cannot increase monotonically with group size. Major's (1978) study of predatory fish shows that average food intake per individual does not increase beyond a group size of three. In lions (Caraco and Wolf 1975) and in wolves (Nudds 1978) also, per capita food availability attains a maximum at fairly small group sizes. In many social species per capita foraging benefits continue to increase until group size is rather large. However, unless food is superabundant, per capita availability must eventually decrease with group size. If there is no other constraint on the size of groups, physiological requirements and competition will eventually constrain foraging group sizes. Furthermore, the division of food according to dominance status may imply that the competitive costs of increasing group size may be borne disproportionately by subordinate individuals.

When dependent offspring have a fixed food requirement, the amount of work done by individual foragers may decrease with group size. Brown *et al.* (1978) find that the per capita foraging effort on behalf of nestlings decreases as the number of helpers in a group increases (also see Kinnaird & Grant 1982). The direct benefit to the parents and offspring is clear, while possible direct and indirect (mediated by non-descendant kin) benefits to helpers are more difficult to establish (Brown 1978; Brown & Brown 1981a; see Chapter 12).

Grouping may decrease metabolic rates in homeotherms. Close proximity retards heat loss in roosting bats (Trune & Slobodchikoff 1976); sociality in active small mammals may reduce an individual's metabolic rate for less obvious reasons (Herreid & Schlenker 1980). In these cases, sociality might enhance survivorship by reducing foraging requirements. Any such per capita benefits probably increase, but ever more slowly, as group size increases.

Some current efforts in foraging theory have begun to focus on risk-sensitive behaviour (Oster & Wilson 1978; Real 1980; Caraco *et al.* 1980c; see Chapter 4). The thrust of these arguments is that foragers should respond to both the mean and the variance of the realized intakes associated with different strategic options. In a patchy environment, increasing group size can reduce the variance in the total time an individual spends searching for a required amount of food, or the variance in the total amount of food acquired during a foraging period of given length (Thompson *et al.* 1974; Caraco 1981b; Pulliam & Millikan 1982). If variance reduction increases an individual's probability of averting starvation, group foraging will be advantageous. If survivorship is instead proportional to variance (a condition occurring when food requirements exceed expected food intake), solitary foraging will be preferred unless mean net benefits increase sufficiently with group size (see Fig. 5.1). Since some animals adjust their food selection on a short-term basis as risk and requirements change (e.g. Caraco 1982), individual propensities to forage in a group may also exhibit variable risk-sensitivity.

5.3.2 Group size and predation

Sociality in a vast number of species must represent, at least in part, a response to predators. A variety of mechanisms may render group membership safer than solitary existence. Grouped prey often detect an approaching predator sooner than do solitary individuals (Pulliam 1973a; Powell 1974; Siegfried & Underhill 1975; Kenward 1978; Lazarus 1979). This occurs even though each individual in a group usually spends less time in vigilance behaviour (Hoogland 1979a; Caraco 1979a; Bertram 1980; Elgar & Catterall 1981) and can therefore allocate more time to other activities (Caraco *et al.* 1980a, 1980b). Earlier detection increases the likelihood of escape; this advantage may be particularly large if group members give an alarm when a predator is sighted (Charnov & Krebs 1975; Seyfarth *et al.* 1980a). If individuals watch for predators independently, the benefits of early warning reach an asymptote fairly quickly as group size increases. However, as the velocity of the predator's attack increases, benefits to individual prey continue to grow over a greater range of group sizes. In a group of a given size, individual members must choose a rate at which they can scan for approaching predators. In some prey groups, survivorship may favour cooperative rates of vigilance behaviour over seemingly more selfish strategies (Pulliam *et al.* 1982).

Prey may aggregate in an attempt to place conspecifics between themselves and an attacking predator (Hamilton 1971). The per capita benefit probably varies both with group size and with an individual's ability to obtain a central position; in nesting colonies, predation can be more intense on the periphery (Horn 1968; Tenaza 1971). When individuals on the periphery of the group experience greater predation, it is difficult to invoke this mechanism as the sole selective force favouring sociality. Peripheral group members might benefit from leaving, leading to the disbanding of the group (Pulliam 1973a). 'Selfish herding' could easily interact with other anti-predator benefits, or factors like the spatial distribution of food, so that the group maintains its integrity. In such cases, greater predation at the group's edge might imply that peripheral group members spend more time in vigilance than do more centrally located individuals (Inglis & Lazarus 1981).

Tightly aggregated prey may confuse an attacking predator, if the predator is prevented from focusing its effort on a particular target. Similarly, a solitary individual of a normally highly social species may be conspicuous to predators simply because it is alone. Aggregated

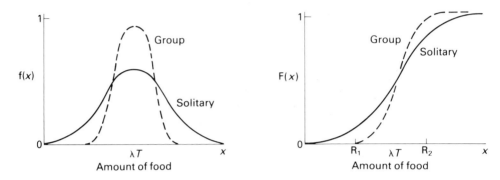

Fig. 5.1 Risk-sensitivity, energetic benefits, and social foraging. Suppose the length of the foraging period (T) is fixed, and the individual must obtain more than R food items during T to avoid starving. The *average* rate at which a solitary discovers food is λ. A solitary consumes all the food it finds. A flock of size G locates food, on *average*, at rate $G\lambda$, and each individual consumes a fraction $1/G$ of the total amount of food discovered. Let x be the random variable representing the amount of food consumed by an individual during the foraging period. Under reasonable conditions, x will have a normal probability density $f(x)$, with cumulative distribution $F(x)$. The average food consumption is independent of G and equals λT for both solitary and social foragers. The variance in the amount of food consumed decreases with group size, as indicated in the diagram on the left.

The probability that the individual starves is simply $Pr(x \leqslant R)$. This probability equals $F(R)$, by definition. Therefore, the individual seeking to minimize its probability of starving will choose the strategy, solitary or social foraging, with the smaller value of $F(R)$. As indicated in the diagram on the right, the choice can depend on a comparison of R with λT, the mean. When the expected amount of food exceeds the required amount (consider R_1; $R_1 < \lambda T$), social foraging should be favoured. This individual behaves in a risk-averse manner, since it prefers the reward distribution with the lower variance about the common mean. When the required amount of food exceeds the expected amount (consider R_2; $R_2 > \lambda T$), solitary foraging should be favoured. This individual behaves in a risk-prone manner, since it prefers the reward distribution with the greater variance about the common mean. Pulliam and Millikan (1982) provide a detailed development including the effects of dominance on survival probability. Just as the random nature of foraging benefits can influence sociality, so can the variance in the costs of searching for food. Caraco (1981b) considers the time required to locate enough food to meet an individual's requirements. Again, when the mean searching costs are independent of group size, risk-aversion favours sociality and risk-prone preferences can induce solitary foraging.

prey also might actually endanger a predator approaching at high velocity (Tinbergen 1951). These hypothetical benefits probably do not continue to increase past a certain group size, and in either case peripheral individuals would again suffer greater risk of predation. Furthermore, confusion effects may wane quickly as the predator learns to handle aggregated prey, as experimentally demonstrated by Milinski (1979a).

If attack rate is independent of group size, and a predator kills only a single prey when successful, an individual benefits from a dilution effect in a group. If the probability of being a victim in a successful attack is simply the inverse of group size, the individual's benefit continues to increase, but at an ever decreasing rate, as the size of the group increases.

Even if prey aggregations attract predators, in particular circumstances individuals may be able to reduce per capita predation on dependent offspring by 'grouping' births in time. If females produce their young during a short time interval, predator appetites might be saturated quickly. As a consequence, individuals might attain a higher probability of keeping their offspring alive than if females gave birth asynchronously.

Grouped prey may be able to mob a predator (Kruuk 1964; Curio 1978), or otherwise defend themselves actively against a predator (e.g. Crook 1972; Altmann 1974). The per capita benefit should reach an asymptote. If prey defend themselves, selection may favour sexual dimorphism, though effects associated with sexual selection are probably often more important. Dominant males in age-structured groups sometimes do the fighting against predators. In doing so, they may incur a cost in terms of their own survivorship, but increase the survival of their offspring.

In some animals, grouping might exact a cost in terms of predation. Groups might be more conspicuous than solitaries to a predator searching at a distance, depending on the predator's sensory mode and acuity (Vine 1973; Treisman 1975; Taylor 1979). For some prey, crypticity and an often solitary existence may be a response to predation (Tinbergen *et al.* 1967; Jarman 1974; Pulliam & Mills 1977).

Whether due to differential conspicuousness or predator preference, attack rate may depend on prey group size. Kruuk (1964), Krebs (1971), and Andersson and Wicklund (1978) have demonstrated that egg predators attack clumped nests of birds more often than they attack dispersed nests. In Krebs' study, egg losses were greater in grouped nests. In the Andersson and Wicklund study, egg losses were lower with clumping, despite the greater attack rate, because approaching predators were mobbed.

Group size also can influence an individual's probability of contracting parasites or contagious disease. Allogrooming may be a benefit of sociality that increases with group size, if that implies greater availability of partners. However, costs probably increase faster than benefits in this aspect of sociality since ectoparasites are more common around larger groups in some species (Hoogland & Sherman 1976; Hoogland 1979b), and individuals may be more likely to contact a diseased conspecific as group size increases.

5.3.3 Group size and reproductive success

In a general sense, group size influences an individual's reproductive success through effects on mating probabilities and the production and survival of dependent offspring. The fecundity-related consequences of variation in group size are intertwined with characteristics of mating systems. Furthermore, within some groups a particular individual might benefit from its own reproduction and/or from reproduction by related individuals. Hence, there can be direct and indirect benefits and costs accrued by group members (Brown & Brown 1981a). Other chapters detail the ecology of mate choice (Chapter 9) and the properties of communal breeding systems (Chapter 12). Therefore, we review only a few circumstances where a group size effect might be strong.

Brown and Brown (1981a) provide direct experimental evidence that the number of fledglings per nest increases with group size in an avian communal breeder. This relationship has been noted elsewhere (e.g. Brown 1978; Emlen 1978; Koenig 1981a), but factors like territory quality sometimes complicate the analysis (Vehrencamp 1978). Non-breeding helpers in such systems presumably benefit in accordance with inclusive fitness arguments (Hamilton 1964; Brown 1978), or because they will later obtain both the territory and breeding status (e.g. Brown 1969; Woolfenden & Fitzpatrick 1978). Sometimes helpers might be better off reproducing themselves, but may not be able to establish a breeding territory, in an environment close to saturation (Brown 1969, 1974; Koenig and Pitelka 1981; see Chapter 12).

Some specific group size effects have been documented in invertebrates. For example, a dominant male dragonfly's mating probability may depend on the number of subordinate males sharing a territory (Campanella & Wolf 1974). In a *Polistes* wasp, nests founded by two sisters more than double the number of offspring produced by a solitary female (Metcalf & Whitt 1977b). Though the subordinate member of two-female nests lays few eggs, benefits through kinship suggest that her reproductive success exceeds that obtained by a solitary female. Bartz and Hölldobler (1982) suggest that in honey ants there is an optimal number of females initially founding a nest, in terms of maximizing per capita survivorship (until workers are produced) and per capita size of the initial brood. Only a single female remains as queen after significant numbers of workers appear. However, the initial advantages of grouping may outweigh the chance of later eviction from the nest by the workers (Bartz & Hölldobler 1982).

In addition to mate competition, reproduction in groups can entail costs which probably increase with group size. Parental investment might be misdirected to others' offspring (Hoogland & Sherman 1976). Offspring might be killed by other group members, or larger groups may attract more predators to the young.

Costs and benefits to individuals in reproductive groups depend on the interwoven complexity of dominance, kinship, life history phenomena, group size, etc. While some aspects of mating systems correlate with ecological variables (e.g. Emlen & Oring 1977), others lack obvious correlations (Brown 1978). In iteroparous species, fecundity effects in breeding groups must often interact with adult survivorship. However, in groups formed outside the breeding season, fitness may be equated with survivorship. This simplification invites both theoretical and empirical investigation of the way individual decisions lead to rules which might predict group size patterns, and we take up this approach in the next section.

5.4 GROUP SIZE AS A CONSEQUENCE OF INDIVIDUAL DECISIONS

Clearly, an individual's inclusive fitness varies as a function of the size of the group in which that individual is found. This observation alone has prompted discussion of the possibility of an optimal group size (Wilson 1975; Pulliam 1976; Caraco 1979b). The most complete treatment to date is by Brown (1982), who considered optimal group size for group territorial animals. Brown (1982) and Bertram (1978) both point out that the question of optimal group size is complicated by the fact that dominants and subordinates may differ in optimal group size. In this section, we argue that there may often be no single optimal group size and that instead the equilibrium distribution of individuals among groups represents a Nash solution to an *n*-person game.

5.4.1 Empirical studies of optimal group size

Using Schaller's (1972) data published in *The Serengeti Lion*, Caraco and Wolf (1975) calculated daily food intake per lion for groups of various sizes. The food available per lion was smaller for larger hunting groups, but capture efficiency increased with group size. The mean edible prey biomass captured per lion per day was greatest for intermediate-sized groups and this 'optimal' group size differed for different habitats and times of year. Caraco and Wolf found that for lions preying on Thomson's gazelle, the mean hunting group size observed matched that predicted on the basis of maximizing daily food intake per lion. However, for lions feeding on wildebeest and zebra, the mean observed group size was considerably larger than predicted. In all cases, the observed group size was the maximum group size consistent with each lion acquiring the daily minimum requirement of 6 kg of prey.

Smith (1980) studied the foraging group sizes of contemporary Inuit (Canadian Eskimos). He tested the hypothesis that foraging party size is adjusted among different hunt types so as to maximize the net rate of energy capture per individual in the hunting party. He calculated the energy expenditure and capture rates for individuals in various parties hunting different prey and using different hunting methods. By calculating the net rate of energy capture per individual for parties of various sizes, he was able to determine the 'optimal' hunting group size for each hunt type. For example, he found that when hunting seals at breaking holes, hunters in groups of two or three averaged over 4000 kcal $\cdot$ hunter^{-1} $\cdot$ hr^{-1} (net energy capture), but in other groups ranging in size from one to seven, they averaged less than 2000 kcal $\cdot$ hunter^{-1} $\cdot$ hr^{-1}. Despite the decreased efficiency of larger groups, the observed modal group size for this situation was more than three hunters per party. In situations where capture rate was highest for solitary hunters, parties of one predominated, but otherwise there was little agreement between the observed group size and the optimal group size.

5.4.2 Combining the components of fitness

From the previous examples, it is clear that whereas foraging efficiency does vary with group size, it is not in itself a sufficient explanation for the observed variation in group size. Many components of fitness vary with group size and combining them into a single metric presents a serious methodological problem. Wilson (1975) suggests that if the components of fitness are measured in the same currency then the net effect on fitness can be calculated as the sum of the individual effects. This approach ignores the possible statistical interdependence of the components, and as pointed out by Smith (1980), the various costs and benefits of group size can only rarely be expressed in a single currency.

A more general approach to the problem of combining the various costs and benefits of group living employs time-budget analysis (Pulliam 1976; Caraco 1979a, b; Krebs & Davies 1981; Brown 1982). This approach has been applied with some success to foraging groups in the non-reproductive season, under the assumption that all fitness components can be reduced to the common currency of increments in survival probability. A simple example concerns the over-winter survival of sparrows in foraging flocks (Caraco 1979b; Caraco & Pulliam 1980).

Individual sparrows foraging in groups on the ground divide their time between feeding (T_F), scanning (T_S), and interference (T_I), including time spent in aggression. The total time (T_T) available is taken as the sum of the time spent in each activity (i.e. $T_T = T_F + T_S + T_I$).

Since expected food intake is directly proportional to the time spent feeding, the probability of starvation declines monotonically with T_F. Knowing the feeding rates and energy requirements of the sparrows, we can calculate the probability of starvation during the night as a function of time spent feeding during the day (T_F). Scanning is thought to be an anti-predator behaviour: if a sparrow spots an approaching predator in time, it can retreat to the safety of cover. Scanning rates are higher for birds far from cover and smaller in larger groups where vigilance is shared (see section 5.3). Pulliam *et al.* (1982) have proposed a quantitative model relating time spent scanning to the daily probability of surviving predation.

If there is no aggression ($T_I = 0$), the probability of daily survival can be calculated as the product of the probability of surviving predator attacks and the probability of not starving given that predation has been avoided, under the constraint $T_T = T_S + T_F$. Doing this for any particular group size usually admits an intermediate optimum with some time devoted to feeding and some devoted to scanning. The optimal allocation of time to feeding versus scanning depends on group size since the risk of predation varies with group size (Pulliam *et al.* 1982). The optimal group size for an individual can be said to be that for which the allocation of time between scanning and feeding maximizes survival probability for that individual.

A problem with this approach is that it assumes that over-winter survivorship is maximized when *daily* survivorship is greatest. In fact, a dominant individual can increase its over-winter survival by allocating some time to aggression, even if this means spending less time scanning or feeding than is otherwise optimal. Aggression by dominants may reduce the rate of food depletion by subordinates and, therefore, time spent in aggressive resource defence on one day may increase the dominant's probability of avoiding starvation in the future. The optimal allocation of time to resource defence can be calculated using optimal control theory, but only if the relationship between time spent in aggression and future feeding rates is known.

The need to consider all aspects of a time budget simultaneously is well illustrated by the authors' studies of the influence of predation on group size in yellow-eyed juncos, *Junco phaeonotus* (Caraco *et al.* 1980c). T_F, T_S, and T_I were calculated as a function of group size from field observations, in both the absence and the presence of a trained hawk. As discussed above, we can assume that the probability of overnight starvation is lowest when time spent feeding (T_F) is greatest. However, in both the presence and absence of the hawk, the mean observed group size differed from the group size for which T_F was greatest. In the absence of the predator, mean group size was lower than expected on the basis of maximizing time spent feeding. We attribute this to the long-term benefit to dominant individuals from reducing mean group size by aggressively defending the best feeding areas. In the presence of the predator, observed group sizes were larger than expected on the basis of maximizing T_F. We attribute this to the benefit to all birds of shared vigilance when predation risk is great.

5.4.3 A game-theory approach

For many reasons, the group size that maximizes one individual's fitness may not be the optimal group size for other individuals. We have already mentioned the case where dominants and subordinates differ in optimal group size (Brown 1982; Bertram 1978). Even when there is no dominance and all individuals have the same apparent optimal group size, the realized group size may be suboptimal.

Consider, for example, a population of identical individuals for all of whom fitness is maximized in groups of size G. If the local population size N is exactly equal to G, or to a multiple of G, each individual may live in a group of the optimal size. But suppose N is six and G is five. If five individuals are in a single group of the optimal size, the sixth faces the choice of living alone or joining a group that will then be suboptimal for it and for all others in the group. The equilibrium distribution of group sizes depends, in such a case, not on the optimal group size *per se*, but on the decision rules used by individuals deciding whether to join or leave groups. If, in the case under discussion, the sixth individual uses an optimal decision rule, it will join the group if, and only if, doing so maximizes its inclusive fitness, given the options available.

An ESS model of habitat selection

Since joining or leaving a group necessarily implies a change in location, habitat selection and group size selection can be considered simultaneously. Furthermore, we feel that models of territorial behaviour which presume that habitat quality declines with density of occupants and models of sociality which presume that habitat quality is enhanced (up to a point) by the presence of conspecifics should be treated together. We argue that in all such models the critical element is that individuals always choose habitats and group sizes on the basis of fitness maximization.

Brown (1969) described a model relating habitat selection to territorial defence. In his model, habitats are assumed to vary in quality, and initially all individuals choose the best habitat. Territories are assumed to have a fixed minimal size, so when the best habitat is full, individuals begin to occupy the second best habitat. Finally, when all habitats are saturated, non-territorial 'floaters' appear. Studies of titmice (Krebs 1971) and red grouse generally support the model.

Fretwell (1972) proposed a model of habitat selection based on the assumption that individuals always choose the best habitat available to them at the time. As the better of two habitats fills up, the quality of this habitat declines until eventually a point is reached at which the two habitats are of equal quality. As density increases further, new individuals continue to settle in both habitats, but always in such a way as to keep the habitats roughly equal in quality. The resulting distribution of the individuals among habitats is referred to as the ideal free distribution, calling attention to the assumption that individuals are free to settle wherever they prefer.

Though it appears not to have been generally appreciated, Fretwell's ideal free distribution represents an ESS in the game of habitat selection (see sections 2.3 and 2.7). The best choice of a habitat always depends on what other individuals are doing and, therefore, habitat selection can be modelled as an *n*-person game (or 'alternative option scramble' in the terminology of Chapter 2). In such a game, a *Nash equilibrium* occurs whenever no individual can increase its personal fitness by changing habitat, given that all other individuals stay put. If we add the requirement, as does Fretwell, that at this equilibrium all individuals have the same fitness, the distribution of individuals among habitats is said to correspond to the ESS (Maynard Smith 1974a; Oster & Wilson 1978; Chapter 2).

In a generalization of Fretwell's model of habitat selection, we assume that N individuals are distributed among the possible patches such that there are n_i individuals in patch i ($\sum n_i = N$). We assume that the quality of patches is highly predictable and that the cost of moving between patches is negligible. Finally, denoting $W_i(n_i)$ as the fitness of any individual in patch i when the patch is occupied by n_i individuals, we assume patch quality declines as n_i increases (i.e. $\partial W_i(n_i)/\partial n_i < 0$).

When patches have been chosen according to the criterion that each individual always chooses the best available place, we denote the resulting vector as $\mathbf{n} = (n_1, n_2, n_3 \ldots)$. Following Fretwell's requirements that (1) all individuals within a patch have the same fitness, and (2) habitats are chosen so as to ensure the approximate equality of all occupied patches, we note that

$$W_i(n_i) = W_j(n_j), \tag{5.1}$$

for any pair of occupied patches, i and j. Furthermore, since $\partial W_i(n_i)/\partial n_i < 0$,

$$W_i(n_i) > W_j(n_j + 1), \tag{5.2}$$

for any pair of patches, i and j. This condition ensures that no individual can increase its own fitness by moving to any other patch as long as all other individuals stay put. Therefore, the vector $\mathbf{n}$ is a Nash equilibrium. Since the model assumes that all individuals within and between patches have the same fitness, the equilibrium distribution of individuals among patches corresponds to an ESS.

A habitat matching rule

In a special case of the ideal free distribution, the fitness of each individual in a patch can be taken to be $W_i(n_i) = K_i/n_i$. This assumption is likely to be valid if a population is resource limited and if the fitness of an individual is directly proportional to the fraction of the total resource consumed or controlled by that individual. In this special case, the equilibrium condition (from equation 5.2) necessitates that

$$K_i/n_i = K_j/n_j,$$

for any pair of occupied patches, i and j. Thus, in the case of food limitation, the equilibrium relative number of individuals utilizing any two patches matches the relative productivity of those same two patches, i.e.

$$n_i/n_j = K_i/K_j. \tag{5.3}$$

The habitat matching rule given by equation 5.3 allows two distinct interpretations (see Maynard Smith 1976a). If group membership in each patch remains constant after an equilibrium is reached, the problem is identical to habitat selection. If, however, group membership frequently changes, the equilibrium ratio implies that each individual matches the ratio of the time it spends in any two patches to the ratio of the productivities of those two patches.

This second version of the habitat matching rule can be considered as the habitat usage analogue to the matching law of psychology (Herrnstein 1974; Herrnstein & Vaughan 1980), according to which the relative number of responses an animal makes at each of two feeding stations (characterized by 'interval' reward schedules) is proportional to the relative rewards obtained at those stations.

Milinski (1979b) appreciated that Fretwell's ideal free distribution corresponds to an ESS. In experiments with sticklebacks choosing between two food patches, he demonstrated that feeding group size approximated the predicted equilibrium (equation 5.3) closely. As he experimentally manipulated the prey density at two feeding sites, he found that the relative numbers of fishes feeding at the two sites quickly reached an equilibrium matching the relative profitabilities of the sites. In similar experiments with ducks, Harper (1982) found that they also distributed themselves between two food patches in close approximation to the ideal free distribution. However, Harper found that because of dominance behaviour of some individuals, not all individuals within a patch achieved the same feeding rate.

In the studies by Milinski (1979b) and Harper (1982) cited above, individual animals moved freely from one feeding patch to the other. Whitham (1980) studied habitat selection in *Pemphigus* aphids feeding on *Populus* leaves. In this situation, once the stem mother aphid chooses a leaf, she is committed to remaining there for the duration of her life. Whitham found that individuals on different leaves had approximately the same mean fitness, as predicted by the ideal free model, but that individuals on the same leaf differed in relative fitness depending on their position on the leaf. Furthermore, he found that dominant individuals secured the better positions.

The line of reasoning employed above for habitat selection can be more generally developed to apply to the problem of simultaneously choosing a patch and a social group in which to live. Fretwell's original model assumed that increased density within habitats always results in reduced habitat quality. He extends the model (Fretwell 1972) to incorporate the Allee principle which posits that habitat quality increases, up to a point, as density increases and thereafter declines. Following Fretwell's lead, we now develop a general model encompassing a wide spectrum of possible social organizations.

In Fig. 5.2 we graph the fitness of each individual in a patch versus the number of individuals in that patch. Each line represents a distinct patch of habitat and in the absence of dominance all individuals within a patch are assumed to have equal fitness. Individuals are assumed to arrive sequentially. (Editors' note: see Sibly 1983 for a similar model.)

Example A represents Fretwell's first model, in which the fitness of all individuals declines monotonically with the number of cohabitants. As previously discussed, for each population size ($N = n_1 + n_2$) the distribution of individuals among patches corresponds to an ESS.

In examples B and C, we assume that all individuals within a patch live in a single organized group, and in some cases there is an advantage, up to a point, of being in a larger group. In both examples, fitness increases to a maximum at six individuals in patch 1. In example B, the first individual settles in patch 1, because $W_1(1) > W_2(1)$. Because fitness initially increases with increased group size, the next few individuals also join the group in patch 1. Even though the optimal group size occurs at $n_1 = 6$, the next few individuals also join group 1 because $W_1(x) > W_2(1)$, for $x = 1, 11$. Finally, individual number 12 settles in patch 2, because $W_2(1) > W_1(12)$. Thereafter, the individuals alternate between settling in patch 1 and patch 2, depending on which is better at the time they arrive. At every point, the distribution of individuals between groups corresponds to an ESS, because all individuals have (approximately) equal fitness and no individual can increase its fitness by changing groups.

In example C, the first 11 individuals settle in patch 1, because $W_1(x) > W_2(1)$, for $x = 1, 11$. The twelfth individual settles in patch 2 because $W_2(1) > W_1(12)$. Once this has happened, it becomes advantageous for one of the individuals in group 1 to leave and join group 2, because $W_2(2) > W_1(11)$. This assumes that the individual deserting group 1 has made no investment in the group or patch, such as building a nest, that would have changed its expected fitness since the time it joined group 1. It also assumes that after making the initial decision to join the group in one patch, individuals continue to monitor the quality of other patches. Given these assumptions, the distribution of individuals among patches still corresponds to an ESS.

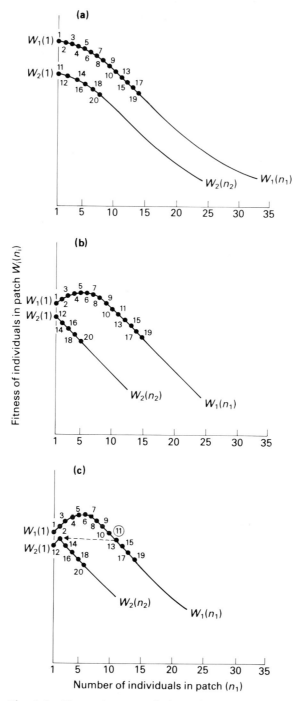

Fig. 5.2 Fitness in groups lacking dominance structure. $W_i(n_i)$ is the fitness of the n group members in habitat i, i = 1, 2. Each graph shows the sequentially ordered assemblage of the two groups. Each individual, in numerical sequence, chooses the habitat where its fitness will be greater. Each example shows a total of 20 individuals assembled into two groups. In (c) the dotted line indicates that one individual might choose to shift from habitat 1 to habitat 2 under certain conditions. Details of decision rules and resulting group sizes are discussed in the text.

To this point we have assumed a one-to-one correspondence between each patch and the group in that patch. This might hold if, for example, the patches are small, discrete and widely dispersed, so that all individuals in a patch necessarily interact with all the other individuals in the same patch and can interact with further individuals only by changing patches. Alternatively, each patch might be much larger than the area occupied by a single group. In this case the patches of the model must be defined as equivalent to the areas used by the various groups. This complicates matters, since when groups merge or split, patch designations may change. Nonetheless, the same analysis can be applied, so long as (1) the groups use distinct, non-overlapping areas, and (2) patch designations are carefully defined and accounted.

In a closed population, the total population size will continue to increase so long as mean absolute fitness exceeds 1.0. Thus, under the assumptions of the ideal free distribution, group sizes will continue to increase until all occupied groups have mean fitness 1.0. This result

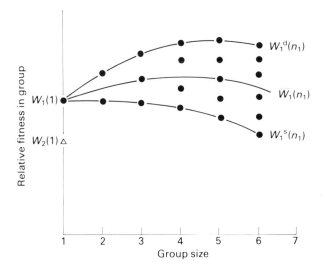

Fig. 5.3 Group size, dominance rank, and fitness. $W_i(1)$ is a solitary's relative fitness in habitat i, $i = 1, 2$. $W_1(1) > W_2(1)$, so a solitary always should prefer habitat 1. As the size of the group increases, a particular individual's fitness depends on its rank in the linear dominance hierarchy. $W_1^d(n_1)$ is the fitness of the most dominant among n_1 individuals in habitat 1. $\overline{W}_1(n_1)$ is the average fitness of the n_i individuals, and $W_1^s(n_1)$ is the fitness of the most subordinate among n_1 individuals in habitat 1. The group forms initially in the better habitat (1). Habitat 2 will first be utilized when the most subordinate individual in habitat 1 would be better off using the poorer habitat. That is, all individuals choose habitat 1 as long as $W_2(1) < W_1^s(n_1)$. When this inequality is first reversed, optimal decision rules require that the lowest ranking individual choose habitat 2. Thereafter, individuals should join groups in a manner depending on their ability to dominate others and the current size of the groups.

ensures that at the demographic equilibrium, groups will be as large or larger than their optimal sizes. In particular, all groups will be as large as possible within the capacity of the patches to support the groups on them.

Dominance-structured groups

We adopt the point of view that the behaviour of a subordinate individual represents an attempt on the part of younger and/or smaller individuals to maximize their individual fitnesses given that dominant individuals have greater resource-holding potential (see section 5.1.2). This perspective does not view subordinate status as a lifetime strategy but rather as a temporary tactic to make the best of a bad situation. It does assume that for the duration of particular dominance relationships, dominant individuals may enjoy higher relative fitness than do subordinate individuals.

The relative fitnesses of dominant and subordinate individuals can be graphed together as shown in Fig. 5.3. For each group size, the fitnesses of all individuals in a group of that size are shown as separate points. As illustrated in the figure, dominant and subordinate individuals may have different optimal group sizes. The equilibrium group size is not determined by these optima but rather by a comparison of the fitness of the lowest ranking member of the group to the fitness the same individual could achieve as a solitary individual in a habitat of poorer quality.

To illustrate how dominance complicates the question of group size selection, we consider the simple case of two individuals choosing between two patches of habitat. The matrices in Fig. 5.4 illustrate some possible relative fitness values for two individuals in each of the following four situations:

1,1—both individuals in patch 1;
1,2—individual 1 in patch 1 and individual 2 in patch 2;
2,1—individual 1 in patch 2 and individual 2 in patch 1;
2,2—both individuals in patch 2.

In each of the matrix cells, the number above the diagonal refers to the relative fitness of individual 1 and the number below the diagonal refers to the relative fitness of individual 2 in that particular situation. Though the matrix 'payoff' values can represent fitness in the general sense, in the examples to follow we assume that the only fitness differences are due to individual differences in food intake and that individuals can assess patch quality on the basis of feeding rates.

The four cases illustrated in Fig. 5.4 represent situations differing in the relative costs and benefits of group living and patch quality. For example, in Fig. 5.4c, the benefits of group membership exceed the costs, as seen by the fact that in either patch each individual does better in a group than by itself. Accompanying each payoff matrix is a transition graph showing changes in patch use and group size as individuals make decisions and move from one patch to the other. For the transition graphs, we have assumed only that (1) each individual has sampled the four possible situations and retains a true estimate of its relative feeding rate in each situation and (2) an individual changes habitats only if by doing so it can expect a higher feeding rate. Thus, the graphs can be said to represent perfect information decision rules.

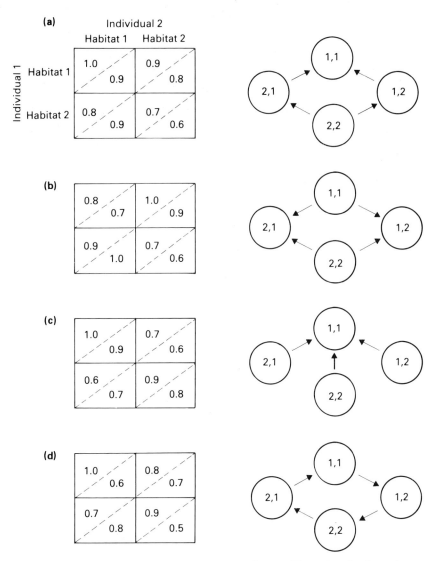

Fig. 5.4 Two-person flocking games. In each payoff matrix, the first player controls the rows, and the second player controls the columns. Within each cell, the payoff to the first player is written above the diagonal, and the payoff to the second player is entered below the diagonal. The diagram to the right of each payoff matrix shows transitions between states that could be expected when individuals employ optimal decision rules. State (i, j) indicates that the first player occupies habitat i and the second player occupies habitat j. Details of the various games are found in the text.

In Fig. 5.4a, an equilibrium occurs only when both individuals are in patch 1. For example, starting with both in patch 2, either individual 1 will move to patch 1 or individual 2 will, because either would benefit from such a move. Once one individual has moved to patch 1, the other will follow, because such a move also increases its expected feeding rate. Finally, once both individuals are in patch 1, any further move by either individual would result in a lower feeding rate for the individual that moves. Contrast this case with that shown in Fig. 5.4b, where the costs of group living exceed the benefits. In this case, whenever both individuals are in the same patch, either benefits by leaving. Thus, the only stable situations are when one individual is in patch 1 and the other is in patch 2.

Figure 5.4c shows a situation like that in Fig. 5.4a, in that the benefits of a group life outweigh the costs. However, in this case, when both individuals are in patch 2, it pays neither to move, so long as the other stays. Nonetheless, both could benefit by a cooperative move, going to patch 1 together. Thus, it can be said that the state 1,1 is a Pareto or 'cooperative' equilibrium and the state 2,2 is a Nash or 'selfish' equilibrium.

Finally, Fig. 5.4d represents a case for which there exists no optimal group size. The dominant individual benefits from feeding in a group but the subordinate would prefer to feed alone in either patch, rather than to feed with the dominant. The result of each individual behaving so as to increase its own feeding rate is a kind of round-robin, in which the dominant individual chases the subordinate from patch to patch (this assumes that the cost of moving is small compared to the benefits to the subordinate from feeding alone and to the dominant from feeding together). Such round-robins are a natural consequence of optimal decision rules whenever one individual always benefits from being in a group and the other always does best feeding alone.

5.5 PROSPECTS

We have argued that, in realistic situations, an optimal group size may not exist and, even if it does, equilibrium group sizes will usually not be of optimal size. Given that this is so, what are the prospects that all of the accumulated facts and ideas concerning the costs and benefits of group life can be organized into a single coherent body of theory?

We argue that natural selection produces individuals that make decisions in approximate accord with the dictates of fitness maximization, given the options available to them. If this is so, the distribution of individuals among available groups and patches of habitat will approximate to a Nash equilibrium to an n-person game. Furthermore, we postulate that individuals will often be in groups of larger than optimal size. In particular, quasi-permanent groups will be of the maximum size consistent with resource productivity.

Ours is not a point of view that would please Pangloss. Indeed, we are postulating the worst, rather than the best, of all possible worlds. Individuals, by always choosing the best of the options available to them, continue to reproduce until the very best option is mere survival and replacement.

Chapter 6
Territory Economics

N. B. DAVIES and A. I. HOUSTON

6.1 INTRODUCTION

In this chapter we will focus on the question of when resources are economically defendable. We will discuss ideas and examples from territorial behaviour, but in principle the same approach can be used to study other social interactions such as dominance hierarchies and group living (Chapter 5). For our purposes we will recognize a territory as 'a more or less exclusive area defended by an individual or group'. Many animals engage in fierce combat along territory boundaries to keep competitors at bay, often sustaining injury in the process. Defence can, however, be more subtle, with individuals maintaining exclusive areas by mutual avoidance of each others' keep-out signals such as scent (Gosling 1982) or song (Krebs *et al.* 1978). Here, just as in combat, the owner has to spend time and energy maintaining the territory and we have to explain why it is prepared to pay this cost.

In some cases defence appears to have minimal costs because all individuals in the population follow some very simple movement rules. In speckled wood butterflies (*Pararge aegeria*), intruding males retreat on perceiving that a territory is occupied, each male following the simple rule 'owner returns to the territory, intruder retreats'. This means that the only way an individual can get to 'defend' a territory is to be the first one to arrive in the area (Davies 1978b). In the dragonfly *Libellula quadrimaculata* there are sometimes physical clashes when males first set up territories around the edge of breeding ponds, but once settled for the day individuals appear to follow the simple rule 'fly until you meet a neighbour, then turn round and fly back again'. By obeying this rule individuals often patrol up and down the same stretch for several hours. Sometimes, however, when a neighbour is perched on vegetation, a male will fly over him unnoticed and carry on until he meets the neighbour beyond. He then turns round and if he now meets his original neighbour on the way back, he again turns, with the result that the two will have swopped territories (pers. obs.). Given this simple movement rule, the only way a male can 'defend' a larger territory is to fly faster; he will then cover a longer stretch before he meets each neighbour and is forced to turn round and go back again.

It is difficult to draw a sharp distinction between territories which are maintained by physical combat and those occupied through individuals avoiding each other by the use of simple movement rules. Whatever the proximate cause of the spacing patterns, we still have to answer the ultimate question of why the maintenance of an exclusive area has been favoured by selection. Why, for example, have certain movement rules evolved rather than others? Brown's (1964) concept of economic defendability states that we would only expect an animal to spend time and energy interacting with others to defend a territory when this yields greater net benefits than an alternative behaviour, for example ignoring others in the population and spending the whole time exploiting the resource.

6.2 ECOLOGICAL FACTORS FAVOURING TERRITORIALITY

Three main factors will influence the economic defendability of a resource. These three factors are: resource quality and distribution in space; resource distribution in time; and competition for the resource.

Resource quality and distribution in space

The range an animal will have to occupy to satisfy its energetic or reproductive requirements will depend on the abundance and distribution of food and mates. At one extreme, if the resource is of poor quality and sparsely distributed (e.g. wildebeest exploiting poor quality grazing on a large plain) the animal will have to roam over a large area and it is unlikely that it will be able to defend this economically. On the other hand, sparsely distributed high quality food might be worth defending (e.g. browse for small antelope, Jarman 1974) (see also section 5.2.1).

If the resource is clumped in distribution (e.g. a pile of nuts under a tree) the animal might be able to restrict its movements to a very small area. Whether this area is economically defendable will depend on competitor density, which in turn will depend on the size of the clump and its quality. In general, large piles of high quality food are not defended (e.g. finches feed in flocks on large clumps of seeds) but smaller piles of food sometimes are. Zahavi (1971) provided a nice experimental demonstration of how food distribution in space can influence social behaviour. When high quality food was presented in small clumps, individual pied wagtails (*Motacilla alba*) defended them, but when the same food was dispersed sparsely, territory defence ceased, presumably because it was uneconomical to defend a large area. Rubenstein (1981b) performed similar experiments with pygmy sunfish in aquaria. When prey were dispersed randomly, males swam freely around the tank, but when prey were predictably located in a central clump, males defended territories near the prey.

A comparative survey of ant territoriality by Hölldobler and Lumsden (1980) illustrates the influence of resource distribution on social organization. The food of the African weaver ant (*Oecophylla longinoda*) is uniformly distributed in space and stable over time (insect prey in vegetation). These ants defend large, fixed, three-dimensional territories and reduce transport costs by distributing nests throughout the territory. Harvester ants, such as *Pogonomyrmex barbatus* and *P. rugosus*, encounter food that is patchily distributed in space but relatively stable over time (large patches of seeds on the ground). Colonies of these ants have one nest with a series of trunk trails leading out to the food supplies. Finally, the food of honey ants (*Myrmecocystus mimicus*) is both patchily distributed in space and unstable over time (e.g. termites under cattle dung). Hölldobler and Lumsden show that adopting a fixed territory in such conditions will result in large fluctuations in gain over time and, furthermore, as resources become scarcer a larger area has to be defended to get enough food. Honey ants adapt to their changeable food supplies by adopting a flexible territorial system. There are no fixed territory boundaries but rather the colony ranges over a large area and defends only the patches in which it is foraging at the time.

Resource distribution in time

Horn (1968) constructed an elegant geometric model to show how the distribution of resources in time might influence animal distribution. If food patches are very ephemeral then individuals might have to roam over a large area to exploit sufficient patches to stay alive. In this case it might pay all individuals to live in a colony at the centre of distribution of the patches in order to minimize their travel time when foraging. If resources are more predictable in time, for example because they are renewed at a sufficient rate for an individual to exploit the same patch for long periods, then it might pay animals to space out in exclusive areas.

Most models of territoriality ignore resource renewal and so we explore this in detail later on. We make one general point here, which is that work on optimal allocation of time to the schedules used in operant experiments turns out to be relevant to the question of how to exploit renewing resources on a territory. The basic property of the concurrent variable interval–variable interval procedure is that the longer an animal has been away from a schedule, the more likely there is to be a reward 'waiting' on that schedule. The variable interval schedule is thus analogous to a renewing resource. We can use the results of Houston and McNamara (1981) to give the optimal time allocation in two areas of a territory as a function of the renewal rates in each area and the travel time between them.

The number of competitors and the individual differences in competitive ability must also influence the economics of defence. In Horn's model, if there were sufficient resources for all competitors then individuals might adopt exclusive areas without any aggression at all. However, resources will usually be limiting and so active defence will be expected in most populations. The costs of defence will be influenced by competitor density and also different individuals may have different thresholds for economic defence. The weakest individuals might never be able to defend territories and may adopt alternative strategies of competition.

6.3 OPTIMAL TERRITORY SIZE AND THRESHOLDS FOR ECONOMIC DEFENCE

6.3.1 A graphical model

The three factors above will influence the costs and benefits of defence. In general we imagine that the benefits of occupying a territory increase with size at first, but eventually reach an asymptote when the resource becomes superabundant in relation to the animal's ability to utilize it (Fig. 6.1a). The costs of defending the area will increase with territory size because more intruders will come on to a larger territory and the owner will also have to patrol over larger distances. Territory defence will only be economical (B > C) between thresholds x and y. Changes in resource quality and distribution will move the benefit curve (Fig. 6.1b) while changes in competitor density will move the cost curve (Fig. 6.1c) and so alter the thresholds

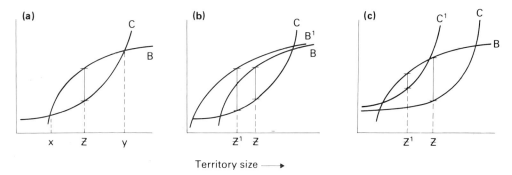

Territory size ⟶

Fig. 6.1 Hypothetical relationships between benefits B and costs C for various territory sizes. **(a)** Defence will be profitable (B ⩾ C) only between points x and y. Maximum net benefit (B − C) is at Z. **(b)** Increase in benefits (B curve shifts to B^1) will change the thresholds at which defence is economical and decrease Z to Z^1. **(c)** Increase in costs (C curve shifts to C^1) will likewise decrease the thresholds and decrease Z to Z^1. (After many authors, including Myers *et al.* 1981.)

for economic defence. In most cases the cost and benefit curves will not change independently. For example, as resource abundance increases, so will intruder pressure.

Figure 6.1 indicates what is optimal for an animal to do but does not specify the proximate mechanisms that might bring about the behaviour. It is possible that an owner adjusts its territorial behaviour by responding not to food *per se* but rather to its internal state (e.g. hunger; see Slaney & Northcote 1974) or some feature of the environment which is a reliable indicator of food availability.

The thresholds at which the animal is selected to defend a territory and the optimal territory size will depend on three conditions: currency, the profitability of alternative options, and the time-scale.

Currency

It is clearly meaningless to ask what is the optimal territorial behaviour unless we define what is meant by 'optimal'. This amounts to finding the currency which relates an individual's behaviour to its inclusive fitness. Consider, for example, a male great tit (*Parus major*) defending a territory in late winter. The territory is apparently not essential for feeding or mate attraction (Perrins 1979) and the main benefit does not come until the summer, when spacing-out reduces predation of nestlings (Krebs 1971). Thus the male spends time and energy setting up a territory in winter, presumably at some risk to its own survival, so that its young have increased survival chances in the summer. The optimal trade-off between the male's own risk and the risk to its young can be found by dynamic optimization theory (see Chapters 4 and 11).

If the trade-offs can be correctly calculated, then our costs and benefits will be in terms of inclusive fitness and the optimal size of territory to defend will always be where (B − C) is a maximum, that is point z in Fig. 6.1a. Most studies, however, consider only certain components of fitness, for example mate attraction, predator avoidance or food intake. Although the concept of optimal territories can be applied widely (see Chapter 10), we will concentrate on those used for feeding and assess benefits and costs in terms of energy (for examples of other benefits, see Table 6.1). For an animal selected to put on fat prior to migration, the optimal size of feeding territory will be that which maximizes (B − C), i.e. point z in Fig. 6.1a. On the other hand, for an animal selected to maintain a constant body weight and to minimize foraging costs, the optimal territory size will be where B = C and C is minimized (point x in Fig. 6.1a). The same net benefit is obtained at point y, but at a higher cost. Point y is the largest territory that can be defended without a net loss. Between points z and y the owner may reduce its net energy intake but defence of a larger territory may exclude more potential rivals. We return to this idea later (section 6.4).

Even without any increase in defence costs with increased area, there can still be an optimal territory size. Andersson (1978) considers the exploitation of a territory by an animal that brings food back to a central place. The currency he considers is maximization of the food obtained over a foraging period during which the food does not renew. Territory size is limited by the increase in time taken to bring food back from great distances. Andersson shows that the forager should spend less time searching in a given area as it gets further from the central place. The distance from the central place at which the optimal search time becomes zero sets the limit to the economic territory radius.

Table 6.1. Examples of benefits other than food of an increase in territory size. In each case the costs are an increase in intruder pressure.

Species	Benefit of larger territory	Reference
Western gull (*Larus occidentalis*)	Decreased cannibalism of chicks by neighbours	Hunt & Hunt 1976 Ewald *et al.* 1980
Three-spined stickleback (*Gasterosteus aculeatus*)	Decreased predation by conspecifics on eggs in nest	Black 1971
Belding's ground squirrel (*Spermophilus beldingi*)	Decreased cannibalism of young by conspecifics	Sherman 1981
Arctic skua (*Stercorarius parasiticus*)	Attract female earlier in the breeding season	Davis & O'Donald 1976
Long-billed marsh wren (*Telmatodytes palustris*)	Attract more females	Verner 1964
Great tit (*Parus major*)	Decreased predation by weasles (*Mustela nivalis*)	Krebs 1971; Dunn 1977

The profitability of alternative options

The lowest possible threshold must be set by the level where the animal cannot get sufficient energy to exist if it continues to defend the territory (e.g. Carpenter & MacMillen 1976). An individual might abandon its territory before this 'death threshold', however, if its gain rate would be higher from an alternative behaviour. The territory might even be defended for short periods under conditions of net deficit if this nevertheless yields better rewards than other options (Wolf 1978).

The time-scale over which benefits and costs are measured

This can be illustrated by contrasting three bird studies. Nectar feeders (hummingbirds and sunbirds) defend feeding territories for short periods such as a day or week and often abandon them as soon as nectar levels become unprofitable (Gill & Wolf 1975). Pied wagtails defend their territories throughout the whole winter, even on days when other alternatives are temporarily more profitable or levels of food are so low that the owner cannot spend the whole day on its territory (Davies & Houston 1983). Ural owls (*Strix uralensis*) spend their whole lives in the same territory and continue to defend them even through periods when prey are so scarce that there is no chance of breeding. Many pairs have to wait five years between breeding attempts and one pair has been waiting for 10 years for prey to increase to a level which would permit breeding (Lundberg 1981)! In these three examples, the time period over which we need to measure benefits and costs are respectively, a few hours, a winter and several years.

Dill (1978) develops a model to compare the maximization of daily versus seasonal net energy intake. He shows that, over a whole season, it sometimes pays an animal to hold a territory larger than that required to maximize daily intake. This is because large territories result in emigration of potential intruders and hence reduce defence costs late in the season. Although Dill looks at the best fixed size throughout the season, it is possible that the optimal strategy is to change territory size as the number of intruders is reduced.

The shapes of the curves in Fig. 6.1 can be varied indefinitely and there is a temptation to invoke different benefit and cost functions, together with various degrees of resource 'abundance', 'quality', and 'patchiness', without ever actually measuring anything. We will now discuss four case studies where measurements have been made in an attempt to understand how behaviour varies with resource distribution and competitor density.

6.3.2 Rufous hummingbirds

Rufous hummingbirds (*Selasphorus rufus*) breed in north-west North America and then migrate along inland mountain ranges to their wintering grounds in southern Mexico. Their migration consists of a series of leap-frog flights between alpine meadows where they defend feeding territories for a few days or a few weeks to fatten up for the next stage of the journey.

Birds moved location from day to day and appeared to track sudden changes in resources. For example, in one meadow there were 25 flowers in bloom and no hummingbirds on one day, and then, just one week later, there were 3200 flowers and 15 territories being defended (Gass 1979). Individuals also made daily adjustments in their territories, defending smaller areas where flowers were dense, with the result that although territory size varied 100-fold in area, there was only five fold variation in the number of flowers defended. If territory size was adjusted to maintain a constant amount of resources, and flowers were uniformly distributed, then we would expect a negative hyperbolic relationship between territory size and resource density. The data of Kodric-Brown and Brown (1978) from a study site in eastern Arizona fit this prediction very well. On a log–log plot the slope of the relationship between territory size and density of flowers defended is not significantly different from -1, which means that the birds were changing their territory size to defend always about the same number of flowers (Fig. 6.2a). Experiments also supported this idea; when flowers were removed, individuals expanded their territories to include sufficient extra flowers such that nectar levels returned to their previous values.

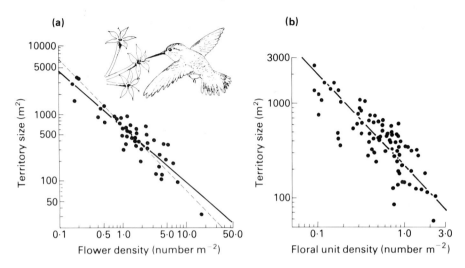

Fig. 6.2 (a) Inverse relationships between flower density and territory size in rufous hummingbirds in eastern Arizona. The slope of the regression is -0.82 (solid line), which is not significantly different from the slope of -1 (dotted line) which would be obtained if the same average number of flowers were defended regardless of territory size (From Kodric-Brown & Brown 1978.) (b) A similar plot from a study in north-west California (Gass *et al.* 1976). Here 'floral unit density' is a measure of nectar production on a territory, taking account of the fact that different flower species produce different amounts of nectar. Again, the slope (-0.95) is not significantly different from -1.

In a study in north-west California, Gass *et al.* (1976) found that the birds did not simply defend a constant number of flowers but varied their territory size depending on both flower density and flower species. The two commonest plants were the Indian paintbrush (*Castilleja miniata*) and the red columbine (*Aquilegia formosa*), which produced four times as much nectar per day as the paintbrush. Measurements showed that territory sizes were adjusted to provide about the same total amount of energy regardless of flower density and species composition (Fig. 6.2b).

Both these studies show changes in territory size as benefits change, as depicted in Fig. 6.1b, but there were no data on how costs of defence varied with territory size or quality, so we cannot answer the question about which currency the birds were maximizing by their behaviour.

6.3.3   Golden-winged sunbirds

Gill and Wolf (1975, 1977) studied golden-winged sunbirds (*Nectarinia reichenowi*) near Lake Naivasha, Kenya, where the sunbirds were defending winter feeding territories consisting of patches of *Leonotis nepetifolia* flowers. Territory size varied enormously, from 6 to 2300 m^2, whereas the number of flowers defended varied less, from 1000 to 2500. Like the hummingbirds, the sunbirds showed short-term changes in territory size, for example increasing their territories when flowers died or where flowers were less dense.

Gill and Wolf recognized that the advantage of territory defence depended critically on the renewal of nectar in the flowers. The key advantage of defence was that, by excluding competitors, nectar levels could increase to higher levels. As predicted, nectar levels in defended areas were higher than in undefended areas, because in the latter there was more depletion. Owners also increased the rewards they obtained by systematic foraging around their territories, so avoiding visits to flowers which they had recently depleted (Fig. 6.3a; see also Kamil 1978; Bibby & Green 1980). The advantage of exploiting high nectar levels was that the bird could satisfy its daily energy requirements in less time.

The benefits of defence, however, depended on the absolute levels of nectar in the flowers and the birds only defended their territories between two thresholds (as in Fig. 6.1a). When nectar levels were very low (e.g. in the case of slow renewal), the birds abandoned their

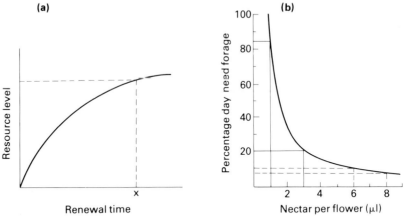

Fig. 6.3   **(a)** Hypothetical relationship between resource level and renewal time. By defending a territory, an owner can prevent depletion by competitors and so crop the resource at profitable renewal times, e.g. x.
(b) Relationship between amount of time a sunbird must forage (percentage of a 10 h day) to maintain its energy requirements (13 kcal nectar) when it exploits various nectar levels in the flowers. At low nectar levels defence can be very profitable (e.g. if defence increases nectar levels from 1 to 3 μl, foraging time is decreased substantially). Defence at very high nectar levels brings little saving in foraging time (e.g. increase from 6 to 8 μl). (From Gill & Wolf 1975.)

territories altogether because, even with defence, nectar would not increase to levels where they could balance their daily energy requirements. At intermediate levels of nectar, defence brought large benefits because small increases in nectar levels due to prevention of depletion by competitors decreased the owner's foraging time considerably (Fig. 6.3b). Calculations showed that under these conditions the energetic costs of defence were more than offset by the energy savings from foraging at higher nectar levels in the territory. As nectar levels increased to very high values, however, the benefits of defence decreased because depletion by others brought little cost (e.g. when nectar renewed rapidly). As predicted, owners then ceased defence and tolerated intruders.

As with the hummingbird studies, no detailed measurements were made on how costs of defence varied with territory size and resource quality. In a model, Pyke (1979) assumed that intruder pressure was an accelerating function of the average energy content of the flowers. He was then able to predict the optimal territory size (number of flowers defended) for various hypotheses of currency. The predicted value for maximization of net energy gain (B − C) was 7070 flowers, while for minimization of costs, subject to the constraint that the owner got enough energy to stay alive (i.e. B = C), the prediction was 1595 flowers, remarkably close to the average observed number of flowers defended which was 1600. Pyke concluded that the currency of cost minimization makes good biological sense for the sunbirds. They were not preparing to migrate or breed and winter feeding conditions were reasonably stable, so little would be gained from accumulating energy reserves (i.e. maximizing B − C). Indeed, putting on reserves unnecessarily might increase predation or metabolic costs, and so minimization of costs might well be the best goal to maximize their overwinter survival.

6.3.4 Sanderlings

Many studies have found that, like the nectar feeders above, territories are smaller when resources are more abundant (e.g. in lizards, Simon 1975; birds, Stenger 1958; Cody & Cody 1972). Myers *et al.* (1979) pointed out two proximate explanations for this relationship. First, the owner could assess prey density and adjust its territory size to include sufficient resources for its needs. As food increased in abundance, it would decide to defend a smaller territory (Fig. 6.1b). Alternatively, the individual could always defend as large a territory as possible, but size would be constrained by competition. As food abundance increases, intruder pressure will increase and so the territory will become more costly to defend. Owners may then be forced to defend smaller territories (Fig. 6.1c). Here, the inverse relationship between food abundance and territory size comes about indirectly through variations in intruder pressure.

It is difficult to tease apart these two proximate explanations for variation in territory size, because food abundance and intruder pressure will usually vary together. Myers *et al.* (1979) were able to do this, however, in a study of winter feeding territories in a shorebird, the sanderling (*Calidris alba*) on the sandy beaches of California. The birds defended 10–120 m linear territories between the tide marks. Some territories were defended for just a day, others for several months, and some birds even defended the same stretches in successive winters. Non-territorial birds (30–90% of the local population) roamed the beaches and often intruded on to the territories. Prey were sampled by sieving core samples and prey density was found to vary fourfold between different parts of the beach (the main prey was a marine isopod, *Excirolana linguifrons*).

Simple correlation coefficients showed that territories were smaller in areas of higher food density, and because more intruders were present where prey was most dense, territories were also smallest where intruder pressure was greatest. By using a multiple regression to unravel the interactions between the three variables it was shown that the relationship between prey density and territory size became insignificant when intruder density was controlled for, whereas the relationship between intruder density and territory size was still significant when prey density was controlled for. This suggests that decreased territory size in areas of high prey density was brought about proximately by increased intruder pressure, which increased the costs of defence (Fig. 6.1c). Further evidence for this view was that in one season intruder pressure declined when non-territorial birds left the area. Territory size of the owners on the beach then increased suddenly, even though prey density remained the same as before.

Why do sanderlings respond to intruder pressure rather than prey density? One possibility is that because prey density changes markedly during the winter, depending on reproductive activities of the prey and on wave action, it may be costly for the birds to keep changing their territory size to track some optimum. A simpler solution may be to defend as large a territory as possible, given intruder pressure. If there are resources in excess of daily requirement then these will act as an insurance in case prey density suddenly declines.

6.3.5 Pied wagtails

We studied the winter feeding territories of an insectivorous bird, the pied wagtail, on a meadow in the Thames Valley, southern England. Some birds defended territories along a river while others fed in flocks on flooded pools nearby. The territory-owning birds exploited a renewing food supply, namely small insects which were washed up by the river on to the muddy banks. Owners typically walked a regular circuit around their territories, up along one bank and then

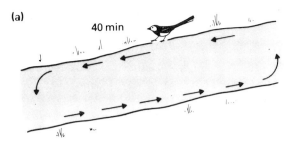

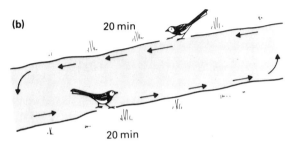

Fig. 6.4 **(a)** Pied wagtails exploit their territories systematically. The circuit of the river bank takes on average 40 min to complete. **(b)** When a territory is shared between two birds, each walks, on average, half a circuit behind the other and so crops only 20 min worth of food renewal.

back down the other side again (Fig. 6.4a), with the result that they cropped the resource at profitable renewal times (Fig. 6.3a). As with the sunbirds, territory defence can be seen as the way an owner protects the prey renewal times on its territory, by preventing competitors from depleting the resource.

Indeed, by systematic depletion of the food supply on the territory, the owner can, in effect, make it unprofitable for intruders to land. The most profitable place to feed (renewal time greatest) is just ahead of the owner. If an intruder landed there it would be easily spotted. If, on the other hand, the intruder landed some way behind the owner it would be feeding over stretches the owner had recently depleted. Measurements showed that intruders often fed at an unprofitable rate, even if they managed to sneak on to the territory undetected, precisely because they fed over depleted stretches. This probably explains why intruders were usually very noisy when flying over a territory, calling loudly 'chisick'. If an owner was in residence it replied 'chee-wee', whereupon the intruder usually retreated (Davies 1981). We suggest that the intruder's noisy calls are an enquiry as to whether an owner is present. If one is, then this signals a depleted food supply and so it pays the intruder to move on.

In theory, the optimal territory size to maximize daily feeding rate must vary with the renewal pattern of prey along the river. As food abundance and renewal rate increase, the wagtail can maintain a profitable feeding rate by walking around smaller and smaller circuits (equivalent to the B curve shifting to the left in Fig. 6.1b). Intruder pressure increases with food abundance, so the costs of defence will also increase (cost curves shift to the left as in Fig. 6.1c).

Unlike the nectar feeders, however, which changed their territory size rapidly to follow changes in resources, the wagtails maintained the same size of territory, a circuit of on average 600 m, throughout the whole winter. It seems likely that continuous adjustments in territory size to track daily optima would be costly for the wagtail. At the start of the winter, neighbours spent a lot of time in boundary disputes but once settled these boundaries were respected and were maintained simply by short displays at intervals throughout the day. Changing boundaries from day to day would be costly, especially when food supply decreased and a bird needed to expand its territory again. Instead of daily changes in territory size, therefore, the owners maintained a fixed territory throughout the winter and adjusted to variations in food abundance by changes in their behaviour towards intruders and changes in the amount of time they spent on their territories.

Four levels of territory defence and time budget can be recognized.

(a) When food was very scarce on the territories, the owners left and fed elsewhere but they kept returning at intervals to announce ownership and evict intruders. They would spend some time on the territory even on days when they did not feed there at all. This suggests that defence was a long-term investment and that the territory was worth maintaining even through periods of low prey abundance. Over the winter as a whole, territory owners were less likely to suffer long periods of severe food shortage than the flock birds. Further evidence that owners were not maximizing short-term (daily) feeding rate was that on some days the flock birds enjoyed much higher food intake rates, yet territory owners still spent much of their time on their territories, forsaking the higher feeding rate they could have achieved by abandoning them and spending the whole day with the flock. The time allocation of owners between their territories and the flock under these conditions suggested that the periodic visits to the territory were to prevent intruders from cueing into the prey renewal pattern and taking it over (Davies & Houston 1983).

(b) At intermediate levels of food abundance, the owners spent all day on territory and evicted all intruders.

(c) As food increased further, however, an owner began to share its territory with a satellite, usually a first-winter juvenile or a female from the flock. Sharing the territory brought costs to the owner because the satellite depleted the food supply on the territory. Owners and satellites usually shared the territory by each walking half a circuit behind the other, so that the owner now cropped the food at only half the renewal time it would have enjoyed if alone (Fig. 6.4b). Sharing also brought a benefit, however, because the satellite helped with defence. We were able to quantify how these costs and benefits combined to influence the owner's feeding rate and showed that owners tolerated a satellite only when the benefits of help with defence outweighed the costs of sharing. On days when the food levels decreased again, and an owner achieved a higher feeding rate by remaining alone, it evicted the satellite from the territory (Davies & Houston 1981).

(d) At very high levels of food abundance, for example when there was a sudden emergence of small insects in the spring, owners abandoned all attempts at defence. This made good economic sense because there was little benefit from evicting intruders. With high levels of food abundance and rapid renewal, an owner's feeding rate was hardly affected even if it walked directly in the footsteps of another bird.

The wagtail's behaviour in long-term occupancy of a territory over the whole winter makes a nice contrast to the sudden daily changes of the nectar feeders in territory size and location. When food supply doubles for a sunbird, the bird simply halves its territory. When a wagtail shares its territory with a satellite it is, in effect, also halving its territory (Fig. 6.4b), but with an important difference. The owner still patrols over and defends its entire territory and maintains the same boundaries with neighbours. The advantage of this strategy over a decrease in real territory size is that when food becomes scarce again, the owner can quickly reclaim the whole area for itself simply by evicting the satellite. It is interesting that owners, who are usually adult males, only tolerate juveniles or females as satellites, birds that they can evict most easily. By maintaining the same territory and reducing defence costs by sharing, as opposed to a decrease in real estate, the wagtail avoids having to set up new territory boundaries every time the food supply changes.

It is worth emphasizing that in both sunbirds and wagtails the birds do not necessarily maximize net daily energy gain, which at first sight may seem the most appropriate currency. The sunbirds appear to minimize daily costs and although the wagtail's decision of whether to share its territory or not seems to be based on daily gain, the allocation of time to the territory even on very poor days clearly does not maximize daily gain. For both birds the appropriate currency may be maximization of overwinter survival, and McNamara and Houston (1982) show theoretically that this is not equivalent to maximization of daily intake.

6.4 SUPERTERRITORIES AND SPITE

The studies we have discussed so far suggest that territory sizes are adapted to maximize the owner's short- or long-term efficiency of access to resources. Verner (1977) has put forward the intriguing alternative hypothesis that individuals will be selected to defend larger territories than they need for their own requirements simply out of spite, in order to prevent others in the population from using the resource. Although defence in such a superterritory will bring no extra absolute benefit to the owner, its relative success will be increased because if it excludes others from the resource then fewer competitors will survive and breed.

As several authors have pointed out, the problem with the evolution of such spiteful behaviour is that although the superterritory holder certainly enjoys increased relative success, so too do other individuals in the population, including those who do not pay the cost of defence of a superterritory (Rothstein 1979; Tullock 1979; Colgan 1979). In effect, the superterritory holder bears all the cost for a public benefit. Theoretical models suggest that although super-territoriality may be an ESS, it is unlikely to spread in large populations and where the costs of defence of the extra resources are high (Knowlton & Parker 1979; Parker & Knowlton 1980).

There are practical difficulties in showing that a territory has resources surplus to an individual's own requirements. If territories are defended as a long-term investment then in some short-term periods they will appear to have excess resources (e.g. sanderlings and wagtails, above; see also MacLean & Seastedt 1979). In theory, territory size could also have been sexually selected in the same way as tail length in some birds (Andersson 1982b), so that males might defend huge territories simply because females find them attractive. Finally, individuals may space out more than expected from the sufficient-resource hypothesis to make it more difficult for newcomers to settle close by. In this case, the defence of a large territory brings direct benefits to the owner because it decreases the time which has to be spent in aggression against neighbours (Getty 1981; Burger 1981).

Harris (1979) attempted to test Verner's hypothesis experimentally. Tree swallow (*Iridoprocne bicolor*) populations are often limited by availability of nest sites, so that numbers of breeding pairs in an area can be increased by the erection of nest boxes. Harris placed boxes very close together and found that a pair would prevent others from using nest boxes very close to the one they used for breeding, even though they never actually used these extra boxes themselves. This appears to be defence of resources in excess of the pair's own requirements. Data showed that the additional aggression did not affect the pair's own reproductive output but it did prevent others from breeding. No tests were made, however, of the possible costs of very close neighbours and so there is still a possibility that some direct benefits were involved, such as decreased cuckoldry or predation.

6.5.1 Sneaks, satellites and the payment principle

Owners often lose resources to intruders who sneak onto their territories unnoticed. 'Sneaks' decrease their chances of detection by creeping silently around the territory and their crypsis is sometimes enhanced by dull colouring. Sneaks impose costs on the owners and are evicted whenever spotted. There are, however, other cases where owners tolerate subordinates on their territories. These 'satellites' also impose costs but in addition they bring benefits, often helping to defend the territory (Table 6.2). (Here we distinguish, for convenience, between 'sneaks' who bring no benefit to the owner, and 'satellites' who do, although elsewhere in the literature these terms have often been used interchangeably.) The transition from sneak to satellite occurs when the subordinate begins to bring a benefit to the owner, which can be regarded as a payment for permission to share the territory (Gaston 1978b). In the pied wagtails mentioned above, we showed that owners only tolerated satellites when the payment (help with defence) offset the costs of sharing (depletion of food).

Brown (1982) has developed a graphical model to show the thresholds at which owners should decide to share their territories. In the graphs (Fig. 6.5) the benefits of sharing are regarded as lower defence costs per individual while the costs of sharing are shown as a decrease

Table 6.2. Examples of sneaks and satellites on territories.

(a) Sneaks: impose only costs on territory owner		
Hummingbirds	Dull-coloured individuals, especially immatures, steal nectar	Kodric-Brown & Brown 1978; Ewald & Rohwer 1980
Frogs and toads	Silent males intercept females attracted by calling male on his territory	Howard 1978; Perrill, Gerhardt & Daniel 1978; Arak 1983
Fish	Dull-coloured intruders steal matings with females in owner's territory	Constantz 1975; Fernald & Hirata 1977; Wirtz 1978
Dragonflies	Steal matings	Campanella & Wolf 1974
(b) Satellites: bring benefits and costs to territory owner		
Waterbucks (*Kobus ellipsiprymnus*)	Steal matings but also help defend territory	Wirtz 1981, 1982
Ruff (*Philomachus pugnax*)	Steal matings but also increase attractiveness of territory to females	Hogan-Warburg 1966; van Rhijn 1973
Pied wagtail (*Motacilla alba*)	Deplete food on territory but also help in defence	Davies & Houston 1981

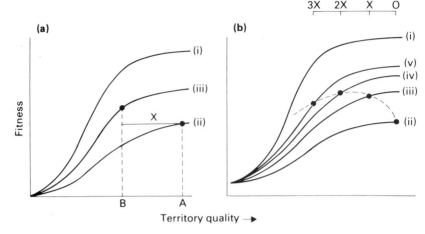

Fig. 6.5 (a) Curve (i) is the maximum fitness an individual can obtain if there are no defence costs. Curve (ii) is the fitness an individual would obtain alone, including defence costs. Curve (iii) is the fitness if an individual shared its territory with another individual and thus only sustained half of the defence costs. The costs of sharing are represented as a decrease in territory quality. For example, with sharing costs of X, it would pay an owner with a territory of quality A, to share it with a subordinate, because its fitness on a shared territory (B) would be increased. (b) An extension of the model to include fitness curves from sharing with two others (curve iv) and three others (curve v). Depletion costs from sharing are represented on the horizontal line above the graph. Sharing with one other individual causes a reduction in territory quality by an amount X, with two others 2X and so on. The dots give the net fitness from sharing and, in this example, an individual gains maximum fitness in a group of three. (From Brown 1982.)

in territory quality, that is to say a decrease in resources available per individual. Other costs and benefits could be considered; for example, depletion costs could be recouped via increased benefits of group living such as better predator detection or nest defence (see also section 5.4 for a discussion of optimal group size).

6.5.2 A model for resource renewal and territory sharing

The cost of depletion depends on how the resources renew. We can investigate the influence of renewal rate by assuming territory size to be fixed and exploring the effects of sharing the territory. We think of the territory as containing a series of identical patches. The amount of food $f(t)$ in a patch depends on the time t since it was last visited by a forager. When an animal has the territory to itself we assume that it returns to each patch after a time t_1, so that it gets an amount of food $f(t_1)$ from each patch. Now assume that another individual arrives and the owner decides to share its territory.

The owner can choose between two ways of sharing. First, both individuals could go round together and exploit the same patches at the same time. In this case the renewal time per patch will stay at t_1 but the food will be shared. Let the owner get a proportion p, so that its food per patch is now $pf(t_1)$. (If $p = 0.5$, the food is shared equally.) The alternative way of sharing is for each individual to forage alone and exploit different patches. They could do this by each having exclusive use of part of the territory or by each using the whole territory and avoiding each other in time (as in Fig. 6.4b). This results in a lower renewal time t_2 for the owner.

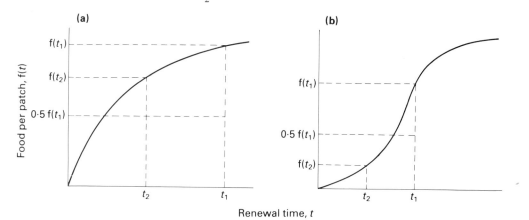

Renewal time, t

Fig. 6.6 The effect of resource renewal on two ways of sharing a territory. (a) The function $f(t)$ is decelerating and so $f(t_2) = f(0.5t_1)$ is greater than $0.5f(t_1)$. This follows directly from the definition of a decelerating function $f(t)$ as one for which $f[at_i + (1 - a)t_j] > af(t_i) + (1 - a)f(t_j)$, $0 \le a \le 1$ for all t_i and t_j. Setting $a = 0.5$, $t_i = t_1$ and $t_j = 0$ establishes the result. Alternatively the $0.5f(t_1)$ condition can be thought of in terms of the owner getting a renewal time of t_1 half the time and of zero half the time. The fact that $f(0.5t_1) > 0.5f(t_1)$ then follows by Jensen's inequality. (b) The function $f(t)$ is first accelerating and then decelerating (the logistic equation is an example). By the same sort of argument as we used in (a), over the accelerating region $0.5f(t_1) > f(t_2) = f(0.5t_1)$.

The question of interest is whether $f(t_2)$ is greater than $pf(t_1)$. A simple case, in which p is 0.5 and t_2 is $0.5t_1$, is illustrated in Fig. 6.6. If $f(t)$ is a decelerating function of t, then $f(0.5t_1)$ will always be greater than $0.5f(t_1)$, and so each individual will obtain a higher food intake rate if they forage in different patches (Fig. 6.6a). In Fig. 6.6b, the renewal function is first accelerating and then decelerating. Over the accelerating region, $f(0.5t_1)$ is less than $0.5f(t_1)$ and so in this case a higher feeding rate will be obtained by both individuals if they go round the territory together.

We cannot necessarily conclude from this that it is always better to go round the territory separately when the renewal function is decelerating, because the resources may not be shared equally between the animals ($p \ne 0.5$) and dividing up the territory may not always halve the return time ($t_2 \ne 0.5t_1$). Furthermore, the renewal function may actually depend on the number of animals present; two animals feeding together may enhance each other's feeding efficiency ($f(t)$ increases) or may reduce it ($f(t)$ decreases). Nevertheless, it is interesting that in the wagtails, where we showed that the renewal function was decelerating and that t_2 was $0.5t_1$ (see Fig. 6.4b), the birds shared by dividing the territory rather than by going round together, a result we would predict from the argument above.

6.5.3 Optimal group size and conflicts of interest

Brown (1982) used his graphical model to predict the optimal group size on a territory (Fig. 6.5b) but this concept may have limited utility because the optimum may vary for different individuals in the group (see also Chapter 5). In pied wagtails, for example, we showed that there was often a conflict of interests between an owner and its satellite (Davies & Houston 1983). Owners had long-term interests in their territories and continued to defend them even when alternative feeding options in a nearby flock were more profitable. Satellites, on the other hand, tended to maximize short-term gain and went to whichever place gave the greatest daily feeding rate. This meant that on some days the owner would have had a higher feeding rate on its territory by sharing, but the satellite did not join the owner because it did better by remaining with the flock. Conversely, on other days the satellite attempted to come on to the territory but the owner evicted it because it did better by remaining alone.

Subordinates could, in theory, increase their payment under conditions of such conflict to make it profitable for the owner to accept them. Owners, in their turn, may be able to manipulate subordinates to get maximum benefits for themselves. Territory sharing might, therefore, involve a subtle interplay of subordinates minimizing payments to make it just profitable for owners to accept them and owners skewing the payoffs to their own advantage but allowing a subordinate just enough benefits to entice it to stay. We illustrate these ideas by considering the effect of the payment made by a subordinate on its own feeding rate and on the feeding rate of the territory owner. The upper graph in Fig. 6.7 shows how the subordinate's feeding rate on the territory decreases as its payment increases. Once the payment exceeds P_s, the subordinate would do better by feeding elsewhere. P_s is thus the maximum economic payment that the subordinate can offer. The lower graph considers the owner's feeding rate. This will increase as the subordinate's payment increases and above a payment of P_o the owner will do better to share its territory rather than remain alone. If, as in the figure, $P_o < P_s$, then the region between them is the 'bargaining' zone, within which both owner and subordinate will agree to share the territory but disagree over the amount of payment.

At first sight we might expect the owner to want a payment just below P_s, but the subordinate to offer a payment just above P_o, which will be the minimum to make it profitable for the owner to

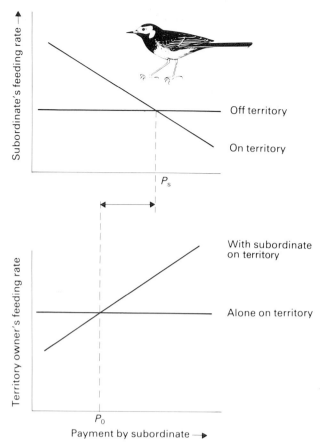

Fig. 6.7 The influence of the subordinate's payment on the subordinate's own feeding rate (top graph) and on that of the territory owner (lower graph). The horizontal lines in each graph represent the payoffs for the alternative options for each bird (i.e. what the subordinate would get if it fed elsewhere and what the owner would get if it were alone on the territory). See text for further explanation.

accept it on the territory. However, the subordinate may not know the value of P_0 and furthermore subordinates may compete with each other to get on to territories, in which case owners could select for the highest bids. Conversely, owners could compete with each other for subordinates, in which case acceptable payments would be less.

The curves in Fig. 6.7 represent a given distribution of payoffs between the owner and subordinate (for example equal shares). Once the subordinate settled on the territory, the owner could increase his own share of the payoffs, in which case the subordinate's feeding rate would decrease. If the subordinate did not change its payment then there would come a point when any further increase in the owner's share would reduce the subordinate's feeding rate to below what it could obtain elsewhere. This will be the maximum skew the owner can impose on the payoffs without the subordinate leaving the territory.

6.5.4 Game theory and territorial behaviour

We have already mentioned the possibility that the owner's decision over acceptance of a subordinate might depend on the behaviour of others in the population. One interesting idea is that if a subordinate does not join an owner as an ally it could join a neighbour as a rival. It could even pay an owner to suffer a net cost from sharing if the alternative was a larger cost via increased intruder pressure if the satellite joined a neighbour (Brown 1982).

Craig (in press) has used exactly this kind of argument to explain the evolution of communal breeding in the pukeko (*Porphyrio porphyrio*), a New Zealand gallinule. Pairs have greater reproductive success than groups (several males and several females on the same breeding territory) but nevertheless most males apparently decide to accept subordinate males on their territory. At first sight the birds seem to be behaving suboptimally because it would pay all individuals to try and breed as pairs. Craig suggests, however, that this is not a stable strategy because if everyone was in pairs it would pay an individual to 'cheat' and accept a subordinate male. By so doing, it would be able to exert greater intruder pressure on its neighbours and so claim a larger territory and enjoy greater reproductive success. The only way a neighbour can prevent this happening is to match the cheat with its own defence forces by accepting a subordinate male itself. Pukekos are therefore apparently caught in a Prisoner's Dilemma (Axelrod & Hamilton 1981); each individual would do best as a pair but under these conditions groups would prosper, so the only stable solution is for all birds to live in groups.

Observation showed that whenever one pair accepted a subordinate, neighbours did likewise, as if they were quickly matching the gambit before any territory loss occurred. An alternative explanation for the apparent matching could be that it is not simply the owner who decides to accept a subordinate, but rather that subordinates force themselves on to territories and settle in an ideal free manner (see Chapter 5); if a subordinate settled on one territory then it may pay the next one to settle on another territory and so on, giving the appearance of 'matching' of defence forces by the owners. Obviously, there could also be conflicts of interest between dominants and subordinates over sharing the territory.

In disputes over territory sharing or the setting up of boundaries between two neighbours, the contestants are competing for a divisible resource. Each may want a larger share than the other is prepared to grant but both may prefer to share rather than to allow negotiations to break down. In the example we analysed in Fig. 6.7 it may pay two individuals to share the territory but in order to obtain maximum benefits the owner risks defection by the satellite and the satellite runs the risk of eviction. Maynard Smith (1982a) has developed some game theory models to explore how a compromise is likely to evolve in such cases. Two main conclusions emerge.

(a) It is important to distinguish games of complete versus incomplete information. In the former, both contestants know the benefits involved and the costs that each can afford to pay in the dispute, whereas in the latter they do not. Games of incomplete information are more likely to end in breakdown even when a compromise would have paid both contestants better.

(b) There are cases where graded signals, which give honest information about the value of the resource to a contestant, can be evolutionarily stable (see Chapter 15). It is tempting to suggest that many of the elaborate displays seen in territorial disputes are concerned with bargaining for a compromise where cooperation pays but where there are also conflicts of interest.

Chapter 7
Learning and Behavioural Ecology

SARA J. SHETTLEWORTH

7.1 INTRODUCTION

The subjects of many chapters in this book are specific problems animals have to solve: what social system to have, how to choose a mate, how quickly to mature, how many children to have, and what sex to make them. One can work out what solutions to such problems ought to have evolved under given assumptions and examine the evidence that animals solve them in the predicted ways. This sort of approach can organize large amounts of data on animals as diverse as bees and monkeys, and generate novel observations and experiments.

A parallel analysis of learning would probably begin with the observation that important details of the environment are inherently unpredictable on an evolutionary time scale (Johnston & Turvey 1980; Plotkin & Odling-Smee 1979; Slobodkin & Rappoport 1974). No individual can know ahead of time exactly what its own mate, young, or territory will be like, where in its living area the most abundant food will be found, what that food will be like, or where predators are. It may be important for animals to know such things, and they may show signs of adjusting their behaviour to these sorts of details of the environment. Selection has equipped animals with learning programs for these sorts of situations (Pulliam & Dunford 1980) rather than with fixed ways of responding.

Seen in this way, the study of learning would be the study of how behaviour is fine-tuned to the details of the individual's environment and how the capacity for this fine-tuning influences fitness. It would include an analysis of when and how animals ought to learn and an examination of life history strategies to see whether they follow the predicted patterns.

This sort of behavioural ecological approach to learning has hardly begun to be developed. While zoologists have long been familiar with a wide variety of examples of learning, the systematic study of learning has traditionally been the province of experimental psychologists. These have been almost exclusively interested in analysing learning processes in an abstract, generalized way, and not in how or what their subjects learn in their natural environments. Calls for more communication between psychologists and behavioural ecologists are increasingly frequent (Johnston 1981; Kamil & Yoerg 1982;

Pulliam 1981; Rozin & Kalat 1971). Nevertheless, the central problems in the two areas remain fundamentally different. Psychologists are primarily concerned with mechanisms of learning, with *how* animals learn. Behavioural ecologists are primarily concerned with evolution and present-day function, with *why* animals learn. In principle, knowledge about the function of learning can inform studies of learning mechanisms, and vice versa, but the relationship between these two sorts of analyses is more complex than it might at first appear (Shettleworth 1983b).

For the moment, we may define learning quite broadly as any change in behaviour due to experience (Rescorla & Holland 1976; see section 7.5). Analysing an example of learning would then begin with describing the environmental events experienced by the animal, the learning paradigm in formal psychological terms, together with the behavioural change that results, the learning phenomenon. If one were concerned only with what learning is for, one could stop here and deal only in learning principles, or relationships between types of experience and types of behavioural change. For instance, if a novel event occurs repeatedly without important consequences, an animal may stop reacting to it. This statement describes habituation, which occurs in many behaviour systems in species from slugs to man. The principle makes intuitive functional sense: time and energy should not be wasted in responding to innocuous stimuli.

However, the psychological approach to learning typically goes beyond learning principles to a study of learning processes or mechanisms. Why, in the causal sense, does the response wane in habituation? Do the stimulus receptors or the muscles that support the response become fatigued, for example, or is there a more central change specific to the stimulus and response? A dishabituation experiment can differentiate between these classes of mechanisms. If a specific central change has occurred, responding will be restored when a novel but similar stimulus is presented. Such experiments are typically done with an either/or answer in mind, but in principle more than one mechanism could be involved in a single outcome, and, more importantly, different mechanisms could be at work in different examples of the same learning principle. This is to be expected because, as D. S. Lehrman put it, 'Nature selects for outcomes, not processes of development.' (Lehrman 1970, p. 28).

The possible processes underlying habituation can readily be tied to actual physiological processes. But psychologists studying learning mechanisms do not always refer so clearly to the physiological substrate. Differences among learning mechanisms are usually defined in terms of differences in behavioural outcomes to be expected from different sorts of experiences. Much of the research on learning within psychology consists of highly sophisticated experiments designed so that the possible outcomes will discriminate one possible learning mechanism from another.

Although there does not yet exist a well-defined and well-developed body of work that could be called the behavioural ecology of learning, recently there have been a number of developments in different areas that would need to be taken into account in elaborating such a discipline. Psychologists have discovered and analysed a number of 'constraints on learning' that may have functional relevance. Others studying learning mechanisms have drawn attention to the possibility that these mechanisms are adaptively specialized, and have called for more experimental analysis of naturally occurring examples of learning. These developments within psychology have led to more attention being paid to the possible importance of understanding the context within which learning mechanisms are used in nature. At the same time, the interest of some behavioural ecologists in foraging behaviour has led to experiments remarkably similar to psychologists' studies of reinforcement schedules. Foraging theory has also led to discussions of how and when animals should learn in order to forage optimally in specific situations. The next three sections of this chapter describe these developments and the final section outlines a behavioural ecological approach to learning.

7.2 CONSTRAINTS AND PREDISPOSITIONS IN ASSOCIATIVE LEARNING

7.2.1 Learned flavour aversions

When an animal is exposed to a consistent relationship between two events and its behaviour changes because of the properties of that relationship (and not, for example, just because of exposure to the events *per se*), the animal is said to have associated the two events. Much of the research done by psychologists on learning aims to analyse the conditions under which associative learning occurs, and to identify its contents and underlying mechanisms (Dickinson 1980 is a recent review). In classical, or Pavlovian, conditioning, the events to be associated are typically a relatively neutral stimulus like a tone and some biologically important stimulus like food. In instrumental (operant) conditioning, the second commonly studied type of associative learning, the predictive event is some behaviour the animal performs; its frequency may be changed by rewarding or punishing consequences.

Learning theorists have traditionally studied basic associative learning mechanisms in a few convenient species, using simple events in controlled, artificial environments. This type of research only makes sense on the assumption that the results reveal a general associative learning process, common to all situations that fit the abstractly defined associative learning paradigm. In the mid-1960s this assumption began to be re-examined. A number of different experimental results and theoretical papers stimulated this re-examination (e.g. Bolles 1970; Rozin & Kalat 1971; Seligman 1970;

Shettleworth 1972a), but the single strongest impetus was the work of Garcia and his colleagues on learned flavour aversions in rats (Garcia & Koelling 1966; Garcia *et al.* 1972).

An experiment on learned flavour aversion fits the classical conditioning paradigm: a flavoured solution, the relatively neutral stimulus, is paired with illness resulting from ingestion of poison. When tested later, a rat consumes less of the solution than it would otherwise. But illness-based aversions have two special properties. First, associations can be acquired when illness occurs hours after the flavour is sampled. This finding contrasts with results obtained with most other pairs of events, where the events must be close together in time, and at first appeared to challenge the assumption that a single general process underlies all forms of classical conditioning. Secondly, long-delay learning about illness also challenges learning theorists' traditional assumption that all events may equally be associated with all other events. For rats at least, flavours are very readily associated with illness, while other events like auditory or visual properties of food are not. In contrast, audio-visual stimuli are readily associated with immediate electric shock to the feet, while flavours are poorly associated with this sort of event (Garcia & Koelling 1966; Fig. 7.1).

Many psychologists were reluctant to accept these anomalous results as reflecting special properties of an associative mechanism underlying learned flavour aversions (Rozin 1977). Perhaps, for example, rats that have recently been poisoned ingest less of any relatively novel flavour than do unpoisoned rats. Such enhanced neophobia might function in nature to help rats avoid poisoned foods, but it is not the same as an ability to associate flavours with illness over long delays. However, this mechanism is ruled out as the sole explanation of conditioned flavour aversions by the simple observation that increasing the delay between flavour and illness decreases later avoidance of the flavour (Kalat & Rozin 1971), since rats poisoned at all delays should have equally enhanced neophobia but different strengths of association. Carefully controlled experiments were also necessary to confirm that associative effects underlie the specificity of illness-induced aversions to flavours (Domjan 1980 and in press, a).

'Biological constraints on learning', as exemplified by flavour aversion learning, quickly became a recognized problem in the field of learning. A number of examples of 'constraints' were identified, and several were thoroughly studied. An excellent basic review is provided by Roper (1983) and a more detailed one by Domjan (in press, b). In the case of conditioned flavour aversions, it seemed obvious that the special properties, or constraints, are related to the natural function of this learning for rats. Many foods can be identified by their flavours, and toxins may act slowly. Therefore, in order to avoid ingesting too much of a harmful food, animals sensitive to flavours, like rats, ought to be able to associate them with illness after a long delay (Rozin & Kalat 1971). These sorts of admittedly

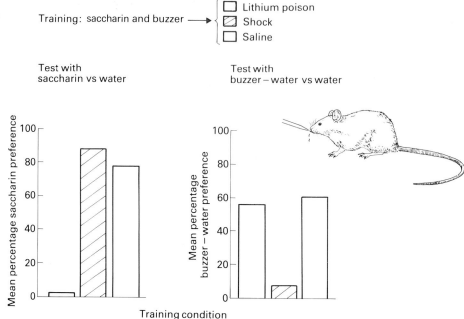

Fig. 7.1 Performance of three groups of rats exposed on a single trial to the sound of a buzzer together with an infusion of saccharin solution into the mouth and then given an injection of lithium, a shock and a saline injection, or just a saline injection. In the tests the next day they could drink from two bottles, one containing plain water and the other either containing saccharin or wired to sound the buzzer when the rats drank. The results are evidence for specificity in what pairs of events are associable, since whether the rats avoided the flavour or the sound depended on whether they had been shocked or poisoned. (From Domjan & Wilson 1972.)

post hoc adaptive considerations suggested to some (e.g. Johnston 1981) that the search for general processes of learning in the laboratory ought to be abandoned or greatly modified to take into account the natural functions of learning for each species. Of course, ethologists had believed this all along (Manning 1976).

In fact, however, most recent research on 'constraints on learning' shows that conditioned flavour aversion and other examples of 'constrained' learning display the same properties as other forms of associative learning, once some degree of specificity in what type of stimuli can enter into associations is taken into account (Dickinson 1980; Domjan 1980 and in press, b; Revusky 1977; Roper 1983). Similarly, the interval between events over which associations can be formed can be regarded as a variable that depends on the particular events and species involved. Rather than weakening it, the investigation of 'constraints on learning' has broadened and enriched the study of general mechanisms of associative learning. This enterprise is often seen nowadays (e.g. Dickinson 1980) as the study of how animals learn about causal relations among events. Both the general learning mechanisms and their specificities can be seen as predisposing animals to learn only about those relationships that reflect true causal relationships in the environment.

7.2.2 Associative learning and mimicry systems

It is an open question whether animals display the same subtlety and sophistication in associating events in the natural environment as in the laboratory, but certainly associative learning does seem to occur in nature. For example, the anti-predator defences of many plants and animals rely on potential predators' ability to associate visual or auditory stimuli with the immediate effects of attempted ingestion. This is particularly apparent in Batesian mimicry systems, which are very detailed adaptations to the selective pressures exerted by predators that can learn. Many insects that are unpalatable to predators are brightly coloured. The black and yellow stripes of bees and wasps are an example of this sort of aposematic coloration. In the laboratory, naive predators attack such prey at first but rapidly learn to avoid them. Concomitantly, they avoid palatable species of similar appearance, the Batesian mimics. This generalization of associative learning is assumed to be responsible for the evolution of mimicry systems in which unrelated species living in the same area may have strikingly similar appearance, behaviour and sounds (review in Rettenmeyer 1970).

For a palatable mimic to benefit from its resemblance to models enough to offset the disadvantages of being conspicuous, its life history must be such that predators can learn about models before encountering their mimics. Models too benefit from being available to predators before these learn to eat mimics, since predators experienced in eating harmless mimics take longer to learn to avoid models, harming or killing more individuals while doing so (Shettleworth 1972c). Even a relatively low proportion of very unpalatable models confers some protection on mimics, especially when alternative food is available (J. Brower 1960; Rettenmeyer 1970). There should be a trade-off among unpalatability of models, resemblance of models to mimics, their relative numbers, and other factors (Arnold 1978).

Models and mimics need not both be available to predators in the same season. In one study in Illinois, wasps and bees were most numerous in mid-summer, when newly fledged birds would be learning what to feed on. Their syrphid fly mimics appeared mainly in late summer and early spring, when most of the predators would have learned to avoid them (Waldbauer & Sheldon 1971). Since models may be killed while predators are learning, conspicuous coloration will probably evolve only where families of models remain close enough together for a predator that has learned by killing one model to avoid its relatives, which thereby benefit (Harvey *et al.* 1982; but see Järvi *et al.* 1981).

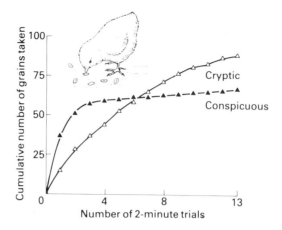

Fig. 7.2 Mean cumulative numbers of distasteful cryptic or conspicuous grains taken by domestic chicks tested for 2 min every hour on one kind of grain. Data are shown for blue or green grains on a blue background but similar results were obtained with a green background. The chicks learn sooner to avoid taking conspicuous grains. This may be because conspicuous grains are taken faster at first, providing a more negatively reinforcing experience, or because they are associated more readily with a bad taste. Both mechanisms could function in nature to favour conspicuous coloration in unpalatable prey (Gittleman & Harvey 1980), but experiments to distinguish them have not yet been done.

In some species, rapid association of a visual stimulus with a nasty taste is facilitated by a specific predisposition to associate sights and tastes of objects (Shettleworth 1972b). Domestic chicks show no sign of learning when sounds signal the presence of quinine-flavoured water, while they learn in a few trials to avoid drinking when visual stimuli are the signals. In contrast, in a fear conditioning experiment, they learn faster with the same sounds than with the visual stimuli. In conventional learning experiments, more salient stimuli enter into associations more rapidly (Kamin 1969), and Gittleman and Harvey (1980) showed that domestic chicks learn faster to stop pecking nasty-tasting grains if these are conspicuous than if they are cryptic (Fig. 7.2). Thus predators' general associative learning mechanisms may have selected for the conspicuous coloration of unpalatable insects. Conversely, formal representations of model–mimic systems (e.g. Bobisud & Potratz 1976) suggest that the existence of unpalatable prey exerts a pressure on predators to learn quickly to avoid them.

7.3 SPECIES-SPECIFIC AND ADAPTIVE SPECIALIZATIONS OF LEARNING

7.3.1 Non-associative learning: imprinting and song learning

The foregoing section contains examples of how animals may use the sorts of associative learning mechanisms discovered in the laboratory to solve problems they encounter in the field. Naturalistic studies of development are better known, however, for showing how experience can affect behaviour in ways that are not readily analysed as instances of associative learning (Johnston 1981). Imprinting and song learning are two of the best-known examples. The most obvious description of filial imprinting is that simply because of being exposed to a certain type of object a young chick or duckling develops a preference for its company and comes to direct filial behaviour toward it (Bateson 1979). Learning theorists have made adroit attempts to show how imprinting really is a form of associative learning, by analysing the imprinting object into two or more separate stimuli (Hoffman & Ratner 1973). However, besides seeming rather forced, these approaches are not completely successful in accounting for the experimental data (Eiserer 1980; Shettleworth 1983b).

The first phase of song learning in many passerines is an even clearer example of learning through simple exposure. Just hearing a species-typical song at the right age may be sufficient to make the bird begin singing it months later (Marler & Peters 1982). In perhaps a similar way, exposure at an early age to the star patterns in the night sky influences migratory orientation of indigo buntings (*Passerina cyanea*) and other birds when they reach adulthood (Emlen 1970; Keeton 1981). In such cases the animal seems to be pre-programmed to attend to certain events at certain times independently of the relationships of these to other events (McFarland & Houston 1981).

7.3.2   Species-specificity in song learning and imprinting

Song learning and imprinting are particularly good examples of species-specificity in learning. Not only do they seem to exemplify a special kind of learning, shown by only a few species, but also they illustrate particularly well how the specific stimuli that are learned, and when and how they have an influence, differ among species (Bateson 1979; Kroodsma 1981). Some of these differences make immediate functional sense. For example, in song learning the young bird must be sensitive to song at a time when adults of its species are singing, and it must be able to identify the appropriate song to learn. This may be done through selective sensitivity to species-specific features of song (the 'auditory template') (Konishi & Nottebohm 1969) or through the bird learning the songs of singers with certain plumages or certain social relationships to itself. Reasonable as this may seem in functional terms, in fact the adaptive significance of the details of the song-learning process is not yet well understood (McGregor & Krebs 1982b).

Given that what is learned in imprinting and song learning is something about a species-specific characteristic, one may ask, 'Why learn at all?'. On the other hand, why *not* learn if the normal circumstances of development invariably provide the opportunity? Learning is often assumed to be costly in terms of neural or genetic material, but too little is known about the neural basis of learning, let alone its genetic basis, for this to be a necessary assumption (Johnston 1982).

Bateson (1979) has recently suggested one possible function for the learning in sexual imprinting. Through sexual imprinting the young bird learns the characteristics of its close kin (its siblings) and later it can mate with optimally discrepant others, thereby avoiding too much in- or outbreeding. Cross-species comparisons of the sensitive period for sexual imprinting show that it occurs when young birds are still in the family group but are acquiring adult plumage. Furthermore, sexually mature Japanese quail (*Coturnix coturnix*) choose to be near cousins rather than siblings or unrelated birds (Bateson 1982). McGregor and Krebs (1982b) state that female great tits' (*Parus major*) learning of their fathers' songs may function similarly to facilitate an optimal degree of outbreeding (see also Chapter 9).

The special features of song learning and imprinting, among other examples, show that learning ability cannot be treated as a single characteristic that varies in a unidimensional way across species (Hodos 1982). Some processes, like associative learning and habituation, may be fairly general across species, although they may not be equally accessible to all events the members of a species can perceive. Other kinds of learning are peculiar to only a few species. These very general observations have suggested to some that animal 'intelligence' is best viewed as clusters of adaptive specializations (Rozin & Kalat 1971; Rozin 1976). The kinds of learning mechanisms a species has and the types of events that have access to these mechanisms are supposed, on this view, to be especially suited to the problems that species has to solve by learning.

This view is well supported by cases like song learning, in which animals have isolated learning abilities that are qualitatively different or quantitatively more spectacular than those of unrelated species. Much of behavioural ecology, however, deals with differences in closely related species facing different selection pressures, or convergence in unrelated species facing similar pressures, and here the evidence for adaptive specializations of ·learning is meagre. Psychologists have provided very little such evidence, since traditional comparative studies of learning have compared very different species, like turtles, goldfish, rats and pigeons (Bitterman 1975). While field studies have provided some suggestive evidence, ethologists have generally not tried to separate learning ability *per se* from other sorts of factors that might be responsible for species differences in behaviour in a situation requiring learning. Of course this kind of distinction is not necessary in studies of behaviour in a functional context, but failure to make it can lead to unsupported conclusions about species differences in learning, as illustrated in the next section.

7.3.3   Species-specificity in nest and egg recognition by gulls and terns

All animals that nurture their young after they leave the mother's body should ensure that they care for their own offspring and not someone else's. This problem can be solved by parents and/or offspring learning to be attracted to the other partner in the relationship and/or learning to recognize a home site to which both are attracted. Among gulls and terns (*Laridae*) there is a great deal of interspecific variation in how this problem is solved, and the different solutions seem to suit the different situations in which the species nest. For instance, royal terns (*Sterna maxima*) make rudimentary nests in dense colonies. Therefore nest recognition is not easy; however, the terns make egg recognition easy for themselves by producing variably patterned eggs. In contrast, herring gulls (*Larus argentatus*), which build more elaborate nests, more widely spaced, recognize their own nests but not their own eggs, and recognize their chicks by the time they are old enough to wander (Tinbergen 1953). Kittiwakes (*Rissa tridactyla*), which nest on small ledges on cliffs, recognize their nest sites but neither their eggs nor their sedentary chicks (Cullen 1957).

These findings and others like them show that there are species-specific ways of solving the problem of caring for only your own offspring, but they do not necessarily illustrate the species-specificity in learning ability that has been attributed to them (Manning 1976; Shettleworth 1972a). A rigorous demonstration of differences in learning ability requires controlling for possible differences in salience of the stimuli and motivation of the subjects. For instance, attributing differences in ease of recognition to differences in learning ability implies that there are no differences across species in the sorts of objects to be recognized. But this is not easy to assess. People would have an easier time recognizing individual eggs from one gull species than those from another, but since this need not be true of the gulls, one cannot tell whether the species that learn to recognize their own eggs do so just because the eggs are more easily discriminated from one another. Cross-species equivalence in salience or ease of discrimination would have to apply to all aspects of the learning situation. For instance, if the nest is easily learned about, this could prevent learning about the eggs that are added later on, a phenomenon known in associative learning paradigms as 'blocking' (Kamin 1969).

Finally, it is necessary to test for learning in a way that discriminates among different reasons why a gull might reject or accept test stimulus objects. Buckley and Buckley (1972) exchanged eggs between neighbouring royal tern nests and found that terns usually chose a neighbour's nest containing their own eggs, showing they did recognize their eggs. However, choice of the eggs did not mean that the birds failed to recognize their own nests, since they acted as though disturbed when brooding on a strange nest.

The point of this discussion is not that there is no species-specificity in learning ability here. The different and ecologically appropriate ways in which different gulls and terns ensure that their parental investment is directed appropriately need not imply any differences in learning abilities, however, and it is not an easy matter to discover whether they do. The entire structure of the situation, much of it created directly and indirectly by the birds, probably conspires with learning ability and perceptual and motivational factors to produce a functionally appropriate outcome. Here, as in other aspects of behavioural development, there may be alternative pathways to a similar end (Johnston 1982).

7.3.4 Memory in food-storing birds: an adaptive specialization?

Despite the difficulties in isolating species differences, or specializations, in learning ability, there is abundant evidence that adaptive differences in what is learned do exist, as illustrated by the species-specificities in what songs birds will learn (section 7.3.1). The specificities in associative learning described in section 7.2.1 are also suggestive of adaptive specializations in what is learned, but they have not yet been closely tied to the requirements for learning in the species involved. Doing this is likely to be no less complicated than the case outlined in the preceding section, as illustrated by a recent attempt to find species differences in the ease with which generalist versus specialist kangaroo rats learn food aversions (Daly *et al.* 1982).

Perhaps the best example of how closely-related species may differ in learning ability in a way that is intimately related to their ecology and life style is in the spatial memory of food-storing birds. A number of birds, particularly tits (*Parus* spp.; Cowie *et al.* 1981; Sherry *et al.* 1981) and corvids (Bossema 1979; Vander Wall & Balda 1981), store large numbers of food items in scattered locations and recover them days or even months later. Four corvids from the American Southwest show particularly striking variations in morphological and behavioural specialization for storing pine seeds over the winter (Vander Wall & Balda 1981). Clark's nutcracker (*Nucifraga columbiana*) has a special structure for carrying seeds to storage sites; it stores thousands of caches in areas that are relatively free of snow in the winter, and it breeds very early, feeding its young on seeds stored months earlier. In contrast, the scrub jay (*Aphelocoma coerulescens*), the least specialized species, has no morphological specializations for harvesting and storing seeds, and makes relatively few caches, placing them in its territory.

Although the experimental study of memory in these birds is just beginning (Vander Wall 1982), it would be surprising if all the other specializations for food storage in birds like Clark's nutcracker were not accompanied by an unusually long-lasting and large-capacity spatial memory. This could be tested by comparing the different species in laboratory tests of recovery of stored food (Shettleworth 1983a). However, just as in the examples in the last section, it is also likely that birds relying on stored food have a number of strategies in addition to memory, like preferences for certain sites, that facilitate their recovering caches in natural conditions, and the most specialized hoarders may have developed these to a high degree.

In storing food and directing parental investment, an animal potentially can control some features of the situation to make learning easier or even unnecessary for itself. A gull could perhaps evolve to produce more individually distinctive eggs, or young that learn to recognize its call or its nest or do not wander. Similarly, a marsh tit could select only certain sorts of storage sites or search its territory systematically for stored food. For the gull's problem, any of several solutions might be equally good, though its physiology, nesting ecology, or perhaps general learning mechanisms must impose some constraints. In food storage, however, there is one best strategy— memory for individual sites. Any systematic pattern of storage could be learned by cheats preying on the stores, and random search would not favour the hoarder. In either case, there would be no selective advantage in investing time and energy in storing food (Andersson & Krebs 1978). Thus the two types of problem differ in the degree to which they favour a solution involving learning. Arguments like these could perhaps be made more rigorous in future developments of a behavioural ecology of learning.

7.4 LEARNING AND FORAGING THEORY

7.4.1 Introduction

Most of the investigations of learning described in the preceding sections start from observations of behaviour assumed to involve learning and proceed to analyse the mechanisms underlying that learning. While the results of such investigations could be relevant to understanding how learning is used to solve problems in nature, this has not been their main emphasis. Optimal foraging theory (see Chapter 4) provides something much closer to the kind of behavioural ecology of learning outlined in section 7.1, but for only one aspect of behaviour. Its starting point is a formalization of the things animals need to 'know' to maximize rate of net energy intake while foraging and ultimately, it is assumed, to maximize fitness. Since these include details of food availability in the individual's environment, optimal foraging may require learning of some kind.

From this starting point, optimal foraging theory has developed in two complementary directions. On the one hand, there have been numerous investigations of whether animals actually follow the predictions of foraging models. Such studies have revealed something about the sorts of mechanisms animals use to solve foraging problems. On the other hand, theoreticians have explored the performance of simple decision rules that animals might use. Such models make assumptions about what animals can learn or lead to conclusions about what they need to learn, in given conditions. Often, of course, these two types of investigations have been combined, as in the work of Krebs *et al.* (1978); see section 7.4.4.

In the past few years a number of authors have discussed the interaction between learning psychologists and behavioural ecologists that has been created by developments in optimal foraging (Houston 1980; Kamil & Sargent 1981; Kamil & Yoerg 1982; Lea 1981, 1982; Staddon 1980). As Lea (1981) has argued, the processes underlying instrumental behaviour on schedules of food reinforcement in psychology laboratories are probably the same as those used in foraging. With this in mind, learning psychologists and behavioural ecologists are beginning to raise new questions for each other as well as to help each other answer old ones. We can expect exciting developments in this area in the next few years. Here I will simply survey the sorts of issues about learning that are raised by optimal foraging theory. The discussion is organized around the four basic foraging problems that have been most extensively analysed (Pyke *et al.* 1977).

—How should the optimal forager move in a patch?
—What items should it eat?
—What patches should it choose?
—When should it leave a patch?

Krebs *et al.* (1983) have partitioned foraging problems potentially requiring learning in terms of three types of information a predator needs: information about the nature of prey items, information about the quality of the patch it is in, and information about the overall quality of the environment. Development of this approach would probably lead to a way of looking at the learning and discrimination mechanisms in foraging somewhat different to that presented here, which follows more established lines (e.g. Lea 1981; Pyke *et al.* 1977; Staddon 1980).

7.4.2 Movement rules and spatial memory

Animals that remove prey from identifiable locations as they forage should be able to avoid revisiting those locations until food there has had time to be replenished. The solutions different species have evolved to this sub-problem of 'how to move' illustrate a wide range of cognitive complexity. Bumblebees (*Bombus appositus*) use one of the simplest sorts of solutions to avoid revisiting flowers on a single inflorescence (Pyke 1979). They have a 'non-cognitive' movement rule, roughly, 'start at the bottom flower and move successively to the nearest flower not yet visited' (Cheverton 1983). (See also section 4.5 for discussion of rules of thumb.)

The amakihi (*Loxops virens*), like other nectar-feeding birds, also faces the problem of avoiding revisits to depleted flowers. The amakihi appears to solve it by remembering which flowers it has depleted and avoiding revisits for an hour or so, long enough for nectar to be replenished (Kamil 1978). Unlike bees (Pyke 1981), amakihis seem to have no systematic movement pattern around their territories, nor do they appear to mark visited flowers, since other amakihis readily visit these.

That animals are indeed capable of remembering large numbers of locations has been shown with rats searching for food (Olton & Samuelson 1976). A rat is placed in a maze with eight or more arms radiating out from a central platform and one piece of food at the end of each arm. Rats quickly learn to collect all the pieces of food in close to the minimum possible number of trips, one per arm. Experiments controlling for hypothetical odour trails and other possible aids to performance have shown that rats do this by remembering which arms they have visited most recently. In fact they can keep track of quite a large number of spatial locations over periods of several hours. In one experiment (Roberts 1981) rats were allowed to collect four of the eight items in one maze, removed and given four choices in each of one or two other eight-arm mazes, and then allowed to search for the remaining food on the first maze. They performed as well as rats with no intervening experience, although three intervening mazes did impair performance significantly. Marsh tits searching for hoarded food also seem to have a large capacity for remembering locations recently visited and avoiding searching the same place twice (Shettleworth & Krebs 1982).

Comparisons of the degree to which nectar-feeding birds and bees use simple movement rules or perhaps marking systems rather than spatial memory in solving spatial problems might suggest what conditions favour evolution of particular types of solutions to these problems. Some conditions that may influence the benefits from learning the exact details of food distribution are: how likely feeding is to be disrupted; what is to be gained by going back to the 'correct' place; how far apart food sources are; how long before a depleted source recovers; and whether this is the same for different foods. The relative ease with which energetic costs and benefits of different foraging decisions can be calculated for nectar feeders (cf. Pyke 1981) makes these sorts of developments a distinct possibility.

In such comparisons of widely differing species, it cannot be assumed that phylogeny predicts what kind of mechanisms are available to a given species. Vertebrates, for instance, do not invariably solve problems in a complex, 'cognitive' way. In winter, pied wagtails (*Motacilla alba*) have territories along riverbanks. Food is continuously washed up along the bank, and the best strategy for exploiting it is to patrol along the edge of the territory, keeping out intruders while periodically revisiting each stretch of the bank (Davies &

Houston 1981). This could be done by 'mindlessly' walking around and around the territory, and apparently this is what the wagtails do. When a wagtail leaves off foraging to chase intruders, it does not always resume feeding at the place it last fed (Davies & Houston 1981). This does not prevent it, however, from having nearly optimal return times on average (see Chapter 6).

7.4.3   Choosing the optimal diet: rules of thumb and reinforcement schedules

In order to select prey items optimally, animals should rank prey types according to profitability (energy yield per unit handling time, or E/h) and should be sensitive to the overall abundance of all but the worst prey types. When the abundance of profitable prey is low they should generalize; as it increases they should specialize (see Chapter 4). In general, predators do roughly this (Pyke *et al.* 1977; Lea 1981; Chapter 4) but they do not need complex learning rules or micro-computers to do so. Under some conditions both profitability and abundance can be assessed by simple rules of thumb. Big prey items may be preferred to small ones, for example. Within the size range where single items can be snapped up instantly, like grains by a hungry pigeon, size may be a good guide to profitability. However, in artificial conditions, animals with these sorts of rules, such as shrews, can be fooled into taking larger but less profitable prey (Barnard & Brown 1981). Other species are more sensitive to actual variations in E/h (Lea 1981).

Assessing prey abundance similarly evokes images of complex cognitive mechanisms, but it too can be done by simple rules of thumb, as vividly demonstrated by Charnov's (1976) analysis of the effect of time since last feeding on the readiness of a mantid to attack houseflies. The fuller the mantid's gut (i.e. the greater its 'estimate' of overall energy availability in the environment), the nearer flies had to be before it would strike (i.e. the larger E/h, since nearer flies have smaller h). In this case diet choice is achieved by a simple reflex mechanism in which responsiveness to visual stimuli is modulated by gut fullness.

The shore crab (*Carcinus maenas*) also uses a simple motivational mechanism for diet selection, but it has shorter-term fluctuations (Elner & Hughes 1978). The crab picks up mussels as it encounters them and after a 'recognition time' either accepts them or drops them uneaten. Large mussels are preferred to small ones. After a crab eats a large mussel it is likely to reject the first one or two small ones it encounters, but if it does not encounter another large mussel it is quite likely to take the third small one (Fig. 7.3). This pattern of prey selection suggests a simple threshold mechanism like that developed by Waage (1979b) for patch selection by parasitoid wasps. Each encounter with a large mussel raises (or maintains at a ceiling) the size

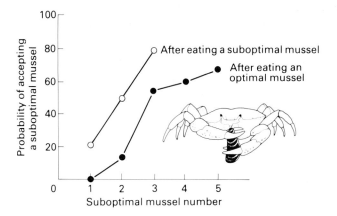

Fig. 7.3 Probability of shore crabs accepting the first, second, etc., suboptimal mussel encountered after eating either a mussel of the optimal size class available in the test, or one from the suboptimal class. Prey selection reflects short-term changes in a threshold for acceptance. (Data from Elner & Hughes 1978.)

threshold for accepting the next mussel, but the threshold decays rapidly to the level where small mussels are accepted. Elner and Hughes suggest that such short-term fluctuations are appropriate for a predator which cannot assess prey abundance visually from a distance and a prey whose abundance varies unpredictably even within small areas.

These examples show how a problem requiring integration of several items of information about local conditions can be solved without any long-term learning about prey abundance. However, simple mechanisms like those used by the crab and mantid may be inappropriate for dealing with prey with different sorts of distributions. If prey distributions are relatively stable, governing behaviour by learned expectations of future reward may be more appropriate than being very responsive to local fluctuations. Noting that the variable of food (energy) per unit time foraging is the same one whose effect is studied using schedules of food reinforcement in the laboratory, Lea (1979) devised an experimental situation that simulates the optimal diet problem with two prey types for pigeons in a Skinner box (Fig. 7.4; see also Abarca & Fantino 1982; Collier & Rovee-Collier 1981; Snyderman 1983). Behaviour in this situation is similar to that obtained with real prey (e.g. Krebs *et al.* 1977): as the item that is better in E/h terms becomes less abundant, the less profitable item is included in the diet, but in a gradual rather than an all-or-none manner (section 4.2; Fig. 4.1).

One basic assumption of Lea's simulation is that decreasing E/h can be simulated by increasing delay to a fixed reward. Although one may question whether this is a reasonable equation for pigeons, it has led to the explicit suggestion (Abarca & Fantino 1982) that optimal

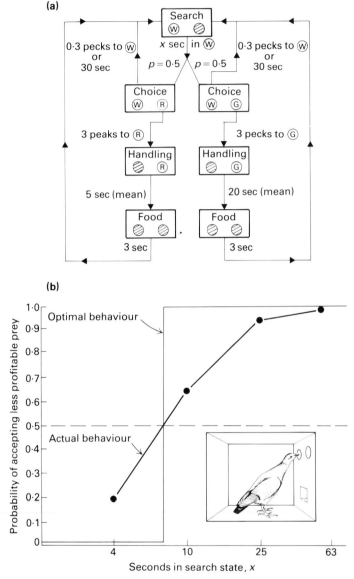

(a)

(b)

Fig. 7.4 **(a)** Flowchart of the simulation of a prey selection problem used by Abarca and Fantino (1982), modelled after that devised by Lea (1979). Rectangles represent different states of the schedule, during which the pecking keys (circles) could be red (R), white (W), green (G), or dark (hatched). The requirements for transition from one state to the next are indicated by the legends. An additional requirement was that to enter the choice state or to obtain food the pigeon had to peck at least once after the indicated time had elapsed. The lighting of the white key signified the search state. Red and green lights on the second key were correlated with different delays before 5 sec access to grain could be obtained. The delays are formally equivalent to handling times. In the example given, the red 'prey' is better in E/h terms.

   (b) Results from a group of six pigeons tested in the procedure shown in a when x, the duration of the search state, was varied. They are typical of results obtained in laboratory tests of diet selection (see also Fig. 4.1).

diet is attained by the same mechanism that governs choice in other reinforcement schedules: choose the alternative associated with the greatest average decrease in time to reward. When the reward involving short delay ('handling time') occurs often enough, food can be expected sooner if the animal continues searching than if it chooses the reward with long handling time (see Fig. 7.4).

At least some animals can discriminate relative delays to reward quite accurately. In Gibbon and Church's (1981) 'time left' paradigm, rats or pigeons work for a food reward which is always given a fixed time after a lever or pecking key is made available, say 60 seconds. At various times during this interval they may be presented with a second lever to press or key to peck. Reward can be obtained after a shorter interval, say 30 seconds, by responding on the second lever or key. In the example given, the optimal behaviour in E/h terms is to choose the shorter interval in the first 30 seconds and stick with the longer one when the choice is presented later on. Animals do tend to switch their preference in the middle of the interval, but not in an all-or-nothing manner. Their behaviour can be generated by a model which assumes that animals compare remembered times to reward and that these have a variance proportional to their mean. A similar approach, in which some variance in the system is assumed so that all-or-nothing choice is virtually impossible, is beginning to prove fruitful in analyses of actual diet choices. A signal detection model with intrinsic variance in perception of each prey type can account for 'errors' in prey selection by great tits (Rechten *et al.* 1983). If such a system can perform close enough to optimally in natural conditions, a more accurate one might not be expected to evolve (Orians 1981).

The simple form of the optimal diet equation assumes that animals know the abundances and profitabilities of prey items. Unlike the case of patch choice discussed in the next section, few models have been developed to deal with how animals should acquire this information. However, some theoretical developments have shown how, with fixed prey types and abundances, it may pay a predator to learn various things. When time is required to recognize prey items, animals can improve their rate of energy gain by reducing recognition time through learning, as by developing a search image, or they can learn to handle prey more efficiently (Hughes 1979). Implicit in Hughes's model is a demonstration of the selection pressures for certain well-known kinds of learning. The presence of such learning can alter the pattern of prey selection from that predicted by the simple optimal diet model (McNair 1981).

7.4.4   Patch choice and optimal learning rules

A troop of monkeys searching through the jungle for fruit must learn which trees have how much fruit and how far apart they are. They must decide which trees to visit and, while foraging on any one, they must keep track of how its yield declines and decide when to leave and head for another tree. It is not yet entirely clear how predators facing problems of patch choice like these should (McNair 1982) or do (Cowie & Krebs 1979) solve them. Nevertheless, analyses of the various problems involved in patch choice are rich in material for a behavioural ecology of learning.

Learning about new patches

The first discussions of optimal patch choice assumed an omniscient predator which knows the abundance of food in all patches in its environment (see Chapter 4). Since such knowledge is obviously impossible in all but the most stable environments, models have been developed to deal explicitly with optimal ways of learning or of using information available over the short term. One example is Krebs, Kacelnik and Taylor's (1978) analysis of how animals should behave when faced with two patches of unknown density. They suggest that, in the manner of a Bayesian decision maker, the predator should sample both patches equally and then fixate on the one it estimates to be the better patch. The longer the 'time horizon', the total time available for feeding, the longer should be spent in sampling before the decision to exploit one patch is made. The optimal amount of sampling is also greater with smaller differences between patches (see see Fig. 4.4).

Krebs, Kacelnik and Taylor tested their model with great tits (*Parus major*) working at two feeders that provided food on variable ratio schedules (i.e. an average of one mealworm piece for *n* hops on a perch). Their report gives the impression that the birds did switch abruptly from equal sampling of the two ratios to exploiting the better, but detailed data are not presented, and in fact the change was gradual (Krebs, pers. comm.). The preferences of pigeons in a similar experiment also changed gradually rather than abruptly (Dow & Lea 1979, quoted in Lea 1981).

The experimental arrangement in these studies would be termed a concurrent schedule by psychologists: two reinforcement regimes are available at once and the animal is continuously choosing between them. However, there are few data comparable to those obtained by Krebs, Kacelnik and Taylor, since until recently conventional concurrent schedule studies have focused on behaviour after long exposures to schedules. An exception is the work of Herrnstein and Vaughan (1980). They suggest that animals continuously change their behaviour according to a 'melioration' rule: switch toward the alternative that has the higher local rate of reward (reinforcements per unit time in the patch). Harley (1981) and Lester (in press) have developed detailed models which show how animals do something like this (see also section 4.2.3). In Lester's model, time is distributed among patches in proportion to estimates of food availability in the patches.

The estimate of food availability is a weighted average of recent and past rates of obtaining food (see Killeen 1981):

$$\hat{r}_i(n) = (1 - b)R_i(n) + b\hat{r}_i(n - 1)$$

where $R_i(n)$ is reward in patch i in the nth time bin and the $\hat{r}$'s are estimates of reward rates. The parameter b is a measure of the 'memory window' (Cowie 1977), the weight given to recent experience in relation to past experience. Such a model predicts a gradual shift to the better of two patches. It also predicts quite well the behaviour of goldfish and pigeons foraging in and depleting two unknown patches. At first they gradually increase the proportion of time they spend in the richer patch, but once their feeding depletes it below the level of the other patch they begin to shift toward the alternative patch, which is now richer (Lester, in press; Lester, Beeby & Shettleworth, in preparation).

The line of research begun by Krebs, Kacelnik and Taylor is important because it makes specific predictions about situations requiring learning: animals should sample, and there may be an optimal rate or amount of sampling that depends on the amount of variability in the environment generally and on the time horizon. Although the notion that animals increase fitness by persistent sampling even to the extent of sacrificing short-term gain (Orians 1981; Pyke *et al.* 1977) pervades discussions of foraging, neither sampling nor time horizon has been much considered in psychological studies. At present more detailed data are needed on the roles of 'sampling' versus incremental reinforcement mechanisms in adjustments to unknown patches and more theoretical work needs to be done on the relative payoffs from different possible strategies. Krebs *et al.* (1983) have discussed how theoretical work can take into account the value of discriminating among different patch or prey types.

Allocation of time to patches

In laboratory studies with two 'patches' offering reward on variable interval schedules, animals match the proportion of time spent in a patch to the proportion of all rewards received in that patch (Herrnstein & Vaughan 1980). For instance, if one schedule provides reward twice as often as the other, two-thirds of the animal's time will be spent there. This behaviour implies that time is allocated so as to equalize rewards gained per unit time in the different patches. Matching provides a higher reward rate than the exclusive preference for variable ratio schedules described in the last section, because with interval schedules reward on both schedules becomes more probable as a function of time, whichever schedule the animal is working on. The optimal policy is to spend most of the time on the schedule offering more frequent rewards and make frequent brief visits to the other schedule to collect any rewards that have 'set up' (Houston & McNamara 1981). With most schedule values usually studied, matching is close to optimal, however. Variable interval schedules have at least one natural analogue, in the accumulation of food on the banks of the Thames described by Davies and Houston (1981) in the study of wagtails discussed in section 7.4.2. Staddon (1980) has suggested that concurrent variable interval schedules are like two patches subject to depletion, because as time passes in one patch the relative benefit of the alternative patch increases. However, it does not follow that animals necessarily use the same mechanisms to solve the two problems.

Both matching and optimization refer to the overall, or in psychologists' terms, molar, outcomes of behaviour. A separate question is what short-term, or molecular, mechanisms produce this outcome. In the case of pigeons on reinforcement schedules it is apparently not a mechanism sensitive to overall reinforcement rates. On complex schedules designed so that the most food is obtained by not matching, pigeons still match (Herrnstein & Vaughan 1983 ; Mazur 1981). They apparently use a rule of thumb like those described in the last section. Presumably the rules of thumb revealed in laboratory studies do yield close to optimal outcomes in natural foraging conditions for the species under study (Houston & McNamara 1981; McNamara & Houston 1982; Staddon 1980).

The preference for matching over maximizing means that the most strongly reinforcing events are not necessarily those correlated with the greatest rate of net energy intake. Nowhere is this more vividly illustrated than in studies of 'self-control' in animals (Green & Snyderman 1980). Pigeons, at least, are very loath to choose a condition in which scheduled rewards are delayed for a few seconds, even if in E/h terms that condition is better than its alternative. The pigeon's behaviour reflects something like our rule of thumb 'a bird in the hand is worth two in the bush'. Perhaps animals that often deal with delayed reward in the form of prey items requiring long pursuit or handling times would behave differently.

When to leave a patch

The time allocation problem in concurrent schedules is peculiar because the animal is thoroughly familiar with the 'patches' and they remain the same over time. As envisioned in optimal foraging theory, however, the problem is how to allocate time among a number of patches, each of which can be depleted by the forager (Pyke *et al.* 1977; see Chapter 4). McNair (1982) has recently pointed out that the original solution to this problem, the marginal value theorem (Krebs *et al.* 1974), implicitly assumes that the predator knows in advance how much time will be spent in each patch. Only under this assumption can the overall rate of intake in the habitat be compared with the current rate. Clearly it is more reasonable to assume that a predator is comparing current intake rate with the intake rate in the recent past, over some sliding 'memory window' (Cowie 1977; Cowie & Krebs 1979). Several models have shown how animals might do this (Green 1980; Ollason 1980; McNair 1982). These models are of interest in the present context because they are formal investigations of how animals should learn about patch quality in order to make close to optimal decisions.

A simple example is Green's (1980) discussion of what animals should do when foraging in patches with prey in discrete locations that can be searched systematically. In this case the probability of finding a prey item in the next bit remains constant. The forager should adopt a criterion for the number of prey items required in the first t bits searched, and leave if it finds fewer. The criterion depends of course on the travel time between patches and the overall quality of the environment, though Green does not describe in detail how the latter should be assessed. Nevertheless, this model constitutes an hypothesis about what mechanisms an animal foraging in these types of situation uses. It should be able to count food items, with no memory loss for more distant items, at least within the bounds of the criterion search time. This contrasts with the decaying influence of past food items assumed by Lester (in press) and Ollason (1980) among others, and with the giving-up time mechanism suggested by Krebs, Ryan and Charnov (1974). Models like Lester's and Ollason's are most plausible psychologically, but in any case the intake rates generated by different decision rules can be compared quantitatively for different habitats. That different decision rules are best under different conditions (Iwasa *et al.* 1981; McNair 1982) suggests comparative studies of what mechanisms different sorts of species actually use.

7.5 TOWARDS A BEHAVIOURAL ECOLOGY OF LEARNING

Having reviewed the varied approaches to studying learning in an ecological context, I conclude with some suggestions for what a behavioural ecology of learning as discussed in section 7.1, parallel with other areas of behavioural ecology, would be like. What phenomena would it deal with and what questions about them would it try to answer?

7.5.1 Memory, motivation, learning and development

A behavioural ecology of learning would deal with the ways in which individuals adjust their behaviour to local conditions in response to experience. It would therefore cut across other functional categories of behaviour like feeding, defence, and mate selection. It would also cut across mechanisms of behavioural change and would include many kinds of behavioural change not usually regarded as 'learning'.

Rescorla and Holland (1976) classify kinds of learning in terms of the structure of the experiences that change behaviour, the learning paradigms. The durability of any effect of experience is not, according to them, a useful criterion for distinguishing 'learning' from 'non-learning'. This caution certainly applies to any attempt to develop a behavioural ecology of learning. The present chapter includes examples ranging from short-lived and generalized effects of eating one mussel on accepting the next (section 7.4.3) to long-lived and highly specific effects of eating an unpalatable insect on attacking others (section 7.2.2). Being associative, the latter would be classified as learning, but the former would usually be regarded as an effect on short-term memory or motivation. As these examples suggest, the criteria for classifying effects of experience implicitly include how long the effect lasts and how specific it is (Johnston & Turvey 1980). An analysis of learning mechanisms might separate phenomena differing very markedly in specificity or durability, but a behavioural ecological approach classifies together different kinds of solutions to the same functional problem, such as different strategies for deciding when to leave a patch (section 7.4.4).

Although any definition that restricts 'learning' to long-lasting associative changes may be too narrow, a definition of learning as adjustments to local conditions on the basis of past experience may be too general. The stipulation that the adjustment be on the basis of past experience is intended to distinguish learning from problems of perception or moment-to-moment control of action, but clearly the lines between any of these sorts of categories are not easy to draw (Johnston & Turvey 1980). In spite of being very broad, a definition of learning as a solution to certain sorts of ecological problems may exclude some effects of experience on development, for example the way in which prehatch auditory exposure influences posthatch auditory preferences (Gottlieb 1975). Johnston (1981) advocates a different sort of 'ecological' approach from that outlined here, and this includes a wider range of 'developmental' effects. He takes as a starting point a description of all the ways in which experience affects behaviour in the natural lives of animals. However, an approach to learning modelled on other areas of behavioural ecology promises a more satisfactory organizational framework in terms of some obvious functional questions about 'learning'.

7.5.2 Questions about learning

Why learn?

'Learning' may be more or less required by different species to solve a given functional problem, like locating offspring (section 7.3.3). Other situations, such as food storage, may absolutely require learning of some kind (section 7.3.4). And there are some very well studied examples of learning, such as song learning and imprinting, of which the function is not yet well understood (section 7.3.2).

193

Learning

What to learn?

Given that learning is required, what aspect(s) of the environment should the animal learn about? Should you recognize your chick by its voice or by its plumage? Should food availability be identified by gut fullness or in some more cognitive way (section 7.4.3)? Some of the aspects of optimal foraging theory discussed in section 7.4 provide examples of formal predictions about what aspects of the environment animals should be sensitive to if they are to solve a specific type of problem optimally, or adequately close to optimally, within some realistic constraints on their information-gathering capacities. The information-gathering capacities themselves may of course have been shaped by the specific requirements of the species (Orians 1981).

How to learn?

The best example of a formal attack on the problem of how to learn is the work of Krebs, Kacelnik and Taylor discussed in section 7.4.4. Informally, discussions of constraints and predispositions in associative learning (section 7.2) have assumed that it is better to learn about some aspects of a situation than others and that animals' learning programs (Pulliam & Dunford 1980) are attuned to the best predictors.

How fast to learn and how long to remember?

It is often assumed that the stability of memory for an event will be matched to the stability of the event itself. Thus the song to be sung in adulthood or the appearance of an unpalatable species of butterfly should be remembered for a long time, while memory for locations of stored food should decay just as rapidly as the stores are depleted by predators (Shettleworth & Krebs 1982). The interval over which recent food availability is averaged should be related to the variability of the environment (Cowie 1977). Learning may not occur at all in a variable environment or in a very short-lived animal (Manning 1976). Some of the work mentioned in section 7.4 represents the beginnings of a more formal attack on this problem, in such notions as sampling, memory window, and time horizon.

7.5.3   Conclusion

Functional questions about learning can in principle be answered in the same way as questions in any other area of behavioural ecology. The questions can be made more precise in the sorts of ways illustrated in section 7.4 and the answers can be sought in comparative studies. Of course any attempt to do this must recognize all the same problems as beset other studies of function (see Chapter 1).

One reason why the first edition of this book contained no chapter on learning and why this chapter can be viewed only as suggestions for a behavioural ecology of learning may be that information about learning in the field is more difficult to gather than information relevant to some other questions in behavioural ecology. Group size or types of items eaten can in principle be observed fairly directly, but observations of learning necessarily require studying the same individuals over time. Furthermore, to answer most of the questions posed above, it is probably necessary to test animals under controlled conditions in the laboratory. Thus the behavioural ecology of learning can be expected to remain at a rather primitive level for some time to come. Perhaps some of the readers of this book will be able to contribute to its development.

PART 3
SEX, MATING SYSTEMS
AND LIFE HISTORIES

Introduction

The four chapters in this section are all concerned with reproductive strategies. Most animals studied by behavioural ecologists engage in sexual reproduction. Females, who make larger gametes, put more reserves than do males into each zygote and in animals with parental care it is usually the females who do most of the work. Why do females accept this parasitism by males of their reproductive investment? In Chapter 8, Maynard Smith considers the problem of why sex has evolved. He shows that in a population of sexual animals, an asexual female variant would have a twofold advantage in terms of gene propagation. Only half of a sexual female's offspring invest in their own children while every one of an asexual female's offspring does so. Of course, if males contribute to parental care such that the reproductive output of a sexual female is doubled compared to an asexual, then this will offset the disadvantage. In animals without male parental care, however, sex has a twofold cost, that of male laziness.

Maynard Smith discusses two types of advantage of sexual reproduction which might outweigh the cost of producing males. In the long term, sexual species may be less likely to go extinct because they can evolve more rapidly to meet changing conditions, and recombination prevents the accumulation of deleterious mutations in the population. This explanation invokes group selection and so we still have to explain why, in the short term, asexual variants do not increase and replace the sexuals in the population. One possibility is that asexual variants may not arise very frequently and a single clone may be unlikely to sweep through the population. The other is that there are short-term advantages of sex. In a changing environment, where the optimal genotype changes between each generation, the greater variability of sexual offspring could more than double the chance of one of them surviving compared to that of asexual offspring. The 'environment' of any species consist not only of physical factors but also of other species. Van Valen's view of evolution, coined as the Red Queen hypothesis, is that each species must evolve simply to keep up with improvements in competitors, predators and pathogens. One of the main short-term advantages of sex, therefore, may be that it enables animals to produce offspring able to compete successfully in an environment where other organisms are changing all the time.

Chapter 8 also considers the evolution of sex ratios, hermaphroditism and parental care. Fisher proposed that the 1 : 1 sex ratio, found in many populations, is the stable outcome of competition between individuals to maximize reproductive success: in a population with an excess of one sex, individuals will be selected to produce an excess of the rarer sex. An alternative view is that the 1 : 1 ratio is not a product of natural selection but simply a mechanical consequence of meiosis. Whatever the cause of the sex ratio, a consequence is that in most animal populations there is intense competition between males for females. A male can father many more offspring than a female can produce. The result is that whereas a female's reproductive success is limited by the rate at which she can convert resources into offspring, a male's success is usually limited by the rate at which he can find and mate with females. In many populations some males are very successful while others fail altogether, and the equal sex ratio exacerbates the competition between males for mates.

In Chapter 9, Partridge and Halliday discuss the causes of variance between individuals in reproductive success. They point out the difficulties of distinguishing non-random mating due to the outcome of male–male competition and that due to female choice. Are some males very successful because they are good at outcompeting rivals or because they are chosen by females? In many cases the female gains obvious material benefits from mating with some males rather than others, for example access to an area with good food or nest sites. Theoretical and practical problems arise, however, when female choice is presumed to have evolved for particular male characteristics, such as bizarre plumage, solely because of the genetic consequences for the offspring. Chapter 9 summarizes some recent theoretical models of how male characters evolve under the influence of female choice. There is also a critical discussion of the kind of empirical data that are needed to show female choice in nature. 'Active choice' implies that the behaviour of the female has been modified during evolution so that she mates with a particular kind of male rather than another. In many cases, however, the 'choice' may simply involve passive attraction to the most conspicuous male advertisements. The evolution of male adornments may represent the outcome of an arms race between males to present the females with supernormal stimuli. Partridge and Halliday also emphasize that many elaborate male displays are directed to other males rather than towards females. Indeed the evolution of male advertisements, such as long tails or loud croaks, may be influenced by male–male competition rather than female choice. Perhaps bizarre vocalizations and plumage have evolved for psychological warfare against rivals in the same way that weapons, such as horns and antlers, have evolved for physical combat.

Chapter 10 places the behaviour of males and females into an eco-logical setting and examines how the distribution of resources and competitors influences the evolution of mating systems. Bradbury and Vehrencamp begin the chapter by listing the factors which influence male and female reproductive success and then go on to describe in detail some models of mating systems. They discuss critically the assumptions and predictions of Orians' polygyny threshold model which has been invoked to explain polygyny in many cases where males have territories of different qualities. The chapter then exam-ines the eco-correlates of polyandry and monogamy and concludes with leks, which are perhaps the most extraordinary and least under-stood of all mating systems. Bradbury and Vehrencamp challenge the conventional wisdom that males in the central territories gain the most matings, and emphasize some of the problems of female choice introduced in the previous chapter. It is clear that more observational and experimental data are needed on male and female behaviour and on the distribution of mating success. Frog choruses may provide one of the best opportunities for studying male aggregations because with external fertilization it is easy to score the reproductive success of different males and furthermore their vocal displays are more amenable to experimental manipulation than the visual displays of birds.

Throughout the chapter, the authors emphasize that classification into strict categories such as monogamy, polygyny and polyandry may obscure the variability of mating strategies within populations. It is perhaps worth stressing that the optimal mating system from a male's point of view is usually very different from that for the female. This is shown by some recent elegant work on the pied flycatcher (*Ficedula hypoleuca*). A female has greatest reproductive success by mating monogamously because she then gets the male's undivided help with feeding of the chicks. A male, however, can increase his reproductive success by attracting a second female, even though he eventually deserts her and leaves her to raise her brood unaided. Polygyny is best viewed here as an example of deception of females by males (Alatalo *et al.* 1981, 1982).

The final chapter, by Horn and Rubenstein, looks at life histories, the ways in which individuals allocate their resources to repro-duction, self maintenance and growth. They show how the optimal life history, in the sense of maximizing gene contribution to future generations, depends on how ecological factors affect adult and juve-nile mortality. They discuss the conditions favouring iteroparity (repeated breeding) versus semelparity ('big bang' reproduction), the allocation of resources between quantity and quality of offspring and the trade-offs between fecundity and survival. It is shown how the optimal solutions often depend on environment predictability.

This chapter points to the links between behavioural ecology and population ecology. Population ecologists explore the consequences for population growth of given birth rates, death rates and dispersal, for example. Behavioural ecologists are interested in understanding how these parameters in the population models have evolved, given that individuals are selected to maximize their reproductive success. Horn and Rubenstein point out the areas where more data are needed, particularly age-specific schedules of mortality, fecundity and dispersal. Because dispersal patterns determine the genetic structure of populations, they provide the setting for the evolution of different social systems, which is the subject of the next section of the book.

Chapter 8
The Ecology of Sex

JOHN MAYNARD SMITH

8.1 THE FUNCTION OF SEX

Although sexual reproduction is almost universal, its functional significance is still a matter of controversy. In eukaryotes, its essential features are the production of gametes by meiosis (halving the chromosome number) and the production of a new individual by syngamy (the fusion of two gametes to form a zygote). In higher animals and plants, there is a division of labour between male and female gametes, egg and sperm. In many protozoa and green algae there is no morphological differentiation between the gametes ('isogamy', as opposed to the 'anisogamy' of higher forms), but there is usually differentiation into 'mating types', such that a gamete can fuse only with one of a different type. Finally, in the ciliated protozoa no free gametes are produced, but there is a process of conjugation in which haploid nuclei are exchanged, with genetic consequences similar to the mutual cross-fertilization of two hermaphrodites (see section 8.3).

The consequence of this process is to bring together in a single cell genes from two parental cells. The process of genetic recombination which occurs during meiosis also ensures that genes from different but homologous chromosomes can be combined in a single chromosome. This has the result that a population reproducing sexually can evolve more rapidly to meet changed circumstances. This point was first made by Fisher (1930) and Muller (1932). There is some controversy about how great the effect is (Crow & Kimura 1965; Maynard Smith 1968), but it is certainly substantial, particularly in large populations. The reason is easy to see qualitatively. Suppose that in an asexual population two favourable mutations, $a \to A$ and $b \to B$, were to take place in different individuals. There would be no way in which an AB individual could arise, except by the occurrence of a second B mutation in a descendent of the original A, or of an A mutation in a descendent of B. In a sexual population an AB individual could arise by recombination. The reason why sex is unimportant in a small population is that each favourable mutation will be fixed by selection before the next occurs.

Muller (1964) pointed out a second advantage of recombination, which has been called 'Muller's ratchet' by Felsenstein (1974). Consider an asexual population. It is inevitable that slightly deleterious mutations will occur. We can classify the individuals according to whether they have 0, 1, 2 ... deleterious mutations. Suppose the number of individuals with no deleterious mutations is small. Then, although these individuals are fitter than average, there is a chance that in one generation they will fail to reproduce. In the absence of sex, there is no way in which an individual free of deleterious mutations can arise again (save by back mutation, which can be shown to be unimportant). The 'ratchet' has clicked round one notch. In this way, deleterious mutations will accumulate in the population. But if sexual reproduction occurs, an individual with no deleterious mutations can be produced by recombination between two individuals with different mutations; the ratchet would cease to turn.

A quantitative analysis of Muller's ratchet (Maynard Smith 1978b) suggests that it could have been important in the origin of recombination very early in the evolution of life. Two individuals, each carrying a deleterious mutant, could produce one perfect individual (and, of course, one doubly damaged one) by recombination. Recombination is then seen as simply a form of DNA repair. Muller's ratchet may also be important in causing deterioration of parthenogenetic strains of higher plants and animals.

So far, the advantages I have suggested for sex (acceleration of evolution, preventing the accumulation of deleterious mutations) are advantages to the population as a whole. This has lead to the 'long-term' or 'group selection' explanation of sex. Species are, for the most part, sexual because those which abandon sex fall behind in the evolutionary race, and go extinct.

There are good grounds for being cautious about group selection explanations in biology (Williams 1966, 1975; Maynard Smith 1964, 1976b). The crucial difficulty is that a gene increasing individual fitness will spread through a population, even if it reduces the fitness of the population; if populations are not completely isolated, a gene which has spread in one population can 'infect' others. There are, however, two reasons why group selection cannot be ruled out as an explanation of sex, even though it is probably unimportant in most other contexts. First, parthenogenetic mutants (which would increase individual fitness) are rare; secondly, species really are reproductively isolated (see also section 8.1.1).

I will discuss the maintenance of sex rather than its origin, because we only have evidence concerning the former. One crucial point is that in a sexual population a parthenogenetic mutant would, other things being equal, have an immediate twofold advantage. Thus if, on average, every female can produce two surviving offspring, a sexual female produces one female offspring like herself, and a parthenogenetic female produces two female offspring like herself. The point is illustrated in Fig. 8.1. Sexual reproduction entails the twofold disadvantage of producing males.

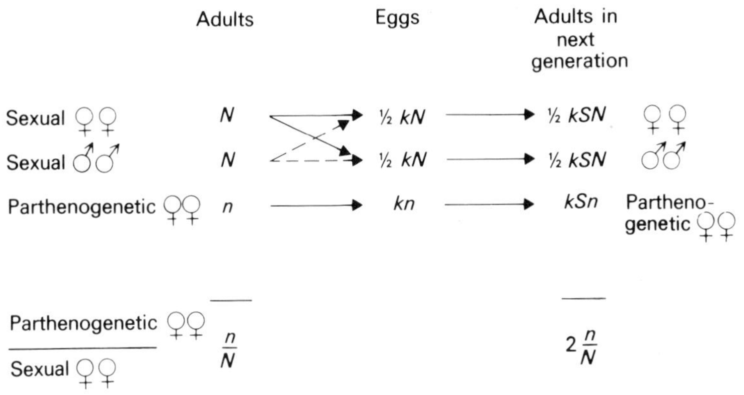

Adults	Eggs	Adults in next generation

Sexual ♀♀ N ⟶ ½ kN ⟶ ½ kSN ♀♀

Sexual ♂♂ N ⟶ ½ kN ⟶ ½ kSN ♂♂

Parthenogenetic ♀♀ n ⟶ kn ⟶ kSn Parthenogenetic ♀♀

$$\frac{\text{Parthenogenetic ♀♀}}{\text{Sexual ♀♀}} \quad \frac{n}{N} \qquad\qquad 2\frac{n}{N}$$

Fig. 8.1 The twofold disadvantages of producing males. k is the number of eggs laid by a female, and S the probability that an egg will survive to become an adult.

There has been discussion about whether the cost of sex is best thought of as the cost of producing males, as just outlined, or as a cost of meiosis—i.e. as the cost resulting from discarding half the genes in a nucleus, and accepting in their place genes from an unrelated individual. One way of tackling this question is to ask whether there is a cost of sex in an isogamous species. Figure 8.2 considers a gene A which suppresses meiosis. Gene A would gain some advantage through not wasting time; this may be why meiosis in protists often occurs at times of food shortage, when growth is in any case impossible. However, the gene A does not obtain a twofold advantage, which therefore need not be taken into account when discussing the origin of eukaryotic sex (it is safe to assume that isogamous sex preceded anisogamy).

This argument suggests that the cost of sex arises from producing males, and not from meiosis as such. However, consider the fate in an isogamous species of a gene A' suppressing meiosis and producing a diploid gamete which, after fusing with a typical haploid gamete, was able to expel the genes in the fusing gamete but to retain its cytoplasm. Such a gene would double in frequency, although there are no males. This example suggests that what matters is the nature of the mutant which is supposed to occur in the typical, sexual, population. In this particular case, mutant A' is somewhat implausible, so the 'cost of males' view seems more helpful than the 'cost of meiosis'. But consider a second example, pointing the other way. Imagine a bird species in which both parents care for the young. Mutant A causes females not to pair or mate with a male, and to produce female offspring without meiosis. Such a mutant would not have a twofold advantage, because an unpaired female could not raise so many offspring. In contrast, a mutant A' which paired and mated normally, but which suppressed meiosis and did not use the sperm, would have

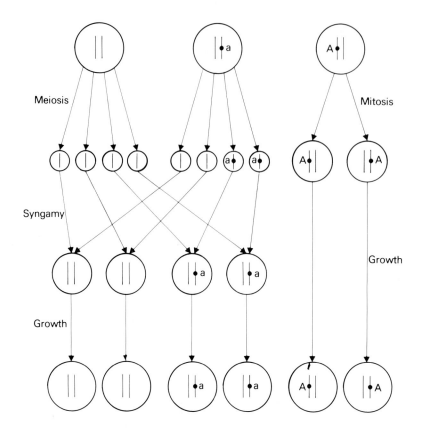

Fig. 8.2 Does meiosis confer a twofold disadvantage in an isogamous population? The gene *A* suppresses meiosis. After mitosis, the number of copies of *A* is the same as the number of copies of an allele *a* after meiosis and syngamy. Therefore, there is no necessary twofold disadvantage, but in a growing population there would be some advantage to *A* through time-saving.

the twofold advantage. In this example, mutant *A'* is more plausible than *A*, so a 'cost of meiosis' interpretation is favoured.

The sensible conclusion is that there is little point in arguing about which phrase is best. What is needed is a clear description of the typical reproductive system, and of the alternative, mutant, system being considered. However, for any anisogamous system with separate sexes, it is easy to imagine a mutant which, if we ignore other selective forces, would have a twofold advantage.

The difficulty of the group selection explanation should now be clear. In its simplest form (and we shall see later that this simple form may be misleading), it implies that each time a parthenogenetic variety arises, it will, because of its twofold advantage, replace the sexual species which gave rise to it. Later, the parthenogenetic species will go extinct in competition with more rapidly evolving sexual species. Thus each parthenogenetic origin is balanced by a species extinction. Can this be true?

Three arguments which support the group selection theory will be considered first.

The origin of parthenogenetic strains may be a rare event

In animals, parthenogenesis can occur in one of two ways, automixis and apomixis. In automixis, the egg undergoes meiotic reduction in the normal way, and diploidy is restored either by fusion of two of the four pronuclei, or by fusion of two genetically identical cleavage nuclei. Most cases of occasional spontaneous parthenogenesis in typically sexual species are of this type. Since the process leads to wholly or partially homozygous offspring, usually of low viability, it is unlikely to give rise to a successful parthenogenetic strain, although some naturally occurring strains are automictic (e.g. in enchytraeid oligochaetes).

In apomixis offspring genetically identical to the parents are produced without meiosis. The origin of such strains of animals is probably a very rare event, although it has occurred, for example, in the ancestry of the cyclically parthenogenetic aphids and cladocerans, and the obligately parthenogenetic weevils.

Among the vertebrates, parthenogenesis occurs in nature in several genera of fish, amphibians and lizards. In the whiptail lizards (*Cnemidophorus*), it leads to the production of offspring genetically identical to the mother, but by the elaborate procedure of first doubling the chromosome number, then pairing sister chromosomes and undergoing a 'normal' meiosis to produce a diploid egg nucleus.

Unhappily, the botanical terminology is totally different to the zoological. However, the essential fact is that when higher plants produce seeds which develop without fertilization, those seeds are genetically identical to the parent. It may be that the origin of such parthenogenetic varieties is much commoner in plants than animals. One reason may be that in higher plants a single somatic cell can develop into a whole plant, whereas in higher animals it seems that it may be very difficult to initiate development without meiosis.

The taxonomic distribution of parthenogenetic varieties suggests that they are short-lived in evolutionary time

If the abandonment of sex condemns a species to early extinction, we would expect to find that most existing parthenogenetic varieties are similar to sexual species. We would not expect to find large taxa (e.g. families or orders) consisting wholly of parthenogenetic varieties; nor would we expect to find isolated parthenogenetic varieties unrelated to any sexual form. By and large, this is what we do find. In animals, existing parthenogenetic forms must represent hundreds, and in all probability thousands, of separate origins. Yet with one important exception (the bdelloid rotifers) no major taxon (subfamily or above) consists predominantly of parthenogenetic forms, and there is no taxonomically isolated parthenogenetic species. To give a particular example, Mockford (1971) states that there are 28 known parthenogenetic strains of psocids; they belong to 13 different families, and in 12 of the 28 cases there are sexual and parthenogenetic forms of the same nominal species. A fundamentally similar picture is found in plants.

In case it be thought that this taxonomic distribution is an accident of sampling, it is worth comparing it with the distribution of another sexual system, male haploidy ('arrhenotoky'). Existing arrhenotokous species may be descended from as few as eight ancestral lineages; of these, five have given rise to families or larger taxonomic groups (the Hymenoptera, Thysanoptera, monogonont rotifers, a group of mites and a group of homopteran bugs) and one to a taxonomically isolated beetle species, *Micromalthus debilis*. Thus, despite the problem raised by the bdelloid rotifers, the taxonomic distribution of parthenogenesis strongly supports the view that it leads to early extinction.

An established parthenogenetic variety may not replace its sexual ancestral species

A parthenogenetic variety may arise as a unique event, and contain, at least initially, only a single genotype. Despite its twofold advantage, it is therefore unlikely to replace its sexual ancestor over the whole ecological range to which the latter is adapted. For example, the lizard *Cnemidophorus uniparens* has been shown by skin-grafting to consist of a single clone (Cuellar 1976).

C. uniparens may be of recent origin. Parker and Selander (1976) have found using electrophoretic techniques that another parthenogenetic 'species', *C. tesselatus*, is more variable. This species consists of diploid hybrids between two sexual species, and of triploid hybrids between diploid *C. tesselatus* and a third sexual species. All the triploids examined belonged to a single clone, suggesting a single origin, but 12 distinct diploid biotypes were found. These probably represent five distinct hybridization events, with some further variability being generated within clones by mutation and recombination. The interesting point is that despite a minimum of six separate original clones, with their twofold advantage, the parthenogenetic forms have not wholly replaced their sexual ancestors. These lizard clones do illustrate the fact that a single established variety has too narrow a range of genetic variability to eliminate its sexual ancestors.

The fish *Poeciliopsis monacha-occidentalis* is a parthenogenetic hybrid between the sexual species *P. monacha* and *P. occidentalis*. Moore (1976) has attempted to measure three components of fitness of the hybrids: the advantage from not producing males, the disadvantage arising because the hybrid must mate with male *P. occidentalis* before its eggs will develop (although the chromosomes provided by the male are later eliminated), and the 'primary' fitness when these two factors have been allowed for. He finds that the primary fitness is high in the region where the ranges of the parental species meet, but falls to half that of *P. occidentalis* further north. Although the genetic system of *Poeciliopsis* is bizarre, Moore's experiments do indicate that a parthenogenetic variety with a narrow range of genotypes could not replace the sexual parent over its whole range.

This point is demonstrated on a larger scale by several plant complexes, of which *Taraxacum* (dandelions) is an example (Richards 1973). There are some 1000 asexual 'species' of *Taraxacum*, and some 50 sexual ones. Parthenogenetic varieties probably first arose in the Cretaceous. New ones continually arise either from sexual ancestors, or, more probably, by hybridization between a sexual species and an asexual variety acting as pollen parent and carrying the gene for apomixis. It is a striking fact that the sexual species still survive, albeit restricted in range, despite continuing competition from their asexual descendants. Part of the reason may be that the asexual varieties do not reap the full twofold advantages of parthenogenesis, since they produce large flowers and, in most cases, pollen. The presence of large flowers and pollen is itself a puzzle, since they serve no obvious function in a parthenogen. Even though the pollen may occasionally contribute to a new clone, this is a selective advantage which arises very rarely. A possible explanation is that these clones, which evolve slowly and which are often of relatively recent origin, have not yet had time to adapt to parthenogenesis; but I do not feel much confidence in this explanation.

To summarize, group selection cannot be ruled out as an important force maintaining sex. Parthenogenetic varieties not suffering from homozygosis may arise rather seldom, particularly in animals; their taxonomic distribution suggests that they may have a short future; a single clone is unlikely to replace the whole of a sexual species.

8.1.2 The short-term advantage of sex

There are, however, strong arguments on the other side, which have been particularly stressed by Williams (1975).

The 'balance' argument

Some species reproduce both parthenogenetically and sexually. There must be some genetic variance in the relative frequency of the two modes, and selection would therefore eliminate the sexual mode if it did not confer some short-term advantage.

The argument applies in two contexts, cyclical and facultative parthenogenesis. *Daphnia* and other cladocerans reproduce mainly by apomixis, producing eggs which develop immediately into females. Occasionally, perhaps in response to crowding, males and females are produced apomictically. Fertilized 'winter eggs' are then produced, which sink to the bottom and hatch only after a prolonged period. One might therefore be tempted to argue that sex is maintained in the short run not because of any genetic consequences, but because winter eggs are the main means of dispersal, and of surviving the drying up of ponds. But this will not do, because some strains of *Daphnia* produce winter eggs apomictically. The balance argument for a short-term advantage for sex seems to hold. Similar arguments apply to other cyclically parthenogenetic groups—aphids, cynipid wasps and monogonont rotifers.

An individual plant of the grass *Dichanthium aristatum* can produce both apomictic and sexual offspring. The relative frequency can be altered environmentally (Knox 1967), and one must suppose that it could be altered genetically. Such facultative parthenogenesis is a powerful argument for a short-term advantage for sex. The case is not unique, but it is difficult to decide how common it is. There are many cases of populations consisting of an excess of females, some of which reproduce parthenogenetically. Unfortunately, it is often not known whether these populations are a mixture of sexual and obligate parthenogenetic females (in which case the parthenogenetic variety may contain only a narrow range of genotypes), or whether they are facultatively parthenogenetic, in which case the balance argument applies.

The timing of sexual reproduction

Williams (1975) points out that if a species can reproduce both sexually and vegetatively, the sexual phase is always timed to coincide with dispersal into an unpredictable environment, when recombination is most likely to be advantageous. The winter eggs of *Daphnia* are an example. In plants which reproduce both vegetatively (by stolons, rhizomes, bulbils, etc.) and sexually (by seeds), it is the sexual propagule which is provided with means of dispersal.

It is not easy to evaluate these opposing arguments. Clonal extinction, in competition with more rapidly evolving sexual competitors, plays some role in maintaining sexual reproduction. But the 'balance' argument suggests that there must also be short-term advantages to sexual reproduction. The problem is discussed at length in Williams (1975) and Maynard Smith (1978b).

There is space here to describe only one possible short-term mechanism—that of 'sib competition'. Figure 8.3 illustrates sib competition in a plant. It is supposed that the environment is divided into

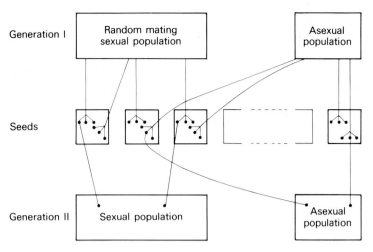

Fig. 8.3 A model of sib competition. Two parents each contribute three seeds to each patch. Only one of these seeds can survive. The sexual individuals are hermaphrodites, and produce half as many batches of seed as asexual ones. To make up for this twofold disadvantage, sexual seeds must be twice as likely to win in competition in the patches. Computer simulation suggests that for this to be the case, a parent must contribute several seeds to a patch (i.e. there must be competition between sibs). Also, there must be at least 30–40 seeds in each patch (i.e. selection must be intense); for simplicity, only six seeds per patch are shown in the diagram. (From Maynard Smith 1976c.)

'patches' (which can be contiguous), each large enough to support a single adult plant. A number of seeds fall in the patch, but only one can survive; in general the seed which is genetically best adapted to the patch will be the survivor. The essential feature (first proposed by Williams 1975) is that a single parent plant contributes a number of seeds to the same patch. Given that the environment in each patch is sufficiently unpredictable, sexual plants will increase in frequency relative to the asexual ones. The reason is best seen by analogy with a raffle. Each patch can be thought of as a raffle with a single prize. An asexual parent is like a man who buys a number of tickets in the raffle and finds that they all have the same number; in contrast, the sexual parent resembles a man who buys fewer tickets, but all with different numbers.

Sib competition can provide a short-term advantage for sex, provided that (i) selection is intense and density-dependent, (ii) sibs compete with one another, and (iii) the environment is unpredictable between generations. The necessity of sib competition can be shown by modifying the model of Fig. 8.3 so that each parent contributes only one seed to each patch; sexual reproduction then confers no advantage, and of course still suffers its twofold disadvantage. It follows that there must be other selection pressures, long or short term, maintaining sex.

The problem of the maintenance of sex is related to that of how, within sexual populations, genetic recombination is maintained; this cannot be explained by group selection, because there is plenty of evidence to show that there is within-population generic variance in recombination frequency. In an unchanging environment, the genome would congeal (Turner 1967); that is, recombination would be abolished. Various models have been proposed which would generate selection for increased recombination. What they have in common is that they assume an ever-changing environment. For example, if there is 'normalizing selection' favouring some intermediate optimal phenotype, but this optimum changes from generation to generation, either continuously in one direction or in a fluctuating manner, there can be strong selection for increased recombination (Maynard Smith 1980a).

This brings me to the one generalization one can make about the selective maintenance of sex and recombination. Such maintenance is explicable only in a changing environment; in a uniform environment we would all be parthenogens. Now the 'environment' of any species consists in the main of other species in the ecosystem. Van Valen (1973) has suggested that each species must evolve because all the other species are doing so. There is evidence (Levin 1975, for plants; Glesener & Tilman 1978, for animals) that sex is more often lost in simple ecosystems than in the complex species-rich systems of the tropics; they suggest that this is because a species which abandoned sex would not keep up with its competitors, predators and pathogens. More recently, Hamilton (1980) has simulated the coevolution of a host and its pathogen, and shown that selection favouring sex can be strong enough to overcome the twofold disadvantage.

8.2 THE SEX RATIO

Females, by definition, produce large gametes containing substantial food reserves, whereas males produce small, mobile ones. If males and females have equal capacities for acquiring resources, it follows that a male can produce many times as many gametes as a female, and can fertilize the eggs produced by many females.

For the population, therefore, the greatest production of offspring would be achieved if there were many times as many females as males. This is rarely the case, because selection acts on individuals, or more precisely on genes, and not on populations. Fisher (1930) explained why, in a random-mating population, the evolutionarily stable sex ratio is 1 : 1. Suppose, he argued, there were more females than males. Then a parent who produced only sons would have more grandchildren than typical members of the population. Therefore the gene which caused the parent to produce sons would increase in frequency. By an identical argument, if there were more males than females a gene causing parents to produce daughters would increase in frequency. Hence in a population with an excess of one sex, selection would favour the production of the other. Only the 1 : 1 sex ratio would be stable. More precisely, Fisher argued that a parent will expend equal resources on offspring of the two sexes. This is a difficult argument to see without a mathematical treatment, but a numerical argument may help. Suppose that females 'cost' twice as much in resources as males, and that a parent can produce 2 ♀♀ or 4 ♂♂ or 1 ♀ + 2 ♂♂. If the population sex ratio were 1 : 1, then on average males and females have equal numbers of offspring. Therefore a parent producing 4 ♂♂ would have more grandchildren than one producing 2 ♀♀, and so the population would not be evolutionarily stable. But if there were twice as many males as females in the population, males would on average have half as many children, and therefore a parent producing 4 ♂♂ would expect the same number of grandchildren as one producing 2 ♀♀. Such a population would be stable; it is one in which twice as many males as females are produced, which implies equal investment by parents in the two sexes.

An obvious difficulty is as follows. The theory predicts that, in most cases, the ratio should be close to 1 : 1, and this is the ratio most easily produced by an X–Y sex-determining mechanism. Hence, when we meet a 1 : 1 ratio, we do not know whether to interpret it as the product of natural selection, or as something which happens because there is no genetic variance of the ratio, upon which selection could act. There are several ways we can tackle this difficulty. One is to ask whether there is any evidence for genetic variance of the sex ratio within species. Some of the most extensive data sets (e.g. Edwards 1970, on man; Toro & Charlesworth 1982, on *Drosophila*) fail to reveal any sign of such variance. Thus, although I feel sure that if the ratio typically favoured by selection was *not* 1 : 1, mechanisms other than the X–Y one would have evolved (I mention two others below), it may be that species sometimes fail to make adaptive changes of the ratio because of lack of the necessary genetic variance. Maynard Smith (1980b) discusses the alternative possibility, that the 1 : 1 ratio is fixed, and that species adapt to changing circumstances by changing the investment they make in offspring of the two sexes.

A second approach to the problem is to ask whether, in species with an X–Y mechanism, the ratio departs from 1 : 1 when our theory predicts that it should. For example, is the ratio distorted away from 1 : 1 when offspring of the two sexes require different amounts of parental investment? In some birds the sexes differ in size at fledging. Howe (1977) looked at sex ratios in the common grackle, *Quiscalus quiscula*, in which males are 20% heavier at fledging. He found 52 males and 83 females at fledging; the difference appears to arise because of heavier male mortality, since he found a 1 : 1 ratio among embryos. These results accord with Fisher's prediction; unfortunately, they also accord with the simpler view that the primary sex ratio is 1 : 1 because that is the ratio produced by a regular meiosis. Thus Howe's results are consistent with Fisher's prediction, but do not prove that natural selection has modified the sex ratio away from

1 : 1. Equally ambiguous results emerged from the more extensive studies by Newton and Marquiss (1979) on the European sparrow-hawk, *Accipiter nisus*. Adult females are twice as heavy as males; they hatch from similar eggs, but a twofold weight difference exists already at fledging. The authors found 1102 males and 1061 females, in 651 broods. These data appear to contradict Fisher's prediction that the parents should equalize investment rather than numbers. However, male and female nestlings ate the same amount. Females put on weight faster, but males feather sooner and leave the nest 3–4 days earlier. So the costs of a male and a female may not be so different as their weights suggest. More data are needed on the evolutionary adjustment of the sex ratio.

Greater success has been achieved by studying haplodiploid organisms, in which a female can determine the sex of each offspring by whether or not she fertilizes the egg. Hamilton (1967) pointed out that if there is inbreeding, Fisher's conclusion is modified. If a female 'knows' that her daughters will be mated by her sons, she can maximize the number of her grandchildren by producing an excess of daughters, and only enough sons to ensure that they are fertilized. He showed that in many haplodiploid arthropods, inbreeding is associated with a great excess of females. The extreme example is a viviparous mite, *Acarophenox*, which has a litter of one son and 10–20 daughters; the male mates with his sisters and dies before he is born.

If more than one female lays eggs in a single host, the expected degree of sex ratio bias is less. Werren (1980) has shown that in the parasitic wasp *Nasonia* a female can tell whether a host is already parasitized and, if it is, lays eggs with the predicted degree of bias.

In the cases discussed by Hamilton, the sons of a given mother are competing with one another for mates. Hence the process is referred to as 'local mate competition'; the females being competed for are usually the sisters of the males, but the bias does not depend on this. A bias is also predicted if a set of sibs of one sex compete for any other limited resource. For example, Clark (1978) suggested that the male bias in bush-babies after weaning may be an evolutionary response to the resource competition between female sibs; the males disperse, and hence are not involved in the competition.

In most turtles, and in some other reptiles, sex is determined by the temperature at which the egg is incubated (Bull 1980). Typically, turtle eggs incubated below a critical temperature develop as males, and above that temperature as females. The choice of a nest site by a female is thus the proximate factor determining the sex of her offspring. It is not easy to understand why such a sex-determining mechanism should have evolved. Charnov and Bull (1977) showed that environmental sex determination could be stable against invasion by sex-determining genes if one sex gained a greater advantage from being raised in a particular environment than the other. It seems unlikely that this could be true of turtles, but it can explain the situation in some parasitic nematodes (discussed in Charnov 1982a), in which eggs laid in favourable circumstances (e.g. few eggs per host) usually develop as females. This makes sense because, in nematodes, fitness probably increases with size more steeply in females than in males.

8.3 HERMAPHRODITISM, SELFING AND OUTCROSSING

It will help to start with the meanings of some words.

GONOCHORISTIC: having separate male and female individuals; for example all mammals and birds.

DIOECIOUS: the botanical equivalent of gonochoristic; for example holly trees (*Ilex*) and stinging nettles (*Urtica dioica*).

HERMAPHRODITE: in animals, an individual which produces both eggs and sperm. It may be 'simultaneous' as in most land and freshwater snails, or 'sequential', when an individual is first male and then female or *vice versa*. In plants, hermaphrodite refers to the flower and not to the whole plant. A hermaphrodite flower produces both seeds and pollen. This condition is commonest in flowering plants and almost certainly primitive. (See Chapter 14 for a more extensive discussion of plants.)

MONOECIOUS: in plants, having separate male and female flowers on the same plant; for example birch trees (*Betula*).

In understanding the functional significance of hermaphroditism, a crucial question is whether an individual is self-fertile. Many hermaphrodite plants (e.g. all members of the pink family, the Caryophyllaceae) are self-compatible. In such cases, the likelihood that pollen will fertilize ovules from the same flower is reduced if anthers and stigma ripen at different times, but since different flowers are not synchronous this does not prevent a plant from pollinating itself. Many groups of plants, however, have evolved more effective mechanisms preventing self-pollination (for a review, see Grant 1958). A variety of genetic self-incompatibility mechanisms have evolved which prevent the growth of the pollen tube down the style of the same individual. Some hermaphrodite species are divided into two (e.g. *Primula*) or even three (e.g. *Lythrum*) morphologically distinct types, such that pollination typically only takes place between members of different types; this phenomenon, known as heterostyly, was first elucidated by Darwin (1877); see also section 14.3.3.

Animal hermaphrodites show a similar range of ability to fertilize themselves. For example, among gastropods *Helix* and *Cepaea* are completely self-sterile; *Biomphalaria* (a freshwater planorbid) typically cross-fertilizes but can fertilize itself if kept isolated; *Rumina*, a land snail, typically self-fertilizes in the wild. In general, however, much less is known of animals than of plants because the matter is harder to investigate; in a plant one can simply put a plastic bag round a flower and see whether it sets seed.

Three main selective forces have been responsible for the evolution of hermaphroditism: the difficulty of finding a mate or being pollinated; the genetic effects of inbreeding; the allocation of resources between male and female functions. These will be discussed in turn.

The 'low density' advantage of hermaphroditism

A single self-fertile hermaphrodite is able to colonize a new habitat, and to reproduce even if the population density is so low that it never meets another member of its species. It therefore makes sense (Baker 1955) that many annual weeds of disturbed soil are self-fertile. Families such as the Caryophyllaceae, which are primitively self-fertile, include many such colonizing species, for example *Stellaria* (chickweeds) and *Arenaria* (sandworts). Many other weeds are apomicts, which have the same advantage as colonizers. Apomicts do not suffer the inbreeding depression which an initially outcrossing population undergoes when it starts selfing, but lose the possibility of genetic recombination.

Ghiselin (1969) has argued that the same selective force has been important in the evolution of hermaphroditism in animals. The case is harder to substantiate because often we do not know whether a particular species is self-fertile. However, as was pointed out by Tomlinson (1966), even a self-sterile hermaphrodite is at some advantage over a gonochorist at low density, because any two individuals are certain to be able to mate. Supporting Ghiselin's view is the fact that hermaphroditism is common among sessile animals, which cannot move around to find a mate, and among parasites (e.g. flukes, tapeworms, and many parasitic genera in predominantly gonochoristic groups, such as crustaceans, nemertines and prosobranch molluscs).

The genetic advantages and disadvantages of selfing

If offspring produced by selfing were on average as fit as those produced by outcrossing, all hermaphrodites would fertilize themselves. Thus imagine a gene *A*, conferring self-compatibility, arising in a population of self-sterile hermaphrodites (Fig. 8.4). An individual with gene *A* would always fertilize itself, and would have the same success in outcrossing as a typical member of the population. If the population size were constant, a typical individual would contribute two gametes to the next generation, one as a female and one as a male; an individual with gene *A* would contribute three gametes, one as a female, one as a male by selfing and one as a male by outcrossing. By a similar but more subtle argument (Bengtsson 1978), if inbreeding had no deleterious effects, all females would mate with their brothers.

The long-term consequences of selfing in hermaphrodites (or inbreeding in gonochorists) would be the loss of the evolutionary

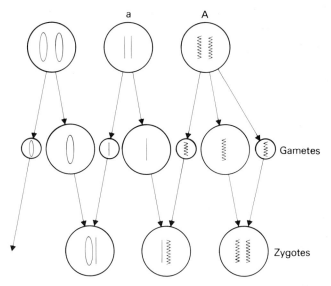

Fig. 8.4 The advantages of selfing. The individual *A* is a self-fertile hermaphrodite; individual *a* is self-incompatible. In the zygote population there are three copies of each chromosome from *A* for every two from *a*.

advantages of sex. Populations which took this road might ultimately be eliminated by inter-population selection. But there is a much more immediate selection pressure countering the spread of selfing and inbreeding: the offspring so produced are usually of low fitness. The exact reasons for this are not immediately relevant, and indeed are still to some extent in dispute. All that matters in the present context is that selfing does in fact lead to a dramatic loss of fitness (for a general discussion, see Lerner 1954).

In plants, it seems that dioecy has often evolved from herm-aphroditism because it prevents selfing, and thus ensures that an individual produces fit, outbred offspring. Monoecy may evolve for the same reason; often the male and female flowers grow on different parts of the plant, which further reduces the likelihood of selfing. This explanation makes sense only if it is in some way easier for a hermaphrodite species to evolve dioecy or monoecy than self-sterility. It is not clear why this should be so, but the taxonomic and ecological distribution of dioecy suggests that it is.

In plants, dioecy tends to occur sporadically in taxa which are otherwise predominantly self-fertile hermaphrodites (Baker 1959); for example, it has evolved at least three times among the Caryo-phyllaceae.

Ecologically, dioecy and monoecy are commoner in trees than in shrubs, and in shrubs than in herbs (Table 8.1). The reason is probably as follows. As the number of flowers on a single plant increases, the likelihood of self-pollination also increases. A small herb can rely on the movements of insects from flower to flower to ensure an adequate frequency of cross-pollination; a large tree cannot.

Resource allocation in hermaphrodites

How should a simultaneous hermaphrodite allocate resources between male and female functions? At what age should a simultaneous hermaphrodite switch sex? Charnov (1982a) points out that these questions, and also the question of what is the stable sex ratio, are all aspects of the same general problem: what is the stable allocation of resources between the sexes? An answer to this question also helps to explain why some organisms are hermaphrodite and some dioecious. I shall discuss here only simultaneous hermaphrodites (see Charnov *et al.* 1976).

Table 8.1. Percentages of forest plants with different habits.

Type of forest	Layer	Hermaphrodite	Monoecious	Dioecious	Reference
Deciduous woodland in Britain (four types)	Tree	50–57	14–30	20–33	
	Shrub	60–89	0–38	0–23	Baker 1959
	Field	86–93	0–9	0–9	
Five North American temperate forests	Tree	0–27	60–83	6–17	Bawa & Opler 1975
Seven tropical forests	Tree	41–68	10–22	22–40	Bawa & Opler 1975

Imagine a plant population consisting of males, females and self-sterile hermaphrodites. Each male produces N 'successful' pollen grains (i.e. pollen grains which contribute to a zygote in the next generation), and each female n successful seeds. Each hermaphrodite produces αN successful pollen grains and βn successful seeds. Obviously, whether hermaphrodites will replace gonochorists, or *vice versa*, depends on the values of α and β. In fact, it can be shown that if $\alpha + \beta > 1$ hermaphrodites will win, and if $\alpha + \beta < 1$ gonochorists will win. The reason can be glimpsed intuitively. Suppose that $\alpha = \beta = \frac{1}{2}$; that is, a hermaphrodite produces half as many successful pollen grains as a male and half as many seeds as a female. In fact, it will pass on as many genes to the next generation as a typical male or female. But if both α and β were greater than $\frac{1}{2}$ (and hence $\alpha + \beta > 1$), the hermaphrodites would pass their genes on to more offspring than males or females, and so would replace them (see Fig. 14.3).

What factors will tend to make $\alpha + \beta > 1$? For plants, three factors can be suggested. First, for any individual the period during which resources must be allocated to pollen production is earlier than the period when seeds are dispersed; consequently, a hermaphrodite may be able to allocate a greater total quantity of resources than either a male or a female. Second, in a hermaphrodite flower, there is a potential saving of resources because the same organs (petals, nectaries) serve both male and female functions. Third, and probably most important, in a sessile organism with limited pollen dispersal, a doubling of the number of pollen grains would not mean a doubling of the number of successful pollinations, because of competition between pollen grains from the same plant; similarly, a doubling of the number of seeds would not mean a doubling of the number of successful seedlings. In both cases, a law of diminishing returns would set in (see Chapter 14).

These factors do not operate with the same force in animals. The 'law of diminishing returns' may operate in sessile animals, but these often have pelagic larvae, so the law will not apply to egg production. It is unusual for the same organ to serve both male and female functions. More often, it seems likely that a doubling of investment in a particular sexual role would more than double success; this is particularly so of males which compete for females.

In gonochoristic animals, there are opposing forces favouring inbreeding and outcrossing, parallel to those favouring selfing and outcrossing in hermaphrodites. In general, the severity of inbreeding depression seems to have favoured outcrossing. There are two ways in which an animal might reduce the likelihood of mating with a close relative:

(i) by dispersing before sexual maturity in such a way that close relatives are unlikely to be encountered;

(ii) by recognizing relatives, and refraining from mating with them.

As an example of dispersal preventing inbreeding, it is common in group-living mammals for males raised in one group to transfer to another before breeding. The olive baboon, *Papio anubis*, affords a striking example. Packer (1975) found, in a 6-year study of three troops, that none of the males which reached maturity during the study remained with their natal troop. Of 41 known transfers, 39 were males. Although in most social mammals it is the males which transfer, it is sometimes the females (e.g. chimpanzees, hunting dogs).

To avoid inbreeding by the second mechanism it is not necessary that animals should be able to recognize their relatives by their genotypes. It is sufficient that an animal should recognize those individuals with which it was raised and should refrain from mating with them. Hill (1974) showed that in the deermouse *Peromyscus maniculatus* there is a delay in reproduction if the two potential parents have been raised together, whether or not they are actual sibs. There is anecdotal evidence of this kind for other mammals and birds. Interestingly, the effect may operate in our own species. A study of Israeli kibbutz (Shepher 1971), in which children of both sexes are brought up communally, showed that of 2769 marriages between second-generation kibbutz adults, not one was between members of the same peer group.

8.4   PARENTAL CARE

Species vary widely in the pattern of parental care; in particular the two sexes often contribute very unequally. Trivers (1972) interpreted these patterns in terms of 'parental investment'. In effect, he suggested that if, at some moment during the breeding period, one parent has invested more in the offspring than the other, then the parent which has invested less will be tempted to desert. He defined 'investment' as anything done by a parent to increase the chances of survival of the existing offspring, at the expense of the parent's ability to invest in future offspring.

This idea was criticized by Dawkins and Carlisle (1976), who pointed out that we would expect animals to behave so as to maximize their reproductive success in the future. To argue that past investment should determine future behaviour is to commit the 'Concorde fallacy', which asserts that if one has already spent a lot of money on a project, it is wise to continue spending money. The criticism is clearly correct, yet Trivers had been able to show that past investment and future behaviour are correlated.

The real difficulty in analysing parental behaviour arises, I believe, because the optimal behaviour for one parent depends on what the other parent is doing. For example, suppose that a female has brought a particular offspring almost but not quite to the stage of independent existence. To abandon the offspring at that stage would be a bad strategy if it resulted in the death of the offspring; but if the male could be relied on to care for the offspring, the best strategy for the female would be to start on a new offspring. In other words, we are analysing a 'game', in which the interests of the parents, although similar, are not identical. The discussion which follows is based on a game theory analysis (Maynard Smith 1977). The idea is to seek for a pair of strategies, M for the male and F for the female, which together form an 'evolutionarily stable strategy' or ESS (see Chapter 2 for discussion of the ESS concept), in the sense that it would not pay a male to depart from strategy M so long as females continue to adopt F, and it would not pay a female to depart from strategy F so long as males continue to adopt M.

The main factors determining whether there will be parental care, and if so which parent will provide it, are:
(i) the effectiveness of parental care by one or two parents;
(ii) the chance that a male deserting a female after mating will be able to mate again;
(iii) the fact that a female which exhausts her food reserves in laying eggs is less able to guard them; and
(iv) whether a male can be confident that a particular batch of eggs were fertilized by him.

The ways in which these factors interact can be most easily illustrated by reviewing some of the patterns observed in birds and fishes,

Table 8.2. Number of families of bony fishes in which (in some or all species) one or other parent cares for the young (data from Breder & Rosen 1966).

Fertilization external			Fertilization internal			No parental care
♂	♀	both	♂	♀	both	
28	6	8	2	10	0	191

which are the best studied groups from this point of view. Some idea of the frequency of different types of behaviour among fishes is shown in Table 8.2 (the data are extracted from Breder & Rosen 1966; more extensive data are now available in the literature). Several points are worth making.

(i) The commonest pattern is no parental care.

(ii) If there is parental care, it is usually by one parent only. Fish do not bring food to their young. Care consists of fanning the eggs, cleaning them of parasites, and protection against predators. These are tasks at which one parent may be almost as effective as two.

(iii) If there is internal fertilization, care is usually by the female. There are two possible reasons. With internal fertilization the male is usually not present when the eggs are laid; if he is present he may have no guarantee that the eggs were fertilized by his sperm.

(iv) If there is external fertilization, either parent may care for the young, but it is more often the male. The male may construct a nest (e.g. the stickleback, *Gasterosteus*) or protect a particular egg-laying site (e.g. *Cottus, Cyprinodon*) or the male may brood the eggs (e.g. sea horses and pipefishes, family Syngnathidae).

Two theoretical points should be borne in mind. First, for a species with a given basic ecology, either male or female care might be evolutionarily stable, once it had arisen. Thus consider a species with male care. If, in the absence of care, eggs have small chance of survival, a male which abandoned the eggs it had fertilized would have a low fitness, even if by abandoning the eggs it increased its chance of remating. The same argument, *mutatis mutandis*, applies to species with female care. The questions therefore arise, why did evolution originally take one path rather than the other, and why did it most often take the path of male care?

The question is most readily answered if we remember that in fishes uniparental care usually evolved from no care. In the absence of parental care, females will exhaust their food reserves in egg-laying. Since, during parental care, a parent fish feeds little or not at all, a female could only evolve this behaviour if, at the same time, she reduced her fitness by reducing the number of eggs she laid. In contrast, a male can fertilize the eggs and still have reserves left to live on during parental care. Also, in many cases (e.g. *Gasterosteus*) a male which guards the eggs laid by one female does not lose the chance of mating with a second one.

Whereas in fishes the commonest pattern is no parental care, followed by male care, in birds the commonest pattern is for both parents to care for the young, followed by female care. In birds, the parents usually bring food to the young. In almost all such cases, both parents care, more or less equally, for the young. If the young are fed, two parents can raise approximately twice as many offspring as one. Even in species which do not feed their young, it is not uncommon for both parents to protect them from predators. In the waterfowl, for example, if the parents are large enough to drive off most predators (geese, *Anser*, and swans, *Cygnus*) both parents care for the young, whereas in the smaller ducks (e.g. *Anas*, *Aythya*) only the female does so.

There seem to be two reasons why, if only one parent guards, it is almost always the female. First, fertilization is internal. Hence, if we suppose the primitive condition was biparental care (as is almost certainly true, for example, for the waterfowl), a male could desert immediately after mating and seek a second female, whereas a female could not seek a second male to care for a second clutch of eggs until she had laid the first. The remarkable condition in the American jacana (Jenni 1974), in which the female is polyandrous and lays clutches of eggs to be cared for by several males, probably evolved from the habit of 'double-clutching' found in some waders. In the sanderling (*Calidris alba*), for example, it is common for the female to lay one clutch of four eggs which is incubated and reared by the male, and then to pair with a second male and lay a second clutch which she cares for herself. Once males have evolved the habit of caring for a clutch of eggs after the female has laid them, it could pay females to compete in order to acquire more males; this is what has happened in the jacana (see section 10.3.2).

The other feature of bird biology which has favoured female rather than male care is that the productivity of a pair is not usually limited by the number of eggs a female can lay, but by the number which can be incubated and by the number of young which can be fed and protected (Lack 1968). Since a female does not exhaust her reserves in laying eggs, as do fish, she is able to care for the young. In some species, egg production is limiting, and only the male cares for the young. The most striking example is the mallee fowl, *Leipoa ocellata* (Frith 1962). The male constructs a large mound of sand and leaves as an incubator for the eggs. The female lays eggs over a period of three to four months, laying up to 30 eggs, each 10% of her own weight. During the laying period only the male constructs and tends the mound. The pair are monogamous for life. Despite the examples of the mallee fowl and the jacana, it is much rarer in birds than in fish for the female to get the male to look after her eggs. There seems to be two main reasons. First, since fertilization is internal a male bird may not be present when the eggs are laid, and in any case cannot be certain he is the father. Secondly, in most birds a female could not greatly increase her fitness by abandoning parental care in favour of laying more eggs, because breeding success is limited not by the number of eggs laid but by the number of young birds which can successfully be raised.

Chapter 9
Mating Patterns and Mate Choice

LINDA PARTRIDGE and TIM HALLIDAY

9.1 INTRODUCTION

It has long been obvious that the gametes produced in natural populations do not pair up at random. Leaving aside the obvious restrictions imposed by species and gender, some individuals may obtain more fertilizations than others, or particular types of pairings may be more common than others. Such non-random mating is of fundamental evolutionary importance because different matings may have different fitness consequences. These in turn will produce powerful selection pressures that can bring about evolutionary changes in the morphological or behavioural characters on which the non-random mating is based.

For mating patterns to evolve, two things are necessary. First, there must be (or have been) genetic variation affecting which individuals pair for mating. We know that such genetic variation can occur in extant populations (Majerus *et al.* 1982; Partridge 1983). The extent of genetic variation in the past can only be a matter of conjecture, but its existence is a universal assumption in theoretical discussions of the evolution of non-random mating. We rarely see evolution occurring, and it is therefore not surprising that some theories invoke genetic events of kinds not yet demonstrated in natural populations. In this chapter there are frequent discussions both of the realism of evolutionary theories in the light of our knowledge of genetic variation in modern populations, and of the implications of those theories for the genetic variation expected in modern populations.

The second important prerequisite for the evolution of mating patterns is that different matings should have different consequences for the frequencies of alleles producing them, in other words, for fitness. Sometimes these consequences are obvious. A genetic variant leading to morphology or behaviour associated with obtaining a large number of fertilizations is likely to spread, because mating success is likely to be associated with reproductive success which in turn is likely to be related to fitness (unless high reproductive success is balanced by a decrease in some other aspect of fitness or by a decrease in the fitness of a relative). In other cases, the fitness consequences of a particular mating pattern may be much less obvious. For instance, Japanese quail (*Coturnix coturnix japonica*) of both sexes apparently choose to mate with relatives that are neither very close nor very distant (Bateson 1982). At present, it is not clear whether mating with an intermediate as opposed to a distant relative can have beneficial fitness consequences (see section 9.3).

Because ESS (Chapter 2) and optimality (Chapter 4) models have been used elsewhere in this volume, we should briefly mention their value in theories of evolution of non-random mating. Some optimization models seek to define a single theoretically ideal phenotype for a specified set of circumstances, which is then compared with the real phenotype. Unfortunately, in models of non-random mating the fitness of a particular genotype often depends on its own relative frequency, or on that of another genotype. Such frequency-dependence can be accommodated in ESS models, which can define an equilibrium mixture that cannot be invaded by a specified set of mutant strategies. If such models are to be applied to mating patterns, it is important that the effect of starting conditions be examined, because the evolutionary path followed can be greatly affected by them (see section 9.2.4). ESS models deal with strategies, in other words with phenotypes, and they therefore make the tacit assumption that the genetic system is unimportant in determining the outcome of the evolutionary process. This is a problem, because it is quite clear that specific assumptions made about genetics can alter both the rate of evolution and what evolves in models of non-random mating (O'Donald 1980; Lande 1981; Kirkpatrick 1982), so that in these cases explicit genetic models are needed. There have been several excellent recent discussions of these problems (Maynard Smith 1978a, 1981, 1982a; Lewontin 1979; Arnold 1983; see Chapter 3).

This chapter deals with five interrelated forms of non-random mating: variation in individual mating success (section 9.2), inbreeding and outbreeding (9.3), assortative and disassortative mating (9.4), frequency-dependent mating (9.5), and non-random mating and the species (9.6). These different mating patterns can have very different consequences, both for individuals through their effects on the number and genotypes of their offspring, and for populations as a result of changes in gene and genotype frequencies. Making evolutionary sense of these patterns is proving to be a considerable challenge, raising as it does basic but still unanswered questions about such topics as the prevalence of sexual reproduction (Chapter 8) and of genetic variation.

9.2 VARIATION IN INDIVIDUAL MATING SUCCESS

9.2.1 Why does mating success vary?

A common finding in population studies is that some individuals, usually among males, mate more often than others and hence leave more progeny. The generally accepted reason for this variation in male mating success is that male gametes are extremely small relative to female gametes; female gametes are therefore each energetically much more costly to produce and a female's reproductive success is more likely to be limited by her own ability to produce gametes than by the number of her mates. A male, in contrast, can produce enough sperm to fertilize the entire gamete output of many females, and his reproductive success therefore tends to be limited by his ability to obtain fertile matings. This means that there will be scope for higher variance in male reproductive success, because each male is potentially capable of producing a much higher maximum (and other males, therefore, a lower minimum) number of offspring than is each female.

This argument never applies with full force; the potential number of matings obtainable by a male is always lower than is implied above, because the cost to a male of each mating is much higher than that entailed in producing a number of sperm equal to the number of eggs produced by the female. For fertilization to reach full efficiency it is often necessary for the male to produce a large excess of sperm. In those species where sperm competition occurs, either because fertilization is external or because females may mate with more than one male, males will be selected to produce more sperm than are needed to ensure that all eggs would be fertilized in the absence of sperm competition. This may explain the correlation between sperm competition and testis size found across primate species (Harcourt *et al.* 1981). Furthermore, it seems likely that the energetic costs of producing sperm have been underrated; recent evidence suggests that, at least in some species, producing sperm may be quite costly (Dewsbury 1982; Nakatsuru & Kramer 1982; Verrell 1982a). In addition, the cost of gamete production is probably only a minor proportion of the total cost to a male of obtaining a mating. He may have to produce energy-rich accessory fluid, as well as incurring costs in terms of time, risk and energy, both in finding and courting females and in countering the activities of other males.

Parental contributions to the offspring after mating will increase the cost of a mating. Parental investment is defined as any contribution to an offspring that reduces the potential for producing further offspring. Trivers (1972) pointed out that parental investment by either sex will tend to result in the lower-investing sex competing to mate with the higher-investing sex. As we have seen, pre-zygotic investment may modify this conclusion. In general, both forms of reproductive effort are higher in females. Occasionally this is not the case, and females may then compete for males, notably in polyandrous birds (Jenni 1974; Orians 1969; Knowlton 1982; Petrie 1983).

From an evolutionary point of view, we are interested in genetic variation affecting lifetime mating success. Almost all the available data are for various short-term measures of mating success which may correlate poorly with lifetime success. There are two main reasons for this. First, high mating success may incur costs resulting in reduced longevity (Gadgil 1972; Partridge & Farquhar 1981). Secondly, mating success may change with age or condition, so that short-term measures would indicate higher variance than lifetime measures (Clutton-Brock 1983). Nonetheless, such few data as are available do support the view that the lifetime mating success of males is more variable than that of females in species where males show no parental care after mating (Clutton-Brock 1983).

$$S^2_{\male RS} > S^2_{\female RS}$$

9.2.2 Sexual selection

As Darwin was first to recognize, variance in the number of successful matings is the raw material for sexual selection, defined as selection on characters giving certain individuals an advantage over others of the same sex in obtaining successful matings. From the general category of sexual selection Darwin was excluding several sorts of selection that also lead to sexual dimorphism, such as that on characters involved in caring for offspring or on male characters essential for mating and hence not advantageous solely in competition with other males. Some sexually dimorphic characters have an ecological significance, for example when males and females specialize on different diets (Selander 1972). Darwin was in fact trying to explain the evolution of maladaptive and extravagant male characters, and suggested that though they are detrimental to survival, this is outweighed by their benefit in competing with other males for mates.

Sexual selection, like many other forms of selection where competition is involved, is particularly potent because it has the characteristics of a coevolutionary arms race (West-Eberhard 1979; Thornhill 1980; Dawkins & Krebs 1979). Any evolutionary advance in one individual creates selection for a similar advance in its competitors.

Darwin suggested that sexual selection would lead to the evolution of two sorts of male characters; *intrasexual* selection would lead to the evolution of characters involved in winning in combat with other males, whereas *intersexual* selection would bring about evolution of characters important in attacting females. For any particular character, however, it may not always be easy to determine which form of selection has been the more important (Halliday 1983). There have been several recent discussions of sexual selection (Trivers 1972; Halliday 1978; Thornhill 1980, 1981; Wade & Arnold 1980; Arnold 1983). Many fascinating and diverse examples of male combat are coming to light (e.g. Lloyd 1979), and indeed this aspect of sexual selection is better documented and less controversial than intersexual selection.

9.2.3 Intrasexual selection

Numerous studies have shown that males that are successful in combat thereby gain high mating success (e.g. Cox & Le Boeuf 1977;

Appleby 1982). In some cases, male phenotypes associated with success in combat have been identified, a very common one being body size (e.g. Davies & Halliday 1979; Berven 1981; Johnson 1982). These characters have almost certainly evolved partly under the influence of intrasexual selection.

Fighting is by no means the only form of intrasexual competition. Males may interfere with or redirect one another's courtship attempts. For example, male bedbugs of the genus *Afrocimex* have external paragenital structures like those of females, which may direct the sexual attention of rivals into inappropriate channels (Hinton 1961, quoted in Lloyd 1979). Males may also destroy each other's sperm; males of the damselfly *Calopteryx maculata* remove sperm from previous matings from females before mating with them (Waage 1979a). Some males may simply search more actively for females or court more than other males.

Such male characters evolve presumably because the gain to the male in mating success exceeds any cost of the behaviour, and evolutionary advance in sexually selected characters will cease when further increase is opposed by natural selection and decrease is opposed by sexual selection. What implications does such an evolutionary process have for genetic variation affecting male mating success in extant populations? The theory suggests that at evolutionary equilibrium under constant selective forces, if heritable variation exists for characters affecting male mating success, then there will be a negative genetic correlation between the mating success of a male and some other aspect of his fitness, such as his own survival or the viability or fertility of his progeny. This follows from the description of the evolutionary equilibrium; alleles increasing the value of a character that increases male mating success must have a pleiotropic negative effect on some other aspect of fitness if the evolutionary equilibrium is to persist. This argument does not apply only to sexually selected characters; it is relevant whenever heritable variation for characters related to fitness occurs in equilibrium populations (Rose & Charlesworth 1981a, b; Rose 1982).

The pleiotropic negative effect may or may not be related to the character itself. For example, large males might have high mating success and large adult size might be disadvantageous because of increased attractiveness to predators. Alternatively, alleles increasing adult size might do so as a result of a high growth rate, which might result in the diversion of nutrients away from energy storage compounds and hence in an increased susceptibility to starvation during early life.

Thus one body of theory about populations at evolutionary equilibrium predicts that any heritable genetic variation affecting one fitness component will have exactly balanced pleiotropic effects on some other fitness component.

This theory assumes that there is no net fitness heritability, as would be predicted for an equilibrium population subjected only to constant selection (Falconer 1981). If there is net fitness heritability, then the balance between pleiotropic effects on different fitness components cannot be perfect. There are no data on net fitness heritability for natural populations, but theory suggests that both recurrent mutation (Lande 1976) and variable selection (Felsenstein 1976; Hamilton & Zuk 1982) could produce net fitness heritability. Where fitness heritability occurs, heritable genetic variation affecting male mating success could be associated with high net fitness, either because alleles producing high mating success produce a smaller negative effect on other fitness components or because they have a positive effect on other fitness components. For example, a new deleterious mutation affecting an important metabolic pathway is likely to have a negative effect on several fitness components.

There are virtually no data on the genetic correlation between male mating success and other fitness components. In one experiment, *D. melanogaster* allowed to mate non-randomly produced more viable offspring than flies mated at random (Partridge 1980). If variation in male mating success was responsible for the non-random mating then these results may indicate that these two fitness components are genetically correlated, although the viability of the male parents and the male mating success of the progeny would have to be measured to confirm this.

It is important to realize that phenotypic correlations between different fitness components cannot necessarily be used to deduce genetic correlations, because the effects of the environment may obscure the latter. For example, in one experiment on *Drosophila melanogaster*, males that succeeded in inter-male competition for mates had higher longevity than other males (Partridge & Farquhar 1981). These data cannot be used to deduce the nature of the genetic correlation between these two fitness components; in this study both components were related to male size, and variation in size was associated with the degree of competition experienced during larval life. To deduce the genetic correlation it would be necessary to hold the environment constant and study the inheritance of both fitness components for at least one generation. Recent theoretical models have suggested that a positive correlation might be expected between sexually selected characters and general phenotypic condition, essentially because animals in good condition will be able to afford to channel more resources into reproductive effort of all kinds (Andersson 1982a; Parker 1982). If general phenotypic condition reflects genotype, then the genetic correlation with sexually selected characters may also be positive.

There is thus an important unresolved problem concerning the degree and consequences of heritable genetic variation underlying sexually selected characters and this topic would undoubtedly repay further study.

9.2.4   Intersexual selection

The second mechanism for sexual selection proposed by Darwin stressed the importance of mate choice by females. His argument was primarily concerned with explaining the evolution of elaborate male ornaments, often referred to as epigamic characters, that appear to be detrimental to survival. Ever since its proposal, intersexual selection has been the subject of controversy. In part this controversy revolves around the selective forces that might cause the evolution of female preference for particular male characters.

While female choice for male characters obviously occurs, it is not at all certain that it is a product of selection on female behaviour (Parker 1982). Females have a particular perceptual make-up that makes some sights, sounds or smells conspicuous to them. For example, during the breeding season female natterjack toads (*Bufo calamita*) move towards the loudest source of male calls they can hear (Arak 1982, 1983). This behaviour results in higher mating success for males that call loudly, and it seems likely that selection has acted in the past so as to increase male call loudness to the point where the costs of calling outweigh the benefits. In contrast, the female's tendency to approach loud sound may simply represent a constraint of her nervous system, without any evolutionary implications for her behaviour.

Certain aspects of female choice are well documented and present no theoretical difficulty. These are all cases where the female chooses a mate with a phenotype or a resource that is beneficial to her own survival or to successful rearing of her offspring. For example, several studies have shown that females are more likely to mate with males holding resources such as high-quality territories (Davies 1978a; Chapter 10) or nuptial gifts (e.g. Thornhill 1981). On an evolutionary time-scale, competition between males for resources attractive to females may then evolve (section 9.2.3). Where there are problems, both with theory and inadequate data, is where female choice is postulated to have evolved for particular male phenotypic characters solely because of the genetic consequences for the offspring.

The evolution of female choice: some theory

Although Darwin initiated the idea of female choice as a selective force on male secondary sexual characters, he was unable to explain the evolutionary origin of female preferences. Fisher (1930) produced a theory of how they might evolve; this has been described by Halliday (1978) and will be discussed here only briefly.

Fisher's model started with a heritable male character advantageous under natural selection and a heritable female preference for the new male character. Non-random mating then occurs, females with the preference being more likely to mate with males with the character. Sons from these matings carry their mother's preference genes and their father's character genes and, because of the latter, have higher survival than the sons of males without the character. They transmit both their parents' genes at mating, and the preference genes therefore spread initially because of their association with the male character genes, brought about through non-random mating. At some point the female preference reaches a high enough frequency for a new process to start; females mating with males with the character leave more grandchildren, because their sons are preferred by females and therefore have higher mating success than males lacking the character. This process involves positive feedback, so that what Fisher called a 'runaway process' occurs. Fisher did not produce an explicit genetic model of this process, but presumably visualized the female preference and male character under polygenic control, and he suggested that the male character would become exaggerated to the point where its disadvantage under natural selection was balanced by its advantage under sexual selection if females chose males on the basis of relatively high values of the character.

O'Donald (1967, 1980) subsequently produced two-locus genetic models of Fisher's process which confirmed the spread of the female preference but not the runaway process. His models highlighted the point that the dynamics and outcome of sexual selection are affected by genetic constraints such as dominance and linkage. Following an earlier suggestion by Zahavi (1975, 1977a), a more recent genetic model (Andersson 1982a) has indicated that Fisher's runaway process can be initiated by a deleterious male character, provided only those males that are highly fit in other respects can afford the cost of the character.

Two recent genetic models (Lande 1981; Kirkpatrick 1982) have independently pointed out an important feature of models of female choice. If we consider a disadvantageous male character, then for any degree of disadvantage there is a corresponding strength of female preference which can maintain the male character in the population. In other words, there is a whole range of equilibrium values for a population where a deleterious male trait and the female preference for it are in balance (Fig. 9.1). The line of equilibria in Lande's (1981) model may be stable or unstable in that, depending on the detailed genetic parameters of the population, perturbation away from the line may result in selection pushing the population either back to equilibrium (although not necessarily to the same point on the line) or away from equilibrium at an ever increasing rate, as indicated by the faint lines in Fig. 9.1. Following perturbations from the equilibrium line, evolution back to or away from the line can be rapid, and random factors acting in different isolated populations could cause them to diverge, perhaps causing speciation (section 9.6).

In these models female preference is assumed to be selectively neutral in that the female's preference does not result in any immediate benefits for herself or for rearing her offspring. This selective neutrality means that female preference for a male trait can drift upwards from zero in the absence of the trait. If the trait then appears

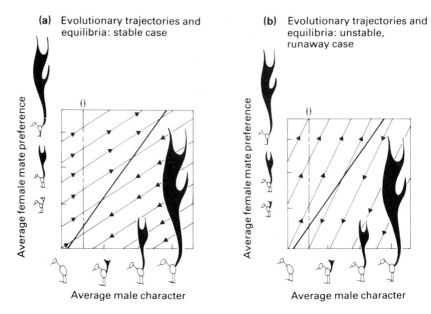

(a) Evolutionary trajectories and equilibria: stable case

(b) Evolutionary trajectories and equilibria: unstable, runaway case

Average female mate preference

Average male character

Fig. 9.1 Lande's model for the evolution of a male character through female choice. In each figure, the heavy line represents a continuum of equilibrium values at which a deleterious male trait and female preference for that trait are in balance. Departures from equilibrium due to random genetic changes may result in selection back towards equilibrium **(a)** or away from it **(b)**. θ represents male tail length in the absence of sexual selection. (From Harvey & Arnold 1982.)

by mutation, and is disadvantageous under natural selection, it can nonetheless become established at low frequency in equilibrium with the previously established female preference. Thus Lande's and Kirkpatrick's models emphasize the role of genetic drift both in establishing female preference for disadvantageous male characters and in perturbing equilibrium populations in a way that can lead to rapid deterministic evolution.

Thus Fisher's (1930) assumption that female preference can only evolve for initially advantageous male characters may be unnecessary. In addition, Lande's (1981) model does not require that females choose males on a relative criterion; choice for an absolute value of the character can have the same evolutionary effect.

Thus several theoretical models give plausible accounts of how female preference for male characters might evolve. What evidence is there that female choice may have been an important force in the evolution of male secondary sexual characters and for the role of female choice in extant populations?

The evolution of female choice: some evidence

There are a number of difficulties in detecting the occurrence of female choice.

Female behaviour that has evolved because it resulted in mating with males with particular genotypes may not be detected as female choice because it may have other causes unrelated to the mate thereby acquired. For example, female *D. melanogaster* often move while being courted. This may have the consequence that small males are more likely than large males to lose the female during courtship. Female movement during courtship may have evolved solely or partly as a consequence of selection for mating with males that are large for genetic reasons. However, this would not be detected as a case of mate choice on the basis of behavioural observations, because the female may move because she is unreceptive to any male or because she is searching for food. Similarly, the type of passive attraction to loud sounds described in natterjack toads (Arak 1982, 1983) may not be entirely the consequence of nervous system constraints; female behaviour may have evolved to make matings with loud callers more likely. Disentangling the evolutionary antecedents of such behaviours is virtually impossible. Many cases of intersexual selection may therefore go undetected as such.

Female choice may also appear to be involved when it is not occurring. For example, cases where the identity of the female has an effect on the identity of the male achieving mating have sometimes been advanced as evidence for female choice. Assortative mating (section 9.4) and inbreeding (section 9.3) are examples. However, these mating patterns may occur if different morphs are active in different habitats or at different times, or if there are interactions between male courtship and female receptivity (Bryant 1979). These mating patterns are therefore not in themselves evidence for female choice.

Studies of female choice for male secondary sexual characters can be problematic also because there are often confounding variables that may mask the choice. The two most common confounding variables are non-genetic benefits to the female or her progeny and inter-male competition.

Masking by non-genetic benefits to the female or her progeny

In many species, males obviously do provide for their mates or progeny. It is not easy to be sure, even in those species where females associate with males only for mating, that choice does not have non-genetic consequences. For example, males may vary in fertility, and females may be selected to discriminate between males of different fertility. In the lemon tetra (*Hyphenobrycon pulchripinnis*) male fertilization rate falls as a function of spawning frequency, and females show a preference for males who have not recently spawned and whose sperm supply is therefore large (Nakatsuru & Kramer 1982). The potential for such discrimination exists in other species; male courtship intensity is related to the amount of sperm available in newts (Halliday & Houston 1978) and butterflies (Rutowski 1979). Choice by males for larger, more fecund females occurs in several species (Manning 1975; Gwynne 1981; Downhower & Brown 1981). In addition, male provision for the female may take a very subtle form. In several insects seminal fluid contains nutrients that are absorbed by the female and metabolized in her own body maintenance and in egg production (Boggs & Gilbert 1979; Boggs & Watt 1981; Marshall 1982). It is not known if this type of male provision is widespread, but where it occurs some males may produce more nutrients than others and females will be selected to recognize and mate with more productive ones. There is as yet no evidence that females can recognize such males prior to mating, but there are data suggesting that female butterflies remate more quickly after mating with a male who produced a small spermatophore (Rutowski 1980; Boggs 1981).

Mate choice may also incur costs in terms of wasted time, failure to mate, predation and injury by males (Parker 1982). These risks will mean females may be selected either not to be choosey or to prefer low-risk males. For example, a common finding in lek species is that females tend to mate with males in the centre of lek (e.g. Buechner & Roth 1974, but see Chapter 10); they may simply be less vulnerable to predation in the central area. Similarly, passive attraction to conspicuous males may minimize the costs associated with finding a mate of the correct species.

In cases where non-genetic benefits occur, to demonstrate choice that has evolved solely because of genetic consequences it is necessary somehow to rule out the non-genetic benefits. If, for example, non-genetic benefits could be manipulated so that females could be shown to choose particular male characters irrespective of the non-genetic benefits, then the evidence would be good. More experimental studies of this kind are needed.

Masking by intrasexual competition

Competition between males can obscure the effects of female choice and this has led to some confusion in the literature as to what exactly constitutes female choice, because 'mate choice' is often used loosely (and incorrectly) to describe any instance of non-random mating, even when intrasexual competition has not been ruled out.

Consider a virgin female fruitfly (*Drosophila melanogaster*) that is courted by several males. She apparently ignores some, moves away from others, kicks one or two, and eventually mates with one. In such a situation, there is a higher probability of her mating with a large male than with a small one (Partridge & Farquhar 1983). Is this female choice? Though similar observations are often attributed to choice, the evidence is insufficient for such an interpretation. It is quite possible that each female has a fixed probability of accepting *any* courting male. Large males might then have higher mating success because they are better at searching for females or at tracking them during courtship, or they may be more strongly motivated to court than small males. One could eliminate these forms of male competition by studying the courtship of large and small males and seeing if, when courtship time was controlled for, females were more likely to mate with larger males. One would then have to rule out passive attraction; fruitfly males produce courtship songs which have been shown to affect male mating success (Ewing & Bennet-Clark 1968). These songs are produced by the wings, and the larger wings of large males probably produce louder songs than those of small males. The high mating success of large males may therefore be due to their being audible to the female from a greater distance. This possibility could be ruled out as the sole explanation by a demonstration that females actively reject small males more often than large ones; this could not be explained on the basis of male detectability. In those species where females approach males for mating, it may be possible to score the number of times different classes of males are approached by females and to rule out passive attraction as an explanation. All of these types of evidence for choice require that behaviour be observed directly; it is not possible to infer choice simply by scoring matings, even where males are tested individually so that combat can be ruled out.

Thus to make a study of active female choice we ideally require cases where passive attraction, intrasexual competition and non-genetic benefits can be ruled out. Only then do we have clear evidence that female behaviour has been modified during evolution solely so as to cause mating with males possessing particular genotypes.

Evidence for active female choice

The existence of elaborate species-specific male courtship patterns throughout the animal kingdom, together with many demonstrations that these induce females to mate (section 9.6), must mean that active mate choice has occurred in the past and that it plays a major role in reproductive isolation between species. One would otherwise have to seek explanations for the selective responsiveness of females to conspecific males in terms of characteristics of their perceptual systems evolved in contexts other than mating. This seems highly unlikely in view of the complexity and extreme species-specificity of courtship patterns. Thus, selection for reproductive isolation from closely related species may be an important cause of the evolution of female preference (Halliday 1978). Is there any evidence for active mate choice within a species population? The obvious place to look might be those species generally believed to show epigamic male characters.

In the literature on evolution in general and on sexual selection in particular, it is common for the consequences of intersexual selection to be exemplified by the peacock and birds of paradise. Evidence that females actually choose their mates in these species is, however, slight or non-existent. Indeed, some recent studies suggest that elaborate male plumage in these birds may be, at least in part, the evolutionary result of inter-male competition; males may be intimidated by the elaborate plumage of rivals in aggressive encounters.

The mating system of birds of paradise has been described by Le Croy and co-workers (Le Croy *et al.* 1980; Le Croy 1981). In *Paradisaea decora*, only some males have elaborate plumes. Mating occurs in a confined area, similar to a lek, containing a few tall trees. Each tree is occupied by two plumed males who are visited frequently by other plumed and unplumed males. The resident males display to one another, perform loud duetting songs and chase other males away. When a female visits a tree, the resident pair of males switch to performing silent displays to her. At this stage, any other plumed males in the tree leave, but unplumed males remain. The display reaches its peak when one of the two resident males moves aside, leaving his partner to display alone. The attending unplumed males then all copulate briefly with the female before the remaining plumed male starts a prolonged mating bout with her. Le Croy found no evidence for active female choice. Within a tree, the same male performed all of the long final copulations that occurred, and was clearly dominant over both his resident partner and other males that visited his tree. The male dominance hierarchy is the result of display interactions between males that occur throughout the year.

It may be a feature of many lekking species that male competition is an important determinant of male mating success. This will be true where female choice is directed towards males holding particular territories, usually central ones, and where possession of those territories is achieved by male competition. Studies of manakins, *Manacus manacus trinitatis* (Lill 1974), sage grouse, *Centrocercus urophasianus* (Wiley 1973), and Uganda kob, *Adenota kob thomasi* (Buechner & Schloeth 1965), provide evidence that females choose particular positions in leks rather than particular males (but see section 10.7.2). There could be a variety of purely phenotypic benefits involved here, such as protection from predators during mating. There is a clear need for studies attempting to unravel the relative importance on leks of female choice and inter-male competition.

It is perhaps in the Phasianidae that the most extreme epigamic plumage occurs. The evolutionary pressures that brought about the peacock's tail are something of a mystery because no field studies of peacocks have been made (Ridley 1981). However, Davison (1981) has studied a related and equally striking species, the argus pheasant (*Argusianus argus*), in its forest habitat. Some males hold display sites which they largely clear of foliage; other adult and sub-adult males are non-territorial and move about widely. Davison found no evidence that females sample several males and exercise choice on the basis of male plumage, but suggests that they only mate with males that hold territories.

Such field studies as have been carried out on species in which the evolution of elaborate male plumage has classically been attributed to female choice generally fail to support that hypothesis unequivocally. There is a need for studies where male plumage is experimentally altered in an attempt to study female choice.

Widow birds

Active mate choice has recently been studied experimentally in widow birds (*Euplectes progne*) by Andersson (1982b). Males of this species defend breeding territories within which the inconspicuous females nest. The males have a very long tail which is used in display. Andersson experimentally altered the tail length of territorial males, lengthening some tails, shortening others and simply cutting and restoring others as a control. As a result of this manipulation, the long-tailed males had more nesting females in their territories than the short-tailed and control males. Thus females exerted a preference for the experimental long-tailed males.

It is not entirely clear what evolutionary significance should be attached to this result. It could be a case of passive female attraction, since the long-tailed males may be more conspicuous, but females visit several males before settling. Ideally, evidence that females actively reject some males in favour of others is needed to establish the occurrence of active choice.

Elephant seals

One of the few cases where active rejection of some males by females has been demonstrated is in the elephant seal (*Mirounga angustirostris*) (Cox & Le Boeuf 1977; Cox 1981). In this species, females are more likely to make a vocal protest when mounted by a low-ranking male, and the effect of their protest is to attract other males who attempt to displace him. The net effect of this female behaviour is to make mating by a dominant male more likely.

In this example, females are effectively choosing to mate with larger, older males. The evolutionary significance of this behaviour is uncertain. It is possible, for example, that older males are more fertile or that by mating with them females avoid injury in fights between males. Cox and Le Boeuf suggest that genetic consequences for the offspring have been responsible, the larger, older males having indicated their ability to survive.

Selection on female choice in extant populations

The evolutionary models suggest that female choice can evolve for male characters with a variety of initial effects on other fitness components. As with intrasexual selection (section 9.2.3), evolution will cease when further change in the male character, or possibly in female preference, is opposed by an adverse effect on other fitness components. When discussing intrasexual selection we pointed out that sexually selected characters might be expected to show a negative or positive correlation with other fitness components. The same comments apply to epigamic characters. Parker (1983b) has suggested that fixation of the alleles producing the preferred male character might be expected to be followed by a decline in the frequency of the alleles producing the female preference if choice is costly. Clearly this will depend partly on whether there is heritable genetic variation for the preferred male character, and empirical studies on this point would be extremely valuable.

9.2.5 Alternative mating strategies

One effect of the recent increase in studies using individually marked animals has been to show that different individuals may use very different behaviours during competition for a limited resource such as mates. These different behaviours have come to be known as alternative strategies. One example has been described by Thornhill (1979a and b, 1981) for scorpion flies (*Hylobittacus apicalis*). In this species, males can only obtain matings by offering a nuptial gift of an arthropod prey item that is consumed by the female during copulation. Males use three different behaviours to obtain prey items. They may simply catch their own, they may steal prey from other males by force, or they may steal them by mimicking female behaviour so that they are offered a nuptial gift by another male.

The existence of a mixture of behaviours in a single population raises interesting questions. Is the mixture a consequence of fixed individual differences or does each individual produce the same mixture? If there are individual differences, are they the consequence of genetic differences? If so, why is the population polymorphic? If not, what is the causation or ontogeny of the different phenotypes? What functional explanations can we find for the mixture?

It is probably not unfair to say that the theoretical treatments of the functional significance of alternative strategies are considerably more advanced than the data (Davies 1982; Dunbar 1982). This is not surprising. Critical testing of the alternative hypotheses requires the sort of information that is extremely difficult to collect in the field, namely a description of the occurrence, costs and benefits of different behaviours measured over the lifetimes of known individuals. The vast majority of studies so far involve only short-term measurements that suffer from the problems outlined in section 9.2.1. In addition, information about the causation and ontogeny of the different phenotypes so far described is virtually non-existent and more experimental studies are required. What follows is therefore an account of some of the fascinating ideas about alternative mating strategies, with a rather more tentative mention of possible examples (see also Chapter 2).

Behaviour contingent on environment

One of the simplest functional explanations for a mixture of behaviours is that the most appropriate course of action may depend upon the environment. For example, male speckled wood butterflies (*Pararge aegeria*) use two different methods to locate females. Early in the morning or on cloudy days, the males patrol the ground and woodland canopy searching for females, whereas when patches of sunlight form on the ground they are defended by some males that sit and wait for females (Davies 1978b). During the time such sun-patches are available, the males sitting in them encounter more females than do the patrolling males. In this case, individual males can show both behaviours, and removal of a male from a patch of sun results in a patrolling male taking it over. It therefore seems likely that individual males are responding in the very short term to sun-patch availability. When sun-patches are not available, all males patrol because, in functional terms, this is probably the most effective way to locate females. Similarly, male red-spotted newts (*Notophthalmus viridescens*) can adopt two different sexual behaviours depending upon the initial responsiveness of the female; if the female does not move away when the male approaches he performs a hula display, while if the female moves away he captures her forcibly and goes into amplexus (Verrell 1982b). Individual males can show both behaviours. As in the case of the speckled wood butterfly, each behaviour may be appropriate in context; capture and amplexus may be necessary if an initially unresponsive female is to be persuaded to mate, while the hula display is less costly in time and energy and may therefore be more appropriate if the female needs less persuasion. Amplexus is also used, even when the female is responsive, if rival males are nearby (Verrell 1983). Under such circumstances amplexus enables a male to monopolize a female.

Variable environments may also maintain genetic polymorphisms for mating behaviour. One possible candidate is the ruff, *Philomachus pugnax* (Hogan-Warburg 1966; van Rhijn 1973). Most males are dark in colour and defend a small mating territory on a communal lek. A few males have very pale plumage and are found within the mating territories as 'satellites'. Satellite males are sometimes able to mate with females when the territorial male is preoccupied. The relative mating success (short term) of resident and satellite ruffs varies with season and lek size, and it may be that these environmental effects maintain a polymorphism for plumage and behaviour. However, it is by no means certain that the differences between residents and satellites have a genetic basis, because pedigree data are not available and there are anyway other possible (not necessarily alternative) functional explanations for the mixture (see Frequency-dependent equilibrium, below).

Behaviour contingent on phenotype

The phenotype of an animal may have a large effect on the outcome of its behaviour (Parker 1982). For example, in many species fighting ability increases with age up to a point and then declines. Under these conditions, young and old animals may have the same fighting ability, and we might therefore expect them to be equally prepared to fight for access to females. However, fighting in general incurs risks, and young animals with their increasing fighting ability and a potentially long reproductive life ahead of them may be less prepared to incur these risks than old individuals with low reproductive values (Fisher 1930). There is some limited evidence that such a pattern of age-related aggression is seen in primates (Clutton-Brock & Harvey 1976).

Probably the most important phenotypic variable affecting competitive behaviour is what Parker (1974b) has called resource-holding potential (RHP) which is a measure of the competitive ability of an animal (Chapter 2). For example, in red deer, size is closely correlated with fighting ability and both increase with age, at least initially (Clutton-Brock *et al.* 1979). This means that males in the prime of life are successful in fighting to hold harems of females while younger and older males, unable to hold harems, attempt to obtain occasional matings.

Age is not the only variable that can affect RHP. In insects in particular, size is fixed at metamorphosis to the adult stage and is therefore unrelated to age, and large males are often at an advantage in obtaining females. Nutrition in early life is probably responsible for much of this variability in adult size.

Variation in RHP may not always be related to fighting. Parker (1982) and Andersson (1982a) have both produced theoretical (ESS) models of competition for females through differing levels of sexual advertisement. Males produce a signal that makes them conspicuous to females, who exert passive choice (section 9.2.4) between the males. The more signal a male produces relative to his competitors, the larger his personal zone of attraction for females. Advertisement incurs costs (e.g. energetic or predatory) and benefits (matings). Both authors conclude that, under these conditions, individuals with high RHP will attract more males because, for a given cost, they can advertise to a higher level. There have been very few studies of variation in advertisement level, and this is an area that would repay a great deal of further study (see also section 9.2.4). In species where some males advertise and others are satellites, advertisers do sometimes incur extra costs because of increased risk of predation (Howard 1979) or parasitism (Cade 1979, 1980). Male guppies (*Poecilia reticulata*) increase the rate at which they display to females, when other males are present (Farr 1976). In a competitive situation they have to incur higher energetic costs to gain the same benefits.

Frequency-dependent equilibrium

Frequency-dependent selection occurs when the fitness of a phenotype depends upon its frequency (see also section 9.5). Several recent studies have raised the fascinating possibility that selection on different competitive behaviours may often be frequency-dependent because selection on one behaviour will depend upon what the rest of the population are doing. If the frequency-dependence is such that the fitness consequences of a behaviour decline as its frequency increases, this may tend to stabilize the population with a mixture of behaviours each of which has the same fitness consequences at equilibrium.

Possible candidates for this type of equilibrium are cases where some males advertise and others act as satellites. In these cases, the advertisers essentially create a niche for satellites, and it seems intuitively plausible that as the frequency of advertisers increases, satellite behaviour will become more beneficial, while as the frequency of satellites rises, fewer females will be attracted so that advertising may be favoured. In the tree frog *Hyla cinerea* (Perrill *et al.* 1978) the two strategies do appear to have approximately equal mating success.

A mixture of behaviours with approximately equal fitness consequences could arise in several ways. It could be that each individual can produce all the behaviours, and either does so with fixed probabilities or adjusts its behaviour to the payoffs it experiences. In the former case the behaviour is usually known as a mixed evolutionarily stable strategy (mixed ESS). Alternatively, the different behaviours may be a result of a genetic polymorphism, each individual showing only one of the behaviours. The behaviours are then known as pure strategies and the population is said to be in an evolutionarily stable state (Chapter 2).

One possible candidate for an evolutionarily stable state is the ruff. As we have seen, there are resident and satellite males, and although their short-term mating success seems to vary with their environment, there may also be frequency-dependence at work. Another likely case is the bluegill sunfish, *Lepomis macrochirus* (Gross & Charnov 1980). Males show two mating strategies which may be genetic morphs, although this remains to be shown. Parental males construct nests, attract females and provide brood care, and do not reach sexual maturity until seven years of age. 'Cuckolder' males go through a developmental sequence of sneaking fertilizations early in life and mimicking female behaviour to gain by deception access to spawnings later in life. Preliminary measurements suggest that the two strategies have equal lifetime payoffs, but more work remains to be done here.

Mixed ESSs are difficult to identify, although we have seen

several cases where an individual may produce different behaviours. Dunbar (1982) has made the point that there will usually be some selection pressure for individuals to adjust their behaviour to the payoffs they experience, so that genuine mixed ESSs, with each individual producing the same stochastic mixture of behaviours, may be rare in nature. As yet there are too few data to tell.

9.3 INBREEDING AND OUTBREEDING

9.3.1 The genetic consequences of inbreeding

Inbreeding is the mating of individuals related by ancestry. The essential genetic result of the sharing of a common ancestor by two individuals is that they may both carry a replica of a single allele present in the ancestor, and if they mate they may both pass one of these replicas to their progeny. Two such alleles present in an offspring are identical by descent, and the individual is hence homozygous for that allele. Inbreeding thus tends to increase or maintain homozygosity, whereas outbreeding has the opposite effect (Falconer 1981).

The best-documented effect of inbreeding is lowered offspring fitness, known as *inbreeding depression* (Falconer 1981). This effect is very well known in laboratory and domestic animals, and inbred matings in natural populations of normally outbreeding species can have similar consequences (Packer 1979; Greenwood *et al.* 1978; but see Noordwijk & Scharloo 1981). The most likely explanation for inbreeding depression lies in the presence of deleterious recessive mutations in populations. New mutations are nearly always harmful, but those that are fully or partially recessive can accumulate because they are protected from selection in the heterozygous condition (Simmons & Crow 1977). The homozygosity that results from inbreeding leads to such mutations being expressed and exposed to selection.

Perhaps as an evolutionary consequence of inbreeding depression, many animals show behaviour that prevents inbreeding. One such mechanism is differential dispersal of the two sexes. For example, male olive baboons (*Papio anubis*) transfer away from their natal group before breeding, whereas females remain and mate with males that transfer in, to which they are usually not related (Packer 1979). In mammals, it is generally males that disperse further than females from their place of birth to breed, whereas in birds the reverse is usually the case (Greenwood 1980).

Differential dispersal by the sexes means that inbreeding can be avoided without any necessity for relatives to recognize one another. There is, however, evidence that in some species at least, relatives can recognize one another and refrain from mating. In most of these examples, recognition of relatives is a result of experience of them early in life (e.g. Hill 1974; Bateson 1982), but evidence is accumulating that in some mammals (Wu *et al.* 1980; Grau 1982; Kareem

& Barnard 1982), amphibians (Blaustein & O'Hara 1981) and arthropods (Linsenmair & Linsenmair 1972; Greenberg 1979) kin recognition can occur without prior experience; in the mammalian examples recognition occurred between relatives from different litters, so that experience *in utero* could be ruled out, while in the amphibians experience in the egg mass was possible.

Inbreeding depression clearly can produce strong selection favouring avoidance of inbreeding. It is less clear whether selection can act against high levels of outbreeding. While it is obvious that avoidance of hybrid matings with other species will be favoured (section 9.6), it is not clear whether, within the confines of a single species, there is some optimum intermediate level of outbreeding.

Two studies suggest that strong outbreeding may be avoided. Japanese quail of both sexes prefer intermediately related mates to individuals that are very close or very distant relatives (Bateson 1982). This preference is based on early experience of brood-mates. These results may not reflect the situation in the wild perfectly, because the experiments were done with birds from a laboratory colony that may be more inbred than is normal in nature; the birds appear to be using some character of their brood-mates, such as plumage pattern, in forming their mating preferences, and the variation in this character within and between broods may differ between wild and laboratory colony birds. Another study has shown that female great tits (*Parus major*) tend to be found paired to males that sing slightly differently, but not very differently, from the females' own fathers (McGregor & Krebs 1982a). It is not yet known what degree of relatedness between mates exists or how the mating pattern is mediated. There is a clear need for studies, preferably under natural conditions, of the fitness consequences of matings between individuals of different degrees of relatedness. Unfortunately, for more distant relatives to be identified, long pedigrees are needed, so if long-term studies are to be avoided it will be important to pick organisms with a reasonably short generation time.

There may well be a cost to strong outbreeding in animals that are specifically adapted to local ecological conditions. If offspring tend to encounter the same environmental circumstances as their parents, either because they are strongly philopatric (Greenwood 1980; Shields 1982) or because of habitat selection (Partridge 1978), then mating within the same local population will act to maintain the adaptation in the offspring, because it will reduce gene flow from other, differently adapted populations. One would expect, therefore, assortative mating within ecotypes of the type discussed in section 9.4.

9.3.2   Possible benefits of inbreeding

Some recent field studies have shown that animals do sometimes mate with close relatives. Several studies of Hymenoptera (Hamilton 1967; Alexander & Sherman 1977; Cowan 1979; Waage 1982) and one of deer (Smith 1979) have demonstrated close inbreeding in nature. The deer study may not reflect the natural situation because small habitat islands and unnaturally high population densities could have distorted the normal mating pattern.

A large advantage of inbreeding may stem from an inclusive fitness effect which may differ for the two sexes (Maynard Smith 1978b; Smith 1979; Packer 1979; Parker 1979). If males compete for mates, then a female can increase the fitness of a male relative by mating with him. This could raise her inclusive fitness, provided the increase in the male relative's mating success outweighs the cost of inbreeding depression in the progeny of the mating. It is significant that inbreeding occurs in the Hymenoptera, because here males are haploid. In these species, therefore, deleterious recessive mutations, whose presence in diploid populations is a probable cause of inbreeding depression, are unlikely to accumulate because they will very quickly be exposed to selection in haploid males and hence eliminated. Consequently, inbreeding depression may be very low in these species, so that the inclusive fitness advantage of inbreeding may be relatively very important.

If inbreeding leads to competition between related males to mate with their female relatives, there will be selection for a female-biased sex ratio (Hamilton 1967); a parent that produces many daughters will leave more grandchildren than one who produces many sons, because mate competition between male relatives will reduce their reproductive potential (see section 8.2). This association between inbreeding, local mate competition and a female-biased sex ratio is common in the Hymenoptera (Hamilton 1967; Alexander & Sherman 1977; Cowan 1979; Waage 1982).

9.4 ASSORTATIVE AND DISASSORTATIVE MATING

Assortative mating is said to occur when there is a significant tendency for individuals to mate with partners with a similar phenotype to themselves. The lesser snow goose (*Anser caerulescens*) is dimorphic for plumage colour, with blue and white forms. Approximately 90% of birds pair with a partner who is the same colour as themselves, as a result of a preference developed in early life for a bird to pair with a mate the same colour as members of its family (Cooke *et al.* 1976; Cooke 1978). Similar effects have been described for domestic fowl (Lill & Wood-Gush 1965; Lill 1968a, b) and for captive mallard ducks (*Anas platyrhynchos*) (Cheng *et al.* 1978). Assortative mating on the basis of local song dialects may occur in white-crowned sparrows (*Zonotrichia leucophrys*), females showing a strong tendency to pair with a male whose song is characteristic of their natal area (Baker 1982; but see Petrinovich *et al.* 1981). In none of these examples is the adaptive significance of assortative mating at all clear. In the snow goose, there are no differences in reproductive success between assortatively and non-assortatively mated pairs, while for the spar-rows there are not yet published data on this point.

Canadian three-spined sticklebacks (*Gasterosteus aculeatus*) exist in two forms, an exclusively freshwater type and a second type that breeds in fresh water but otherwise lives in the sea. In choice experiments, both sexes showed a significant tendency to mate with a partner of the same type (Hay & McPhail 1975). This assortative mating between ecotypes appears to have obvious adaptive value; hybrid progeny would probably be less well adapted to either the anadromous or the purely freshwater way of life than pure-bred progeny. There are, however, no empirical data that this is so.

The most commonly reported form of assortative mating is that based on body size, whereby individuals pair with a partner of comparable size to themselves. In no case has this been shown to be due to individuals exercising mate choice. Usually, all members of one sex share the same preference but only some are able to exercise it. Thus, all males may prefer larger, more fecund females, but only the bigger males, because they have an advantage in male–male competition, are able to gain access to them. This kind of effect has been reported in the freshwater crustacean *Asellus aquaticus* (Manning 1975; Ridley & Thompson 1979) and the weevil *Brentus anchorago* (Johnson 1982). In *Gammarus pulex*, size-assortative mating appears to involve neither choice nor competition, but to result from a non-random distribution of individuals within the habitat with respect to body size (Birkhead & Clarkson 1980).

The evolutionary consequences of assortative mating, both at the individual and at the population level, will depend on the heritability of the character on which it is based. If the heritability is high, the progeny of assortatively mated parents will be very likely to inherit genes for the character. Within a population, assortative mating will have a variety of consequences (McLain 1982), including a tendency to maintain genetic variation between individuals. If assortative mating is based on a non-heritable character, such as age or breeding experience, then the consequence will be immediate rather than genetic. For example, a union between two very experienced breeders may produce many more offspring than one between novice breeders, but those offspring cannot inherit that experience. They could, however, inherit the genes that control the assortative mating preference, which would then evolve.

A possible consequence of assortative mating, if it were a very strong effect and were maintained over a long period, would be sympatric divergence of phenotypes, eventually leading to complete reproductive isolation between them. Thus, assortative mating can in theory provide a mechanism for sympatric speciation (section 9.6).

Disassortative mating, in which individuals tend to mate with partners with a different phenotype to themselves, has been described only rarely. Females of certain inbred strains of house mice can show a preference for males of different inbred strains (Yanai & McClearn 1972, 1973) although such experiments can give contradictory results (d'Udine & Partridge 1981). Mice show disassortative mating in respect of genes in the region of the Major Histocompatibility Complex (Yamazaki *et al.* 1976, 1978, 1982). This preference appears to be mediated by the smell of urine; any functional significance of the behaviour is unclear at present. It has been suggested that such a preference will be adaptive because it will tend to prevent inbreeding, and there is some indication that such a system may have a long evolutionary history (Jones & Partridge 1983). The rare-male effect (section 9.5) can be regarded as disassortative mating, and has been suggested to be a form of inbreeding avoidance. Relatively subtle disassortative mating in Japanese quail has been described (section 9.3), and again may be related to inbreeding avoidance.

9.5 FREQUENCY-DEPENDENT MATING

Frequency-dependent mating occurs when the mating success of a particular phenotype depends on its frequency within a population in one of two ways.

1. Positive frequency-dependent mating: the mating success of a phenotype increases as its frequency increases (Fig. 9.2a). Such a mating pattern will usually eliminate genetic polymorphism for a trait very rapidly and is thus not a process likely to be seen operating in nature, but rather one that has produced changes in the evolutionary past. It is implicated, for example, in Fisher's runaway process (section 9.2.4).

2. Negative frequency-dependent mating: the mating success of a phenotype increases as its frequency decreases (Fig. 9.2b). This is a process of considerable evolutionary and present day interest because of its role in the maintenance of genetic polymorphism. It is often known as the rare-male effect.

Numerous examples of the rare-male effect have been reported, for example in *Drosophila* species (e.g. Petit 1958; Spiess & Ehrman 1978), the parasitic wasp *Nasonia* (Grant *et al.* 1974) and the guppy (*Poecilia reticulata*) (Farr 1977). These are all laboratory studies which can be problematic, partly because it is easy to introduce experimental artefacts (Bryant *et al.* 1980; Kence 1981; Goux & Anxolabehere 1980; Anxolabehere *et al.* 1982), but primarily because a phenomenon only has evolutionary significance if it occurs in the field. Only one study, that of two-spot ladybirds (*Adalia bipunctata*) (Muggleton 1979; O'Donald & Muggleton 1979), has demonstrated the rare-male effect in nature (Fig. 9.3). More work is needed to discover how common the effect is under natural conditions.

Why might rare males have high mating success? One possibility is that females might prefer them. It has been suggested that a female preference for male rarity *per se* may be adaptive. Assuming that a

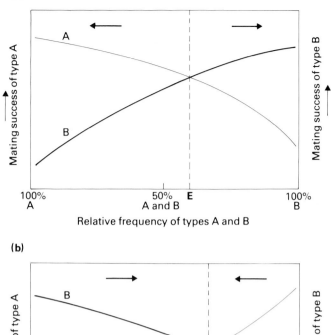

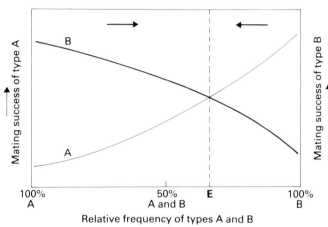

Fig. 9.2 Graphical representations of the two kinds of frequency-dependent mating. **(a)** In positive frequency-dependent mating, the mating success
of each of two types, A and B, in a population increases as they become
relatively more common. This leads to an unstable equilibrium ratio of the
two types, at E. Any departure from this point will lead to selection
(indicated by arrows) that reinforces that departure. **(b)** In negative
frequency-dependent mating, the mating success of each type decreases as it
becomes relatively more common. The equilibrium ratio E is stable in this
case; any departure from equilibrium will be opposed by selection. The
exact shape of the curves is not significant.

female does not know her own genotype and will on average belong
to a common type, she will, by choosing to mate with a rare male,
produce offspring with high levels of heterozygosity and hence,
perhaps, greater fitness (Averhoff & Richardson 1974; Farr 1977;
Lacy 1979; Grant *et al.* 1980). However, there is no empirical evidence that rare males have fitter offspring; indeed, the very fact that
they are rare suggests that the reverse may be true. Rare genotypes
are likely to have low fitness, either because they carry deleterious

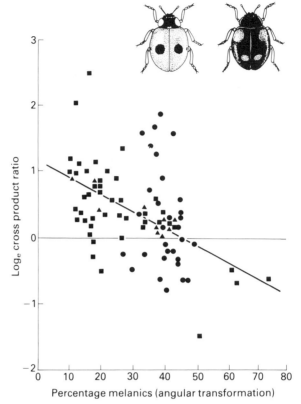

Fig. 9.3 Negative frequency-dependent mating in the two-spot ladybird (*Adalia bipunctata*). The natural logarithm of the cross-product ratio of the two morphs (melanics and non-melanics) in the mating part of the sample and in the whole population is plotted against the angular transformation of the percentage of the melanic morph in the whole sample. A cross-product ratio value of >1 indicates an excess of melanics among mating pairs, a value of <1 an excess of non-melanics. Samples from Moscow and Riga (▲), Potsdam (●) and England (■). (From Muggleton 1979.)

$$\text{Cross product ratio} = \frac{\text{number of non-melanics in sample}}{\text{number of melanics in sample}}$$

$$\times \frac{\text{number of melanics mating}}{\text{number of non-melanics mating}}$$

The natural logarithm is used for reasons of scaling.

mutations, or because they are immigrants from other populations adapted to different ecological conditions. Furthermore, in none of the studies of the rare-male effect is there any evidence that females behave differently towards different male morphs (Partridge 1983). Such evidence is needed to support the hypothesis that female choice is occurring.

Moreover, some sort of frequency-dependent effect may occur if two male morphs of unequal mating ability compete for access to females (Ewing 1978; Hill *et al.*, in preparation). Ewing considered competition between two populations with different distributions of male mating ability. He showed that frequency-dependent mating may occur if truncation selection is applied to the mixed population so that only a certain proportion of the total male population can mate. The direction and extent of this effect will depend on the mean and variance of male mating ability in the two competing strains, and on the degree of truncation.

In other words, the rare-male effect can be a consequence solely of male–male competition, without any female choice necessarily being involved. There is a clear need for more field and experimental studies of the rare-male effect in order that both the importance of the phenomenon in the field and the relative roles of inter-male competition and female choice in producing it can be better understood.

9.6 NON-RANDOM MATING AND THE SPECIES

According to the biological definition of a species, it is axiomatic that across species there is total assortative mating such that individuals avoid hybrid matings. There are, of course, cases where species do hybridize, commonly producing progeny with reduced fitness; a number of such cases have been reported among anuran amphibians (Littlejohn 1965; Kobel *et al.* 1981). Theories of speciation see a role for assortative mating, not only in maintaining the integrity of existing species, but also in the formation of new ones. In allopatric speciation, an initial period of geographic isolation between populations, during which they adapt to different environmental conditions, provides the conditions in which genetic divergence between them can occur. When the two populations come together again, there will be powerful selection favouring assortative mating if there is any loss of fitness as a result of hybrid matings between individuals from different populations. At this sympatric stage, partial reproductive isolation becomes reinforced by the evolution of assortative mating to produce total isolation (Mayr 1963).

Evidence that this process has occurred is provided by cases in which closely related species that are sympatric show more marked differences in reproductive behaviour than other closely related species that are allopatric. Such an association has been shown among *Drosophila* species, for example by Hoikkala *et al.* (1982). Rather more direct evidence comes from studies that have revealed character displacement in reproductive behaviour in the overlap zone between partially sympatric species. For example, two species of frog, *Litoria ewingi* and *L. verreauxi*, have ranges that overlap to a small extent in south-east Australia (Littlejohn 1965; Littlejohn & Loftus-Hills 1968). The advertisement calls of males taken from allopatric populations show only slight differences between the two species. In the overlap zone, however, the pulse rates of the two species' calls are markedly different (Fig. 9.4). It is not only characteristics of male calls that have

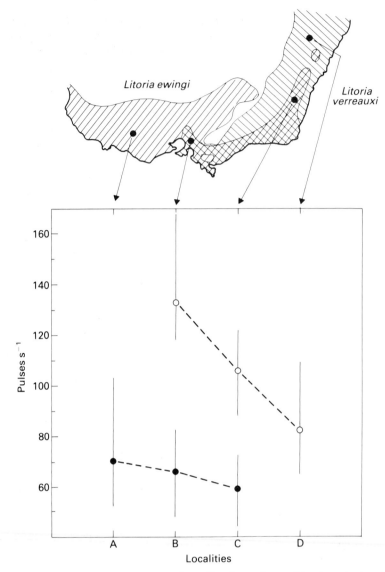

Fig. 9.4 Geographic variation in the calls of two frogs, *Litoria ewingi* and *L. verreauxi*, that are partially sympatric in south-east Australia. The lower part of the figure shows mean values and ranges for the pulse rates of male advertisement calls recorded at four localities (A–D). Whereas calls of the two species from allopatric sites (A and D) are very similar, those from sympatric sites (B and C) show no overlap. (From Littlejohn & Loftus-Hills 1968.)

shifted in the overlap zone; the preferences of females for the kinds of sounds that they will approach are also different between animals from allopatric and sympatric populations (Table 9.1).

Because hybrid matings can incur severe costs in terms of reduced fecundity, there will be powerful selection favouring female choice for any character that clearly identifies a male's species. Such a situation could then be the starting point for intersexual selection,

Table 9.1. Results of experiments to test the responses of female *Litoria ewingi* and *L. verreauxi* to the recorded calls of males of their own and of the other species. All the females were collected from sites within the area of sympatry shown in Fig. 9.4. (From Littlejohn & Loftus-Hills 1968.)

Species of female	Number of tests	Calls presented	Positive responses
L. ewingi	10	Sympatric *L. ewingi*	10
		Sympatric *L. verreauxi*	0
L. verreauxi	9	Sympatric *L. verreauxi*	9
		Sympatric *L. ewingi*	0
L. ewingi	27	Sympatric *L. ewingi*	14
		Allopatric *L. verreauxi*	13
L. verreauxi	7	Sympatric *L. verreauxi*	7
		Allopatric *L. verreauxi*	0

leading to the elaboration of that male character (Halliday 1978). Such a process has been elegantly demonstrated experimentally using *Drosophila* (Crossley 1975).

More controversially, it has been suggested that assortative mating within a species could lead to species formation in the absence of geographic isolation (Maynard Smith 1966). This type of sympatric speciation has particularly been suggested to occur as a result of assortative mating within different host plant races in phytophagous insects (Bush 1974; Tauber & Tauber 1977). Sympatric speciation has recently been severely criticized, on both theoretical and empirical grounds (Futuyma & Mayer 1980). Part of the difficulty concerns the definition of sympatric; the point here is that if movement of individuals is such that members of different ecotypes do not have the opportunity to interbreed, then the two ecotypes are effectively allopatric. The important question is whether speciation can occur in the face of any gene flow and, if so, by what mechanisms. The final stage of allopatric speciation in the hypothesis outlined above occurs in the presence of gene flow, the selective force being provided by lowered hybrid fitness. Whether speciation can be initiated without a barrier to gene flow is uncertain. Some data are very hard to explain with the allopatric speciation model (Gibbons 1979).

Studies of the numerous *Drosophila* species on the Hawaiian islands have led Carson (1982) to suggest that, in many cases, these have evolved by a speciation process in which the widely accepted sequence of events has been reversed. Rather than shifts in behaviour reinforcing genetic differences due to ecological adaptations in founder populations, he suggests that behavioural changes initiated speciation. Certain species differ markedly in their sexual behaviour and in aspects of male morphology associated with courtship displays, but show little ecological divergence and low levels of inviability in hybrids between them. Carson suggests that the first stage in their speciation was that local variants in sexual behaviour became established through coadaptation of male and female behaviour. The differences between these variants then became accentuated by intersexual selection. Finally, populations that were reproductively isolated on the basis of behavioural differences gradually accumulated genetic differences as they adapted to their local ecological conditions.

Lande (1982) has suggested that a similar process may lead to speciation within a cline. Local variations, both in male epigamic characters and in female preferences for those characters, could lead to rapid divergence in mating behaviour and morphology between populations that are not geographically isolated. As in Carson's theory, this hypothesis sees reproductive isolation through mate choice being instrumental at the start of the speciation process, in contrast to allopatric theories which see it as reinforcing already established differences.

Chapter 10
Mating Systems and Ecology

SANDRA L. VEHRENCAMP and
JACK W. BRADBURY

One expects mating behaviour to be a prime focus of selection for all sexual organisms. The ensemble of behaviours and physical adaptations specific to mating, as well as some of the social consequences of these behaviours, are called the 'mating system' of the population. Mating systems exhibit a diversity of forms both between and within populations. One of the early important insights about mating system evolution was the recognition that the form of mating systems is more closely correlated with environmental contexts than it is with phylogenetic heritage. This was first demonstrated in the early 1960s by the pioneering work of Orians (1961), Crook (1964, 1965), Brown (1964), Verner (1964), Willson (1966; Verner & Willson 1966), and Eisenberg (1966) among others. Many of the correlations demonstrated during this period have stood the test of time and the major thrust of subsequent work has been to explain and generalize them.

The classical correlations were made at a gross level of comparison: different mating systems occurred in different habitats. Fitness entered the picture as a third variable since it was assumed that adoption of the observed system in a given site led to higher fitness than the adoption of alternatives. To understand why the observed correlations occur, one must examine all the possible links which exist between habitat, mating system, and fitness. Since each of these gross variables consists of a large number of components, the list of potential relationships is enormous. The hope has been that a wise combination of theory and field work would allow us to sort through this greater list and identify a smaller subset sufficient to explain the general rules.

The overall list of relationships can be thought of as a set of functions. For example, if we divide overall fitness into a set of exclusive components, each component has a function which predicts that component's values when particular mating systems are adopted in a particular habitat. These functions will of course have a large number of arguments, since a mating system consists of a *set* of compatible behavioural patterns and a habitat must be described by a number of ecological variables. The research strategy has been to make educated guesses about which fitness components are under the strongest selection and for the moment ignore the others. Once a component is

251

selected, one makes additional guesses as to which behavioural and ecological variables should be examined and which ignored in subsequent analyses. Next, one derives, measures or guesses at the form of the functional relationship between the retained behavioural and ecological parameters on the one hand and the selected fitness component on the other. Finally, one compares predictions from these simplified functions with the observed distributions of particular mating systems in particular habitats.

There have been three discernible generations of implementation of this paradigm. Early studies took hints from the observed correlations and devised *post hoc* scenarios that would explain the patterns seen. While this approach does not necessarily lead to erroneous conclusions, it was roundly criticized because it only showed that a mating system observed in a given habitat was beneficial, instead of showing that the observed system was better than the alternatives. This ushered in a second generation of studies in which the expected values of specific fitness components were computed for each of several mating system alternatives and the observed patterns compared to those expected if this fitness component were being maximized. A good fit was taken as evidence of progress, and a bad fit as evidence that the wrong fitness component and/or the wrong functional relationship had been selected. We have only recently entered into a third generation of studies based upon the realization that such fitness contrasts must be made with a variety of controls, and interpretations must be made with extreme care. There are at least three reasons for this recent concern.

(1) Predictions made by most second-generation studies are not unique. For example, if the assumption that annual fecundity is the primary selective focus leads to predictions that mating system A should occur in habitat 1 whereas mating system B should occur in habitat 2, a good fit says nothing about whether some other fitness component, such as adult survival, might not also make that prediction. Given that the intent of the research is to identify the primary selective foci, clearly some quantitative test is needed in addition to the traditional qualitative one. Such a test would indicate whether one or the other fitness component accounted for the largest changes in overall fitness as habitat and mating system were varied. It would also ameliorate a second problem. It is entirely possible that the values of two separate fitness components change inversely as habitat and mating system are varied. Were they tightly linked, their combination might be invariant to changes in habitat or mating system even though each by itself was quite sensitive to such changes. Again, to rely on a good fit between prediction and observed patterns when using only one of these components and without controlling for the effects of the other would risk erroneous conclusions. Our quantitative test therefore should also include some way to control for negative correlations between components.

(2) Most second-generation contrasts use mean values for the variables in the function arguments regardless of the nature of these functions or any stochastic variability in those variables. This may be quite wrong. Where the argument of a function is variable, and where the function is non-linear, the expected value of the function is not obtained by using the expected values of the argument in the functions. Instead, the argument means must be supplemented with some measure of the variability of each parameter and of the degree of non-linearity of the function. An approximation based on a Taylor expansion is often used in other fields for this purpose (Meyer 1970), but it has only recently been applied to fitness contrasts in the context of social evolution (Caraco 1980; Rubenstein 1982b).

(3) Finally, whenever fitness values depend on the frequencies with which alternative mating systems are adopted in a population, second-generation contrasts may be entirely misleading. This conclusion derives from evolutionary game theory (cf. Chapter 2 and Maynard Smith 1982a). Second-generation contrasts only compare fitness values computed with the assumption that an entire population adopts the same mating system. If this assumption makes no difference to the calculation, the contrasts are legitimate. If, however, the assumption changes the computations, then game theory is required and one must also compare the fitness values of uniform populations with those in which some individuals adopt one strategy and others an alternative strategy. It is entirely possible that, in this case, the analysis predicts uniform adoption of that mating system which in the second-generation contrast showed the *lower* average fitness. Such an outcome, though paradoxical from the second-generation point of view, becomes quite obvious when proper game theory contrasts are made. It is now clear that a wide variety of mating system analyses must be re-examined with game theory protocols.

Ideally, a third-generation approach would measure, rank and examine correlations for all components of fitness concurrently, would adjust for the propagation of stochastic variation through non-linear functions, and would use game theory where appropriate. Other changes, such as the inclusion of genetic constraints, may also be required. For many species, these may be unreasonable expectations, but for a number of selected forms the general goals should be attainable. Once several systems have been studied in this manner, one would at least have precedents to justify the emphasis *a priori* on particular components of fitness or particular behaviours within a mating system. Our goal in this chapter is to encourage this changing of the 'generational guard'. Because many prior studies ignore important components of fitness, we first provide a sample check-list of these components and make some suggestions on how they might be ranked once measured. We then discuss behavioural options which might be included in a given mating system to increase particular fitness components and outline some of the correlations which are generated following adoption of more than one behavioural option. The main body of the chapter examines current thinking about three of the classical categories of mating system with particular emphasis on how third-generation approaches have altered, or are likely to alter, those views. The chapter concludes with some comments on how these new approaches might force us to revise our current methods of classifying mating systems.

10.1 COMPONENTS OF FITNESS

We have argued that one of the major problems with second-generation studies has been the focus *a priori* on a few fitness components and the consequent neglect of other potentially important components. A first step is thus to provide as exhaustive a list as possible. To guarantee an exhaustive list of components, one simply divides the total up into exclusive sections which should recombine to give the measured values. The total 'pie' could be divided any number of ways. Because of basic asymmetries between the sexes, the male and female pies will usually be divided somewhat differently. We shall divide total fitness for males into the following nine components.

1. The number of females encountered and courted per season.
2. The fraction of courted females inseminated.
3. The probability that any given zygote in an inseminated female is fathered by the male in question.
4. The number of zygotes produced by each female inseminated.
5. The survival probability from zygote to maturity of offspring of each inseminated female.
6. The adult survival rate for the male.
7. A correction summarizing any nepotistic consequences of adopting a given behavioural option.
8. A sexual selection correction summarizing the additional benefit gained by exercising 'epigamic' mate choice above and beyond benefits such as mate territory quality, provisioning by the mate, and parental care by the mate. (This term times the actual fecundity can be considered the number of offspring which would have been needed without sexual selection to generate the same number of grand-offspring.)
9. A demographic correction summarizing the effects of early or late maturity in unstable populations.

The corresponding partitioning for females will utilize components 1 and 4–9 suitably defined for the female point of view. For either sex, total lifetime fitness will be an additive and multiplicative combination of these components according to standard life history methods (cf. Chapter 11). The three correction terms can be considered as multiplicative components which take the value of one when a given strategy adoption does not affect them.

Once the separate components of fitness have been measured for a sufficient sample size, recent methods pioneered by Arnold and Wade allow one to compute both the relative intensities of selection on each component and the importance of any covariances between components (Wade & Arnold 1980; Arnold 1983). These are, of course, 'local' analyses in that they rely on measured variations within a sample and may lead to different values for samples taken from other populations in other habitats. Were we to find that the same few components were the dominant foci of selection in a variety of species and contexts, then one could replace second-generation 'guesses' with actual precedents. If, however, no consistent pattern was obtained, the second-generation approach might have to be dropped entirely in favour of these third-generation methods.

10.2 BEHAVIOURAL OPTIONS AND FITNESS COMPONENTS

For each of the fitness components listed above, there are a number of behavioural options whose adoption would enhance that component according to the functional rules between ecology, behaviour, and components of fitness. In Table 10.1, we have provided sample lists of behavioural options for males and females. Unlike the lists of fitness components, the behavioural option lists are undoubtedly not exhaustive. However, the examples in the table give a reasonable feeling for the diversity of options described so far.

Any mating system is an assortment of some subset of the possible options. Each option present might be emphasized heavily or only in a minor way. One would want to characterize mating systems not only by the presence or absence of different behavioural options but also by their relative emphases. In principle, the scaling of the investment in each option should not be made relative to other options, but each investment should be compared with its own maximum when no other options are adopted. These maxima would vary for different options depending on whether energy, time or anatomical constraints limited overall investments. Properly characterizing a mating system by this method requires access to considerable information.

One wants to know why a population adopts a particular ensemble of options in a given context instead of known alternative ensembles. There are two clear criteria which must have been involved in this selection. First, not all ensembles of behavioural options are possible. Since large numbers of options will share the same upper limits on their implementation (e.g. total energy or time available), a heavy investment in one option automatically reduces investment in others. Environmental factors, anatomy and physiology set the values of the

Table 10.1. Behavioural options for increasing fitness components

Components of fitness	Options for increasing component
	A. MALES
1. Female encounter rate	Stay near resources females need
	Stay near major female traffic nodes ('hotspots')
	Stay in sites selected for signal transmission properties and display to attract females
	Search for females either randomly or using cues emitted by them
	Remain with females once encountered
2. Copulation rate	Accelerate physiological receptivity by causing abortion, killing young, feeding female, or assuming some of her current costs
	Provoke mate choice of receptive female by advertising valent cue
	Force insemination
3. Fertilization rate	Exclude other males by territoriality
	Exclude other males by dominance
	Tolerate other males but control order or timing of insemination
	Ream prior sperm out of female tract before insemination
	Displace other sperm with large amounts of own sperm
	Form sperm plug after insemination
	Guard female after insemination
4. Female fecundity	Select preferred mates using size or age as cue
	Provision female to increase zygotes/breeding bout
	Assume some of female's costs to increase zygotes/breeding bout
5. Juvenile survival	Select females of appropriate age, size, status, resource control or genetic complementarity
	Invest territoriality, parental care, mate survival
6. Adult survival	Defer breeding to later ages
	Reduce investment in any of above which cause increased risk
	Defend territory for own access to resources
	Form social liaisons to dilute or share risks and costs
7. Nepotistic term	Modify amount of investments in above so as to increase relative's benefits
	Direct interaction behaviours preferentially towards or away from kin
8. Sexual selection and demographic terms	Unlikely to be important for males in most species, but there are exceptions
	B. FEMALES
1. Encounter rates	Any of methods outlined for males above in 1
	Force males to aggregate in known or accessible locations by differential mate choice
2. Female fecundity	If sperm are limiting, increase encounter rates as in 1, exclude other females from located males by dominance or territoriality,

Table 10.1—(*continued*)

Components of fitness	Options for increasing component
	control order of access to males, or mate repeatedly with the same male
	If resources are limiting, establish own territories, dominate other females, or trade confidence of paternity in exchange for assumption of territorial defence, food provisioning, or parental care by males
3. Juvenile survival	Invest in maternal care
	Select male with quality territory, higher status, parental competence, or genetic complementarity
4. Adult survival and nepotistic term	Identical to male options
5. Sexual selection term	Invest in careful selection of mate using appropriate cues
6. Demographic term	Modulate age of first breeding to maximize demographic benefits

critical limits and play important roles in determining the nature of the correlations between investments in different concurrent options. The first criterion is thus that the set of adopted options be compatible and feasible.

The second criterion is that the observed ensemble be that alternative which produces the higher overall fitness (controlling for game contexts where necessary). Adoption of any option may enhance a given component of fitness, but because of correlations between options, it may also decrease or increase other fitness components. Because of differences in the functions linking behavioural options and fitness components, it is possible that these secondary effects on other components of fitness are larger than the primary effect gained by adoption of the behaviour. In fact, our assignment in Table 10.1 of specific options to gains in specific fitness components is somewhat arbitrary; without foreknowledge of the magnitudes of all effects of a particular option, any assignments are only 'educated guesses'.

We thus have the problem of ranking the contributions that each behavioural option in the mating system makes to the determination of each fitness component. As with the ranking of the contributions that each fitness component makes to overall fitness, there are now quantitative methods for doing just this (Lande & Arnold, in press). These require that we have some quantitative measure of the relative investment made in each behavioural option and corresponding measures for each fitness component. The analyses indicate intensities of both directional and stabilizing selection and they include controls for correlations between option investments. Once completed, one could then identify which behavioural options are most critical in determining variations in the values of each fitness component. Similar multivariate methods can be used to identify the major ecological parameters which affect specific fitness components. With the most important fitness components, ecological variables and behavioural options identified, one would then have some reasonable grounds *a priori* to derive the functions relating these three variables from existing first principles.

10.3   ANALYSES OF SPECIFIC MATING SYSTEMS

No one has ever performed the kind of analysis proposed above. Perhaps, in fact, no one ever will and it is at best only a mental masturbation. The concept, however, has some clear heuristic value if keeping it in mind helps avoid the pitfalls and short-sighted approaches of many second-generation studies. In the following sections, we shall summarize existing thinking about three of the many known mating systems. In so doing, we shall try to outline the core assumptions in prior work and then re-examine these from a third-generation view. Has focus on only a few components of fitness obscured more complicated consequences of adopting particular mating system alternatives? Does the invocation of non-linear functions and stochastic habitat variables change the expected fitness payoffs? Would game models make different predictions from other existing approaches? We cannot cover all mating systems in this way in this chapter. However, we have tried to discuss three well-known examples in sufficient depth to make our point. The systems we have selected are resource-defence polygyny, polyandry, and lek mating.

10.3.1   Resource-defence polygyny

A common male strategy among both vertebrate and invertebrate species is the defence of a resource that females require. By excluding other males from the vicinity of the resource, the defending male can count on female visitation and thus biased mating access to females that visit while receptive. Typical resources include food, refuges and sites for egg deposition or nesting. Where more than one female is likely to visit such a site, the effective result is polygyny. The primary strategy entails emphasis on two components of male fitness: enhancement of encounter rates by male positioning at a known female resource, and guarantee of fertilization by exclusion of other males from the vicinity of the site.

Male defence of 'territories' near resources during the breeding season occurs in many animals; only some are polygynous. Early correlations between the number of mates obtained by resource defence and ambient ecological conditions suggested to Verner (1964), Verner and Willson (1966), and Orians (1969) that much of the variation in mating pattern might be explained by a simple mapping of male territories on habitats of differing spatial heterogeneity. Imagine a grid of male territories of fixed size. If such a grid is superimposed upon a habitat in which the defended resource is uniformly distributed, all males will have equal amounts of resource. If the same grid is imposed on a habitat in which the resource occurs in small, rich patches, some males may have abundant resource within their territories while others may have very little. If females in turn distribute themselves in proportion to local resource abundances, then males with more resource in heterogeneous habitats will acquire more females and those with less resource fewer. Thus the average number of females per male will be the same as in a uniform resource distribution of equal density, but the variance will increase. Where the average number of females per male is one or less, a large variance is required before one observes any polygyny at all; where the average is greater than one, a higher variance increases the levels of polygyny.

The model determining female settlement

This basic notion was summarized in a graphical model by Orians (1969). The total resource in each male territory was plotted on the abscissa and female fitness on the ordinate. A set of S-shaped functions were drawn with higher curves for monogamous females and successively lower curves for the second, third and later arrivals on each territory. The S-shape reflected a female's need for a minimum amount of resource to breed at all and decelerating increases in fitness at higher values. If females settle only according to perceived expectations of resource access, the shape and proximity of the curves and the distribution of resources among males ought to be sufficient to predict both the sequence of settlement and the degree of polygyny at the end of settlement.

In fact, no one has tried to test this model using these curves and male distributions. To see why, note that it is simpler to draw a single function curve for all females: instead of plotting total male resource for each male territory on the abscissa, one can plot the amounts available for all possible female options, for example as values of $k_j Q_i$ where k_j is the fraction of resource on a territory made available at settlement to the jth female settling there and Q_i is the total amount on the territory of male i (Fig. 10.1). As long as this function is monotonic, the ordering of the fitnesses for the entire set of options will be the same as the ordering of the $k_j Q_i$. Although only some of these options are available to any one female, the model predicts she will take that $k_j Q_i$ which is the largest in the set available to her. In short, one need only know the k_j and Q_i values to predict the sequence of settlement and the final numbers of mates per male; the shape of the curve, if it is monotonic, is not relevant. If Q_1 is the resource available on the best territory, a necessary condition for polygyny is that $k_2/k_1 > Q_M/Q_1$ for at least one territorial male M. The higher the variance in the Q_i values, the more likely it is that at least two of

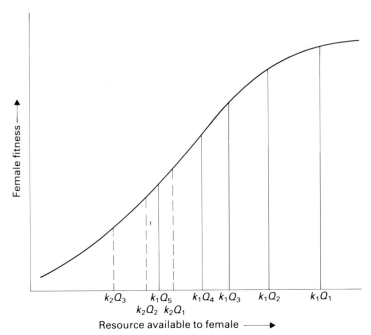

Fig. 10.1 A single utilization function version of the 'Orians–Verner–Willson polygyny model'. Female fitness is presumed to be an S-shaped function of access to resources. The amount of resource made available to a female *at settlement* is $k_j Q_i$ where k_j is the fraction of resource on a male's territory made available to the jth female to settle there and Q_i is the total amount of resource on the territory of male i. In this graph, $k_1 = 1$, $k_2 = 0.5$, and the distribution of Q_i is such that if five or more females settle according to the model, polygyny will occur.

them will be sufficiently different for their ratio to be less than k_2/k_1. This minimal range of differences is known as the 'polygyny threshold' (Willson 1966)

Model testing and selection of fitness components

The obvious way to test this model of female settlement is to identify the critical resources, measure the partitioning of them among the males, estimate the partitioning by females in the same harem, and then see if settlement follows the order predicted. Even this has proved difficult. There have been two major problems: (a) difficulty in identifying the critical resources, and (b) as a consequence, a retreat to more indirect tests, the more popular of which have since been shown to be erroneous tests of the model. The resulting literature is a morass of conflicting claims and counter-claims and, despite recent attempts at clarification (e.g. Wittenberger 1979), adequate tests are still wanting.

To identify properly the critical resource, one must show a causal relationship between one or more environmental parameter and female fitness (correcting for harem size). Parameters correlated with critical resources may be used to rank male territories, but the rankings will only be as good as the correlations and this limits the power of any test. In addition, the relationships, and thus the rankings, will only be valid if the appropriate measures of fitness are used. Preferably one would use all the components of fitness listed earlier; in practice, studies invariably focus on a subset of these. Identifying the correct resources and the correct components of fitness go hand in hand: if major components of fitness are omitted, the wrong resource measures will be used to predict female settlement and any tests will be inappropriate.

As an example, Pleszczynska (1978) in a careful set of studies showed that the critical variable affecting nesting success of lark bunting females is the amount of nesting cover available. Female settlement patterns can even be altered by varying cover in male territories. However, the sequence of settlement does not follow the model predictions: successive female settlers switch to bigamous matings long after the computed threshold favouring bigamy has been reached among mated males (Pleszczynska & Hansell 1980). The explanation provided by these authors is that secondary females do not get parental help from their mates and the weight loss they experience in making up the difference lowers their survivorship expectations. Thus the threshold difference in males expected on the basis of female nesting success alone is too low; overall female fitness depends on both nesting success and future survival. A quantitative demonstration that survival is the key to this mismatch has yet to be performed. The invoking of survival benefits was also suggested by Elliott (1975) as a reason why Downhower and Armitage (1971) did not get a good fit to the model in their study of yellow-bellied marmots. Lenington (1980) found that red-winged blackbird females do not select male territories in which overall nesting success would be highest, but instead choose those in which one component of nesting success, number of young fledged from nests escaping predation, is maximized. She also suggested that adding some omitted components of fitness might make the data fit the model better. Weatherhead and Robertson (1977a, b, 1979) failed to find any good correlates between female settlement and female nesting success in their populations of redwings. Their proposition (later corrected and modified by Heisler (1981)) is that females give up some resource benefits and instead choose males partly for epigamic effects (see above). The importance of this component of fitness is still in considerable dispute (Wittenberger 1981b; Searcy & Yasukawa 1981; Weatherhead & Robertson 1981). Finally, Altmann *et al.* (1977) noted that rather than suffer by dividing up resources, females on a common territory might benefit by cooperating to enhance the fitnesses of all harem members. In these and other studies, the focus on only a subset of the components of fitness has left the interpretation of tests of the model quite ambiguous.

A second cause of mismatch between theory and observation is that one cannot expect females to be perfect judges of available resources. This introduces a degree of stochasticity to the argument of the fitness functions in the polygyny model and, as noted earlier, this must be taken into account when computing expected fitness values. If females do not take this uncertainty into account in selecting mates, the result is simply a poor fit between the model and observation, which makes tests of the model ambiguous. If females do incorporate this uncertainty into their choices, then the model must also incorporate this uncertainty in computed predictions. As discussed at the beginning of the chapter, there are methods which give approximate expected values of a function when the argument is uncertain and the function itself is non-linear (cf. Meyer 1970; Rubenstein 1982b). In this case, we are concerned with estimating the expected fitness $\bar{W}$ for a given combination of k_i and Q_j. The approximation in this case is

$$\bar{W}(k_j\,Q_i) \approx W(\overline{k_j\,Q_i}) + W''(\overline{k_j\,Q_i}) \cdot \mathrm{Var}(k_j\,Q_i)/2$$

where $W(\overline{k_j\,Q_i})$ is the functional expectation given the expected (mean) value of $k_j\,Q_i$, W'' is the second derivative of the function evaluated at the mean $\overline{k_j\,Q_i}$, and $\mathrm{Var}(k_j\,Q_i)$ is the variance in the $k_j\,Q_i$ values. It is immediately obvious that this estimate will produce different values from a model which ignores variation in the function argument. Since the functions in this case are S-shaped, the second derivative has different signs depending on whether the mean, $\overline{k_j\,Q_i}$, is to the left or the right of the argument below the curve inflection point. This makes it impossible without knowing the mean argument value to make sweeping predictions about whether females will avoid or seek out the more variable contexts. Clearly, the larger the uncertainties, the more important proper computations become. With the exception of Wittenberger (1981a), whose treatment was not analytical, uncertainty has typically been left out of tests of the female settlement model.

Erroneous tests of the female settlement model

One problem which has beset the entire history of this model is the invocation of erroneous tests of its propriety. There are three basic errors which recur: (1) assumption of part of the model to facilitate short-cut tests of the entire model; (2) reliance on the prediction that secondary females and monogamous females settling at the same time should have similar fitnesses; and (3) reliance on the prediction that earlier settlers should have higher fitnesses than later ones. We will examine each of these in turn.

   (1) Properly interpreted, the model predicts pairwise correlations between (a) the ranking of males based on controlled resources and

the ranking of males by final harem size; (b) the order in which females who were all the jth female to settle on a territory chose males and the ranking of controlled resources by those males; and (c) the ranking of fitnesses of the jth settling females in all harems of identical final size and the ranking of the resources controlled by their males (Altmann *et al.* 1977). Because the identification and measurement of the critical resource is so difficult, there is a temptation to use one of the variables which is correlated with amounts of controlled resource when the model is true as a substitute for actual measurements of the resource. The usual substitute is final harem size. This is then used to test the balance of the model. This practice is dangerous because the rankings of males by final harem size, order of selection by females, and fitnesses of females of similar rank may all be commensurate and yet have no relationship to amounts of controlled resource *if* the model is false.

(2) The prediction that secondary females and monogamous females settling at the same time should have similar fitnesses is clearly an unnecessary prediction of the model (Altmann *et al.* 1977). The model requires only that females adopt options according to their rank order regardless of their relative magnitudes. It is entirely possible that the first bigamous options are sufficiently removed on the abscissa of Fig. 10.1 from the last monogamous options to generate quite different fitnesses. Even if they are not widely separated, they may produce different fitnesses if the slope of the curve above them is steep. Wittenberger (1979) tried to salvage this prediction by invoking 'continuity' in the distribution of resources controlled by males, but his definition was insufficiently quantitative to be of much use. Rather than impose arbitrary qualitative distinctions, it seems wiser to recognize that a variety of relations between secondary and monogamous females is possible and use other tests of the model.

It is worth noting that the invocation of this prediction (Martin 1974; Crawford 1977; Wittenberger 1978a) was intended as a refinement over the previous procedure of comparing monogamous female fitness with the 'average' fitness of all polygynous females (e.g. Carey & Nolan 1975; Dyrcz 1977; Harmeson 1974; Holm 1973; Price & Bock 1973; Verner & Engelsen 1970; Weatherhead & Robertson 1977a; Zimmerman 1966). Clearly, averaging the fitnesses of all females in a harem obscures the options actually available to each at settlement. The earlier practice should thus continue to be discouraged.

(3) The final prediction that early settling females ought to have higher fitnesses than later settling ones is also unnecessary (Altmann *et al.* 1977). A settling female must choose among available options at settlement. These may not be accurate measures of what she will obtain after all females are settled. Consider the simplest case, in which females within a harem partition the available resource equitably. Certainly within any harem, the early and late settlers will have equal fitnesses, which violates the prediction. More to the point, consider

two separate harems settled on different amounts of resource. It is entirely possible that the best option for the last settler is to join the larger harem on the larger amount of resource. This may reduce the resource per female in this harem below that in the other harem. The first settler in the large harem would then have lower fitness than all females, whether early or late settlers, in the second harem. The prediction would be more reasonable if females took final outcomes into account when selecting mates. Lenington (1980) considered this alternative with red-winged blackbirds but concluded that the data were more consistent with immediate expectations at settlement.

Determinants of resource partitioning by males

The prior model begs the question of when and why there will be sufficient diversity in the controlled resources of males to favour polygyny. Verner and Willson (1966) and Orians (1969) argued that habitat heterogeneity seemed to be correlated with polygyny. As noted earlier, this correlation is easily generated *if* one presumes that a fixed grid of male territories is being imposed on habitats of different degrees of spatial heterogeneity. It is not clear that an assumption of fixed territorial grids is warranted. Certainly territorial settlement can lead to *either* equitable *or* uneven partitionings of resources (cf. Fretwell 1972). While it is true that uneven partitionings are necessary conditions for the polygyny model outlined above, it is not clear that such uneven partitioning can only occur in heterogeneous habitats or that heterogeneous habitats necessarily lead to uneven partitionings.

There are several conditions which would favour uneven partitioning even in homogeneous habitats. These include: (a) high variances in male size or age which are reflected in resource-holding abilities (Carey & Nolan 1975; Searcy 1979a; Yasukawa 1979); (b) asynchrony of settlement (van den Assem 1967; Krebs 1971; Maynard Smith 1974b); (c) errors by males in evaluating controlled resources in their own and adjacent territories (Wittenberger 1981a; Searcy 1979b); and (d) site fidelity between years even when resource distributions change annually (Searcy 1979b; Yasukawa 1979). Recent game theory models have shown how such intrinsic asymmetries can generate stable but uneven partitionings of resources (cf. Maynard Smith 1982a, and references therein). However, they also dictate specific conditions which must be met before stable outcomes are possible. On top of these constraints, the magnitude of variation among male territories must be sufficient to exceed the 'polygyny threshold'. Assembling the game theory and female settlement model constraints into a single set of necessary conditions has yet to be accomplished.

In heterogeneous habitats, it is possible that uneven resource partitionings are more likely but this cannot be simply assumed by invoking a fixed territorial grid. Instead, it must be the case either that a given amount of resource is more easily defended when locally concentrated than when spread out, (e.g. Chapter 6; Brown 1964; Wolf & Stiles 1970; Wolf 1975), or alternatively, that heterogeneous habitats are more likely to show greater age and size variation among males, greater settlement asynchrony, more difficult resource evaluation, or greater site fidelity. Clearly, the entire issue will require much more research before male partitionings of resources are understood.

Summary on resource-defence polygyny

Resource-defence polygyny has received considerable attention, but clear understanding of its evolution has been hindered by a limited focus on a few components of fitness. Only recently have the effects of uncertainty been incorporated into models, and this has not been sufficiently systematic or quantitative to evaluate by how much conclusions might be altered. Game approaches have been invoked for some within-sex contexts, but the more difficult between-sex analyses of this form of polygyny have been ignored. This is an important type of mating system about which we know a considerable amount. Although we have concentrated on the problems with prior work, it is precisely because of this detailed research that subsequent implementation of third-generation approaches ought to be very productive.

10.3.2 Polyandry

Polyandry is a mating system with roles reversed from those of polygyny (section 8.4). Females have several mates, and where there is parental care, it usually is undertaken by the male alone. As with polygyny, polyandrous systems can be subdivided into separate categories. It will be useful in the ensuing discussion to use the following distinctions.

(a) Simultaneous polyandry. Each female has several mates simultaneously and each of these cares for a separate subset of her offspring. Females typically defend a resource within which the several males establish breeding sites; this is thus an equivalent of resource-defence polygyny by males. The behavioural roles of the two sexes are reversed, with the female the larger, more adorned and more aggressively territorial sex. Well-studied examples are jacanas and spotted sandpipers (Jenni & Betts 1978; Oring & Knudson 1972; Oring & Maxson 1978).

(b) Sequential polyandry. Each female provides eggs for a single male, and then moves on to provide additional sets for additional males at later periods. Examples occur in belastomatid water-bugs and in phalaropes (Smith 1980; Hilden & Vuolanto 1972; Schamel & Tracy 1977; Howe 1975a, b). Behavioural roles are again usually reversed, with the larger, more adorned females competing among themselves for direct access to mates (Oring 1982).

(c) Cooperative polyandry. Each female has several mates simultaneously, but instead of dividing up the brood, the female and the males jointly share in the care of a single set of offspring. There is no role reversal in this system. Examples include the Tasmanian native hen and the Galapagos hawk (Ridpath 1972; Faaborg *et al.* 1980).

(d) Polygynandry. This is similar to sequential polyandry in that each female provides eggs for a sequence of males, but differs in that each male collects and cares for the eggs from several females, making them simultaneously polygynous. This is a common system in substrate-nesting, egg-guarding fish and in many ratites (Perrone & Zaret 1979; Bruning 1974).

(e) Non-parental polyandry. Females mate with several males but neither sex engages in parental care. Certainly, mating with different males sequentially is probably the rule in a variety of species with no fixed associations or long-term residences. Pelagic fish are a good example. Less common, but more interesting, are species in which a given female mates simultaneously with several males but neither sex cares for the eggs or young. This occurs in several fish species (Reighord 1920) and possibly in some frogs (Salthe & Mecham 1974).

The study of polyandry has followed a quite different history from that of resource-defence polygyny. Unlike the latter system, the classical studies provided no clear correlates between the presence or absence of polyandry and habitat type. Although a number of first-generation scenarios were proposed to fill the gap (e.g. Jenni 1974; Graul *et al.* 1977; Emlen & Oring 1977), none of these has proved to have the generality or has provoked the extensive field testing that the Orians–Verner–Willson model has for resource-defence polygyny. A few second-generation studies have contrasted values of selected fitness components for polyandrous and monogamous alternatives (Oring & Maxon 1978; Maynard Smith & Ridpath 1972; Faaborg *et al.* 1980). Using these measures, the outcome has been that polyandry has advantages over monogamy for females, but constitutes a loss for males. Such studies are still scarce, perhaps because polyandry is so rare in the more studied taxa (e.g. birds and mammals), and because in most fish, the system is complicated by concurrent polygyny among males.

The absence of conspicuous correlations and second-generation studies with some generality has had the beneficial result of catapulting the study of polyandry into a third-generation perspective ahead of studies on other systems. Because of the general theoretical vacuum on the topic, work turned to the one consistent correlate of polyandry: extensive paternal care. Early studies assumed that if one sex invested heavily in finding additional mates, the other was obliged to undertake any parental care; or alternatively, that if one sex was committed to the parental care, the other was then free to find additional mates (Trivers 1972). (The latter is a basic and unquestioned assumption of the polygyny models discussed earlier.) However, careful reflection shows that this linkage between polygamy and unequal parental care allocation need not be tight at all. For example, neither kiwis nor many cuckoos deviate from monogamy even though males have assumed much of the parental care (Vehrencamp 1982). Similarly, there are many monogamous birds in which the female performs most of the parental care (Kendeigh 1952; Skutch 1959). We have mentioned fish which mate polyandrously but which assume no parental care. These highlight the fact that the only tight linkage to be expected is within a sex: if time and energy are limited, investment by either sex in one activity automatically limits its concurrent investment in another activity. While it is true that the allocations by one sex may affect the expected payoffs of alternative allocations by the other, these correlations need not be either tight or inverse in nature. Clearly, the way to analyse these problems is with evolutionary game theory: the two sexes are the players and the relative allocations to parental care and additional mate encounters are the alternative strategies.

It is the recent introduction of third-generation game theory which has had the greatest impact on the study of the evolution of polyandry. The game can be framed in several ways, but a modified version of the analysis by Maynard Smith (1977, 1982a) will be used here (see section 8.4). Each sex has the same two strategies to choose from: 'Guard' which entails a heavier investment by that sex in parental care, and 'Desert' which entails a heavier investment in additional matings or increased fecundity. The resulting 2-by-2 asymmetric table can lead to any of seven general classes of results, depending on the directions of four separate inequalities (not two, as in second-generation tests). Four of these results predict a single outcome (or ESS) such as 'Female Deserts, Male Guards'. One set of inequalities will allow either of two outcomes to evolve with the two sexes adopting the same strategy, i.e. 'Both Guard' or 'Both Desert'. Another set of inequalities will also allow two ESSs but with the sexes adopting opposite strategies, i.e. 'Male Guards, Female Deserts' or 'Female Guards, Male Deserts'. In all double-outcome cases, which of the two alternatives is seen depends largely on prior conditions. Finally, two separate sets of inequalities will lead to no stable outcome.

The game theory approach explicitly recognizes that there are four possible combinations of male and female options. This leads immediately to two important consequences. First, there is no reason to assume that if one sex reduces its role in parental care the other will automatically increase its investment to maintain the same level of parental care. Thus there are no grounds for expecting that the strategy 'Female Deserts, Male Guards' can only alternate with 'Male Deserts, Female Guards'; a total increase in parental care with 'Both Guard' or a total decrease in parental care with 'Both Desert' could just as easily evolve. Each sex adopts the strategy which maximizes its own fitness given the strategy adopted by the other.

Secondly, this approach identifies the fitness contrasts we must make to understand the significance of a particular mating system. To analyse such an asymmetric game table, the four cell values for each sex must be compared two at a time with each other but in only two of the four possible ways. There are thus two contrasts required per sex and four overall. To achieve any single ESS, there are two specific inequalities that must be met, one for each sex. The directions of the other two contrasts are not critical, as long as they do not produce a second ESS. Even if there is a double ESS, the strategy in question may still evolve if conditions favour it over the alternative.

Consider a second-generation study which compares fitnesses for each sex when monogamous ('Both Guard') with that when poly-androus ('Male Guards, Female Deserts'). The contrast made by comparing female fitnesses in these two situations is a correct one, and it is essential that the deserting female's fitness be higher than that of a monogamous female for polyandry to evolve. The other (less essential) female contrast is a comparison of a female's fitness when polygynous versus the no-care situation. The contrast for males in a second-generation study is an incorrect one: the relative fitness of a mono-gamous versus polyandrous male is irrelevant to the evolution of the system. The proper contrasts for males are (a) polyandrous males versus males in the no-care context (essential), and (b) monogamous males versus polygynous males (less essential). In other words, the loss (or gain) in the male's fitness when females are selected to desert is not important; what is important is whether the male should then stay or also desert. This fact may explain why prior second-generation studies showed that females benefited from polyandry but males did not: the male's fitness was not compared with the correct reference. At least in one study of cooperative polyandry, by Faaborg *et al.* (1980), the proper male fitness contrast was acknowledged.

Several third-generation game models have been applied to the general issue of parental care allocation and mating strategies of the two sexes. The question of greatest interest with respect to polyandry is under what conditions male care/female desertion will evolve rather than the alternative of female care/male desertion. The outcome depends on certain asymmetries in the reproductive energetics of the two sexes, their relative opportunities for additional matings, and ecological conditions. The payoff matrix for a modified version of Maynard Smith's (1982a) model is shown in Fig. 10.2. In this model, males can increase their encounter rate with a probability of p when they desert and with a probability p' when they guard. Deserting females can either increase their clutch size (to V, versus v for a guarding female) or increase their encounter rate with a probability q' (when male guards) or q (when male deserts). Presumably females cannot increase both clutch size and encounter rate, so that q and q' are inversely related to V. Finally, the probability of the offspring surviving is P_0, P_1, and P_2 for no, one, and two parents, respectively,

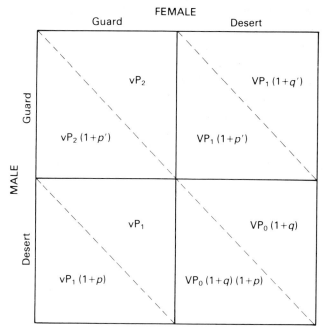

Fig. 10.2 Payoff matrix for the asymmetric male versus female parental care game. Payoffs in the upper right corners of each box are the female's, and in the lower right corners are the male's. P_0, P_1, and P_2 are the probabilities of offspring survivorship with no, one, and two parents, respectively. v is the clutch size of a guarding female, V is that of a deserting female. p is the probability that a deserting male can obtain an additional mate, p' is the probability that a guarding male can obtain an additional mate. q is the probability, or fraction of an additional clutch (of size V), that a deserting female can lay during the time it would take a guarding female to guard. q' is the probability that a deserting female can lay an additional clutch of size V when males guard. There are four ESSs: 'Both Guard', 'Both Desert', 'Male Guards, Female Deserts', and 'Female Guards, Male Deserts'.

where $P_0 < P_1 < P_2$. With this general framework in mind, one can not only establish which fitness contrasts are critical but also predict the conditions that must be met for each type of polyandrous system to evolve.

For polygynandry to evolve, it must be the case that a guarding male has a female encounter rate about equal to or greater than a deserting male's female encounter rate. Whenever p' is significantly greater than zero, q' will also be greater than zero, and females will be able to lay multiple clutches. As long as one parent is about as good as two and much better than none, the two critical conditions for male care

$$(1 + q') > \frac{P_2}{P_1} \quad \text{and} \quad \frac{P_1}{P_0} > \frac{(1 + p)(1 + q)}{(1 + p')}$$

will be met. When the above inequalities are met, it is highly likely that male care will be a single ESS. This is because where $p' \geq p$, the second (less essential) contrast for males favours male guarding rather than deserting if females were to guard. In birds, this system will be found in species with precocial young (enabling one parent to guard) and with a small egg size or clutch size, so that several females' clutches can be incubated by one adult (Lack 1968). In fish and frogs, a similar system will be found in substrate-breeding, egg-guarding species where large numbers of eggs can be accommodated in one nest and where one parent can significantly reduce egg predation compared to no parents (Baylis 1981; McDiarmid 1978).

For the case of female desertion and sequential polyandry, q', q and p' must be zero, and V must be greater than v. As for polygynandry above, one parent must be about as good as two and much better than none, so that these two essential inequalities are met:

$$\frac{V}{v} > \frac{P_2}{P_1} \quad \text{and} \quad \frac{P_1}{P_2} > (1 + p).$$

The most important condition for the evolution of male care in this case is that a deserting female's clutch be substantially larger than a guarding female's. This will occur only if females are energetically limited by egg-laying and cannot produce an optimally sized clutch unless released from parental care. If these essential inequalities are met and a deserting male's female encounter rate (p) is very low, then male care can be a single outcome ESS. However, it is also possible that p is substantially greater than zero, resulting in a double ESS for either male or female care. The critical issue is then the male's ability to increase his encounter rate by deserting versus the female's ability to increase her clutch size by deserting. In the case of a double ESS, the ancestral condition or evolutionary starting point may play an important role in determining which solution evolves. If the starting point is no care, and monoparental care becomes selectively advantageous, then it is more likely that male care will evolve. This is because a female in the no-care situation puts all of her reproductive effort into eggs, so that guarding would result in a large decrease in her fitness (i.e. V >> v). If the starting point is biparental care, as in birds, then parental effort is likely to be the most important component of fitness. Clutch size will therefore be limited well below the number of eggs a female could lay, by the number of young the parents can care for. In effect v ≈ V, so females gain nothing from desertion, and males are more likely to benefit from desertion, via increased encounter rates.

Resource-defence (simultaneous) polyandry is the sexual mirror-image of resource-defence polygyny. In terms of the general model, $p' = 0$ and V ≤ v. Both sexes attempt to increase encounter rates (p for males, q' for females) by despotically defending resources (or groups of the opposite sex). The two essential inequalities that must be met are:

$$(1 + q') > \frac{P_2}{P_1} \quad \text{and} \quad \frac{P_1}{P_0} > (1 + p)(1 + q)$$

The ability to defend more than one's fair share of the resources is primarily dependent on the spatial and temporal distribution of resources, and both sexes have a similar potential. Therefore, the outcome to this game will always be a double ESS, and it is not clear which sex will prevail.

In the case of such a double ESS with male versus female care, initial conditions and ancestral history will play an important role in determining the actual outcome. The relative magnitude of the benefits of deserting for the two sexes does not determine which adaptive peak is favoured. However, in the case of a biparental ancestor, as conditions gradually begin to favour monoparental care, the sex with the greater advantage will meet the conditions for desertion first. Similarly, in the case of a no-care ancestor, the sex with the smaller loss will meet the conditions for guarding first as parental care gradually becomes advantageous. In general, males will derive a greater advantage from deserting, since they do not have to bear the burden of egg-laying and have more time and energy available for mate or territory defence. This may explain why simultaneous and sequential polyandry are so rare. There are, nevertheless, several types of circumstances that may tip the balance in favour of female desertion and male care: (a) external fertilization, as opposed to internal fertilization (Dawkins & Carlisle 1976; Gross & Shine 1981); (b) skewed sex ratio favouring males (Maynard Smith 1977; Grafen & Sibly 1978); (c) preselection for males to perform most of the care in a biparental ancestor (Vehrencamp 1982; Grafen & Sibly 1978); (d) male territoriality in a no-care ancestor (Ridley 1978; Blumer 1979); (e) female territoriality in a biparental ancestor; (f) continuous breeding, as opposed to a single brood (Perrone & Zaret 1979); (g) high female reproductive costs relative to male costs (Knowlton 1982); (h) post-copulatory male guarding of females in insects (Smith 1980).

In summary, early attempts to generate specific hypotheses for the evolution of polyandry were hampered by the failure to find any broad ecological correlates of polyandry, by the wide diversity of polyandrous systems, and by the rarity of these systems in nature. The recent invocation of game theory to model the trade-offs of parental care and desertion tactics between the two sexes has finally brought the study of polyandry back into focus. General game theory models not only provide a synthesis of the different types of polyandry, but they identify the correct fitness contrasts and specific parameters that must be measured in the field.

10.3.3 Leks

Lek mating is one of the more extreme mating systems known. Any parental care is undertaken only by females, and males defend neither resources nor mates. Instead, males establish display territories where they advertise themselves to females. Where male display territories are aggregated in a common area, the assembly is called a 'lek'.

Females typically visit a variety of males, mate with one (or at most a few), and then go on their way. Since most females in an area tend to mate with the same few males, and males in most species do not coerce females into mating, some choice of mate by females is implied.

Leks are considered an 'extreme' system because sexual transactions have been reduced solely to the provision of sperm for females. This makes the lek system an attractive candidate for third-generation studies since the numbers of relevant behavioural options and fitness components are reduced *a priori*. The reductions are more pronounced for males. In lek species, adopted male options are most likely to modulate three fitness components: (a) encounter rates with females, (b) copulation (perhaps in conjunction with fertilization) rates, and (c) adult male survival. For females, the fitness components most likely to depend on sexual transactions are: (a) encounter rates with males, (b) fecundity rates if sperm from preferred males are in short supply, (c) juvenile survival if paternity has any heritable effect on offspring survival, (d) sexual selection benefits of epigamic mate choice, and (e) adult female survival if there are risks during mate choice and copulation.

Existing methods of weighting fitness components and options rely on intrinsic variation. There are two aspects of lek mating which appear to vary widely among lek species and which can be tied *a priori* to the fitness components listed above. One is the pattern of dispersion of displaying males. This can vary from widely scattered clusters of tightly packed displaying males, through less scattered but also less densely packed males, to relatively uniform distributions of displaying males. Male dispersion is likely to affect the following fitness components: (a) encounter rates of males with females; (b) encounter rates of females with males; (c) adult male survival on display grounds; (d) adult female survival on display grounds. One can ask which sex determines the observed dispersions of males and which fitness components are most affected by the resulting pattern. The second aspect which varies with species is the degree of unanimity in mate choice within a lek. As we shall show below, the patterns of mate choice are likely to affect and be affected by: (a) male behaviours enhancing copulation and fertilization rates; (b) female behaviours ensuring adequate insemination; (c) female mate preferences which confer heritable survival benefits on offspring; (d) female preferences which confer epigamic benefits; and (e) survival consequences for males of adopting behaviours which enhance copulation rates. With the exception of some overlap in male survival, these two aspects of lek mating are likely to be associated with independent sets of fitness components. This facilitates subsequent analyses as we shall show below.

There are a number of hypotheses which can be advanced to explain observed variations in lek male spacing. These can be subdivided according to whether males or females would be the primary beneficiaries of adopting a given pattern. Males might benefit from adopting a particular dispersion because: (a) sites safe from predators are so distributed; (b) sites allowing for maximal signal propagation are so distributed; (c) sites near which maximal numbers of females will pass ('hotspots') are so distributed; or (d) signal range was maximized by some optimal group size of males. Females might induce males to adopt a given dispersion by favouring males in one pattern over alternative patterns. Females might benefit if: (a) sites were those with lower predator risk; (b) sites were far from resources reducing intersexual competition; (c) sites were convenient for female access (likely to be compatible with a male 'hotspot' advantage); or (d) males were forced to aggregate in as large a group as possible to facilitate female choice.

A number of these hypotheses are applicable only to a subset of lek species. For example, a variety of forms do not suffer predation risks on the display grounds and thus either female- or male-initiated anti-predator dispersions are unlikely causes for differences between species in male dispersion. Several other hypotheses, such as grouping to increase signal range or to reduce intersexual resource competition, appear to have limited applicability (cf. reviews by Bradbury 1981; Oring 1982; Bradbury & Gibson 1983). The most likely causes of differences in male dispersion appear to be a male-initiated hotspot strategy and a female preference for large male aggregations.

Each of these determinants of male spacing in lek species has been examined with simple models (Bradbury & Gibson 1983). In both cases, variations in male spacing patterns will arise as a simple consequence of variations in female home range size. This latter parameter is assumed to vary with resource dispersion, anti-predator strategies of females, nest visit strategies, etc. Both models predict increasing clusterings of males with increasing female home range size. However, the female-initiated model predicts a final male dispersion with clusters spaced about one female range diameter apart; the male-initiated hotspot model allows clusters to be closer than this distance.

As discussed in a recent review (Bradbury & Gibson 1983), nearly all species studied establish leks that are closer than the diameter of one female home range. The data are thus more consistent with a pure male-initiated hotspot model than a pure female-initiated large-cluster model. It is of course possible that neither sex dominates the final dispersion and that observed patterns are due to a mix of influences by the two sexes. For example, males less favoured by females on leks dispersed by male rules might add enough additional copulations by displaying at female-favoured sites to justify the move. Females with home ranges furthest from a settled hotspot might accept males at interstitial sites. Either type of shift would produce a deviation of male spacing away from a pure hotspot model which might not have been detected with the qualitative tests currently used. Certainly, these tests can be refined and models evaluating the likelihood of shifts can be developed. The appropriate analysis both within and between the sexes would be game models.

Mate choice on leks

Mating on leks is never random: usually, a small fraction of the males present account for the majority of all copulations. As with spacing, this bias in mating success of males could arise from differences in the execution of male strategies, unanimity in female preferences, or some interaction between these two causes. Males might generate differential mating success through asymmetrical interventions between displaying neighbours, or by differences in the location or execution of displaying. Direct male intervention between neighbours does occur in some lek species, but for most forms there is no obvious correlation between the degree of bias and amount of intervention nor any general evidence that disrupters enhance their own probabilities of mating (Bradbury & Gibson 1983).

This leaves locale within the lek and/or display execution (including advertisement of intrinsic male traits such as age or size) as possible cues for female preferences. Despite mixed evidence supporting it, the former is often cited as a major determinant of mate choice on leks (Wiley 1974; Emlen & Oring 1977; Borgia 1979). Several authors have even called lek mating 'male-dominance polygyny' because they accepted the premise that dominance of specific 'mating centres' on a lek was a primary determinant of mating success. The supporting evidence for this idea comes from the findings that the more successful males on a lek are often closer together than expected by chance and that males will compete for more central locations (Buechner & Schloeth 1965; Kruijt & Hogan 1967; Wiley 1973; Lill 1976; Bradbury 1977). The qualifier is that on a per male basis, being close either to the most successful male, the current centre of gravity of successful males, or any absolute reference point only accounts for a small fraction of the variance in male mating success (Bradbury & Gibson 1983). The nearest neighbours of a top male may in fact obtain no copulations at all, and solitary males away from main leks may be very successful (de Vos 1983). Thus, even if one argues that the correlation between proximity to some reference and mating success reflects causation, one must invoke other important factors to explain the observed distributions of mating on leks.

The alternative (or additional) cues that females might use to select mates on leks include age, size, and display performance. Older males do tend to have higher mating success in many lek species, but age is often confounded by position and performance correlates (Wiley 1973; Kruijt *et al.* 1972). Lill (1974) found no correlations between male size and mating success in manakins, and Stiles and Wolf (1979) found no relationships between hummingbird male size and position on leks. Male size is correlated with mating success in some frogs, but this is of less importance than time spent calling (Ryan 1980 and pers. comm.). Display performance *has* been implicated in mating success in a variety of species. Successful blackcock males are those which dance in particular ways (Kruijt *et al.* 1972). Durations of display in indigo birds both early and late in the breeding season correlate with later mating success (Payne & Payne 1977). Strut rate in sage grouse is positively correlated with mating success if proximity to females is controlled for in the analysis (Hartzler 1972). (Wiley's (1973) study did not find this result, but his controls for female proximity were quite different.)

It is thus not yet clear which cues females use to select males on leks, nor which options available to males might have the greatest effect in increasing copulation rates. One might have some insight into the cues if we knew in advance what components of female fitness were being enhanced by this choice. The two likely components have already been listed: heritable effects of paternity on juvenile survivorship (and perhaps fecundity), and sexual selection benefits. There is considerable current debate about the relative importance of these two components of fitness. Lande (1981) and Kirkpatrick (1982) have both shown that female choice can evolve in conjunction with the elaboration of male traits without there being any benefit in offspring juvenile survival or fecundity of daughters. In contrast, a variety of authors have argued that sexual selection through female choice could have some natural selection benefit that offsets the obvious costs of sexual selection (Williams 1966; Trivers 1972; Zahavi 1975; Borgia 1979). The resolution of this issue is not clear: the models of Lande and Kirkpatrick only argue that a natural selection benefit is not needed for sexual selection; they therefore do not preclude the type of effect proposed by the alternative camp. In short, current theory does not help one to decide which fitness components a female enhances by making a choice on a lek (see also Chapter 9). On the other hand, it may be possible in lek species, where paternal care and resource defence can be ignored, to resolve the theoretical dispute by measuring the relative contributions of sexual selection and natural selection to the fitnesses of participating females. Partridge's (1980) recent study on promiscuous (albeit not lek) *Drosophila* is a case in point, although the role of female choice is obscured in these experiments by possible male–male interventions.

The final fitness component that may be related to mate choice is adult male survival. Whichever model of sexual selection one accepts, it remains likely that adult male mortality is one of the factors con-

tributing to any equilibrium value of the trait used in female choice
(Borgia 1979; Lande 1981). The most likely causes of increased mortality in lek males are augmented predation risk and excessive energetic
expenditures during display. In the frog *Physalaemus pustulosus*,
calling is very expensive energetically (Bucher *et al.* 1981) and attracts
predatory bats (Tuttle & Ryan 1981). Although predation on displaying males is known for a number of lek species, the energetic
costs of display have been measured only for the frog. In a few lek
forms, there are indirect clues that males may be limited energetically.
Numbers of displaying males on both hammer-headed bat (Bradbury
1977) and sage grouse leks (J. Bradbury and R. Gibson, unpublished
observations) are reduced after several days of cold weather; this is
true even if females continue to visit leks, and weather *on the day* of
reduced numbers is warm. One might expect such an effect if female
choice relied on a behavioural cue and had pushed this behaviour to
its energetic limits. The indications that strut rate plays a role in sage
grouse mate choice would be compatible with this notion.

Lek male survival may depend not only on the amount of energy
expended in display, but also on rates of energy acquisition. For
males with balanced energy budgets, the amount available for display
would depend on the amount recently acquired minus other costs.
Since mating success is highly biased on leks, the function relating
energy recently acquired, i.e. foraging success, to copulation rate is
probably concave. Because the amount of food obtained in any period
is surely stochastic, it follows that the expected fitness for any given
food source will increase as the variance in amount of food expected
increases (using the same Taylor approximation outlined in earlier
sections). A lek male with a choice of two foods of equal mean energy
value, but differing in the variance of energy value, would benefit on
average by choosing the food of higher variance. A consequence is a
higher likelihood of obtaining too little food and thus dying. In other
words, lek males pushed to an energetic limit might exacerbate their
risks by favouring more risky food sources over less risky ones. This
would explain the apparent differences in diet of some lek species,
such as hammer-headed bats (Bradbury 1977), where males opt for
much riskier food sources than females and suffer much higher adult
mortalities. In short, female choice using energetically expensive cues
may increase male mortality both by pushing display up to existing
limits and by inducing males to adopt even riskier options in foraging.

Summary on lek mating

Although simpler in some respects, leks are still sufficiently complex
to make rankings of relative fitness components and behavioural
options difficult. A few fitness components which are most likely to
be affected by lek male dispersion can be ranked roughly by analysis

of variation in this spatial parameter. We still have only the vaguest ideas of how the components and options related to mate choice are ranked, although the links between a few, such as adult male survival and female choice, are beginning to emerge. Game models and corrections for uncertainty in computing payoffs are likely to be of particular importance in future analyses of lek evolution.

10.4   MATING SYSTEM CLASSIFICATION

Most second-generation mating system classification schemes are based on the same *a priori* focus on particular behaviours or fitness components which characterizes second-generation field studies. Often, the criteria for such taxonomies mix components of fitness and behavioural options: the term 'resource-defence polygyny' refers to a specific set of behavioural options which result in increased encounter and fertilization rates with females. The major risk of using such taxonomies is that they automatically deflect attention away from other fitness components and other behavioural options within the mating system which may in fact be more important from an evolutionary standpoint. Life would be much easier if the lists of fitness components and behavioural options could be subdivided into autonomous sets: then teasing apart the major factors favouring one aspect of a mating system could be attempted without considering the concurrent role of other aspects. But life is not that simple and the potential links between most behavioural options and most fitness components make the interactions within each set very complex and very relevant to the evolution of the entire system.

   Once a variety of different species are analysed from a third-generation point of view, perhaps there will be enough consistency in the rankings of components and behaviours to devise a taxonomic scheme as simple as current ones but justified by actual measures of selective pressures. Perhaps one would even end up with one of the current systems. The risk, as with current systems, is that the precedents will only exemplify some of the possible combinations of weightings. Species with weightings intermediate between two precedent categories will be difficult to place.

   One solution to the latter problem is to classify mating systems according to the relative emphases selection places on each of the components of fitness. This has the advantages that the criteria are exhaustive and, since every species will have some weighting on every component, there will be no awkward intermediates. Using the weighting of each component as a coordinate in some hyper-space, one could use spatial proximity to lump systems after the fact. The risk is that the resultant groupings may have no relationship to habitats. Since the historical aim has been to link sets of behavioural options to environmental factors, invoking a fitness component taxonomy might entail a major change of course. In short, we cannot evaluate the relative merits of switching to pure fitness components, pure behavioural options, or some selected mixture of the two in classifying species until more third-generation results are available. In the meantime, it seems prudent to keep in mind that current systems do rely on assumptions which are far from tested and which may even be misleading.

Chapter 11
Behavioural Adaptations and Life History

HENRY S. HORN and
DANIEL I. RUBENSTEIN

11.1 INTRODUCTION

Natural selection favours those individuals who most abundantly transmit copies of their genes to future generations. This task demands appropriate allocation of limited resources between the conflicting requirements of reproduction and survival, in an environment which is at best capricious and at worst predictably hostile. The result of this allocation is the life history, which is defined by a schedule of fecundity and survival that may show changes with age, environment, individual condition, and social setting. Depending on the time-scale of environmental change, adaptive adjustments of life history may be behavioural, physiological, developmental, or genetic. Of course, behavioural adjustments are of most direct interest to behavioural ecologists, but responses at other levels are also important because they underlie interspecific comparisons that are used as critical tests of many of the ideas of behavioural ecology.

The chapter begins with a précis of the theory of adaptive life histories. Reproduction can be viewed as an individual's investment in the genetic future of its population. The sooner the investment is made, the more time the reproductive rate will have in which to bear compound interest. Yet many animals delay reproduction even after they have reached physiological maturity. This paradox is resolved by exploring the environmental conditions that favour early and exhaustive breeding, and contrasting them with those that favour conservative breeding. If the environment provides many vacant opportunities, then natural selection favours early and exhaustive investment in as many offspring as possible. Conversely, if the environment is crowded and if young must compete for vacancies or fight to create them, then extensive investment in only a few young is favoured, the better to prepare them for competition. Crowded populations favour dispersal, particularly of young individuals, in search of rare openings or uncrowded locales. Our discussion emphasizes the importance of age-specific schedules of reproduction and dispersal as adaptive responses to environmental factors over which an animal has little control. Behavioural choice is often the mechanism by which animals achieve adaptive allocation of resources among reproduction, self-maintenance, and growth.

We discuss more fully those situations in which the choice among current behavioural options affects future opportunities. In these cases the criterion of adaptation is not simply the immediate reproductive output, but rather 'reproductive value', which takes into account current reproduction, likelihood of further survival, and future reproduction, discounted for its delay. In this discussion, our ideas are speculative and our examples tentative. The critical study of behaviour in the context of life history is new enough that ideas are difficult, data are scant, and there are more questions than answers.

11.2 INTRODUCTION TO LIFE HISTORIES

In a classic paper, Cole (1954) showed by numerical simulations that an asexual animal that lived only one year could become reproductively equivalent to a potentially immortal perennial by adding only one further young to its litter. Thus he implicitly raised the question of why all animals do not breed as early and as exhaustively as possible. A more general form of Cole's result has been proved by Charnov and Schaffer (1973), and their proof has been somewhat simplified by Horn (1978) and extended to sexual organisms by Waller and Green (1981).

The simple proof assumes a creature that reproduces annually and that grows to maturity in a single year. The alternatives offered to a parent are conservative reproduction and potential survival to reproduce again versus exhaustive reproduction to produce additional young in its first season but to die in consequence. The self-sacrificial effort is worthwhile if the parent can produce an additional number of young that replaces its own potential future reproductive output. This magic number of additional young must take account of the average risk of mortality for both parent and young, and of the fact that each young is only a partial genetic replica of its parents. If only a fraction Y of all young survive their first year, then in order for a parent to ensure one additional surviving young at breeding time, the number of young added to the litter must be $1/Y$, since $(1/Y)(Y) = 1$. If the probability of survival to the next breeding season for any given parent is P, then the current value of that parent is only a fraction P of the value of one young who ultimately survives to breed. In a sexual and outbred population each young carries half the genetic complement of a given parent; thus two surviving young are needed to propagate a representative sample of the parent's genome. Combining these criteria, we find that a sexually reproducing animal that bears $B + 2P/Y$ young and then dies is reproductively equivalent to a perennial that bears B young each year for as long as it lives. A visual representation of this proof is given in Fig. 11.1.

We can use the result of Fig. 11.1 to generate adaptive relations between various aspects of life histories. Massive, early breeding is favoured, even at the expense of death of the parent, if a mortal

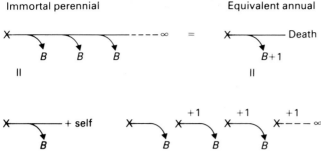

Immortal perennial   Equivalent annual

Fig. 11.1 Reproduction of an asexual, immortal perennial compared with a more fecund annual. An individual is born at X and then bears *B* or more offspring at the times shown by arrows. That the patterns are reproductively identical is shown by rewriting the perennial's pattern in the format of the annual's and vice versa.

reproductive effort can add more than 2 P/Y young to the litter. This addition is made relatively less expensive if the litter is already large and if the offspring are small and require little parental care. Dispersal of the young may augment their survival if the habitat is a mosaic of occupied and unoccupied patches. Low parental survival (P) and high youthful survival (Y) also lower the expense of the addition since they lower the value of 2 P/Y. Massive reproductive effort will further lower parental survival, so that the selection for early and copious reproduction is self-reinforcing. So the following factors favour early, copious and self-sacrificial breeding: large litter, small offspring, little parental care, low parental survival and high survival of juveniles, particularly of dispersing juveniles. By converse arguments the factors that favour restrained early breeding and thus survival for iterated breeding are: small litter, large young, parental care, high parental survival and low survival of young. These conclusions refer to patterns of unavoidable mortality. If the pattern of parental investment in young has a very dramatic effect on the prospects of either adult or juvenile mortality, the predictions might be different (Hirshfield & Tinkle 1975).

For heuristic simplicity we have assumed a beast that grows to maturity in a year and reproduces annually, but the qualitative results can be extended to more general life histories. If development is so rapid that maturity is reached in the interval between two reproductive seasons, then offspring of a young parent are already breeding by the next time that their parent breeds. The reproductive benefit to a parent from its first offspring is compounded multiplicatively. However, if there is a very long period of development and adolescence before actual breeding, the reproductive value of the first batch of young is more nearly a simple addition to that of further young, rather than being compounded. Hence earlier and more exhaustive breeding, when it is favoured, is more highly favoured in species with rapid development to maturity than in those with slow development.

For animals that continue to grow throughout life, the ratio of adult to juvenile mortality still shapes reproductive patterns, but in a somewhat more complicated manner. Although growing larger can simultaneously increase future survival and future fecundity, natural selection will only favour delaying reproduction if the more abundant future output exceeds the compounded success of fewer or smaller young born earlier. Simply put, delayed breeding and growth are favoured if the potential growth of a population of offspring within the body of the parent exceeds the potential growth of a population of offspring deposited earlier in the environment.

An environmental limit to growth may bias some animals toward exhaustive reproduction. Many species of tropical hermit crabs face a shortage of appropriate shells to serve as mobile homes. Bertness (1981) showed that hermit crabs in populations with too few large shells tended to breed at smaller sizes and more frequently than their counterparts in places where large shells were abundant. Environmental limits to growth may also be enforced indirectly through competition for food. Hirshfield (1980) found that individual medaka living in aquaria increased their reproductive effort as food intake decreased. Rubenstein (1981a), working with pygmy sunfish, showed that as the intensity of competition increased so did the proportion of energy devoted to reproduction, despite declines in average size and reproductive output.

To discover the effect of variation in mortality, we must again separate the mortality of adults from that of juveniles. If the survival of parents is variable, the future is further discounted in a way that is closely analogous to lowered average survival of parents. For any survival rate other than 100%, the survival of a given individual is uncertain; if the survival rate is itself subject to unpredictable variation, this uncertainty is compounded. The increased uncertainty of survival to breed again favours early and copious breeding, just as lowered average survival does. Variation in survival of juveniles, especially if they die in batches, favours spreading the risk of reproductive failure among many batches; that is, conservative but iterated breeding. Thus unpredictable variation in juvenile mortality also has an effect analogous to lowered juvenile survival, though the line of reasoning is slightly different from the case of parental mortality.

Perhaps the most important result of this section is that juvenile mortality and parental mortality have opposite effects on the prediction of adaptive tactics. Juvenile mortality biases towards parents' conserving resources and investing them in themselves or in a few well-endowed young. Conversely, adult mortality biases toward investing in many small young, even if the effort hastens the parents' demise. There is really nothing subtle about this difference. Since the only reproductively effective offspring are those who reach maturity and breed, pre-reproductive mortality can be viewed as subtracting from the ultimately effective natality of the parent (Charnov & Schaffer 1973). In a sense, juvenile survival enters the analysis as though it were simply birth rather than survival itself. The effect of variation in mortality also differs for juvenile and parental mortality; and the effect of increased variation is analogous to lowering the respective average survivals. We obtain fully concordant results from an analysis of either average mortality or unpredictable variations in mortality; the paradox discussed by Stearns (1976) disappears. The dichotomy is rather whether environmental mortality falls more heavily on juveniles or adults.

Stearns (1976) tested these ideas by plotting a graph of the average number of seasons in which an individual breeds, against the ratio of average juvenile mortality to average adult mortality, for a variety of birds, mammals, fish and insects. He found a strong positive correlation over about a sixfold range of both parameters; the species with relatively higher juvenile mortality indeed bred more often. However, there was nearly a fourfold scatter of points, and Stearns pleads for more data to decide whether the cause of this scatter is biological or statistical. Bell (1980) argues that the correlation may have a spurious component because an animal cannot have many opportunities to breed unless its adult mortality is low. Southwood (1981) has gathered a more exhaustive set of relations, showing a correlation between size and generation time, both of which are inversely correlated with rate of population increase.

Particularly good examples of adaptive adjustment of schedules of fecundity are found among fish. Reznick and Endler (1982) have found that guppies in Trinidad mature and breed earlier in streams where predators of adults are common, and breed later where predators are rare. They have duplicated this field observation with clever experiments both in the laboratory and in the field. Similar observations have been made for shad and salmon, which are born in rivers, migrate to the sea to mature, and return to their natal rivers to spawn. Those populations of shad that breed exhaustively are found in the rivers of Florida, where juvenile survival is predictably high. In more northern rivers, where conditions are less predictable, spawning takes place in several, more conservative episodes (Gilebe & Leggett 1981). Schaffer and Elson (1975) give the same interpretation of the exhaustive spawning of Pacific salmon in contrast to the repeated spawning of Atlantic salmon, which face the unpredictable dangers of spring ice freshets. Schaffer and Elson also analyse the timing of spawning by Atlantic salmon in different river systems from Maine to Ungava. The mean age of first spawning is greater in the harsher rivers that forebode greater mortality for smaller fish. Rapid growth at sea also favours delayed spawning. Conversely, commercial fishing increases the death rate of larger and hence older fish, and thus favours earlier spawning. A promising subject for further work is the way in which behaviour mediates the adjustment of breeding schedules in these or any other populations that are subject to human exploitation.

11.3   *r*- AND *K*-SELECTION

In the last section we saw how the ratio of inescapable adult to youthful mortality tends to bias a population toward one of two extreme, self-reinforcing selective regimes. A high adult mortality rate favours early breeding, high fecundity, rapid development, dispersal and small size, even at the expense of greater sensitivity to environmental changes and hence a still higher rate of adult mortality. A low adult mortality rate produces a crowded population in which survival and competitive ability are favoured, both of which are made easier by large size and extended parental care, even at the expense of delayed and reduced fecundity which makes parental survival more likely and crowding more severe.

This generalized dichotomy has been called *r*-selection versus *K*-selection (MacArthur & Wilson 1967), in a metaphor that derives from the parameters of the logistic equation of population growth (where *r* measures the growth rate per individual at low population densities, and *K* measures the equilibrium population to which a crowded population tends). The self-reinforcing selective regimes of *r*- and *K*-selection are diagrammed in Figs 11.2 and 11.3.

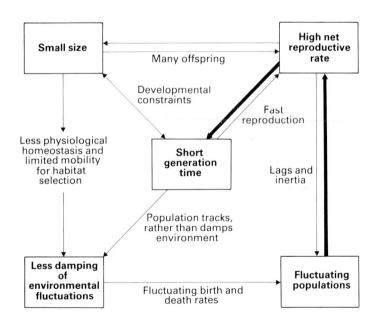

Fig. 11.2   Positive feedback loops reinforce *r*-selection. This figure summarizes the relations among aspects of life history, with arrows pointing from causes to effects that are discussed in the text. The heavy arrows represent natural selection. The idea of making this diagram is pilfered from Southwood (1977).

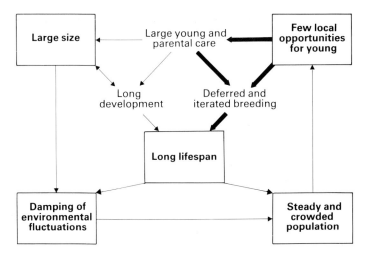

Fig. 11.3 Positive feedback loops reinforce *K*-selection. Some of the effects in this diagram are just the converse of those in Fig. 11.2, but the two diagrams are different in important respects that are discussed in the text.

Comparing Figs 11.2 and 11.3, it is tempting to suggest that small animals are subject mainly to *r*-selection, and large animals to *K*-selection. There is enough truth in this suggestion for it to be a major tenet of modern ecology, but there is enough triviality in it to warrant scrutiny. Southwood *et al.* (1974) have noted the importance of relating the temporal scale of changes in the environment to the scale of the generation time of the animal. If environmental changes occur within a few generations, *r*-selection prevails. If the environment changes slowly over many generations, *K*-selection prevails. In complete ignorance of environmental effects, one can trivially state that most large animals will delay reproduction until they grow to be large. They may also be insensitive to environmental fluctuations that would destroy smaller animals, but they may be just as sensitive as the small beasts to longer term fluctuations that are commensurate with their lifespans. Stubbs (1977) has analysed data from a variety of animals, and found that mammals and birds are indeed dominated by *K*-selection, as befits large animals. However, she found that insects have some species whose life histories are dominated by *r*-selection, and some species dominated by *K*-selection.

This generalized dichotomy, though often broken in interesting ways both in theory and in fact, underlies many of the interspecific adaptive patterns of life history. However, it is important to note that, as with all adaptive patterns, the important point is not to classify a particular species as either *r*- or *K*-selected, but rather to make comparative statements about the relative difference in selective regime in a comparison between two species. An instructive example of this comparative approach is that of Western (1979), who plots for various mammals the size of newborn young against the size of the reproductive adult. He then interprets species with relatively large young for their size as being *K*-selected, and those with relatively small young as *r*-selected. The converse can be done by plotting litter size against adult body size. This approach identifies body size as an important variable for its indirect effects on life history, as well as for its more direct effects on metabolism, biomechanics, and fighting for resources or for mates. Because it explicitly accounts for body size in the determination of selective regime, it eliminates the trivialities mentioned above. Further development of these ideas is provided by May and Rubenstein (1982). In particular, they have plotted a variety of reproductive parameters against weight for many mammals, and they discuss the fine-scale adaptive patterns that are superimposed on a coarser pattern arising from the design constraints of mammalian structure and physiology.

Having made this comparative point between species, we can now extend it to comparisons of different individuals of the same species or even to different ages or bodily conditions for the same individual. Thus one can envision a species in which individuals adopt a relatively *r*-selected pattern of reproduction at one stage of their life cycle, and a relatively *K*-selected pattern at another stage. Perhaps the clearest cases of such switches are found among invertebrate and fish species whose individuals undergo a change of sex during their life history. In some species, reproductive competition is severe among males, and only the largest are successful. Individuals of such species reproduce first as females and then continue to grow. Then when males are in short supply, the largest of the females switch sex and reproduce as males (Warner *et al.* 1975). In other species, there is no reproductive competition in either sex, but growth in size augments female reproduction more than male because eggs are relatively more expensive to produce than the reproductively equivalent amount of sperm. Individuals of these species begin reproduction as males, and only after they have grown much larger do they adopt the female role. A wide variety of examples of adaptive sex change are discussed in detail by Charnov (1982a). There are many other instances among invertebrates of *r*-selected and *K*-selected stages of a single species' life cycle. Parker's (1978b) dung flies, for example, develop rapidly to disperse from their natal cowpat before it becomes an unsuitable habitat, yet they encounter potentially severe competition for mates as adults.

It is at the *K*-selected side of the dichotomy that variations in crowding provide alternative social environments, to which different behaviours by individuals may be adaptive, including dispersal in search of either denser or less dense populations. It is also in crowded populations, or at crowded stages of the life cycle, that the activities of neighbours become an important part of the selective environment. What is adaptive behaviour for a given individual then depends on how its neighbours behave, and game theoretic considerations are crucial (Maynard Smith 1976a; Chapter 2). Games with neighbours

often produce alternative evolutionarily stable strategies and hence
behavioural polymorphism within a species.

287

Behaviour and
Life History

Examples of individuals adjusting their behaviour in response to
the actions of others are common among males competing for mates
(Dunbar 1982; Rubenstein 1980). They are less common for females.
Where females choose among males who control resources in a territory, females may have a choice of mating with a bigamous male on a
rich territory or a monogamous male on a poor territory (Chapter 10).
Orians (1969, following Verner 1964) has argued that if differences
among male territories are sufficiently large, then polygyny, with
females sharing resources with each other, may be favoured over
monogamy. Pleszczynska (1978) has elegantly tested these ideas with
lark buntings (*Calamospiza melanocorys*) nesting in North American
grassland. The major cause of nestling mortality is overexposure to
the sun, and the males that defended territories with the most shaded
nesting sites were chosen for bigamous matings, despite the fact that
males would assist in raising only the offspring of their first nest.

Direct competition for nest sites has been observed in digger
wasps by Brockmann *et al.* (1979). Reproduction requires a burrow,
provisions gathered by a female, and of course an egg. A female may
either dig her own burrow or enter someone else's. Digging costs time
and energy, but a non-digger may provision another female's burrow
only to be displaced by the rightful owner. Thus either behaviour has
costs and benefits that depend on how many individuals in the vicinity are choosing the opposite tactic. Brockmann *et al.* could calculate
at what frequencies of occurrence both behaviours would be equally
successful. For a population of digger wasps in New Hampshire, the
observed frequencies closely matched the prediction. Bertram's
(1980b) observations of female ostriches laying eggs in each others
nests are open to a similar interpretation. More and closer observations are needed to determine whether the 'egg dumping' often
observed in birds is casual accident or strategic parasitism.

Behavioural decisions about life history are most important at the
K-selected end of the spectrum. Behavioural options should play a
critical role at relatively crowded and limited periods of life, when
resources, be they habitat, shelter, food or mates, are in short supply
relative to the number of individuals seeking them. Alternative
behavioural patterns should be displayed less often in tiny, fecund
beasts than in large ones with moderate reproductive output. Behavioural decisions should also be especially important for those individuals in relatively good condition; that is those with large amounts of
stored nutrients and energy available to be divided among various
behavioural options.

11.4 DISPERSAL

A further lesson from the diagrams of Figs 11.2 and 11.3 is that they
are not simply the converse of one another. This shows that *r*-

selection and *K*-selection are not simply opposite ends of a one-dimensional spectrum (Southwood *et al.* 1974). In particular, the introduction of dispersal into population models has dramatically different consequences for *r*- and *K*-selection. Dispersal further reinforces *r*-selection because an abundance of small offspring is favourable for dispersal, and the mortality incurred by dispersal of young further increases the value of early and copious breeding. However, the introduction of dispersal into the diagram for *K*-selection may change its topology entirely. If dispersal provides youngsters with opportunities for colonization at some distance from their place of birth, then the fact that there are few local opportunities for them is less important, and the positive feedback loop that reinforces *K*-selection is broken.

Dispersal is obviously adaptive in an environment that changes in a spatially uncorrelated way, such that when one place deteriorates, better conditions are available elsewhere (Gadgil 1971), but Hamilton and May (1977) argue that it may be adaptive for parents to enforce dispersal of some of their offspring even from a stable and favourable environment. Their argument is developed with game theoretic considerations. In a crowded population of sedentary adults, non-dispersive offspring compete among themselves for a limited number of openings near the parent, where they compete as well with the dispersive offspring of other parents. By dispersing some of its offspring, a given parent has a chance to establish progeny in both near and distant openings. The increased number of opportunities for rare and distant establishment may outweigh the cost in mortality due to dispersal itself, as well as the cost of lowered numbers of non-dispersers competing for the very rare opportunities near the parent.

Among insects there are many cases in which some individuals of a cohort are winged, and thus suited to dispersal, and others are not. Jarvinen (1976) speculates about the adaptive significance of such a dispersal polymorphism in water striders, and Hamilton (1979) does the same for male fig wasps. Such situations would be ideal for a critical analysis of the costs and benefits of dispersal because comparisons may be made between dispersers and non-dispersers of the same species from the same locality.

Dispersers from a crowded population may land in an uncrowded environment, in which case the adaptive life history may include elements favoured by *r*-selection and *K*-selection. However, adaptive tactics if the dispersers land in a crowded population in another locality have yet to be thoroughly explored by behavioural ecologists. The model of Hamilton and May (1977) suggests that dispersal will be favoured even in globally crowded populations of motile species. In general, in a crowded population of sedentary, long-lived adults, vacant territories due to recent deaths will be rare and scattered. Dispersal, especially in the form of itinerant adolescence, could expose individual youngsters to a large number of potential vacancies, even at the expense of encountering young competitors who are also looking for those same openings.

Faced with the difficulty of finding a suitable breeding site in a crowded environment, males and females may adopt different behaviours. In general, dispersal by either males or females may ensure genetic interchange among established populations, but dispersal by females is necessary to colonize unoccupied habitat. Greenwood (1980) has noted that dispersal between crowded populations is often predominantly by one sex, usually the female in birds but the male in mammals. He suggests that the dispersing sex is the one that has the more restricted access to mates or resources, and that this access is in turn determined by the social system. Woolfenden and Fitzpatrick (1978) have done an exemplary analysis of sexual patterns of dispersal in Florida scrub jays, mapping the dispersal of young of known parentage in a patchy and crowded environment. Male and female offspring both remain on their parents' territory, help to defend it, and assist in the feeding of the next batch of siblings. Females tend to spend a year or two helping while exploring widely in search of suitable mates and territories. Males stay at home for as much as three to five years, often extending the family's territory at the expense of adjacent families. A male helper then tends to 'bud off' his own territory from his father's, or to move quickly into any nearby vacancy (see Chapter 12). Thus in the face of the same crowded environment, males and females find breeding territories by very different patterns of dispersal.

There may be a conflict of interest between parent and offspring in the models of Hamilton and May (see Comins *et al.* 1980). It is in the parent's best interest to disperse some of its offspring, even when the mortality during dispersal is high enough to make successful dispersal and establishment of a given offspring less likely than successful establishment near its parent. Further conflicts of interest between parent and offspring arise in species in which grown offspring remain with their parents (Emlen 1982b and Chapter 12). The adaptive outcome of these cases must be a compromise whose resolution varies in favour of parent or of offspring as one or the other is morphologically and behaviourally suited to disperse and to establish a family at a new location.

The resulting patterns of dispersal produce different patterns of genetic structure in a population. In particular they have a strong influence on the likelihood that neighbouring individuals will be close kin, which in turn affects the evolution of behavioural interactions among neighbours. If adults are dispersive with batched young, and pre-reproductive young are sedentary, then the population consists of associations of pre-reproductive siblings who are outbred, and casual mixtures of adults. This population structure provides the opportunity for incest among siblings, simply by the attainment of reproductive maturity prior to dispersal of the sibling associations. Such incest would produce a population of inbred sibling associations of youth, and incestuously inbred adults of mixed parentage. Hamilton (1967) discusses the potential advantages of such incestuous behaviour, and the consequences for the evolution of extreme sex ratios and seemingly bizarre social behaviour. The degree of local inbreeding influences the advantages to be gained from sexual recombination (Horn 1981). It also determines the amount of additional dispersal necessary to balance inbreeding and outbreeding optimally.

Each of these types of population structure has a different implication for the form that natural selection takes, particularly selection for social and altruistic traits (Horn 1981, 1983). Inbred sibling associations favour cooperative behaviour toward neighbours and competition with strangers and immigrants. Associations of outbred siblings favour altruistic behaviour within the sibling group and coherence of the associated siblings until reproductive maturity, to take advantage of each other's cooperation. Associations of stable composition, whether individuals are related or not, favour the evolution of reciprocal altruism (Trivers 1971), where altruistic behaviour is differentially expressed on behalf of those individuals who are likely to reciprocate in the future. The casual mixture of outbred individuals is the only setting in which purely Darwinian selection for maximal individual reproductive success is relatively unaffected by other selective machinery.

Thus the detailed pattern of dispersal in life history is of critical importance to nearly all aspects of a species' ecology and behaviour, including even the nature of its social interactions (Horn 1983). Dispersal poses a series of problems and opportunities for adaptive matches between ecology and behaviour. Should many small young be broadcast at random? Should fewer young be sent out to seek their fortunes with a heavy grubstake from their parents? Should offspring remain at home and engage in territorial feuds with neighbouring clans? Should offspring look for vacancies or fight to create them? How should the adaptive tactics of dispersal differ for males and females? These and other adaptive questions about dispersal are explored further in a volume edited by Swingland and Greenwood (1983).

11.5   REPRODUCTIVE VALUE

Reproductive output is an appropriate parameter for exploring the broad qualitative trends in the adaptation of life history to environment. However, a more detailed analysis requires some elaboration of the notion of 'reproductive value' and its role as a technical criterion of an optimal life history.

Growth of the population as a whole accentuates the reproductive value of the first batch of young over later batches. For a population that is growing at a per capita rate of r, young born T years in the future should be discounted by e^{-rT} because they will face e^{rT} times as many competitors as those born in the present. Reproduction that is discounted for population growth is completely analogous to what economists call 'discounted present value' (Clark 1976). It was an economic analogy that prompted Fisher (1930) to define the reproductive value of the future offspring of individuals of age x as:

$$\frac{e^{rx}}{l_x} \int_x^\infty b_t\, l_t e^{-rt}\, dt,$$

where r is the per capita growth rate of the population, b_t is the birth rate to individuals of age t, and l_t is the probability of survival from birth to age t. It is useful to rewrite Fisher's formula in a readily interpretable form:

$$\int_x^\infty b_t (l_t/l_x) e^{-r[t-x]}\, dt.$$

In this form reproductive value is clearly the sum into the future from age x of the product of (reproductive output at a future age) times (probability of reaching that future age from age x) times (discount on future reproduction for growth in number of competitors in the population between age x and that future age).

The effect of a declining population is slightly more complex, though the criterion of natural selection is still for an individual to leave a disproportionately large number of copies of its genes in succeeding generations. If the population decline is due primarily to juvenile mortality, delayed breeding may be favoured since future offspring face either a lowered death rate or fewer competitors than present offspring. If the decline of the population involves heavy mortality among adults, then any competitive value of delayed reproduction will be offset by the low probability of the potential parent's surviving a delay. If the population is steady, then age of reproduction *per se* is irrelevant, though any parent whose offspring carry some other competitive advantage like higher tolerance of crowding can augment this competitive advantage by breeding early.

Pianka and Parker (1975) use reproductive value in a graphical analysis of life histories. They partition reproductive value at any given moment into current and future components and evaluate the qualitative effects on lifetime reproduction of various schedules of reproductive effort. This approach has been carried further and tackled more analytically by Schaffer (1974), Taylor *et al.* (1974), Bell (1980), and Goodman (1982). They suggest that an optimal life history, that is one which makes a maximal contribution to the future of the population, can be achieved by behaviour that maximizes reproductive value at each age. However, the choice of a particular reproductive option now may have many and complex effects on survival and on the reproductive options that are available later, both of which are ingredients of reproductive value and hence are important in evaluating current reproductive options. This feedback of current options on the criteria of their evaluation makes the mathematics of optimal life histories difficult, though Bell (1980) has given a complete and clear account of the more restricted question of the optimal age of first reproduction.

Goodman (1982) gives a partial answer to these difficulties by showing that for a life history that is already optimal, any lowering of reproductive value at any age will lower the total contribution to the next generation. This amounts to a proof that maximizing reproductive value at each age ensures a life history that is unbeatable by slightly different life histories. This leaves open the question of whether such a life history is a global optimum over the full range of variation that an individual might face. Schaffer (1974), Bell (1980), Caswell (1981) and Ricklefs (1981) all give examples of plausible theoretical conditions that produce multiple optima. In addition, Caswell (1981) has argued that a crucial constraint on the general optimality argument is that the modelled effect of changes in fecundity at a given age be confined to resultant changes in probability of survival to the next stage of reproduction. Ricklefs (1981) gives a counterexample in which trade-offs between age classes lead to inconsistent results. Goodman (1982) provides a somewhat wider range of constraining assumptions on the principle that an optimal life history can be attained by maximizing reproductive value at each age. The principle is clearly true when all trade-offs between fecundity and survival are confined to a single age class, but it is also true for any other monotonic relations between current fecundity and future reproductive value that could be mathematically mimicked by the more restricted assumption. Ricklefs (1981) suggests that even the more restricted assumption may be true for many species that grow to reach a determinate size at sexual maturity and then grow no further. Other complications are added if parental care can increase the likelihood of survival of offspring at the expense of decreased parental survival. Charnov (1982b) has shown that the optimal degree of parental care differs depending on whether it is the parent's or the offspring's reproductive value that is maximized. This particular instance of parent–offspring conflict (Trivers 1974) may prevent the attainment of an optimal life history for the parent.

The two major challenges in applying these ideas to real animals are testing to see when the simplifying assumptions are true, and posing more realistic forms of the theory if they are not true. Even the empirical tests are fraught with difficulties, which can be illustrated with recent attempts to discover the relation between clutch size and parental survival in birds. Smith (1981) showed that song sparrows with large clutches survived at least as well as those with small clutches. De Steven (1980) added eggs to clutches of tree swallows and produced relative decreases in the weight and survival of parents, but the effects were not statistically significant. Bryant (1979) reduced large clutches of the house martin until they contained the average number of eggs, and found that the females tending experimental nests actually gained weight over the nestling period. Högstedt (1980) suggested that clutch size in magpies is positively correlated with subsequent survival of parents. None of these studies shows clear evidence for the conventional assumption that increased fecundity leads to decreased subsequent survival, though the tendency for greater weight loss by parents tending larger clutches is suggestive. Under natural conditions it is possible that variations in quality among territories affect both clutch size and survival. It is also possible that females with differing resources lay clutches that are as large as possible without reducing the likelihood of survival below some minimum level. Either mechanism would reduce or reverse any direct evidence of an underlying trade-off between fecundity and survival, making it difficult to detect without carefully designed manipulative experiments.

Despite both theoretical and empirical difficulties, reproductive value is clearly a more appropriate parameter than current reproductive output for evaluating the adaptive significance of variations in life history. Reproductive value is particularly appropriate as the prime currency in which to express the evolutionary costs and benefits of behavioural acts because it measures the reproductive effect of a behavioural option at the precise moment that a behavioural choice is available.

In particular, there is a growing recognition of the importance of reproductive value in evaluating the costs and benefits of behaviour in the context of kin selection or inclusive fitness (Milinski 1978; Charlesworth & Charnov 1980; Gadgil 1982). Even youngsters from the same litter will differ in reproductive value, and hence in their genetic value to relatives that might give them aid. In general, aid should be concentrated on the relative that will have the highest reproductive value after the assistance is performed, even if it had a lower initial value. Furthermore, aid should be given by those individuals who will lose little, to those relatives who have much to gain. Hence differential aid should be concentrated upon a class of relatives, like youngsters, that has high variance in reproductive value (Rubenstein 1982a).

An example of animals behaving in accordance with these criteria comes from the study of rhesus macaques by Schulman and Chapais (1980). Macaque society is governed by dominance hierarchies among females. Mothers rank above their daughters; but among sexually mature daughters, rank varies inversely with age. As a daughter reaches sexual maturity, around the age of four, she rises in rank above her next older sister, and typically she retains this rank until a younger sister matures and surpasses her. The cause of this rise in

rank is a mother's support of her daughter in contests, and the
support is given at the time when the daughter normally attains her
highest reproductive value.

11.6 ADAPTIVE LIFE HISTORY IN AN UNPREDICTABLE ENVIRONMENT

So far we have discussed life history as though animals faced a predictable environment. Heterogeneity, both in space and in time, has
been subsumed into averaged expectations from reproductive behaviour that is adaptive to an appropriate average of environmental
heterogeneity. We now investigate how risk and uncertainty might
affect adaptive patterns of behavioural investment in reproduction
(Rubenstein 1982b).

Figure 11.4 shows how reproductive output might be expected to
vary with the amount of a resource necessary for reproduction that an
individual has available at a given time. The sigmoidal function is
chosen to give a generally rising reproductive output as supply
increases, but to account at the high end for diminishing returns as
investment increases toward the point at which this resource is no

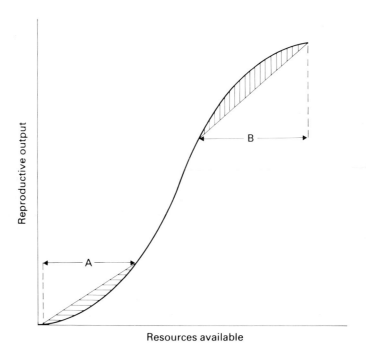

Fig. 11.4 Reproductive output attainable by expending available resources.
The horizontal axis may be interpreted directly as the amount of critical
resources provided at a given time by the environment, or indirectly as the
amount of resources stored by an animal, and thus as its condition for
breeding. Variations over the range A will produce an average reproductive
output that lies somewhere within the horizontally shaded region; over the
range B, within the vertically shaded region. See text for discussion.

longer limiting, and at the low end for the initial cost of entering the reproductive condition. These limits and costs at either end may be the result of nutrient requirements, physiological or biomechanical constraints, energetic demands of behaviour, or some combination of these. Figure 11.4 can be used to explore several aspects of an individual's adaptive response to an unpredictable environment, by representing several kinds of variation in available resources as shifts in resource levels on the horizontal axis, and then reading the resulting expected reproductive output on the vertical axis.

First let environmental variation directly cause changes in the amount of resource available, and examine two alternative strategies. A beast may respond directly to the environmental resource levels at a given time, producing a reproductive output appropriate to the resources available at that time. Alternatively, it may store up the resources through periods of variation, and produce a reproductive output appropriate to the average supply of resources rather than to the variations. The former is a prodigal strategy; the latter, if it is attainable, is a more provident strategy. The provident strategy produces a reproductive output that lies on the sigmoidal curve at the point of average resource supply. The prodigal strategy produces a reproductive output that will lie within the shaded area near the average resource supply. If the average resource level and the variations around it are confined to the lower, upwardly concave section of the curve, then the shaded area is above the curve and the prodigal strategy produces a larger expected reproductive output than the provident strategy. The opposite result, an advantage to the provident strategy, is found when variations of the resource levels span the upper section of the curve, where diminishing returns accompany increased investment. Thus we derive a limited version of a more general result proposed in part by MacArthur (1968) and extended and made rigorous in the context of life history by Schaffer (1974). Individuals in relatively poor condition or in relatively poor environments should devote whatever resources are available to immediate reproduction, while those in good condition and good environments may profit from storing resources so as to average the effect of environmental variation on their reproduction (see also Chapter 4).

A parallel result follows from a similar analysis of risky versus conservative strategies in the face of an uncertain environment. Let the conservative strategy produce a reproductive output that depends on environmental conditions at a later time according to Fig. 11.4. Let the risky strategy depend on environmental conditions in the same way, but with some added variance in the individual's condition caused by the taking of the risk. For example, suppose that a territorial beast can increase its territory, and thus its resources for reproduction, by fighting with its neighbours, but that if any fights are lost, some of the original territory is lost with them. Then the conservative strategy is to keep the original territory and not to fight. The risky strategy is to fight, having the possibility of the resulting territory being either larger or smaller, with the average expectation being the same size as the original territory. The alternatives are evaluated as before, using Fig. 11.4. The conservative strategy yields a reproductive output that lies on the sigmoidal curve at the point of average available resource. The risky strategy produces a reproductive output that will lie within the shaded area near the average level of resource. The shaded area lies above the curve at low resource levels, favouring the risky strategy, but below the curve at high resource levels, where the conservative strategy is favoured. Again, individuals in relatively poor condition or in relatively poor environments should adopt the risky strategy. This may entail experiencing reproductive failures in order to take advantage of the possibility of reproductive gains. Conversely, those in good condition and good environments should behave more conservatively. This result is derived and discussed more fully by Rubenstein (1982b).

The preceding analyses have maximized reproductive output in a single season. The analysis becomes more complex when the whole life history is considered, so that current reproduction affects the probability of survival to the next season or the possible reproductive output in that season. These effects are discussed by Rubenstein (1982b), but we shall only give the qualitative rudiments of his analysis here. By forgoing reproduction in the current season, an individual might enter the next season in better condition or with more resources. Conversely, current reproduction may decrease condition for next season. These effects may be represented in Fig. 11.4 by vectors from current condition and potential reproductive output to the alternative future conditions and reproductive outputs (see Fig. 11.5). If the vectors are large, that is if there is a severe effect of this season's reproduction on next season's condition, the analysis is complex and messy. However, for effects that are small enough to remain within a given region of the sigmoidal curve, the results are straightforward and intuitive. Individuals in poor condition or in poor environments will generally be expected to forgo breeding, either because their expected output is very low if they are in the initial part of the sigmoidal curve, or else because reproduction would slide them down the steeply descending part of the curve. Conversely, individuals in good condition or in good environments should generally attempt to reproduce at every opportunity that their condition allows. Tests of these ideas often involve contrasts between young and old individuals in a population, because the young are often smaller, in worse condition, and relegated to less preferred habitats than their elders.

Note that this prediction, conservation in poor condition and profligacy in good, is just the opposite of the earlier predictions of this section. This is in part due to the special conditions involved, but it

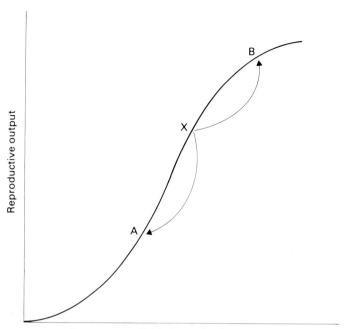

Fig. 11.5 Effect of this season's choice on next season's reproductive output. Point X represents the reproductive output attainable in this season with the amount of resources currently stored. Point A represents the output attainable next season after reproduction in this season. Point B represents next season's potential output after forgoing current reproduction.

also dramatizes the difference between ideas that are based on maximizing current reproductive output and those that take account of the full effect of reproduction on subsequent life history.

We do not emerge from this discussion with a definitive optimal life history, but rather with a list of the ingredients for an appropriate analysis. These include: the shape of the curve relating reproductive output to available resources, the mean and range of environmental quality, the mean and range of individual condition, the predictability of environmental quality and individual condition, and the effect of current effort on future condition. The appropriate common currency in which to evaluate the adaptive effect of behavioural options is reproductive value. A complete evaluation has yet to appear.

11.7 THEORIES OF LIFE HISTORY AND DATA OF BEHAVIOURAL ECOLOGY

The lessons of this chapter concern the interplay between theory and field studies. In studies of life history, it is of crucial importance to quantify not only average mortality, fecundity, and dispersal, but also their age-specific schedules. The most fully developed theories about life histories explicitly use the age-specific pattern of one of these variables to predict the age-specific pattern of another. To date, these theories have been tested predominantly by contrasts among the life histories of different species inhabiting different environments. In these tests, adaptive differences are often achieved by behavioural mechanisms, but these are confounded by large differences in genetics, development, mechanical design, and physiology. The role of behaviour as an adaptive aspect of life history appears more clearly in analogous contrasts among life histories of individuals of a single species.

Among individual differences in behaviour, age-related and sex-related strategies of life history have long been of interest, and they are the subjects of a considerable body of theory. Polymorphisms that are not related to age or sex have yet to receive the attention that they deserve, both in theory and in the gathering of facts. Such polymorphisms are likely candidates for game theoretic analysis, in which an important part of an individual's adaptive behaviour is an accurate assessment of what other individuals are doing. Parent–offspring conflicts also deserve more explicit attention than they have received thus far.

Nowhere is the interplay between fact and theory more important than in the case of reproductive value. Theoretical predictions depend critically on details of the functional dependence of both reproductive output on effort and current reproduction on future reproductive value. Few detailed data exist, and gathering them is a considerable challenge. In those field studies in which reproductive value has actually been calculated, it has led to valuable insights into discriminatory behavioural interactions.

Particularly for reproduction and social behaviour, adaptations to environment feed back on themselves to become part of the environment for further adaptations. Examples include the self-reinforcing cycles of *r*-selection and *K*-selection, and the effect of current reproduction on future opportunities for reproduction. Such feedback adds interesting challenges to both theory and analysis.

At a grander level, there is an important feedback that links major features of life history and of social behaviour. Age-specific schedules of mortality and reproduction produce variations in crowding from place to place and from time to time. Different patterns of dispersal are adaptive to the variations in crowding, and dispersal determines the local genetic structure of a population. The genetic structure, particularly the relative likelihood that close neighbours are close kin, provides the setting for the evolution of different social systems. The social system in turn influences the age-specific patterns of mortality, reproduction, and dispersal. Thus dispersal links life history to the genetics of social behaviour. The theoretical analysis of dispersal is in its infancy, as is the empirical analysis of individual differences in dispersal. The study of dispersal is worth special attention since it promises to draw together life histories and social behaviour, two more mature areas of inquiry into behavioural ecology that have until now had largely separate intellectual histories.

PART 4
COOPERATION AND
CONFLICT

INTRODUCTION

The final section of the book looks at how cooperation between individuals on the one hand, and underlying conflicts of interest on the other, have influenced the structure of societies and the evolution of communication.

In Chapter 12, Emlen considers cooperative breeding systems in birds and mammals, which he defines as cases where more than two adults provide care for the young. Sometimes there are only two breeders in the group and the others are 'helpers', often previous offspring of the breeders (e.g. jackals, scrub jays). In other examples, several males and females breed together communally (e.g. lions, mongooses, anis). The neatest way to test whether helpers really help is to perform a removal experiment. So far only a few of these have been done, probably because field workers are understandably reluctant to disturb their long-term studies of marked populations.

Emlen then asks why the helpers do not leave home and breed. The answer is often that there are ecological constraints which force them to remain on the parental territory; these include a full habitat, strong advantages from group living and prohibitive costs of dispersal and independent reproduction (e.g. due to poor food supplies). Given that helpers remain at home, the next problem is why they help. Three main advantages are considered, namely help as a payment to the breeders for permission to stay on the territory, future inheritance of the territory, and increased inclusive fitness from helping to raise collateral kin. Finally, the chapter outlines the theory for conflicts of interest in cooperative breeders, for example conflicts over the amount of help and over who breeds. Although there is abundant descriptive evidence for conflict (for instance fighting and the killing of nestlings) it is clear that a lot of good long-term field data are needed before we can test these ideas quantitatively.

In Chapter 13, Brockmann considers the evolution of cooperation and conflict in insect societies, concentrating on the social Hymenoptera, which show the greatest diversity of social systems. Their breeding systems show some of the characteristics of cooperative breeding in birds but in many Hymenoptera the 'helpers-at-the-nest' are permanent worker castes rather than transient stages in an individual's life history. Brockmann tackles two problems in her chapter.

First she uses the comparative approach to unravel the possible evolutionary pathways from simple to complex social systems. Two different routes to sociality are identified. In the 'subsocial' route, daughters stay at home to help their mothers reproduce (equivalent to Emlen's 'helper-at-the-nest' category of cooperative breeding), while in the 'parasocial' route, females of the same generation nest together (equivalent to Emlen's 'communal breeders'). Brockmann identifies the ecological pressures which favour the evolution of complex sociality along both these pathways.

The second problem considered in the chapter is how attributes such as worker sterility can evolve when these individuals do not reproduce. The evolution of cooperation by mutualism, kin selection and reciprocation is discussed. Haplodiploidy in Hymenoptera means that there is a high degree of relatedness between sisters and so favours the evolution of altruistic behaviour, but Brockmann emphasizes that there are also often conflicts of interest in a colony, between queens (breeders), between workers and queens (helpers and breeders) and between the workers themselves.

Chapter 14, on plants, may come as a surprise in a behavioural ecology text, but Charnov shows that many of the concepts developed in previous chapters apply equally well to plants, for example sexual selection and life history theory. He considers three topics to give a flavour of this fast developing new field of research: sex choice, breeding systems and conflicts of interest in plant reproduction.

Many plants can function both as a female (producing seeds) and as a male (producing pollen). The chapter considers the ecological factors influencing sex allocation and the theory for its evolution. Plants often change sex with age or size (sequential hermaphroditism) because older or larger individuals are often relatively better at being one sex, for example better at attracting pollinators or producing seed. In other cases, sex choice may depend on the microenvironment in which the plant grows. Charnov then applies sexual selection theory to the evolution of breeding systems in plants. Bateman's principle, that a male's reproductive success is limited by access to females while a female's reproductive success is limited by resources for eggs (Chapter 9), applies to plants as well as animals. Furthermore, reproductive success through male or female function depends on what other individuals in the population are doing and so the concept of an ESS (Chapter 2) is needed to derive optimal sex allocation. Charnov derives the ESS sex allocation under various possible conditions for trade-off between male and female fitness and applies the theory to the evolution of heterostyly.

Finally, Charnov suggests various cases where there may be conflicts of interest over mating decisions in plants. He suggests that the theory of kin selection (Chapter 3) is relevant to the evolution of double fertilization in angiosperms. In this, the pollen nucleus divides into two identical copies, one of which fertilizes the egg to form a

zygote while the other goes to form the endosperm, a tissue which may provide nutrition for the developing zygote. When various endo-sperms compete for resources each endosperm can be viewed as a sterile worker, aiding the reproduction of a relative, namely its
. zygote. By means of double fertilization, therefore, the pollen nucleus in the zygote may provide itself with a helper!

The book ends with a chapter on animal signals. Krebs and Dawkins define a signal as a means by which one animal (the 'actor') exploits the muscle power of another animal (the 'reactor'). They view signals as a result of coevolution between mind-reading and manipu-lation. Individuals can benefit if they are able to predict the future behaviour of others, such as competitors or predators. They do this by mind-reading, picking up signals from others which predict what they will do next. Once this occurs, the stage is set for the evolution of counter-measures to make mind-reading more difficult, including concealed signals and active deception. Victims of mind-reading can exploit the fact that others are using their signals, in order to manipu-late the behaviour of the mind-readers.

Two kinds of coevolution are recognized between mind-reading and manipulation. In cases where actors and reactors have conflicts of interest we would expect an arms race. Krebs and Dawkins suggest that many of the exaggerated conspicuous signals used in animal dis-plays are the end product of a coevolutionary arms race for improved mind-reading and manipulation. Sometimes, however, communication may be cooperative in the sense that actors gain a benefit from having their minds read by reactors. Coevolution here is expected to lead to a reduction in the conspicuousness of displays, to minimize signal costs. The chapter ends with a discussion of ecological and social factors which influence signal detection and a critical look at the kinds of information transmitted by animal displays.

Chapter 12
Cooperative Breeding in Birds and Mammals

STEPHEN T. EMLEN

12.1   INTRODUCTION

Among higher vertebrates, young typically are born (or hatched) in a highly dependent state and require large amounts of adult care during their development to independence. In most species, such care is provided either by the mother alone or by both members of the breeding pair. As a result of the near ubiquity of this parental care pattern, we often overlook the fact that not all animals are restricted to 'parents-only' systems of child rearing. Among some rodents, a number of mammalian carnivores, and over 300 species of birds, additional adults play major roles in the raising of dependent young. Such individuals are termed *alloparents* (see Table 12.1).

In this chapter, I follow the terminology of Emlen and Vehrencamp (1983) and use the term *cooperative breeding* to refer to any situation where more than two adults provide care in the rearing of young. I further distinguish between two forms of cooperative breeding, forms which more accurately should be considered as representing ends of a continuum of cooperative child-rearing possibilities. (1) *Helper-at-the-den (or helper-at-the-nest) systems* are those in which auxiliary (non-breeding) adults contribute physically, but not genetically, to the young being reared. These extra adults serve as helpers, but they do not engage in sexual activity with the breeding pair. (2) *Communal breeding systems* are those in which parentage of the offspring is shared. In the case of shared paternity, more than one male in the social group has a significant probability of fathering some of the offspring. With shared maternity, several females in the group have a significant probability of giving birth, after which the young are communally nursed (mammals), or several females have a significant probability of contributing eggs to a communal clutch, which is then jointly incubated (birds).

12.1.1   Helpers-at-the-den/nest

The silver-backed jackal (*Canis mesomelas*) of the plains of East Africa exemplifies the helper-at-the-den situation. Moehlman (1979, 1983), studying this species in Tanzania, found that roughly one-third of the young produced in a given year remained with their parents through

Table 12.2. Factors influencing dispersal from natal groups

Species	Type of cooperative system	References
Mammals		
Mustelidae		
European badger (*Meles meles*)	Communal	Neal 1977
Canidae		
Red fox (*Vulpes vulpes*)	Helper-at-the-den	Macdonald 1979, 1980; Macdonald & Moehlman 1982
Bat-eared fox (*Otocyon megalotis*)	Communal	Lamprecht 1979
Silver-backed jackal (*Canis mesomelas*)	Helper-at-the-den	Moehlman 1979, 1983
Golden jackal (*Canis aureus*)	Helper-at-the-den	Moehlman 1983
Coyote (*Canis latrans*)	Helper-at-the-den/Communal	Bowen 1978; Camenzind 1978; Bekoff & Wells 1980
Wolf (*Canis lupus*)	Helper-at-the-den/Communal	Mech 1970; Zimen 1976; Fentress & Ryan 1982
African wild dog (*Lycaon pictus*)	Helper-at-the-den/Communal	van Lawick 1973; Frame & Frame 1977; Frame *et al.* 1979
Felidae		
Domestic cat (*Felis catus*)	Communal	Macdonald & Apps 1978
Lion (*Panthera leo*)	Communal	Schaller 1972; Bertram 1975; Bygott *et al.* 1979
Hyaenidae		
Brown hyaena (*Hyaena brunnea*)	Communal/Helper-at-the-den	Owens & Owens 1979a, 1979b; Mills 1982
Viverridae		
Dwarf mongoose (*Helogale parvula*)	Helper-at-the-den	Rood 1980
Banded mongoose (*Mungos mungo*)	Communal	Rood 1974, 1978
Birds		
Grallidae		
Tasmanian native hen (*Tribonyx mortierrii*)	Communal	Ridpath 1972
Pukeko (*Porphyrio p. melanotis*)	Communal	Craig 1979, 1980
Cuculidae		
Groove-billed ani (*Crotophaga sulcirostris*)	Communal	Vehrencamp 1977, 1978
Alcedinidae		
Kookaburra (*Dacelo gigas*)	Helper-at-the-nest	Parry 1973
Pied kingfisher (*Ceryle rudis*)	Helper-at-the-nest	Reyer 1980
Meropidae		
Red-throated bee-eater (*Merops bullocki*)	Helper-at-the-nest	Fry 1972

Table 12.1. (*continued*)

Species	Type of cooperative system	References
White-fronted bee-eater (*Merops bullockoides*)	Helper-at-the-nest	Emlen 1981; Hegner *et al.* 1982
Phoeniculidae		
Green woodhoopoe (*Phoeniculus purpureus*)	Helper-at-the-nest	Ligon & Ligon 1978a; Ligon 1981
Picidae		
Acorn woodpecker (*Melanerpes formicivorous*)	Communal/Helper-at-the-nest	MacRoberts & MacRoberts 1976; Stacey 1979a, b; Koenig & Pitelka 1979, 1981
Malurinae		
Superb blue wren (*Malurus cyaneus*)	Helper-at-the-nest	Rowley 1965
Splendid wren (*Malurus splendens*)	Helper-at-the-nest	Rowley 1981
Timilinae		
Arabian babbler (*Turdoides squamiceps*)	Helper-at-the-nest	Zahavi 1974
Common babbler (*Turdoides caudatus*)	Helper-at-the-nest	Gaston 1978c
Jungle babbler (*Turdoides striatus*)	Helper-at-the-nest	Gaston 1978a
Grey-crowned babbler (*Pomatostomus temporalis*)	Helper-at-the-nest	Councilman 1977; Brown & Brown 1982
Meliphagidae		
Noisy miner (*Manorina melanocephala*)	Communal/Helper-at-the-nest	Dow 1977
Grallinidae		
White-winged chough (*Corcorax melanorhamphus*)	Communal/Helper-at-the-nest	Rowley 1978
Corvidae		
Florida scrub jay (*Aphelocoma coerulescens*)	Helper-at-the-nest	Woolfenden 1975; Woolfenden & Fitzpatrick 1978, and in press
Mexican jay (*Aphelocoma ultramarina*)	Helper-at-the-nest	Brown 1963; Brown & Brown 1981b

the following breeding season. Groups of three to five adults were formed, and the non-breeding members served as helpers in rearing the next year's litter of pups.

Such assistance can take many forms. Jackal pups are born under ground, and first venture out of their den at 3–4 weeks of age. They remain highly dependent upon the lactating female until weaning at 8–9 weeks and continue to depend upon adults for food until three months old. Non-breeding helpers play an important role both in provisioning the nursing female and in regurgitating food directly to the pups once they are old enough to take solid food. In breeding groups of three, Moehlman (1979) found that 18–32% of all pup feedings were provided by the helper.

Helpers also guard the pups when the parents are absent. The

307

result is that pups are left unattended at the den less frequently than is the case when parents must rear their pups alone. Jackal guards alert the pups to approaching danger (by giving alarm calls to which the young respond by returning underground), and also chase off many potential predators (such as spotted hyaenas, *Crocuta crocuta*). The increased attendance at the den thus provides greater protection for the pups. Finally, all adult members of the jackal groups play with and groom the young as well as helping in teaching them to hunt.

The Florida scrub jay (*Aphelocoma coerulescens*) provides an avian analogue to the jackals described above. These birds, studied in detail by Woolfenden and his colleagues (Woolfenden 1973, 1975; Woolfenden & Fitzpatrick 1978, and in press) live on year-round territories in relict oak-scrub habitat of peninsular Florida. The basic social unit consists of a monogamous breeding pair together with some of their young from the previous one or more years. These mature offspring often forage with their parents, and they assist in group defence of their natal territory. During the breeding season, they play a major role in providing food for the nestlings as well as in tending and guarding the nest when others in the group are off foraging. In the event of the approach of a predator, the bird in attendance utters an alarm note which has the effect of summoning the rest of the group, all of whom then mob the potential predator.

After the young fledge from the nest, all group members continue to provide food and protection for the fledglings during the latter's transition to full independence.

The social organizations of other avian species with helper-at-the-nest cooperative breeding systems become more complex. Groups become larger, extended rather than nuclear families predominate, and individuals from several overlapping generations may be involved.

The white-fronted bee-eater (*Merops bullockoides*), which I and my co-workers are studying in Kenya, provides such an example (Emlen 1981, 1982a; Hegner *et al.* 1982). These birds live in large colonies of several hundred individuals but each colony is substructured into a number of smaller interacting units termed clans. It is within these smaller groups that helping behaviours occur.

Clans have a kin-structure based largely upon extended family relationships. Each clan is composed of from one to five monogamous pairs plus a smaller assortment of young and widowed individuals. All pairs are potential reproductives in any given year, but the number that exercise this option varies greatly across the years depending upon the local environmental conditions. Individuals that do not breed are likely to serve as helpers at the nest of a clan member that does. Additionally, birds that start a season as breeders, but whose nesting attempts fail, are likely to join as helpers with other clan members at nests where eggs or young are still being tended. Such redirected helping results in adult bee-eaters shifting back and forth between breeder and helper status. As a consequence, reciprocal exchanges of helping between donor and recipient, or donor and the offspring of the original recipient, are not uncommon.

Bee-eater helpers take part in all phases of breeding activity, including excavating and defending the nest, allofeeding the breeding female, incubating the eggs, feeding the nestlings, and caring for fledglings during the 6 week period after they leave the nest.

12.1.2   Communal breeders

The banded mongoose (*Mungos mungo*) exhibits a communal breeding system. These small members of the viverrid family have been studied in East Africa (Neal 1970; Rood 1974, 1978), where they live in packs of from four to 40 individuals. Packs excavate dens in abandoned termite mounds which are used as nocturnal sleeping chambers. Breeding also takes place in such dens. Many dens may be located within the 1 km^2 home range, and the pack frequently moves between them.

Several females in each pack are reproductives, and they produce their litters synchronously. As a result, the young are the same age and they suckle indiscriminately from any lactating female. (In a related species, the dwarf mongoose (*Helogale parvula*) even some non-reproductive females develop teats, lactate, and communally nurse the young (Rood 1980). It is not yet known whether this is true also of *M. mungo*.)

Male banded mongooses also care for the young. During the first 3-4 weeks following birth, the young remain below ground. Each morning most of the pack depart to forage for food, but one or more adults remain behind, guarding the den. Such guards are able to defend the young against snakes as well as against many other ground predators. Rood (1974) found that three-quarters of the guarding was performed by adult males and that lactating females never acted in this capacity. This partial division of labour freed the breeding females to spend more time foraging and replenishing the nutrient reserves needed during lactation.

Avian species also provide clear examples of communal breeding systems. In the groove-billed ani (*Crotophaga sulcirostris*) from one to four monogamous pairs of unrelated adults plus an occasional unpaired helper form a breeding unit. Vehrencamp (1977, 1978), who studied this species in Costa Rica, found that all members of a group contribute to the building of a single nest into which all females communally lay their eggs. The dominant male, as well as all females, share the task of incubation. Once the nestlings have hatched, all the adults of both sexes play roles in feeding and defending the young.

Behavioural systems such as those described above prompt us to ask numerous questions. Why is cooperative breeding behaviour so rare among higher vertebrates? What determines when it will develop, and what form it will take? And, if helpers actually postpone or forgo personal reproductive opportunities to serve as helpers, then how can such seeming altruism evolve at all?

In the following article, I will attempt to address these questions, looking at the roles which ecological, genetic and social factors play in the development and maintenance of cooperative breeding behaviour. But, before proceeding, it first is necessary to ask whether helpers really do help. Is it possible that such behaviours have been mis-named as a result of anthropomorphic interpretations of alloparental care? Or does the presence of alloparents lead to an increased production of young?

12.2 DO HELPERS REALLY HELP?

Typically one asks this question by comparing the production of young from dens/nests tended by the breeding parents alone with that from dens/nests tended by three or more adults. When more than one helper may be present, we examine the regression of reproductive success upon group size. If the number of young reared is correlated with the presence or number of helpers, we surmise that the latter contributed significantly to the rearing effort—that is, they 'helped'. Simple correlational analyses of this sort are plagued with problems of interpretation (see below), but they can provide a first indication of the effect of helpers upon the reproductive success of breeders.

Moehlman (1979, 1983), in her studies of silver-backed jackals, examined the relationship between reproductive success and the presence of non-breeding auxiliaries. Her results, compiled over five years, are plotted in Fig. 12.1a and show a linear increase in pup survival with increasing numbers of helpers at the den. A similar trend is shown in my own results from white-fronted bee-eaters. Figure 12.1b again shows a linear increase in numbers of young fledged per nesting attempt with increasing numbers of bee-eater helpers tending the nest (Emlen 1981, and unpublished results). Woolfenden's (1981) studies of the Florida scrub jay furnish a third example. His results are plotted in Fig. 12.1c. As with the jackals and bee-eaters, the presence of Florida scrub jay helpers is correlated with an increase in the survival of young.

In these and other studies, helpers contribute to the survival of offspring in two principal ways, by providing (1) better protection from predators and (2) better food provisioning than can be supplied by the breeding parents alone.

Silver-backed jackal pups apparently benefit in both ways. They are protected more continuously as a result of the helper(s) guarding the den while the breeders are away foraging. They also profit by the extra food brought back and regurgitated by the helpers. The amount

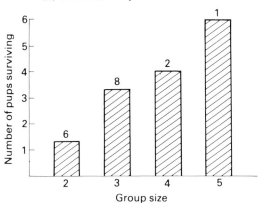

(a) Silver-backed jackal

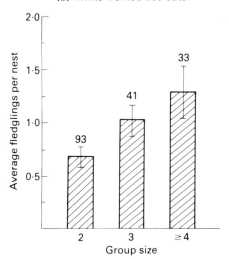

(b) White-fronted bee-eater

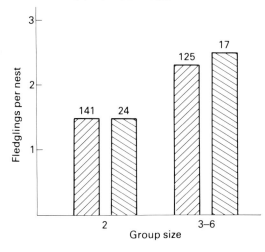

(c) Florida scrub jay

= all pairs

= previously successful pairs

Fig. 12.1 The effect of extra-parental helpers upon reproductive success in three species of cooperative breeders: **(a)** silver-backed jackals (from Moehlman 1979, 1983); **(b)** white fronted bee-eaters (from Emlen *et al.*, unpublished data); and **(c)** Florida scrub jays (from Woolfenden 1981).

(a) Silver-backed jackal

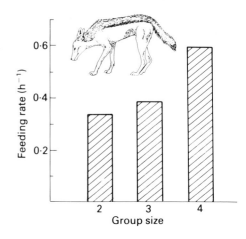

(b) White-fronted bee-eater

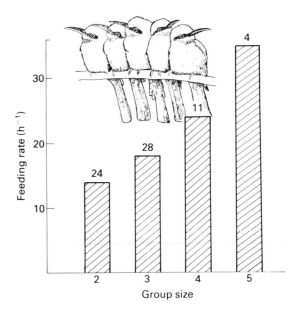

Fig. 12.2 The effect of helpers upon the amount of food brought to the young. **(a)** Data for the silver-backed jackal is taken from Moehlman (1983), and include 414 hours of observation taken on five family units. Feeding rate is calculated as the number of regurgitations and/or nurses provided to the litter of pups per hour. **(b)** Unpublished data from white-fronted bee-eaters are provided by Emlen *et al.*, and include data from 67 nests. For each nest, feeding rate is calculated as the total number of insects brought to the young per hour based upon an 8 hour sample of observations collected when the nestlings were between 14 and 22 days old. **(c)** The relative contributions of different group members to the feeding of nestlings in the grey-crowned babbler (from Brown *et al.* 1978). Note the increase in the proportion of total visits made by helpers as group size increases.

(c) Grey-crowned babbler

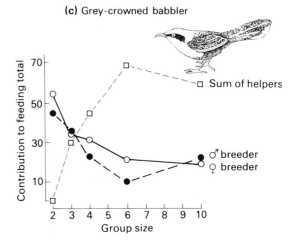

of food brought to litters by groups of varying sizes is shown in Fig. 12.2a, adapted from Moehlman (1983). Note the increase in food provided as helpers are added to the nuclear family.

Among white-fronted bee-eaters, nests are partially protected by being excavated deep into the vertical sides of cliffs and gulleys (Emlen 1981). Predation accounts for only 6% of breeding losses (egg and chick), but starvation of nestlings accounts for fully 30%. Studies show that helpers play full contributory roles as feeders of nestlings, with the result that the quantity of food delivered to a nest increases directly with the number of helpers (Fig. 12.2b, from unpublished results of Emlen *et al.*).

In contrast, starvation of young is uncommon in the grey-crowned babbler (*Pomatostomus temporalis*), an Australian passerine that lives in groups of 8–12 individuals. Only one pair in each group breeds, and the remaining individuals behave as helpers-at-the-nest (Councilman 1977). Although babbler helpers do bring food to the nestlings and fledglings, since the young generally are not food stressed the result is not an increase in total food provided; rather, the same amount of food is delivered, but now it is divided among a larger number of provisioners (Brown *et al.* 1978). Each adult makes fewer foraging trips (Fig. 12.2c). Stated another way, when the food needs of the young are being met, the presence of helpers enables breeders to slacken their work load. Similar findings have been reported from other avian cooperative breeders (e.g. common babbler, *Turdoides caudatus* (Gaston 1978c); green woodhoopoe, *Phoeniculus purpureus* (Ligon & Ligon 1978a); Florida scrub jay, *Aphelocoma coerulescens* (Stallcup & Woolfenden 1978).

Woolfenden (1978) believes that predation is the major mortality factor facing newly hatched Florida scrub jays and larger groups of scrub jays are more effective in mobbing and deterring several predators (especially snakes) than are simple pairs. Stallcup and Woolfenden (1978) argue that the value of helpers in providing food to the nestlings is that it leads to a reduced energetic stress and predation risk upon the *breeders*, and calculations show that the annual survival rate of birds breeding with the assistance of helpers is greater than that of birds breeding alone. These data demonstrate the important demographic point that one must consider long-term survival as well as short-term fecundity gains whenever calculating the effects of a helper upon the lifetime fitness of a breeder.

Correlation analyses of group size against reproductive success form a weak basis for making inferences about helping behaviour. Field manipulations involving additions or removals of extra-parental adults would allow a more direct assessment of the effect of helpers upon reproductive success. Recently, Brown and Brown (1982), in collaboration with D. Dow, performed the first experimental test of the question of whether helpers really help. The Browns selected 20 groups of grey-crowned babblers for their study, picking groups of approximately the same size which occupied equivalent territories.

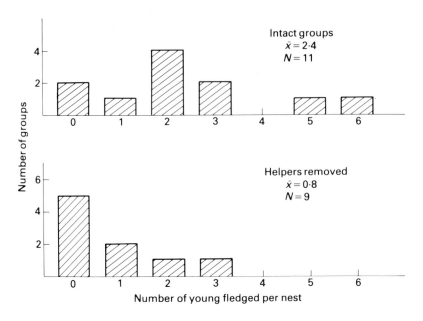

Fig. 12.3 The effect of artificial removal of helpers upon reproductive success in groups of grey-crowned babblers (from Brown & Brown 1982).

They then artificially removed all but one helper from nine breeding groups. The 11 control groups averaged seven members, with helper numbers ranging from four to six; the nine experimental groups each had three members, the breeding pair plus one helper. The nesting success of all groups was monitored in their subsequent breeding. The results were unambiguous (Fig. 12.3). The success of nests with their full complement of helpers was more than twice that of nests tended by the artificially depleted groups.

To summarize the available evidence, both helper and communal breeding systems typically lead to an increased production and/or survival of young by the breeding pair(s). Thus we can answer the rhetorical question raised at the beginning of this section in the affirmative. Helpers *do* really help, and the breeders are the obvious beneficiaries. But what of the helpers themselves? Are they also benefiting or would they do better to leave their groups and attempt to breed independently?

Let us make the oversimplified assumption that all adults, whether breeders or helpers, can provide an equal amount of parental (or alloparental) care. An auxiliary then would be predicted to remain with its parental group as a helper only if by so doing it increased their reproductive success to the point where the *per capita* reproductive success remained constant. If the effect of its contribution was less, the auxiliary would better its fitness by channelling its contribution into its own breeding effort.

Koenig (1981a) recently analysed the productivity data for avian cooperative breeders from this perspective. He examined data from 15 species and found that although reproductive success of the breeding pair(s) increased with group size in 13 of the cases, there was no consistent trend between *per capita* reproductive success and group size. On a *per capita* basis, pairs were the most productive units in seven species, pairs were roughly equivalent to larger groups in five, and groups were more successful in only four species. If our assumption is valid, we must conclude that helpers often would do better to cease acting as alloparents and to initiate their own independent breeding efforts.

Charnov (1981) presented a numerical argument that leads to the same conclusion. When a single auxiliary remains as a helper with its parents, the result is a group of three adults providing care to young to which the auxiliary is related by 0.5. If the same auxiliary pairs and breeds on its own with its new mate, the result is a total of four adults providing care to two sets of young, both of which are related to the auxiliary (now a breeder) by 0.5. Other things being equal, this simple numerical advantage should always favour independent breeding over helping behaviour.

Why then do we find cooperative breeding systems at all?

12.3 WHY DON'T HELPERS BREED ON THEIR OWN?

12.3.1 Routes to sociality

There are two primary reasons why animals might live and reproduce in social groups larger than simple pairs. They may gain directly from group living, or they may be 'forced' to remain in groups as a result of high costs or risks associated with departure.

Normally, we think in terms of the benefits that accrue to individuals by virtue of living gregariously (Chapter 5). The two most often cited benefits are increased alertness and defence against predators and increased capabilities for detecting and harvesting food resources that are difficult to locate (Alexander 1974; Hoogland & Sherman 1976; Bertram 1978). In such instances, the average fitness of individual group members ($\overline{W}$) will increase as some function of increasing group size (k) up to some optimum size and decrease thereafter (Fig. 12.4a).

Alternatively, individuals might be 'forced' to remain in groups because of ecological or other constraints that restrict their option of dispersing and breeding on their own. Consider a species in which grouping leads to a net *decrease* in *per capita* fitness compared to solitary breeding (shown as the solid line in Fig. 12.4b). But consider further that openings for breeding are scarce and competition for them is severe. The intrinsic benefit of independent breeding must be devalued by the probability of dispersing and becoming established successfully, here denoted as s. When s is low, a 'hump' is created in

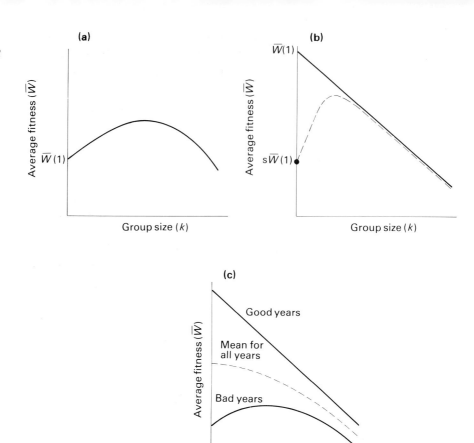

Fig. 12.4 **(a)** Fitness curve for a hypothetical species in which increasing group size leads to a direct benefit to group members. $\overline{W}(1)$ denotes the fitness of a solitary individual. **(b)** Fitness curve for a hypothetical species in which increasing group size leads to a decrease in average individual fitness (solid line), but the possibility of leaving the group is severely constrained (the dashed line, where s denotes the probability of successfully dispersing and becoming established independently). **(c)** Fitness curves for a hypothetical species in which grouping is disadvantageous but the intensity of constraining factors limiting dispersal varies greatly across years. 'Good' years denote times when constraining factors are minimal; 'bad' years denote times when constraints are severe.

the fitness curve and we speak of individuals being 'forced' to remain in their natal groups (dashed line in Fig. 12.4b).

The magnitude of the constraint, s, for any given species may vary across different geographic areas and across years. Such changes in s lead to predictable differences in the size and complexity of the social groupings, with concomitant changes in the potential for development of helping behaviour. Fitness curves for a hypothetical species residing in an unpredictably varying environment, where the severity of constraining factors changes greatly from year to year, are

shown in Fig. 12.4c. In 'good' years, all individuals will breed independently, groups will not form, and helping behaviour will be non-
existent; in 'bad' years, many individuals are forced to postpone
dispersal, groups develop and the potential for helping interactions is
present.

12.3.2 The constraints model of helping

Regardless of the route to group living, every non-breeding individual is constantly faced with the choice of either remaining in its
current group as a non-breeding auxiliary or leaving and attempting
to become established as an independent reproductive. At least four
factors will enter into this decision: (1) the risks associated with dispersal itself, (2) the probability of successful establishment on a suitable territory (area), (3) the probability of securing a mate, and (4) the
likelihood of successfully reproducing once 'established' (Emlen
1982a). Any one of these factors can operate as a constraint, tipping
the cost–benefit balance against the choice of personal reproduction
(Table 12.2).

According to the constraints model of the evolution of helping,
most social groups of cooperatively breeding birds and mammals
form by the second, forced-retention, route. If true, then the selective pressures favouring group living in many cooperative breeders
would be fundamentally different from those favouring grouping in
such aggregations as blackbird roosts or seabird colonies, pinniped
pods or ungulate herds. Cooperative breeding, according to this
hypothesis, develops when mature offspring have their 'backs against
the wall', figuratively speaking—when the possibility of their becoming successful breeders is exceedingly low.

There are two reasons for proposing such a model. First, if a
cooperatively breeding group has formed because of direct benefits
accruing to its individual members, then the potential for auxiliaries
to leave the group and initiate personal reproduction always is
present. The *average* fitness of group members may increase with
group size (k), as in Fig. 12.4a, but a large discrepancy normally exists

Table 12.2. Factors influencing dispersal from natal groups

Risk of dispersal	Probability of obtaining a vacant territory	Probability of obtaining a mate	Probability of successfully reproducing once established	Resulting behaviour
Low	High	High	High	Disperse and breed independently
High	Low	Low	Low	Remain in natal group as an auxiliary

between the fitnesses of a dominant breeder and a subordinate auxiliary in the same group. The fitness of the former will be greater than
$\bar{W}$, that of the latter, less than $\bar{W}$. If the fitness of subordinate non-
breeders falls below that which they would achieve as members of
solitary breeding pairs (denoted as $\bar{W}(1)$ in Figs 12.4 and 12.8), the
group will dissolve and members will breed independently. Consequently, the direct gains received by auxiliaries must be large (the
fitness curve must be steep), or the bias in the fitnesses of group
members must be low (benefits must be shared equitably) for cooperatively breeding groups to form in this manner (see section 12.5).
These conditions are expected to be uncommon.

In contrast, any of the four constraining factors mentioned above
can effectively eliminate the possibility of an auxiliary becoming a
breeder. For example, if no mates are available or no territories are
vacant (for a territorial species), the breeding option simply is
unavailable. No matter by how much $\bar{W}(1)$ exceeds $\bar{W}(k)$, the average
fitness of a group member, $\bar{W}(1)$ cannot be realized and auxiliaries
will be retained in their natal groups.

Second, the two routes to sociality produce groupings of fundamentally different genetic composition. When there are intrinsic
benefits to sociality, groupings will form irrespective of kinship links
between the members. But when groups form due to constraining
factors limiting dispersal, such groups will always consist of close
genetic relatives since they form by retention of natal members. This
substructuring of the population into nuclear or extended family kin
groups creates a favourable genetic environment for the development
of cooperative behaviour (Hamilton 1964; Wilson 1977). It is not mere
happenstance that the vast majority of groupings of cooperatively
breeding birds and mammals are based upon genetic family ties
(Brown 1978; Emlen 1978; Macdonald & Moehlman 1982).

The importance of ecological constraining factors has dominated
ornithological thinking about helping for much of the last decade
(Brown 1974, 1978; Ricklefs 1975; Woolfenden & Fitzpatrick 1978;
Gaston 1978a; Koenig & Pitelka 1981; Koenig 1981a; Emlen 1982a).
These ideas are just beginning to extend into the mammalian literature, however, and it may be premature to speculate too broadly
concerning their applicability. For example, it is probable that direct
benefits of group protection against predation has led to obligate
sociality in many species of diurnal mongooses. Similarly, among
wild dogs (*Lycaon pictus*) and perhaps some populations of wolves,
the foraging gains realized from cooperative hunting and group
defence of kills undoubtedly provide direct benefits to group living
(see Kleiman & Eisenberg 1973). Nevertheless, I will attempt to argue
that the obligate nature of sociality in such species leads to analogous constraints upon the options of younger or more subordinate
members to disperse and become independent reproductives. In
essence they too are 'forced' to remain in their groups (see section
12.5).

12.3.3 Types of constraining factors

(a) High risks of early dispersal

The mortality risks of dispersal depend upon the relative accessibility of food and shelter, and the pressure from predators and competitors in the habitats available to dispersing and resident individuals, respectively. When the risks are high, an individual may increase its probability of surviving to the following breeding season by continuing to reside in the security of its natal area. Regardless of the specific ecological factors of importance for the species, survival in an established and familiar breeding area probably will exceed that in the unoccupied, marginal areas available to dispersing individuals. In such a situation, the short-term cost of postponing dispersal and reproduction may be more than compensated for by improved longevity, and consequently the lifetime reproductive success of a delayed reproducer may be higher than for a non-delayer.

(b) Shortage of territory openings

Many cooperative breeders are permanently territorial species that inhabit stable or regularly predictable environments. Further, many have specific ecological requirements such that suitable habitat is restricted. All available high-quality habitat becomes filled or 'saturated'. Unoccupied territories are rare and territory turnovers are few. As the intensity of competition for space increases, fewer and fewer individuals are able to establish themselves on high-quality territories. Assuming that occupancy of a suitable territory is a prerequisite for reproduction, the option of breeding independently becomes increasingly limited. The non-breeder is trapped into a waiting game as a 'hopeful reproductive' (West-Eberhard 1979). It must wait until it attains sufficient age, experience, and social status to enable it to obtain and defend an independent territory. As I have argued elsewhere (Emlen 1982a), such waiting is best done at home, on an area of proven quality, and in the company of close kin.

Territorial limitations are the most common form of ecological constraint among cooperatively breeding birds. Shortages of breeding openings have been stressed as causal factors underlying helping in Tasmanian gallinules (Ridpath 1972), New Zealand pukekos (Craig 1979), acorn woodpeckers (MacRoberts & MacRoberts 1976; Stacey 1979a; Trail 1980; Koenig & Pitelka 1981), green woodhoopoes (Ligon & Ligon 1978a), campylorhynchus wrens (Selander 1964; Wiley & Rabenold 1981); Arabian, common, and jungle babblers (Zahavi 1974; Gaston 1978a), and numerous species of jays (e.g. Hardy 1961; Hardy *et al.* 1981; Brown 1974; Woolfenden 1975; Woolfenden & Fitzpatrick 1978).

When the death of a breeder creates a breeding vacancy in such species, non-breeding auxiliaries from numerous nearby territories usually converge on the 'vacated' territory within hours. There generally follows a period of intense vocalizations, agonistic challenges, and aggressive contests. These power struggles, as they have been termed in acorn woodpeckers, are vividly described by Koenig (1981b). Contesting auxiliaries that fail to secure the breeding slot return to the security of their natal territories where they continue the waiting game, and continue acting as helpers.

Among mammals, similar constraints probably operate in red fox (*Vulpes vulpes*), silver-backed jackal (*Canis mesomelas*), wolf (*Canis lupus*), wild dog (*Lycaon pictus*), and brown hyaena (*Hyaena brunnea*), and possibly operate in most of the canids which exhibit cooperation in breeding. In red foxes, females often postpone dispersal and remain in their natal areas (Jensen 1973; von Schantz, 1981). For those that do disperse, their 'chances of surviving at all, much less finding a new territory, are low' (Macdonald 1980, p. 168). Those that remain occupy small, suboptimal areas within the larger territory of the dominant female (von Schantz 1981). Only the dominant or alpha female has access to the food-rich portions of the territory, and time-budget studies indicate that only she can obtain sufficient food resources for reproduction. As a consequence, subordinate females are constrained from becoming breeders, and Macdonald's studies (1979, 1980) indicate that such females frequently act as helpers at the den of the dominant pair.

Habitat saturation by itself is insufficient to explain why auxiliaries simply do not disperse into more marginal habitats where competition is less. Koenig and Pitelka (1981; see also Brown 1969) proposed that a second factor was necessary for the evolution of cooperative breeding in permanently territorial species. Not only must optimal habitat be saturated, but marginal habitat must be rare. When this is the case, a maturing individual is severely constrained from either establishing itself as an independent breeder in the optimal habitat, or successfully surviving and breeding in an outlying area. Territory constraints thus are predicted to lead to the retention of non-breeding auxiliaries in: (1) species where ecological requirements are sufficiently specialized that marginal habitat is rare (e.g. Tasmanian gallinules), and (2) species which occupy habitats that are physically restricted or relict in distribution (e.g. Florida scrub jays). To these may be added a third: (3) species which modify their habitat, thereby magnifying differences between occupied and unoccupied areas (e.g. acorn woodpeckers, which artificially increase the food available on occupied territories by constructing storage granaries).

(c) Shortage of sexual partners

An excess of males has been reported in the population sex ratio of many cooperatively breeding birds (e.g. Rowley 1965; Fry 1972; Ridpath 1972; MacRoberts & MacRoberts 1976; Dow 1977; Reyer

1980). Among wild dogs, there is a strong male excess in the secondary sex ratio (Mech 1975). The reason for such skewing is poorly understood, but its effect is to increase competition for mates, leading to a demographic constraint on the option of becoming established as an independent breeder.

(d) Prohibitive costs of independent reproduction

Many species of cooperative breeders reside in areas where territory vacancies and marginal habitat are common. This is especially true of arid and semi-arid environments in Africa and Australia, where many cooperatively breeding birds are either nomadic or inhabit areas subject to large scale, unpredictable fluctuations in environmental quality (Rowley 1968, 1976; Harrison 1969; Grimes 1976). The carrying capacity in such environments changes markedly and erratically from year to year. Avian and mammalian populations, with their relatively low intrinsic rates of increase, cannot track these changes. Consequently, the degree of habitat saturation (if any) changes dramatically across seasons. One cannot speak of any consistent shortage of territory openings as a driving force in the evolution of cooperative breeding, but one can still speak of constraints on the options of independent breeding.

In variable and unpredictable environments, erratic changes in the carrying capacity create the functional equivalents of breeding openings and closures (Emlen 1982a). As environmental conditions change from year to year, so too does the degree of difficulty associated with successful breeding. In benign seasons, abundant food and cover decrease the costs to younger, less experienced individuals of dispersing from their natal groups and breeding independently. In harsher seasons, the costs associated with such reproductive ventures increase, eventually reaching prohibitive levels. As conditions deteriorate, breeding options become more constrained and the constraints first hit the younger, more subordinate individuals. The predicted outcome is the continued retention or the joining of such individuals in cooperatively breeding groups. This process may be important in the alloparental systems of such species as pied kingfishers (Reyer 1980), white-fronted bee-eaters (Emlen 1981, 1982a), and wild dogs (Frame et al. 1979).

12.3.4 Tests of the constraints model

The ecological constraints model predicts that the frequency of occurrence of non-breeding auxiliaries will vary directly with (1) the degree of difficulty in becoming established as a breeder (for permanently territorial species residing in saturated habitats), (2) the degree of skew in the population sex ratio (for species facing a shortage of mating partners), and (3) the level of environmental harshness

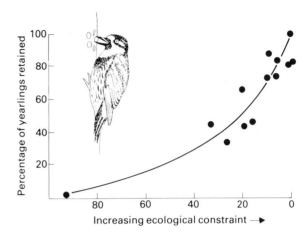

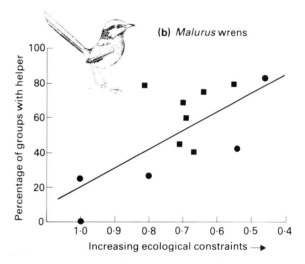

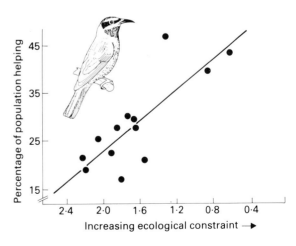

Fig. 12.5 **(a)** Retention of yearlings versus territory shortage in the acorn woodpecker. **(b)** Occurrence of male helpers versus availability of female mates in malurid wrens. ● = *M. cyaneus* and ■ = *M. splendens*. (From Rowley 1981.) **(c)** Incidence of helping versus environmental harshness in the white-fronted bee-eater.

(for species in erratic, unpredictable habitats). Data currently are available for three species of cooperatively breeding birds, and these have been analysed and plotted in Fig. 12.5a–c. (Unfortunately, no mammalian databases are yet available which are sufficient for testing the constraints model.)

The first graph shows the effect of territory constraints on the retention of offspring in the acorn woodpecker (*Melanerpes formicivorous*). This species lives in permanently territorial groups of 2–15 individuals. During breeding, only a single nest is tended at any one time, and most or all group members help to incubate the eggs and feed and defend the young. Yearling individuals either emigrate and attempt to become independent breeders, or remain with their natal groups. Those that remain postpone breeding themselves and play full roles in their group's cooperative breeding attempts. Figure 12.5a plots the annual incidence of yearling 'helpers' as a function of the proportion of all territories that became vacant during the pre-ceeding year (the annual turnover rate). The figure incorporates data reported by MacRoberts and MacRoberts (1976), Stacey and Bock (1978), and Stacey (1979a), as well as unpublished results kindly pro-vided to me by P. Stacey and W. Koenig, R. Mumme and F. Pitelka. Acorn woodpecker populations from Arizona, New Mexico and coastal California are pooled in the diagram.

The second example comes from studies by Rowley (1965, 1981) on two species of cooperatively breeding malurus wrens in Australia. These are classic helper-at-the-nest species, and breeding units consist either of simple pairs or trios (the helper virtually always being a male). Figure 12.5b graphs percentage of breeding groups with a (male) helper as a function of the shortage of sexual partners (using the ratio of females to males in the population as the index of demo-graphic constraint).

The third example is taken from my own work (Emlen 1982a) on the white-fronted bee-eater. This is an example of a colonial, coopera-tive breeder that inhabits the erratic, unpredictable environment of the Rift Valley of Kenya. The survival of nestlings is strongly influ-enced by the food supply, and reproductive success varies greatly across different seasons (Emlen, unpublished data). The food supply, in turn, is highly dependent upon the pattern of local rainfall. Conse-quently, Fig. 12.5c plots the percentage of the population serving as helpers in each of 13 colonies over five years, using a measure of rainfall as the index of ecological constraint (the log of total rainfall occurring in the month preceding breeding).

Although the proximate factors responsible for the constraint upon independent breeding differ among the three species, in each case the intensity (magnitude) of the constraint is a good predictor of the incidence of helpers.

The retention of non-breeding auxiliary individuals in breeding groups is a necessary first step in the evolution of cooperative breeding. By itself, however, it is insufficient to explain the development of actual helping or communal rearing behaviour. I have addressed the questions of why helpers remain in their natal groups and why helpers do not breed on their own. I now consider the question of why such auxiliaries help in rearing offspring that are not their own.

12.4 WHY DO HELPERS HELP?

In formulating a testable hypothesis for investigating the selective advantage of helping, it is necessary to establish an alternative behavioural strategy as a reference point for comparison. The alternative strategy considered here is one of not helping, but rather dispersing from the natal territory at the age of sexual maturity and breeding independently thereafter. The evolutionary hypothesis specifically predicts that the fitness of individuals that help during their lives will be equal to, or greater than, the fitness of individuals that attempt early dispersal and breeding.

The choice between these two strategies will depend upon the fitness gains associated with each. From the perspective of an auxiliary, the fitness realized by helping in a given season can be estimated as:

$$(N_{RA} - N_R)r_{ARy}$$

where $(N_{RA} - N_R)$ is the increase in the number of young successfully produced by the breeding pair as a result of the activities of the helper, and r_{ARy} is the coefficient of relatedness between the auxiliary and the young it helps to rear (R referring to the recipient of helping = the breeder; A referring to the donor = the auxiliary).

Similarly, the fitness realized by dispersing and breeding independently in that season is given by:

$$sN_A r_{Ay}$$

where s is the probability of successful dispersal and establishment as a breeder, N_A is the number of young produced by an auxiliary that leaves its group and breeds independently, and r_{Ay} is the coefficient of relatedness between the auxiliary (now a breeder) and its own young.

Helping will be favoured only when

$$(N_{RA} - N_R)r_{ARy} > sN_A r_{Ay} \qquad (12.1)$$

This condition rarely will be met. The coefficient of relatedness r_{Ay} always will equal or exceed r_{ARy} unless the pair being helped are close relatives *and* the auxiliary's certainty of parentage of its 'own' offspring is low. Further, the efforts of the auxiliary plus its mate (N_A) usually will exceed the contribution of the auxiliary alone $(N_{RA} - N_R)$ as outlined in Charnov's (1981) argument. The equation will be tipped to favour retention and helping by auxiliaries only when very harsh environmental conditions cause the success of new breeding attempts (N_A) to be very low, or when other constraints cause s to be sufficiently low that many dispersers fail to become established as breeders at all.

Equation 12.1 provides an inadequate statement, however, because it uses only a single breeding season as its time reference. Most cooperative breeders are long-lived, and it is lifetime fitness comparisons which are important in any test of evolutionary hypotheses. Since the helping strategy affects not only the immediate production of offspring, but also the chances of auxiliary survival, both fertility and survivorship must be considered. We have already seen how postponement of risky dispersal can enhance the survival probability of an auxiliary.

Helping in a current season also can improve an individual's chances of becoming a successful breeder in the next. Such long-term or delayed benefits of helping can occur in numerous ways.

(a) Breeding experience

It is well-known that inexperienced breeders are less successful in rearing offspring than experienced ones. If the difference between experienced and inexperienced breeders is large (N_A is small), then young individuals could clearly benefit from helping by gaining alloparental experience. Macdonald and Moehlman (1982, p. 456) question whether helpers are 'gaining practice at parenthood, by making their mistakes at someone else's expense' (see also Rowley 1965). Lawton and Guindon (1981), for example, found that among brown jays, first-year helpers made 'mistakes' much more frequently than did breeders. Such mistakes included arriving at the nest with food items that were totally inappropriate for the nestlings, arriving with food but eating it themselves, and arriving with food but becoming distracted and leaving without feeding the nestlings.

It is not clear how much experience an auxiliary needs before it reproduces effectively, but the duration of helping in some species greatly exceeds any reasonable notion of what is required. Gaston (1978a), for example, has estimated that in jungle babblers a typical individual acts as an alloparent for many years and fewer than 25% of individuals ever attain breeding status. Thus gaining breeding experience may be one benefit of helping behaviour, but it is not sufficient to explain its occurrence in most species.

(b) Inheritance of the parental territory

Consider a species in which the shortage of vacant territories is the primary ecological constraint. An auxiliary could become established by challenging and defeating the current owners of an occupied territory, or by waiting for vacancies to occur in nearby areas. There is a

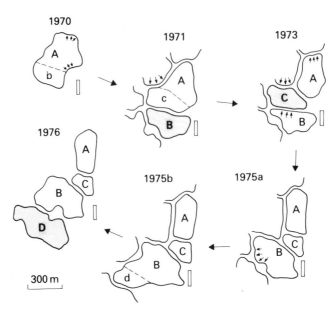

Fig. 12.6 The process of territory fission and territory inheritance as illustrated by a nuclear family of Florida scrub jays over a six year period. The shaded areas represent land inherited by two sons (B and C) and one grandson (D) of the original pair which occupies territory A. Broken lines show the incipient point of budding off and small arrows show how territorial boundaries expanded. (From Woolfenden & Fitzpatrick 1978.)

third possibility—remaining in the natal group until the death of the same-sex breeder and then inheriting the parental territory itself. This is a widespread practice, having been described in jackals (Moehlman 1979, 1983), red foxes (Macdonald 1980; von Schantz 1981), wolves (Mech 1970), lions (Bertram 1975), babblers (Zahavi 1974; Gaston 1978a), jays (Brown 1974; Woolfenden & Fitzpatrick 1978); wood-hoopoes (Ligon 1981) and acorn woodpeckers (e.g. MacRoberts & MacRoberts 1976), to name a few. An example of this process is illustrated in Fig. 12.6, taken from Woolfenden and Fitzpatrick's (1978) study of the Florida scrub jay. The territorial boundaries of an original pair of breeders, two of their sons, and one grandson are shown as they appeared across a span of six years. As the size of the group increased, the territorial boundaries expanded at the expense of smaller, neighbouring groups. Both sons 'budded off' portions of the parental territory for themselves, and the same process was repeated later by the offspring of one of these sons. By such an inheritance process, ownership of a high-quality territory can be passed down through generations, remaining in the same family lineage well beyond the lifespans of individual members.

When several helpers are members of the same cooperative unit, a dominance hierarchy normally forms with the oldest, most dominant member being first to inherit the territory. Consequently, the magnitude of this advantage to helping will depend upon the severity of the territorial shortage, the average residency time required to gain inheritance and the number of other competing helpers 'in line' to inherit (Gaston 1978a; Ligon 1981).

(c) Group dispersal

When competition for reproductive vacancies is intense, auxiliaries may gain by dispersing in groups. Among cooperatively breeding birds and mammals, non-reproductives typically use the natal territory as a base from which to prospect and challenge for breeding openings. And, in many species, groups of dispersers have a greater probability of territory takeover than do solitary dispersers.

Among lions (*Panthera leo*) most females remain and reproduce in their natal groups and female lineages endure for many generations (Schaller 1972; Bertram 1975). Males, however, disperse from the natal pride when approximately three years old, and wander nomadically until they are able to take over another pride by challenging and defeating the resident male(s). Lionesses in a pride frequently come into oestreus simultaneously, with the effect that several litters of cubs are born synchronously. When the young males reach dispersal age, they emigrate together, forming small coalitions of from one to seven individuals. Data collected in the Serengeti over a period of 12 years showed that single males were rarely successful in taking over prides and becoming reproductives at all. In contrast, two males roaming together were able to take over prides in 23 of 39 instances, and large coalitions (three or more males) were virtually always successful (20 of 21 groups) in expelling the previous residents and assuming control of a pride. Furthermore, the length of time that males were able to hold control of a pride of females was proportional to their coalition size. Figure 12.7 graphs the combined effect of these two benefits of group dispersal calculated in terms of individual and inclusive fitness per male (from Bygott *et al.* 1979; but see Grafen 1982). Group dispersal or transfer between packs also is typical of wild dogs. In this species, however, it is coalitions of females (generally sisters) that emigrate together (Frame & Frame 1977; Frame *et al.* 1979).

Among cooperatively breeding birds also, group dispersal is commonplace. Among green woodhoopoes, unisexual groups of different aged individuals explore, locate, and fight for breeding vacancies together (Ligon & Ligon 1978a). Typically, the older members of these coalitions helped to raise the younger ones, and the Ligons (1978b) were the first to hypothesize that one advantage of helping was that it cemented a tight social bond between helper and helped (nestling). When the older, more dominant auxiliary emigrates and attempts to secure a territorial position, it 'takes along' several younger, more subordinate members that it helped rear in previous breeding seasons. As with the lions, these coalitions are better able to secure breeding openings than are solitary individuals.

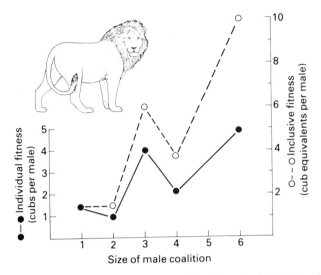

Fig. 12.7 The advantage of group dispersal in lions. Individual and inclusive fitness achieved by males in different sized coalitions. For discussion see text. (From Bygott *et al.* 1979.)

Analogous cases in which former helpers and the young they helped to rear disperse together have been documented in Arabian and jungle babblers (Zahavi 1974; Gaston 1978a) and acorn wood-peckers (Koenig 1981b). To the degree that the social bonds formed during helping itself are important for the development of these coali-tions, we may conclude that helping in one season may increase a helper's breeding prospects in future years.

(d) Reciprocity

In a system based on reciprocity, an individual forgoes breeding and helps in one season on a probabilistic expectation of future receipt of helping at some later date. In direct reciprocity (Trivers 1971) the initial recipient of aid is itself the later donor; in delayed reciprocity (Wiley & Rabenold 1981), the helper gains the services of the young that it previously had helped to rear.

Models of reciprocal altruism have been considered by many to be evolutionarily unstable because they are prone to invasion by cheaters (e.g. West-Eberhard 1975; Dawkins 1976). Axelrod and Ham-ilton (1981), adopting a game theoretical approach, recently showed that once reciprocity is established, it is resistant to cheating provided there is a sufficiently high probability of repeated encounters between the same participants. And cheating ceases to be an in-surmountable problem if the species possesses sufficient individual recognition and memory capabilities. Individuals then are predicted to maintain a knowledge of the behavioural attributes and past behav-ioural actions of other members of their group and to dispense their own behaviours accordingly.

I believe that cooperatively breeding species fulfil many of the conditions considered by Trivers (1971) and Axelrod and Hamilton (1981) to be essential for the development of any system of reciprocity. Populations are divided into small groups whose membership is highly stable; individuals are long lived; and individual recognition capabilities appear to be universal. Perhaps most important, by living in relatively closed social groupings, individuals repeatedly interact with the same small set of close acquaintances throughout much of their lifetimes.

Cases of helpers recruiting the young that they helped to rear have been reported in Tasmanian native hens, *Tribonyx mortierrii* (Ridpath 1972), jungle babblers, *Turdoides striatus* (Gaston 1978a), Arabian babblers, *Turdoides squamiceps* (Zahavi, pers. comm.), green woodhoopoes, *Phoeniculus purpureus* (Ligon & Ligon 1978a, b), Helmut shrikes, *Prionops plumata* (C. Vernon, cited in Brown 1978), and acorn woodpeckers, *Melanerpes formicivorous* (Koenig 1981b). In other instances, when a helper inherits its natal territory it also gains the services of the group's more subordinate members as its helpers.

Instances of more direct reciprocity are rare among cooperative breeders (but see Brown & Brown 1980) but redirected helping, in which a breeder that fails in its own nesting attempt switches to become a helper at another active nest of the group, occurs in pinon jays, *Gymnorhinus cyanocephalus* (Balda & Bateman 1971; J. D. Ligon, pers. comm.); long-tailed tits, *Aegithalos caudatus* (Gaston 1973); white-fronted bee-eaters, *Merops bullockoides* (Emlen 1981; Hegner *et al.* 1982) and Mexican jays, *Aphelocoma ultramarina* (Brown 1978). Among white-fronted bee-eaters, such role reversals are sufficiently commonplace that most adults repeatedly act in both breeder and helper capacities within their clans (Emlen 1981; Hegner *et al.* 1982; Emlen & Demong, unpublished observation).

In communally breeding species, the distinction between reciprocity and mutualism depends on the timing of reproductive activities of the different members of the group. Consider a hypothetical mammal in which most females are reproductives and suckle the communally produced young indiscriminantly. In a given year, some females may lose their pups, yet continue to suckle young; in the following season a different female might lose her pups or be barren, and yet suckle the young of our original female. In the terminology of Hamilton (1964), a female might be a donor one year and a recipient the next.

It will be difficult to tease apart the relative importance of mutualistic, nepotistic and reciprocal factors to the evolution of aid-giving behaviours. But I suspect that reciprocity is an underrated force that will be found to be increasingly common as more detailed studies of cooperatively breeding species are undertaken in the decade ahead.

(e) Increased inclusive fitness via collateral kin

All of the benefits of helping mentioned above increase the fitness of a helper via the personal component of inclusive fitness, i.e. lifetime production of offspring. A second type of genetic gain inherent in helping behaviour involves that component of inclusive fitness attributable to gene copies through collateral kin (the 'kin-component' in the terminology of West-Eberhard 1975; the 'indirect component' of Brown & Brown 1981a). Unfortunately, an unproductive controversy has arisen concerning the importance of kin selection to the evolution of cooperative breeding systems.

It is true that for the vast majority of helper species for which data are available, the helpers are predominantly grown offspring that remain with their parental groups and help to rear full or half sibs. Thus the kin component of inclusive fitness is large and could prove to be a major factor in the evolution of helping behaviour. Some workers, however, have mistakenly taken this correlation of group relatedness with helping as the sole evidence to build a case that kin selection is both necessary and sufficient to explain the evolution of alloparental behaviours. Other workers have erred in the opposite direction. By finding ways in which individual helpers gain personally through helping (i.e. (a)–(d) above) they have concluded that kin selection is either unimportant to understanding of helping behaviour, or that kin selection does not even exist.

Both approaches are too narrow in outlook and miss the point. Evolution is the process of changing gene frequencies in a population through time. Natural selection does not distinguish between a gene copy produced by a direct descendant and one fostered by a collateral relative. All gene copies are tallied, irrespective of who or what was responsible for their survival in the population. The interesting question then is not whether kin selection exists or not, but rather whether collateral kin interactions (as opposed to personal offspring production) have been an important or essential component in the evolution of helping behaviour.

To take this debate out of the realm of semantic argument and into the realm of quantitative biology, Vehrencamp (1979) has devised a simple index for determining the relative importances of individual and kin selection to the current evolutionary maintenance of a behavioural trait. I_k, the kin index, calculates the proportion of the total gain in inclusive fitness that is due to the kin component for any type of cooperation among kin, compared to the non-cooperative situation:

$$I_k = \frac{(W_{RA} - W_R)r_{ARy}}{(W_{AR} - W_A)r_{Ay} + (W_{RA} - W_R)r_{ARy}}$$

where $(W_{RA} - W_R)$ is the change in lifetime reproductive success of the recipient, R, when it is aided by the donor A;

$(W_{AR} - W_A)$ is the change in lifetime reproductive success of the donor A, when it provides aid to R;

r_{ARy} is the relatedness of A to R's young;

r_{Ay} is the relatedness of A to its own young.

The index is applicable only to situations in which the net change in inclusive fitness of A (i.e. the denominator) is positive. When I_k is greater than 1, there is a net cost to A's personal reproduction and pure kin selection is acting. When I_k is less than zero, then A is manipulating R at a net cost to R's personal reproduction. When I_k is between 0 and 1, both the personal component and the kinship component are increased by the cooperation, and the value of the index gives the proportion that is due to the kin component.

To date, three estimates of I_k are available from cooperatively breeding vertebrates. Vehrencamp (1979), re-analysing the published data for lions, calculated the index of kin selection to be 0.51. Similarly, utilizing published information available for Florida scrub jays, she estimated I_k as 0.55. More recently, Rowley (1981) calculated I_k for the splendid wren, *Malurus splendens* (an Australian helper-at-the-nest species), to range from 0.35 to 0.51 for a typical male helper. These three analyses suggest that both the kin and personal components of fitness are increased via helping behaviour, and each contributes about equally to the overall fitness gain realized by exhibiting alloparental behaviour.

In conclusion, helping does not seem to present the evolutionary paradox that it first appeared to. Severe ecological constraints often eliminate or greatly restrict the option of independent breeding. When viewed against the backdrop of this constraint, a number of possible adaptive functions can be proposed for helping behaviour. These range from direct improvements in a helper's lifetime reproductive success to improvements in the kinship component of fitness by aiding close relatives. The differing importance of such factors for each type of cooperative system remains one of the principal challenges of future work on cooperative breeders.

12.5 WHEN DO HELPERS BECOME COMMUNAL REPRODUCTIVES?

In the helper-at-the-den/nest systems emphasized so far, the dominant breeding individuals realize a much greater fitness than do subordinate helpers. We can speak of there being a large bias in the personal reproductive benefits accruing to the different group members. This degree of bias is the key factor distinguishing helper-at-the-den/nest from communal breeding systems. Among communal breeders, several individuals of one or both sexes are reproductives, and the fitness benefits as well as the parental care costs are more equitably distributed among the various group members. What then determines the degree of sharing of reproductive activities? Why do helpers not always reproduce communally in their groups?

Whenever individuals live in groups, social competition will occur. Questions of who breeds and who will breed in the future are of crucial importance to each individual, no matter how seemingly

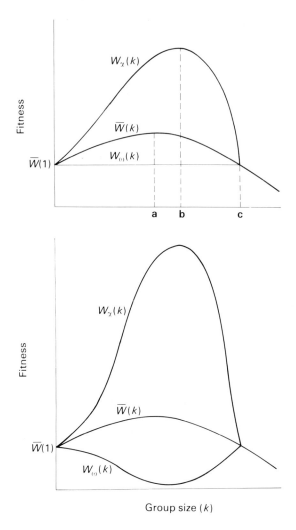

Fig. 12.8 The maximum degree of bias in fitness that can be generated between dominants and subordinates in groups containing (top) unrelated individuals and (bottom) individuals related by a coefficient of relatedness $r = 0.25$. $\overline{W}(k)$ is the average or *per capita* fitness of group members, W_{α} is the fitness of the dominant, and W_{ω} is the fitness of a subordinate. The distance between the W_{α} and W_{ω} curves represents the amount of bias possible at that group size. (From Vehrencamp 1983.)

cooperative the group may be. This view forces us to recognize that any cooperatively breeding group is composed of competing individuals with conflicting strategies. At one extreme, a dominant member may monopolize breeding and recruit the services of others in the group to be its helpers; at the other, differences in leverage between dominant and subordinate may be slight, neither can manipulate the other for its own benefit, and reproduction will be shared. It is these differences in leverage among the various members of groups which determines whether the breeding system will be one of helping or communal reproduction and care of offspring (Emlen 1982b; Vehren-camp 1983).

Vehrencamp (1979, 1983) has developed an algebraic model for determining how much bias a dominant can force from a subordinate before the latter will do better by leaving the group and attempting to breed independently. Consider the case of a group of unrelated indi-viduals where grouping benefits all group members on average. The function relating the *per capita* reproductive success and group size k is given by $\bar{W}(k)$ (Fig. 12.8, top). For each unit of fitness that the dominant can usurp from a subordinate, one unit of fitness accrues to the dominant. For simplicity, we also assume that there is only one dominant in the group and that all subordinates are affected by the same amount. The dominant cannot lower the fitness of any one sub-ordinate below the fitness of a solitary breeder, $\bar{W}(1)$, otherwise sub-ordinates will be selected to leave and breed solitarily. The lowest possible fitness of a subordinate in a group, $W_\omega(k)$, is therefore depicted as a horizontal line intersecting $\bar{W}(1)$. The fitness increment of the dominant, $W_\alpha(k)$, is the difference between $\bar{W}(k)$ and $\bar{W}(1)$ times the number of subordinates. The bias in fitness is therefore reflected in the degree to which the fitness of dominant and subordi-nate diverge. Clearly the greater the average benefit to grouping com-pared to solitary living, the greater the bias can be. Note also that as the magnitude of the bias increases, so too does the optimal group size viewed from the perspective of the dominant (compare points a versus b in Fig. 12.8; see also Rodman 1981).

For groups of related individuals, a similar graph can be derived (Fig. 12.8, bottom), but in this case the subordinates leave the group when their *inclusive fitness* would be greater if they bred solitarily. Here, the personal fitness of a subordinate can go well below the personal fitness of a solitary breeder, and the higher the degree of relatedness, the greater the bias can be. As in the case of unrelated individuals, grouping and biasing can only occur when there is a net benefit to grouping, when $\bar{W}(k) > \bar{W}(1)$.

The Vehrencamp model assumes that it is always beneficial for a dominant to extract as much fitness as possible from a subordinate—up to the point where the subordinate would do better to leave the group altogether. However, dominant and subordinate or, more gen-erally, breeder and helper, will disagree over the location of that departure point. In equation 12.1 on p. 324, I expressed the cost/benefit trade-offs of the two strategies of helping versus dispersing *from the perspective of the helper*. The equation must be modified if it is to express the trade-off as seen *from the viewpoint of the breeder*. The breeder benefits by retaining the auxiliary as long as:

$$(N_{RA} - N_R)r_{Ry} > sN_A r_{RAy} \qquad (12.2)$$

Note that the same terms enter into both equations 12.1 and 12.2, but they are weighted by different values of r. A current breeder will always be more closely related to its own offspring than it will to the offspring of its helper $(r_{Ry} > r_{RAy})$. This difference sets the stage for conflicts of interest between breeder and helper over the point at which the auxiliary should cease helping and strike out on its own.

I (Emlen 1982b) have modelled these conflicts between breeder and helper using the approach of Trivers (1974) to define precisely the areas where conflict will occur. In so doing, I used fitness equations incorporating two successive breeding seasons since the gain achieved via retention often is realized as an increase in survival or in reproductive output the following year. The results of my analysis showed that the direction of the conflict, i.e. whether the breeder or the helper has the greater leverage, changed depending upon two critical (and by now familiar) variables: (1) the severity of the ecological constraints on younger, subordinate individuals, and (2) the magnitude of the advantage (or disadvantage) of group living.

The results of this analysis are summarized graphically in Fig. 12.9a and b. Each figure is a two-dimensional schematic representation of the fitness consequences of continued retention of an auxiliary. The fitness of a breeder increases from left to right while that of an auxiliary increases from bottom to top. The vertical and horizontal lines passing through the centre of each figure represent positions at which retention of the auxiliary results in a zero change in

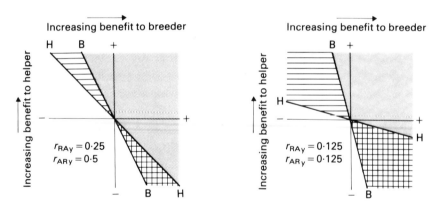

Fig. 12.9 Fitness space diagrams for breeders and helpers of two degrees of relatedness. The inclusive fitness of a breeder is increased through the retention of a helper in all areas to the right of the line marked B–B; the inclusive fitness of an auxiliary is increased by such an association in all areas lying above the line marked H–H. Shaded areas indicate zones where continued association is mutually beneficial. Hatched and cross-hatched areas represent zones of conflict, in which one party gains but the other loses by the association. In the hatched zone, helper-at-the-den/nest breeding systems are expected with large differences in the fitnesses realized by breeder and helper. In the cross-hatched zone, the fitnesses realized by different group members will become more equitably distributed, and communal breeding systems can develop. See text for further explanation. (From Emlen 1982b.)

individual fitness of breeder and helper respectively. Thus the individual fitness of a breeder is increased in all areas to the right of the vertical line while its *inclusive fitness* is increased to all areas to the right of the line marked B–B. Similarly, the individual fitness of an auxiliary is increased by the association with a breeder in all areas above the horizontal line, while the auxiliary's *inclusive fitness* is increased in all areas above the line marked H–H.

When retention and helping benefit both helper and breeder, there clearly is no conflict and helping will evolve (shaded zone). Similarly, when retention and helping are disadvantageous to both parties, helping will not evolve (unshaded zone). But, because of asymmetries in the inclusive fitness equations, there are two zones of breeder/helper conflict over the issue of helping, and these have important implications for the type of cooperatively breeding society that will develop. The size of these conflict zones increases as the degree of relatedness between breeder and helper decreases (compare Figs 12.9a and b).

In the first type of conflict (depicted as the hatched zone in the fitness space of Fig. 12.9), severe ecological or demographic constraints effectively prohibit independent breeding by auxiliaries, yet retention of such individuals results in a decrease in the personal fitness of the breeders. This decrease could be brought about through increased intra-group competition for resources located within the group territory, interference in the courtship of the breeders by sexually active auxiliaries, incompetence of inexperienced auxiliaries in helping rear young, or heightened risks of predation caused by the increased activity (conspicuousness) of the enlarged group (Zahavi 1974, 1976; Gaston 1978b).

At any point in this conflict zone, an auxiliary gains, but a breeder loses, from continued retention. The auxiliary is demanding more investment (in the form of continued access to the security and resources of its natal area/group) than the breeder is selected to provide. If we assume that breeders are dominant over helpers, one resolution would be the expulsion of the subordinate from the group. But since the subordinate is the individual with more to gain by remaining, the dominant has extreme leverage and is in an excellent position to extract fitness from the helper. A second stable resolution is achieved if the auxiliary increases its alloparental contribution to the level where the fitness of the breeder is raised above what is termed the 'breeder tolerance line' (labelled B–B in Fig. 12.9). Anthropomorphically speaking, the breeders are in a position to demand a helping payment in return for allowing the subordinate to remain (Gaston 1978b). The result is a helper-at-the-den/nest breeding system with a strong bias between the fitnesses of dominant and subordinate (breeder and helper).

Communal breeding systems develop when a dominant loses its leverage over subordinates yet still benefits through intrinsic gains associated with group living. At least two factors will contribute to the erosion of such leverage. First, as an auxiliary becomes older, it gains experience and improves its dominance position. Eventually, this will translate into an increased ability to compete successfully for breeding positions. Secondly, environmental conditions may change in such a way that constraints are relaxed (s is increased). Either or both of these factors can raise the value of $sN_A r_{Ay}$ above the point at which it is in the best interests of the auxiliary to cease helping and initiate breeding on its own.

If the $\overline{W}(k)$ curve has a negative slope (i.e. grouping is 'forced', as in Fig. 12.4b), auxiliaries will disperse at this point and no further cooperative breeding is to be expected. But if the breeder would *gain* by the continued retention and contribution of the auxiliaries (the $\overline{W}(k)$ slope is positive), then a second form of conflict arises whose resolution can lead to the evolution of communality (Emlen 1982b).

In this second form of breeder–helper conflict (denoted as the cross-hatched zone in the fitness space of Fig. 12.9), the breeder gains from the continued presence of an auxiliary but the auxiliary loses from retention. However, the breeder has little leverage to prevent auxiliaries from dispersing.

If the conflict is minimal (the distance between the B–B and H–H lines is slight) and breeders are dominant over auxiliaries, we would expect to observe manipulation of the subordinates to induce them to remain as helpers. Forcible retention is impractical but breeders might achieve the same effect more subtly by disrupting the breeding opportunities of auxiliaries. Such disruptions could be indirect, taking the form of harrassment of the would-be breeders, or direct, involving (in the case of mammals) the suppression of breeding in subordinate females or the killing of pups born to such individuals, and (in the case of birds) the destruction of the nests or eggs of subordinates.

Among birds, both types of behaviours have been reported among cooperative species (e.g. Vehrencamp 1978; Trail *et al.* 1981; Emlen *et al.*, unpublished data; Mumme, Koenig and Pitelka, unpublished data) and I have suggested (Emlen 1982b) that their adaptive significance lies in the manipulative recruitment of subordinate individuals for whom the cost/benefit ratio of personal reproduction was initially favourable.

Among many species of cooperatively breeding mammals, subordinate females conceive but later abort or otherwise lose their young. One major cause of these breeding 'failures' is social harrassment from more dominant group members (Macdonald & Moehlman 1982). For example, in the dwarf mongoose, beta females became pregnant in synchrony with the alpha females in approximately 20% of the packs studied by Rood (1980). Yet his evidence suggested that only one subordinate may actually have given birth. Similarly, in red fox (*Vulpes vulpes*), a variable number of subordinate females conceive, but most abort or abandon their young and then become helpers to the dominant pair (Macdonald 1979, 1980; von Schantz 1981). Macdonald and Moehlman (1982) speculate that the cause is the continual subjugation of the lower ranking female by the dominant.

Studies of wolves (*Canis lupus*) also show that the beta female rarely reproduces, instead serving as a helper to the alpha pair. Explanations of the proximate causes for this suppression have included repeated harrassment by the dominant female and lack of sexual interest by pack males in any female other than the dominant (Mech 1970; Zimen 1975, 1976, 1981).

Finally, in the dingo (*Canis dingo*), Corbett (reported in Macdonald & Moehlman 1982) describes how a dominant female killed the pups born to a subordinate female, who then became a helper and suckled the cubs of the dominant. A similar incident in wild dogs was described by van Lawick (1973). It thus appears that various forms of behavioural suppression and manipulation are widespread among the social canids. I suggest that so long as leverage is possible, dominant individuals will attempt to monopolize reproduction within their social groups.

As the magnitude of the conflict of interest increases, however, the leverage of the dominant individuals decreases. Interference strategies will become more costly in terms of time, energy and risks of retaliation. If the fitness gain associated with increasing group size is sufficiently great, current breeders will be selected who yield a portion of their fitness to the auxiliaries (Alexander 1974; Emlen 1982b). The most direct way to achieve equity is through shared parentage of offspring. This will have the effect of reducing the conflict space (B–B and H–H lines in Fig. 12.9 will converge), reducing the fitness inequalities among group members, and enticing subordinates to remain in the social units. Such tendencies toward multiple paternity or maternity will evolve only to the point where the fitness gained by the current breeder via the retention of other group members at least equals the fitness forfeited to ensure that such individuals remain.

An intermediate level of communal breeding is achieved when breeders tolerate, or even solicit, sexual interactions with other members of their group. Female breeders of many species of cooperatively breeding birds and mammals copulate regularly with auxiliary males (see, for example, Robinson 1956; Ridpath 1972; deVries 1975; Mader 1975, 1979; Dow 1977; Stacey 1979b; Craig 1980; Faaborg *et al.* 1980; Koenig & Pitelka 1981). Such promiscuous matings lead to an increased probability of paternity for the auxiliary males. Whether or not such paternity is achieved, the promiscuity in itself could operate as a behavioural tactic to induce additional males to remain with the group and contribute alloparental care (Dow 1977; Craig 1980; Emlen 1982b). Stacey (1979b) provided the first empirical evidence in support of such a hypothesis. In a study of acorn woodpeckers in New Mexico, he reported high levels of female promiscuity and further recorded that those males who copulated with the breeding female served as helpers in bringing food to the nestlings, while those males who had not copulated did not.

The most extreme form of communality results when multiple females in the same group become reproductives simultaneously. This form of breeding system is extremely uncommon among vertebrates (a list of such species can be found in Table 12.1). From the dominant's perspective, true communality is adaptive only when (1) the dominant breeder gains more from continued association with subordinates than the subordinates gain by remaining, yet (2) it has little ability to induce subordinates to stay. It behoves the subordinate to remain in such groupings only if it can share in the heightened reproductive success of the group. Ecologically speaking, this translates into situations where (1) there is a strong *per capita* benefit from grouping, *and* (2) factors constraining independent reproduction by subordinates are minimal. The interplay of these two factors in shaping the different forms of cooperative breeding systems is shown in Table 12.3.

The trade-off of communality for a dominant individual is the loss of monopolization of reproduction versus the retention of auxiliaries (now reproductives) in the social group. This cost/benefit balance is expected to shift as ecological and social factors change. As the positive slope of the $\overline{W}(k)$ curve becomes larger or the constraints upon subordinates increase, so too does the leverage ability of the dominant. As a consequence, communal breeding systems are predicted not only to be rare, but also to be unstable. Whenever conditions permit, dominant group members should attempt to bias reproduction in their favour. The result is that considerable variability is expected in both the presence and the magnitude of communality exhibited by a given species across different years or in different geographical areas.

In the first edition of this book, I concluded my chapter by stating that 'Studies of cooperative behaviour remain in their infancy. . . Hopefully, a decade from now, we will be in a much better position to provide concrete answers to such questions as "What ecological conditions favour or promote the expression of cooperative tendencies?" and "What behavioural strategies are followed by individuals that live within such seemingly 'cooperative' societies?".' In the

Table 12.3. Ecological conditions promoting different types of cooperative breeding systems

| | | Effects of group breeding: | |
		Advantageous	Disadvantageous
Intensity of constraining factors	Strong	Biased breeding Helping-at-the-den/Nest	Biased breeding Helping-at-the-den/nest
	Weak	Equitable breeding Communality	Solitary breeding

short time since the publication of that edition, considerable advances have been made both in the development of theory pertaining to cooperative breeding and in the collection of empirical field data. We now have before us a preliminary set of models and testable hypotheses. The decade ahead should be an exciting one as we begin to see vigorous testing of these various hypotheses.

Chapter 13
The Evolution of Social Behaviour in Insects

H. JANE BROCKMANN

13.1 INTRODUCTION

The evolution of eusocial behaviour in insects involves two issues: (1) the phylogenetic origins or probable evolutionary steps by which adaptive modifications accumulated over time, and (2) the adaptive significance or how such behaviour has been maintained in populations when it appears to decrease the fitness of the individuals possessing it. Both of these problems worried Darwin (1859). He described the complexity of cell-making behaviour in honeybees and wondered how anything short of this perfection could have been favoured by natural selection. He also worried about the adaptiveness of social insect behaviour, since sterile workers often possess behavioural and morphological traits which are quite distinct from those of the reproductive members of the colony. How can such characters be adaptive when the individuals possessing them do not reproduce? Since Darwin's time a large literature has accumulated on both problems. Some of it has been quite controversial. In this chapter I attempt to resolve some of these issues through examining their logical and empirical bases. I concentrate on the Hymenoptera because this group shows the greatest diversity of social behaviour.

13.2 NATURAL HISTORY OF EUSOCIAL INSECTS

Highly social, or eusocial, insects are characterized by four attributes: (1) a number of adults living together in a group, (2) overlapping generations, (3) cooperation among adults in nest-building and brood care, and (4) reproductive dominance or in some species sterile, structurally distinct castes (Wilson 1971). Such extreme forms of social life are rare, being found in only two orders of insects, Isoptera (termites) and Hymenoptera (wasps, bees and ants). Eusocial insects are, however, surprisingly diverse in their behavioural and life history patterns.

There are two quite different patterns of colony founding among the eusocial Hymenoptera (Hölldobler & Wilson 1977; Fig. 13.1). (a) A colony may be initiated by one or more reproductive females who rear the first brood alone (independent founding), constructing the nest, laying the eggs and feeding the larvae. When the first brood

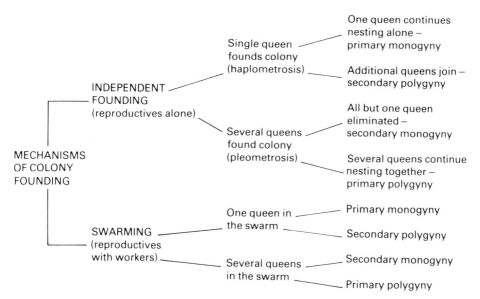

Fig. 13.1 A classification of methods of colony founding and subsequent colony composition for eusocial insects. (From Hölldobler & Wilson 1977.)

emerges, they take over the work of the colony and the queen then specializes in egg production. (b) Alternatively, a new colony may be founded by one or more queens and a group of workers from the original colony (swarming), in which case the queen specializes in egg production from the start. In either case the colony may be founded by a single queen (haplometrosis) or by several reproductive females of the same generation (pleometrosis). If the colony is founded by several queens, then it may remain polygynous or become secondarily monogynous through queens fighting or through workers killing off all but one queen. Later, workers may adopt newly mated sisters, leading to secondary polygyny. Termite nests are founded independently by a reproductive pair or occasionally by several queens (and kings), but if polygynous they apparently become secondarily monogynous at later stages (Thorne 1982a, b). Even social spiders (which are apparently not eusocial) show the same general patterns of colony founding. A single female may initiate a web and her offspring may remain with her, or alternatively a number of females may found a communal web (Kullmann 1972; Buskirk 1981), sometimes even by 'swarming' (Lubin & Robinson 1982).

After the colony has built up sufficient numbers (called the ergonomic stage of colony development) and when the environmental conditions are right, reproductive males and females are produced. In some species this occurs at the end of the growing season when all colonies simultaneously release large numbers of reproductives which engage in nuptial flights and mating. The inseminated females may then overwinter by themselves in a protected location or with a group of workers who then help to initiate the growing phase of the colony

341

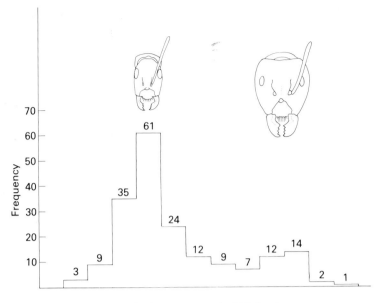

Fig. 13.2 In ants there is a division of labour among the non-reproductive members of the colony based on a size polymorphism. The polymorphism in the workers of the carpenter ant is depicted here. The size distribution within a colony is bimodal, with most individuals being the minor workers and some being majors. (From Wilson 1953.)

in the following spring. In tropical species, colonies may produce reproductives more or less continuously.

The most characteristic feature of eusocial colonies is the presence of castes. Living together in one nest are one or a few reproductives and many individuals that are either not reproducing or show greatly reduced fecundity. In some species, workers have vestigial gonads, in others they do not mate and are thus permanently sterile or produce only male offspring. The non-reproductive individuals in a colony usually show a further division of labour, based either on morphological differences or on age (Brian 1979; Oster & Wilson 1978; Fig. 13.2). In honeybees, for example, young workers generally remain in the nest, caring for the brood and queen, whereas older individuals guard the nest and forage for food (Lindauer 1961). In some ants and termites, the queen becomes grossly distended by eggs and unable to move on her own. Although this does not normally occur in wasps and bees, there are some species of vespine wasps and stingless bees (Meliponinae) in which the queen loses mobility because of her greatly enlarged ovaries and increased weight and by the wearing away of her wings (Spradbery 1973; Michener 1974). One common thread underlying the diversity of eusocial patterns in Hymenoptera is that castes are almost always the result of environmental influences. Whether an individual is a queen or a worker is determined by the amount or quality of her food and the chemicals to which she was exposed during development, her age or morphology, and current nest conditions (Crozier 1977; Brian 1979). In a few ants and some meliponine bees a specific gene apparently influences whether a female becomes a queen or not (Velthius 1976; Crozier 1979; Buschinger 1978).

Termites differ from Hymenoptera in a number of very important respects (Wilson 1971). (1) They are hemi- rather than holometabolous, which means that the immatures are not helpless grubs as in ants, wasps and bees. (2) Unlike drones, reproductive males among termites take an active role in brood and colony care and remain with the queen throughout colony life. (3) In termites, workers and soldiers may be either male or female or both depending on the genus, whereas among Hymenoptera they are always female. (4) In termites both sexes are diploid whereas in Hymenoptera females are diploid and males haploid. In primitive termites, immatures perform most worker tasks, undergoing several moults and continuing to grow.

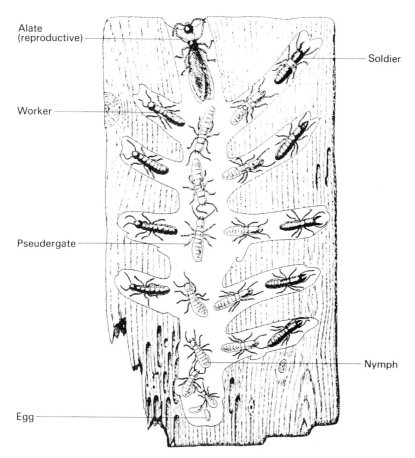

Fig. 13.3 The development of castes in a primitive termite, *Kalotermes flavicollis*. Eggs hatch into nymphs which serve as workers in the nest. They develop and moult into second and third instars before beginning to differentiate into one of the adult castes. A termite may remain in the juvenile form and continue to moult but not grow, a pseudergate. Also, once a nymph has begun to differentiate into one or another of the castes, it may revert into a pseudergate stage. (From Wilson 1971.)

During the later instars, they differentiate into one of the adult forms: either growing wings and becoming a reproductive or developing into a soldier with its enlarged head and mandibles, or remaining relatively unchanged and becoming a worker (Fig. 13.3). Among higher termites, the adult worker caste performs most of the colony labour and immatures contribute little.

13.3 TAXONOMIC DISTRIBUTION OF EUSOCIALITY

Although eusocial Hymenoptera are enormously abundant and conspicuous, they are nonetheless taxonomically quite restricted (Evans 1977; Crozier 1977). The Hymenoptera are divided into two major suborders, the Symphyta (sawflies and horntails, which feed on plants) and the Apocrita. The Apocrita are separated into two divisions, the Parasitica (whose larvae are generally endoparasites of arthropods) and the Aculeata, which are distinguished by having the ovipositor modified into a sting. Of the 250 000 or so known species of Hymenoptera, roughly two-thirds are parasitic or feed on plants (including all the Symphyta and Parasitica and the Bethyloidea of the Aculeata) (Krombein *et al.* 1979). There are no known cases of eusocial behaviour among any of these species, and even extended maternal care, which is so characteristic of higher taxa is rare (Crozier 1977; Eickwort 1981). However, there are a number of examples of group living and apparent cooperation among kin which are at least reminiscent of more complex forms of social behaviour (Malyshev 1968). For example, in a polyembryonic parasite of caterpillar pupae (Encyrtidae), a precocious, non-reproductive larval form is produced which defends the host from other endoparasites, thus protecting its siblings (Cruz 1981).

It is only among two rather similar looking, non-parasitic superfamilies of Aculeata that eusocial behaviour arises, the Sphecoidea, which includes digger wasps and bees, and the Vespoidea, which includes the familiar social wasps and ants (there is much controversy over higher categories of Hymenoptera; I am using the taxonomy of Brothers 1974). These two superfamilies have a pattern of maternal care that is very different from that of all other Hymenoptera: the female constructs a nest and provisions it with food for her offspring (except the brood parasitic species). However, of the 22 families of Sphecoidea and Vespoidea (Table 13.1), only six have eusocial species (Snelling 1981). These include one small group of tropical Pemphredoninae sphecid wasps (Matthews 1968a, b; Matthews & Starr, in press) and some bees in the Sphecoidea, and two subfamilies of wasps and all ants in the Vespoidea. Among bees, eusociality apparently evolved once among the anthophorids, twice among the apids (Winston & Michener 1977) and repeatedly among the Halictinae, where closely related genera contain both social and asocial species (Michener 1969, 1974; Knerer & Schwarz 1976; Sakagami & Maeta

Table 13.1. Taxonomic distribution of eusociality among the aculeate Hymenoptera (taxonomy based on Brothers 1974). In this table, nest-sharing is used as a general term to refer to regular nesting in groups with no known reproductive dominance. These associations may be subsocial or parasocial. (Modified from Snelling 1981, with additions from Michener 1974 and Wilson 1971.)

Sphecoidea
  Spheciformes (digger wasps, sand wasps and mud-daubers)
    Sphecidae EUSOCIALITY EVOLVED ONCE
      Ampulicinae SOLITARY
      Sphecinae SOLITARY
      Pemphredoninae MOSTLY SOLITARY; ONE EUSOCIAL GENUS, *Microstigmus*
      Astatinae SOLITARY
      Laphyragoginae SOLITARY
      Larrinae SOLITARY
      Crabroninae SOLITARY OR NEST-SHARING
      Entomosericinae SOLITARY
      Xenosphecinae SOLITARY
      Nyssoninae SOLITARY
      Philanthinae SOLITARY
  Apiformes (bees)
    Colletidae SOLITARY OR NEST-SHARING
    Andrenidae SOLITARY OR NEST-SHARING
    Oxaeidae SOLITARY
    Halictidae (sweat bees)
      Halictinae SOLITARY, NEST-SHARING OR EUSOCIAL; EUSOCIALITY EVOLVED REPEATEDLY
        Halictini (*Lasioglossum, Halictus*)
        Augochlorini (*Augochlorella, Augochlora*)
      Nomiinae SOLITARY OR NEST-SHARING
      Dufoureinae SOLITARY
    Melittidae SOLITARY
    Megachilidae (leaf-cutting, resin or mason bees) SOLITARY OR NEST-SHARING
    Fideliidae SOLITARY
    Anthophoridae EUSOCIALITY EVOLVED ONCE
      Nomadinae BROOD PARASITIC
      Anthophorinae SOLITARY OR NEST-SHARING
      Xylocopinae SOLITARY, NEST-SHARING OR EUSOCIAL
        Ceratinini
          allodapines (*Allodapa, Allodapula, Exoneura*) NEST-SHARING OR EUSOCIAL
          Ceratina SOLITARY
        Xylocopini (carpenter bees) SOLITARY OR NEST-SHARING
    Apidae EUSOCIALITY EVOLVED TWICE
      Euglossinae (orchid bees) SOLITARY OR NEST-SHARING
      Bombinae (bumblebees) BROOD-PARASITIC OR EUSOCIAL *Bombus*
      Apinae (honeybees) EUSOCIAL *Apis*
      Meliponinae (stingless bees) EUSOCIAL *Trigona, Melipona*

Vespoidea
  Vespiformes
    Tiphiidae SOLITARY
    Sapygidae SOLITARY
    Mutillidae SOLITARY

Table 13.1—(continued)

Sierolomorphidae SOLITARY
Rhopalosomatidae SOLITARY
Pompilidae (spider wasps) SOLITARY OR RARELY NEST-SHARING
Bradynobaenidae SOLITARY
Scoliidae SOLITARY
Masaridae SOLITARY
Eumenidae (potter wasps) SOLITARY OR NEST-SHARING
Vespidae (social wasps) EUSOCIALITY EVOLVED THREE TIMES
    Stenogastrinae SOLITARY, NEST-SHARING; A FEW EUSOCIAL *Parischnogaster*
    Vespinae (yellowjackets and hornets) EUSOCIAL *Vespa, Vespula*
    Polistinae (paper wasps) EUSOCIAL *Polistes, Ropalidia, Metapolybia, Mischocyttarus*
Formiciformes
  Formicidae (ants) EUSOCIALITY EVOLVED ONCE OR TWICE
  1. Myrmecioid complex ALL EUSOCIAL
    Myrmeciinae (*Myrmecia; Nothomyrmecia*)
    Aneuretinae
    Pseudomyrmecinae (*Pseudomyrmex*)
    Dolichoderinae (*Iridomyrmex, Conomyrma*)
    Formicinae (*Oecophylla, Formica, Lasius, Camponotus*)
  2. Poneroid complex ALL EUSOCIAL
    Ponerinae (*Ponera*)
    Cerapachyinae
    Leptanillinae
    Dorylinae (*Dorylus, Eciton*)
    Myrmicinae (*Myrmica, Leptothorax, Myrmecina, Pheidole, Solenops Pogonomyrmex*)

1977; Eickwort 1981). There are two quite different taxonomic lineages among the Formicidae: the Myrmecioid Complex and the Poneroid Complex (Wilson 1971). It is unclear whether eusociality arose once in their common ancestor (a tiphiid-like wasp) or separately in the two lines.

13.4   TWO ROUTES TO SOCIALITY

The study of phylogenetic origins of hymenopteran sociality is made considerably easier by the diversity of nesting patterns found in closely related species. There is an almost continuous series from entirely solitary to highly eusocial. This sequence provides clues to the evolutionary steps by which eusociality evolved. Two different (but not mutually exclusive) sequences have been identified (Fig. 13.4). The 'subsocial' route is a continuum of nesting habits from brief encounters between mother and daughter to long-term cooperative associations (Wheeler 1928; Evans 1958; Malyshev 1968; Spradbery 1973; Michener 1974). In most solitary wasps and bees, the female provisions her nest with a store of food, lays an egg, seals the brood cell and goes on to the next (mass provisioning). In some solitary

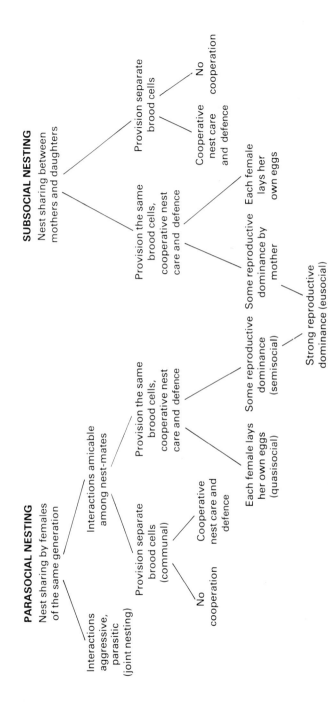

Fig. 13.4 Two routes to eusociality

species, however, the female lays her egg early in provisioning and then continues to feed the larva after it hatches (progressive provisioning). She may actually contact the offspring by feeding it whole or partially macerated food (Evans 1958; Michener 1969, 1974; Wilson 1971). In certain species the newly hatched young use the maternal nest and if the mother is still alive they cooperate with her in the sense that they defend, repair and extend their common nest, although each lays her own eggs. In most of these species, however, there are clear differences in the reproductive abilities of mother and daughter, with the daughter specializing in brood care and the mother in egg-laying. This subsocial route to sociality resembles the pattern of development in an independently founded haplometrotic eusocial colony of ants, bees or wasps.

There is a second, quite different parasocial route to social living (Lin & Michener 1972), which involves increasing levels of association among females of the same generation (Fig. 13.4). In many solitary species, two females may jointly occupy the same nest. Often these associations are clearly parasitic, with strong aggression among the females. However, in some species two or more females may nest amiably side by side, each provisioning her own brood cells and aiding the other in the sense that they both repair and defend their common nest (Brockmann & Dawkins 1979). In a few species these associations go a step further, with joint construction and cooperative provisioning of the same brood cells (particularly common among bees). At times, a few females in these nests appear to be more reproductively active than others. Whether this is due to reproductive dominance or environmental variation is not known, but it is nonetheless similar to the pattern of behaviour observed in some pleometrotic eusocial species during the early stages of colony founding.

Termites are thought to be close taxonomic relatives of cockroaches and extended maternal care is not uncommon in this group (Wilson 1971; Eickwort 1981). They seem particularly similar to the woodroaches, Cryptocercidae (Cleveland *et al.* 1934; Wilson 1971). Like termites, they live in logs, chewing out galleries and digesting the wood with the aid of intestinal protozoa, which are nearly identical to those found in termites. In an elegant study of one species of *Cryptocercus*, Nalepa (1982, 1984) found that woodroaches live in family groups consisting of a mated pair and a number of nymphs. The young remain with the adults for at least three years and during this entire period the parents do not produce additional offspring. They feed the young on hindgut fluids (where the nymphs incidentally pick up their gut fauna) and possibly on accessory gland secretions. This pattern of intensive, long-term parental care in woodroaches suggests that eusociality in the closely related termites may have originated from a similar pattern of subsocial nesting. Polygynous termite nests occur (Thorne 1982a) but only among higher termites, which suggests that this is a derived condition.

Understanding the circumstances that lead to social living and castes requires that one know not only the costs and benefits for solitary and social nesting, but also the degrees of relatedness among those involved and the options open to each individual. For example, an animal that emerges early in the season may have a choice between using the maternal nest and building a new nest of her own, whereas a later individual may only have one available option. What favours the evolution of group living and the development of reproductive dominance? The factors can be categorized according to their influence on different features of sociality: (a) those that tend to throw females together and increase their opportunities for helping, (b) those that tend to make group living more profitable under a wider range of conditions, (c) those that tend to increase the relatedness among individuals within a group, which is crucial when considering the evolution of sterility and (d) those that tend to alter the reproductive options available to females. Most of the factors favouring sociality contribute to more than one of these categories and each has many ramifying effects.

The nest, prey size and parasitism

The taxonomic distribution of sociality within the aculeate Hymenoptera reveals one of the most important preconditions for social life. Group living occurs only among taxa in which females construct a fixed nest to which they return repeatedly with food (arthropods in wasps or many loads of pollen and nectar in bees) for their offspring (Eickwort 1981). This is in contrast to most Hymenoptera, which parasitize hosts or provide their young with only one large prey item (Richards & Richards 1951). The nest and this form of behaviour greatly improve the opportunities for helping as well as the benefits. It is difficult to imagine what one female parasitic wasp might do to help another, but if a female must leave her nest repeatedly to provision, as in wasps and bees, then there are considerable advantages to the presence of an extra guard at the nest. Furthermore, a helper can actually work to accumulate food for the offspring, something which would not be possible in a species that parasitizes or hunts down only one prey item for each egg laid. A fixed nest also means that there is likely to be competition among females for nesting sites or materials, which will result in selection to use old nests, to remain in or near the maternal nest and to guard nests from the intrusions of conspecifics. This will have the effect of aggregating wasps, often close relatives, at specific favourable locations. However, these aggregations will tend to increase the frequency of parasites and predators. Indeed, all the aggregating solitary and social wasps and bees are plagued by a wide range of flies, brood-parasitic aculeates, parasitic Hymenoptera, moths, beetles, ants and mites (Michener 1958; Lin 1964; Lin & Michener 1972; Evans 1977; Gibo 1978). High levels of parasitism and predation should further improve the benefits derived from the joint defence of one nest.

Provisioning behaviour and maternal manipulation

The total dependence of larval aculeates on the provisioning behaviour of the adults provides further opportunities for the evolution of sociality. Among most insects, the size and fecundity of the adult is strongly affected by the amount of food it was able to accumulate as a larva or juvenile before metamorphosing. This is particularly true of insects (such as Hymenoptera) which ingest relatively little protein as adults. Unlike most insects, the aculeate mother provides the larva with nearly all the food it will ever get. This fact means that the size and fecundity of an individual is not simply a reflection of its genes and environment, but also of specific investment decisions made by its mother (or by workers). The reproductive options open to small females may be quite different from those available to larger individuals. They may on occasion derive greater inclusive fitness by remaining in the maternal nest and helping rather than by attempting to nest alone. The immediate result would be reproductive division of labour, i.e. a 'caste'. If the presence of a helping daughter improves the mother's fitness, then selection will favour continued maternal manipulation. The asymmetry is maintained because the offspring will not be able to counter investment decisions made during the mass provisioning of its brood cell, before it has even hatched (which occurs in most bees and sphecid wasps). Even in progressively provisioning species, the asymmetry may be profound, with the larger, more mobile mother feeding the helpless grub.

Generations and overwintering stage

Subsocial nesting can only occur among species in which there is more than one generation per year (i.e. bi- or multivoltine species) and a relatively long-lived mother (Michener 1969; Eickwort 1981). Clearly, this is more likely to occur in subtropical or tropical areas than in temperate regions. There is an interaction between overwintering stage and the number of generations per year. Most solitary wasps live but one season, overwintering as larvae. In the spring they must first pupate (which takes 3–5 weeks) before emerging and beginning to nest. However, in those species that overwinter as mated adults, females can begin to nest with the first warm weather in the spring. If a bivoltine animal overwinters as a larva, then the only opportunity for helping a parent is second-generation females assisting their first-generation mothers to produce the first generation of the following year. However, if a bivoltine animal overwinters as an

adult, then the first-generation females too have an opportunity to
help their mothers produce the second generation of the same year.
This proves to be an important factor since there are usually sex ratio
differences between the first and second generation in a year (see
below).

Haplodiploidy and sex ratio control

All Hymenoptera have a haplodiploid form of sex determination (in
which males develop from unfertilized eggs and are haploid, whereas
females develop from fertilized eggs) and an associated control over
the sex of the offspring (Bull 1981). Haplodiploidy alters the calculations of relatedness when compared with diploid organisms (Fig.
13.5; Hamilton 1964, 1972; Crozier 1970). In particular, a female is
more closely related to her full sisters (3/4) than she is to her own
offspring (1/2). But this does not mean that a daughter should inevitably give up her own reproduction in favour of helping her mother
(West Eberhard 1981). First, she must be able to contribute significant
benefit to her mother's fitness. If she can rear 1.5 times more surviving offspring on her own than she is able to add to her mother's
brood, then selection will favour nesting alone. Secondly, the genetic
benefits from helping drop rapidly if a daughter is not helping to rear
full sisters. Female wasps and bees often mate more than once, which
means that daughters are frequently helping to rear half ($r = 1/2$)
rather than full sisters (Alexander & Sherman 1977; Wade 1982).
Thirdly, in many species large numbers of male offspring are produced, which also greatly reduces the genetic benefits of helping
(Hölldobler & Michener 1980; Crozier 1979). This is because with
haplodiploidy females are not equally related to brothers and sisters

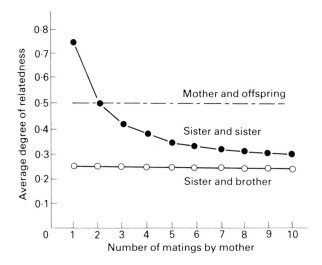

Fig. 13.5 Degrees of relatedness in relation to the number of matings by
the mother for a haplodiploid organism.

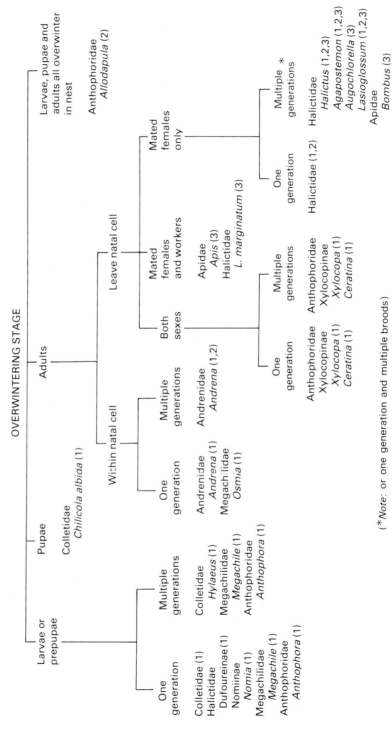

Fig. 13.6 The generations and overwintering stages of temperate bees (from Stephen, Bohart & Torchio 1969, with additions from Michener 1974 and Eickwort 1981) and their degrees of sociality (1 = solitary; 2 = nest sharing; 3 = eusocial).

(Fig. 13.5). In fact, a female is always more closely related to her own offspring than she is to her mother's sons. Thus, the payoff for helping to rear sisters is likely to be higher than that derived from helping to rear brothers. If daughters can bias the sex ratio of their mother's offspring (Trivers & Hare 1976; Alexander & Sherman 1977; Stamps & Metcalf 1978; Metcalf 1980), or produce males by laying eggs of their own (Crozier 1977; Charnov 1978), or if they can preferentially assist their mother at times when she is producing daughters rather than sons (Seger 1983), then helping is likely to be more profitable.

Seasonally variable sex ratios

In general, a female allocates her available resources among male and female offspring so as to maximize her long-term reproductive success (Charnov 1982a; Trivers & Willard 1973). Seger (1983) points out that in bivoltine species, sons of the first generation are of greater reproductive value to their mother since they can, if they live long enough, mate with females of both generations (see also Werren & Charnov 1978). (If males are short-lived, then sons and daughters are of equal value since they both produce one generation and hence selection should favour equal maternal investment in the two sexes.) This means that selection will favour male-biased sex ratios in the first generation of a year and female-biased sex ratios in the second. If a daughter is going to help her mother, then clearly she will gain more from helping to produce the second generation than in helping with the first. As already mentioned, the only opportunity for helping to produce this female-biased second generation occurs when the mother overwinters as a mated adult. This means that the payoffs for subsocial nesting are increased by adult overwintering. It must be emphasized that subsocial nesting is by no means an inevitable consequence of these or any other conditions. The female-biased sex ratio during the second generation (relative to the overall population sex ratio) merely increases the cost ratios over which a daughter is prepared to help her mother. However, if her help is of little consequence relative to what she could produce on her own, then selection will favour solitary or communal nesting rather than subsociality. Is there, in fact, any evidence that eusociality is particularly likely to occur among bi- or multivoltine species in which adult females overwinter?

The taxonomic distribution of social nesting in Hymenoptera largely confirms Seger's (1983) prediction (Fig. 13.6). Among the Sphecidae where overwintering occurs, the vast majority hibernate as larvae. The only eusocial sphecids are a very few tropical species with continuous life cycles. Among the vespid wasps, most eumenids are solitary and they overwinter as larvae, whereas all the temperate social vespine and polistine wasps overwinter as adults (Spradbery

1973). Bees show the greatest diversity of life history patterns. There is a clear separation between those species that overwinter as larvae and show solitary nesting and those that overwinter as adults and show a great diversity of social nesting habits. The only exceptions are some solitary xylocopine and halictid bees. Some of these differ from the usual pattern in that both males and females overwinter and mating occurs in the spring (Stephen *et al.* 1969). The exceptions among the halictids, where sociality is a particularly labile trait, would make an interesting study of the costs and benefits for social nesting. Many of these are undoubtedly cases of mutualistic, parasocial nesting (Michener 1974), where Seger's model does not apply.

As with all correlations, one must be wary of the cause–effect relationships. It seems likely, in fact, that the same factors which select for overwintering as an adult may select for social nesting (J. R. Lucas & A. Grafen, pers. comm.). The advantage to overwintering as an adult would seem to be that nesting could begin with the first warm weather in the spring. This would be important only if nests founded a little later in the season were less successful than those founded at the very beginning. But if this were true, then one would expect that females who emerged later in the season would similarly find themselves in an unfavourable situation for founding a new nest. This would favour joining an established nest rather than initiating a new nest on their own. The literature on social wasps such as the paper wasp *Polistes* confirms that nests which are founded late in the season have higher mortality and produce fewer reproductives than those founded early in the season (Strassman 1981a, c; Noonan 1981; Gibo & Metcalf 1978; Gamboa 1980).

Kin recognition

Among the primitively and highly eusocial bees (Kukuk *et al.* 1977; Buckle & Greenberg 1981; Greenberg 1979) and the *Polistes* wasps (Klahn 1979; Noonan 1981; Ross & Gamboa 1981; Shellman & Gamboa 1982), it has been convincingly demonstrated that individuals recognize and discriminate kin from non-relatives (Crozier & Dix 1979). It is not known whether kin recognition is found among solitary Hymenoptera, but it seems likely that such an ability may well exist (Hölldobler & Michener 1980). Kin recognition may be an outgrowth of the ability of animals to discriminate their own nests, eggs, and young from those of other females (Michener 1974). Such an ability, based on chemical cues, has been demonstrated for some solitary bees (Steinmann 1976). The high incidence of intraspecific parasitism among solitary wasps and bees suggests that there may be selection for females to recognize whether their nest has been marauded or brood parasitized by conspecifics. If kin recognition were found commonly among solitary Aculeata, then this would be an important pre-adaptation for social nesting, since it would mean that a

mechanism by which close relatives could recognize and preferentially join one another was already in use (if such joining were
mutually advantageous).

13.6 ORIGINS OF EUSOCIAL NESTING

We have now identified a number of factors which contribute to or
are at least correlated with the appearance of social nesting. Eusociality arises in multivoltine, aculeate Hymenoptera which provision
their nests with numerous prey, suffer high rates of parasitism, and
overwinter as adults. But these ecological and taxonomic correlates
merely set the stage for the appearance of sociality. Why does social
nesting pay in any particular case?

The evolution of eusociality involves two separate but related
problems. First, why do animals live in cooperative groups and, secondly, why do castes evolve in some of these groups? There will be
selection for an animal to give up its individual reproductive future
under two, not mutually exclusive, conditions: (1) when the inclusive
fitness gained by an individual from not reproducing is greater than
that gained from trying to reproduce on her own (this can occur only
if a female spends her time contributing fitness to a relative), and (2)
when one individual has manipulated another into helping rather
than reproducing (Charnov 1978; Crozier 1979). The same arguments
hold for group nesting: selection will favour joining when the fitness
gained by nesting socially is higher than that gained from nesting
alone (but here aiding relatives is not required) or when one individual has manipulated another into joining (Richards & Richards 1951;
Hamilton 1964; Lin & Michener 1972; Alexander 1974; West Eberhard 1975).

There are two broad categories of benefit to cost ratios which will
favour an individual joining rather than nesting on her own. First, the
association may be mutualistic, i.e. both founder and joiner may have
higher individual fitness when nesting together than when nesting
alone. For example, in one univoltine species of bee, communal nests
were continuously guarded and thus avoided parasitism whereas solitary nests were heavily parasitized (Abrams & Eickwort 1981). Secondly, asymmetries may exist which favour joining by some females
in the population and not others. There are several possible explanations for such asymmetries.

(1) *Delayed reproduction.* Among most solitary wasps and bees, there is
a period of several weeks between emergence and the start of nesting.
In a few species these young adults remain in the nest until fully
mature, where they incidentally guard the nest from intruders (Evans
1973) or may even help with some provisioning (Elliott & Shlotzhauer
1980). For species that overwinter as adults, such as carpenter bees,
this period may be greatly lengthened (Gerling & Hermann 1978) and
young adults may engage in considerable brood care (Michener &

Lange 1958a), thus gaining inclusive fitness while they wait to reproduce later.

(2) *Hopeful reproductive.* A female joining another may care for the nest and brood until the primary female's death, whereupon she inherits the nest and workers she helped to rear. The work she does increases her inclusive fitness in two ways: she is usually caring for close relatives and she is helping to ensure that she will inherit a strong nest when the queen dies (West-Eberhard 1978b).

(3) *Making the best of a bad situation.* Group nesting may also be better than nesting alone if a joining female has little prospect of rearing offspring in her own nest (Bartz & Hölldobler 1982). Possible reasons for this include: (a) she may be small, unmated, of low fecundity or unlikely to live long enough to produce offspring, due to what her mother provided for her (maternal manipulation) or due to other sources of environmental variation; (b) it may be impossible for her to establish a nest elsewhere; (c) parasitism and predation may be so high that nests with single females may not survive, so by joining a nest she gains the benefit of additional females helping to guard. Given these problems, if the queen she joins is her mother or sister, then it is easy to see how her inclusive fitness could be higher than if she nested alone (particularly if her mother has not been mated by many males). However, some females may actually lay more eggs and have more surviving offspring if they join nests of other females than if they nest alone (Michener & Brothers 1974). This can occur despite adaptations on the part of the queen for preventing such 'queen-like' behaviour by subordinates (these adaptations include aggression and eating the eggs).

Conditions favouring mutualistic communal nesting among nonrelatives may arise, but for a number of reasons nest-sharing generally occurs among relatives (Wrangham 1982). (a) Kin groups and philopatry to the natal nesting area are the most likely mechanisms by which individuals locate one another. (b) Selection would favour sisters joining one another if this were the only mechanism by which they could be sure that they, and their sister, had a mutualistic association in which to nest. For example, Noonan (1981) has demonstrated that in *Polistes fuscatus* two co-foundresses on a nest produce more than twice the number of offspring of each nesting alone. She was able to show that females who came from small families (had few sisters) were more likely to nest alone than those who came from large sibships. (c) If any altruism is involved in a nesting association, then clearly the fitness gained from nesting with relatives will exceed that from nesting alone. If relatives were not available, then a female might join non-relatives if her individual fitness were raised by doing so. However, if intraspecific parasitism were involved, one would expect selection against nesting with relatives. (d) If an individual's only source of fitness is through aiding relatives, then clearly she should seek out and join the closest kin she can find. In fact, most

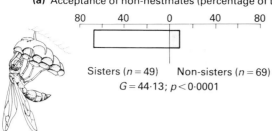

(a) Acceptance of non-nestmates (percentage of trials)

Sisters (n = 49)   Non-sisters (n = 69)

G = 44·13; p < 0·0001

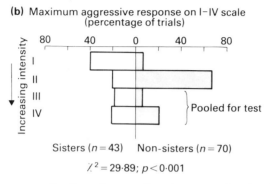

(b) Maximum aggressive response on I–IV scale
(percentage of trials)

Increasing intensity

Pooled for test

Sisters (n = 43)   Non-sisters (n = 70)

$\chi^2 = 29·89; p < 0·001$

Fig. 13.7 The response of *Polistes fuscatus* foundresses to females that are placed on their nest (not nest mates). The bar represents the percentage of trials in which the intruder was allowed to remain on the nest for 15 min. **(a)** The response to sisters and non-sisters was significantly different. **(b)** The level of aggressiveness toward sisters and non-sisters was also significantly different. (From Noonan 1981.)

known foundress associations are between closely related females (Craig & Crozier 1979; Crozier 1977, 1979; Strassmann 1981b; Metcalf & Whitt 1977a, b; Lester & Selander 1981; Pamilo 1982). Females do not simply depend on finding relatives at the natal nesting site, but rather they actively search for and choose nests with sisters (Fig. 13.7; Noonan 1981; Klahn 1979). The founder female also responds differently toward intruders that are sisters than she does toward non-relatives, viciously attacking the latter and driving them from the nest (Greenberg 1979; Noonan 1981). The behaviour of both the foundress and the joiner suggests that nesting with relatives is more advantageous than with non-relatives.

Both theoretical models (Crozier 1979; Charlesworth 1978; Charnov 1978; Craig 1979) and empirical studies support the view that maternal manipulation plays an important role in the evolution of social nesting. In a study of the primitively eusocial, ground-nesting bee *Lasioglossum*, queens were found to inhibit the ovarian development of their worker daughters and manipulate them behaviourally into remaining in the nest (Michener & Brothers 1974). Similar observations have been made on wasps (West Eberhard 1977; Metcalf & Whitt 1977b; Hunt 1982) and ants (Brian 1980; Brian et al. 1981; Hölldobler 1976). It is possible that manipulation may also be a factor in the evolution of sociality in termites. In the cryptocercid cockroaches, the adults are capable of controlling the food source by walking away when a nymph approaches or by closing off the proctodeum, thus denying access to hindgut fluids (Nalepa 1984). Even among some social spiders, maternal feeding occurs (Buskirk 1981) and may result in a degree of maternal manipulation (Y. D. Lubin, pers. comm.). Unlike Hymenoptera, however, juvenile cryptocercids, primitive termites and spiders are not completely at their mother's mercy, since they are capable of feeding and, in fact, gain considerable nourishment on their own.

13.7 THE MAINTENANCE OF EUSOCIALITY: CONFLICT AND COOPERATION

Throughout we have been concentrating on the advantages of social nesting, but a eusocial insect colony is just as likely to be characterized by a high degree of reproductive competition and conflict as it is to show cooperation. The line between mutualistic and parasitic associations is often indistinct (Eickwort 1975). For example, in *Polistes* wasps, the advantage to multiple-foundress associations clearly drops after a few females have joined (Noonan 1981). In a communal sphecid wasp, *Trigonopsis cameronii*, up to four females occupy a joint nest, provisioning their own brood cells and working together in constructing the common mud structure (Eberhard 1972, 1974). However, when provisioning is difficult, they also steal prey from one another. If an association is clearly and consistently mutualistic, then we expect the evolution of specific adaptations which further increase each individual's benefit. Cooperation could arise through mutualism (Lin & Michener 1972), kin selection (Hamilton 1964, 1972) or reciprocation (Trivers 1971). In a social insect colony, the repeated encounters and generally high degree of relatedness make cooperation likely even when defections by an individual might lead to a short-term advantage (Axelrod & Hamilton 1981; Eshel & Cavalli-Sforza 1982). But there are always clear sources of genetic conflict in a colony and in order for eusociality to be maintained these must either be suppressed or the advantages must outweigh the disadvantages. What is the nature of these conflicts of interest?

(1) *Between colonies.* Eusocial insects commonly show intense competition among colonies for control of nests, foraging areas and workers (Hölldobler & Lumsden 1980). For example, Bartz and Hölldobler (1982) found that in the honey ant *Myrmecocystus mimicus*, colonies were generally initiated by small groups of two to four foundresses. Single gynes were incapable of initiating a colony alone, because of raids from neighbouring conspecific colonies which stole their brood.

(2) *Between queens.* There is often intense competition among queens for control of a nest during the early stages of colony develop-

ment. For example, when *Polistes metricus* nest in high densities, nest usurpations of single foundresses are frequent and account for the more productive multiple-foundress associations (Litte 1979; Forsyth 1975; West Eberhard 1969, 1978b; van Bendegem *et al.* 1981; Gamboa 1978). The usurped female may either remain with the dominant female and become a subordinate or she may leave, attempting to take over another nest (Gamboa & Dropkin 1979) or she may be killed by the workers, as in *Vespula* (Matthews 1982).

(3) *Between workers and supernumerary queens.* One of the most serious sources of conflict in a eusocial nest is the presence of queens other than the mother. Polygyny is more serious than multiple mating in reducing relatedness between workers and the brood. Noonan (1978) argues that in a polygynous colony the replacement of a lost queen by her sister results in a greater loss in inclusive fitness to the workers (which are functionally sterile) than if she were replaced by a daughter (i.e. the worker's sister). Among ants, this is in fact exactly what happens. Although pleometrosis is common, it never seems to lead directly to primary polygyny (Hölldobler & Wilson 1977). Rather, the co-foundresses are killed after the first brood matures and new queens are later added (secondary polygyny) by adopting the mother's reproductive daughters from the same nest (usually after a nuptial flight) (Bartz & Hölldobler 1982; Hölldobler & Wilson 1977). In many species of eusocial wasps, subordinate gynes are driven from the nest or killed by workers (West 1967; West-Eberhard 1978a; Gamboa *et al.* 1978; Wilson 1971; Forsyth 1980; Matthews 1982). These observations suggest that in some species workers play a role in reducing the incidence of polygyny.

(4) *Between workers and their queen.* In some species of eusocial insects there is visible competition between the queen and workers over who is to lay eggs. This may occur when the colony is very large and the workers lose contact with the queen (Wilson 1971; Smeeton 1981; Brian *et al.* 1981), but there are many species in which workers regularly sneak in a few eggs. In *Bombus*, queen-like workers dominate other workers and compete with the queen for opportunities to lay (van Honk *et al.* 1981). In many species 'queen-like' workers are subject to vicious attacks by the queen and are soon coerced into more acceptable behaviour (Velthuis 1977; West Eberhard 1977). Worker-laid eggs are normally eaten by the queen or by other workers (Brothers & Michener 1974), but there is no doubt that some survive (West-Eberhard 1978b).

Since workers usually have not mated, the eggs they lay are unfertilized and hence develop into males. However, in a few ant species and in one subspecies of honeybee, queenless workers can produce diploid eggs parthenogenetically and these develop into viable females (Tschinkel & Howard 1978; Ruttner 1977). Sometimes, however, the workers do mate. In a ponerine ant, workers with special glands advertise for males, mate with them and then return to their natal nest, where they lay fertilized eggs (Hölldobler & Haskins 1977). In nests that are founded through pleometrosis, mated secondary queens often continue to live in the nest, dominated by the queen and generally behaving like a worker, but on occasion laying eggs (Wilson 1971). In many species it is not clear which individuals are queens and which workers (Strassmann 1981d).

(5) *Between workers*. Workers may compete with one another for laying opportunities or for a position which increases their chance of taking over should the queen die. Cole (1981) has found that in a species of tiny *Leptothorax* ant that lives in grass stems where colony orphaning is common, there is a permanent dominance heirarchy among workers, with the most dominant females receiving more food than the subordinate individuals. Perhaps because of this, the dominant females have better developed ovaries than the subordinates and even with a queen present they account for 22% of the eggs laid. Even in the most highly eusocial species with generally sterile workers, such as honeybees and ants, worker-laying occurs (Michener 1974; Wilson 1971; Tschinkel & Howard 1978). In one subspecies of honeybee, *Apis mellifera capensis*, a laying worker is capable of producing the queen pheromone from her mandibular glands. This results in the laying worker being treated like a queen and inhibiting egg-laying in other workers (Crewe & Velthuis 1980).

The degree of cooperation and the organization of coordinated colony life vary markedly among eusocial species. In many the workers apparently take charge, as in honeybees, building brood cells of different sizes for workers, queens and males and controlling the use of colony resources. In these highly social species, the colony seems to run itself with little obvious interference from the queen. In more primitively eusocial species, such as the ground-nesting halictid bee *Lasioglossum* (Fig. 13.8), the queen takes a direct role in daily operations, prodding the workers into action or leading them to a cell that needs to be provisioned; in her absence, little work gets done (Brothers & Michener 1974; Breed & Gamboa 1977). In *Polistes* the queen attacks inactive workers and forces them off the nest to forage (Pardi 1948; West Eberhard 1977). In all eusocial species, the removal of the queen results in rapid changes within the colony and often new queens appear from among the workers. If the queen is removed in *Vespa orientalis*, the workers neglect larvae, fight and lay eggs (Kugler *et al.* 1979). This suggests that the queen's position is maintained by suppressing reproduction in the workers around her. In many species this is accomplished through pheromones, but in some the queen dominates the other females through aggressive behaviour. A particularly powerful effect has been found in a weaver ant (*Oecophylla*) (Hölldobler & Wilson 1983). If the queen is removed, workers readily produce male eggs, but as long as the queen is present (or even if a dismembered piece of their queen is in the nest), no viable eggs appear. Rather, workers produce trophic eggs (also found in other

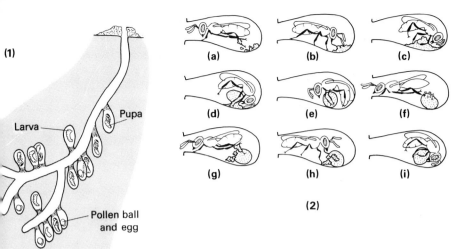

Fig. 13.8   The nest and provisioning behaviour of *Lasioglossum*.
1. The nest and brood cells with pollen ball and egg, larva and pupa.
2. Provisioning of the cell. Depositing of the first loads of pollen and nectar
(**a, b**), forming the pollen into a ball (**c, d, e**), depositing additional loads
(**f, g, h**) of pollen and nectar and perfecting the nearly complete pollen ball
just prior to egg laying (**i**). (From Michener & Large 1958b and Batra 1964.)

highly eusocial species) which are an important source of food for the
queen. The chemicals produced by the queen do not alter ovarian
development in workers, but control whether a trophic or a functional male egg develops. In some species the queen dominates the other
females through aggression. In some polygynous paper wasps
(*Polistes*) for example, there is a linear dominance hierarchy among
the larger females and when the queen dies or disappears the next in
line takes over her role (Pardi 1948; Turillazzi & Pardi 1977; West
Eberhard 1969, 1977, 1981; Gamboa & Dropkin 1979; Hermann 1979;
Strassmann 1981b). Even in species in which workers appear to
control the day-to-day functioning of the colony, there is evidence
that the queen and the chemicals she produces are the controlling,
coordinating agents in colony life (Brian 1980).

Differences in colony organization are often considered to be
associated with 'highly' as opposed to 'primitively' eusocial species
(Wilson 1971). Recalcitrant workers and primitive levels of eusociality
are characteristic of species in which workers are capable of laying
eggs and taking over the role of the queen should she die. Completely
sterile workers are generally more cooperative. The reason for this is
likely to be that fertile workers have other reproductive options available to them. Sterile workers are trapped. Their only source of fitness
is through cooperating with other workers and the queen. The evolution of the 'superorganism' qualities of eusocial colonies are surprising
when one considers the high levels of conflict which are likely to be
present (Starr 1979; West Eberhard 1982). It is not surprising that the
most highly specialized eusocial species are monogynous (or secondarily polygynous) with very long-lived and extremely fecund queens.

Chapter 14
Behavioural Ecology of Plants

E. L. CHARNOV

14.1 INTRODUCTION

While plants may not have behaviour in the strict sense used by ethologists or psychologists, their inclusion in a behavioural ecology text serves to point out that many of the concepts developed in this animal-oriented field apply equally well to plants, even if behaviour is not involved. Furthermore, plants differ in some important ways from mobile animals and present some unique opportunities for testing evolutionary theory. By way of example, most animals are diploid, producing haploid gametes. Plants, particularly algae, show many variations. Some species show the dominant life form (i.e. the plant we commonly see) to be haploid, with the diploid reduced to a transient zygote. Other species are like the animals, while still others show two dominant life forms: one a haploid, the other a diploid. Only recently have questions concerning natural selection been seriously posed about the meaning of such variation (Willson 1981; Bell 1982, also for review of earlier thought). As a second example, phylogenetic evolution in animals may be due to the alteration of a single lineage, or to speciation, the splitting up of a lineage. But plants have another trick; many speciation events among higher plants are due to hybridization between two existing species. That is, two separate evolutionary lines may produce hybrids which then form a third line (Grant 1971). Hybridization appears to be of negligible importance in animals. Plants have photosynthesis, while animals do not. And finally, some plants have flowers and use animals to carry pollen. I know of nothing comparable in animals.

Natural selection thinking has been widely used by students of plants, in viewing topics as diverse as (for example) plant–pollinator relations, fruit dispersal, leaf and canopy structure, photosynthetic physiology, leaf movements (solar tracking), plant–herbivore relations, foraging for and allocation of nutrients (also water and carbon dioxide) and life history strategies. Indeed, the old natural selection adage of 'nature red with tooth and claw' is perhaps more truthfully replaced with a statement of 'nature shadier with leaf and limb'. Obviously, this brief introduction will just scratch the surface of this literature. Readers may wish to further consult the books by Harper (1977) and Grime (1979), or the volumes edited by Solbrig, Jain,

Johnson and Raven (1979), and Townsend and Calow (1981). Also of interest are the review articles by Ehleringer and Forseth (1980) on solar tracking; Mooney and Gulmon (1982) on herbivores as a source of carbon loss; Chapin (1980) on mineral nutrition of plants; and Givnish (1980) on ESS methods applied to canopy structure. This chapter will concentrate on three aspects of the evolutionary ecology of higher plants, aspects which touch base with themes discussed elsewhere in this volume and which to me represent some new and exciting directions. In order of consideration these are (i) labile sexuality (sex choice), (ii) evolution of breeding systems, (iii) some aspects of conflict of reproductive interest shown by males and females in plants. This last section will also discuss aspects of intragenomic conflict of interest, particularly conflict between nuclear and chromosomal 'genes' over pollen production. The reader should be forewarned that some of these topics are controversial, and the discussion here is undoubtedly far from the last word.

14.2 LABILE SEXUALITY (SEX CHOICE) IN PLANTS

Most higher plants are fairly immobile as adults and an individual must literally breed where it is rooted. As a consequence, plants are usually very plastic in their growth (Harper 1977). Since opportunities to reproduce through male versus female function (pollen versus seeds) might be expected to vary greatly with micro-habitat or plant size, we should expect many plants to show the ability to adopt the sexuality appropriate for the individual's particular condition, provided the individual knows some things about itself and the others around it. This section reviews the evidence for such responses, discusses some of the theory proposed to account for them and, finally, briefly reviews data on two of the best-known systems.

14.2.1 Evidence

Botanists have long been aware that, at least for some species, sexuality is responsive to environmental or growth conditions (Heslop-Harrison 1957, 1972). Many publications support the fact that various environmental factors can alter sex expression in dioecious species. Likewise, the ratio of male to female flowers on individuals of monoecious species has often proved manipulatable through alteration of the plant's environment and (less often) the ratio of male to female organs of perfect-flowered taxa. The terms 'monoecious', 'dioecious' and 'perfect flowers' are defined in Table 14.2. As summarized in Table 14.1, age, injury and disease have all been shown to alter the sexual expression of individuals of some species. Similarly, physical and nutritional characteristics of the environment (e.g. light intensity, soil

Table 14.1. Some factors known to modify the sexual expression of vascular plants under controlled conditions. (Table modified from Freeman, Harper & Charnov 1980.)

Factor	Direction of shift
1. Increasing age or size	Male to female
2. Cold weather	Female to male
3. Dry soil (wet soil often the reverse)	Female to male
4. High light intensity (low gives the reverse)	Male to female
5. Manure	Male to female
6. 'Rich' soil	Male to female
7. High temperature	Female to male
8. Trauma—removal of leaves, flowers or crown pruning	Male to female
9. Removal of storage tissue	Female to male

fertility, soil moisture) are reported to affect the sexual expression of a variety of species.

Some variables predispose individual plants of labile sex expression towards femaleness, while others favour maleness (Table 14.1). The data reveal a strong tendency for stress (broadly defined) to induce maleness. By 'stress' I mean conditions which reduce growth, a plant's ability to allocate resources to reproduction or an individual's chances for survival. For example, small size, cold weather, dry or infertile soil, low light intensity and removal of storage organs all tend to incline the individual towards maleness (Freeman, Harper & Charnov 1980).

Most of the work described above was done under controlled laboratory conditions. However, recent work shows that sex choice is a usual feature of the life history of several species. Barker *et al.* (1982) studied the maple (*Acer grandidentatum*), which is a small tree showing either monoecy or purely male individuals. They surveyed the floral sex ratios and showed that trees on xeric sites were relatively more male, while trees on mesic sites were more female. Natural sex change has also been documented for two species of juniper by Vasek (1966). He showed that over a 2–5 year period 7% of the individuals of one species, and almost 25% of individuals of the other changed sex. In other recent work, McArthur and Freeman (1982; McArthur 1977) have demonstrated that environmental factors are strongly correlated with the sexual expression of individuals of the subdioecious species *Atriplex canescens*. They document sex switching following three environmental stresses: (1) an unusually cold winter, (2) drought and (3) heavy seed set by females. During the flowering period following each of the stressing events, there was a pronounced shift towards maleness. Over 40% of the plants involved in the study changed their sexual state during the seven years of observation.

Their data also support the idea that the switching towards maleness during stressful years enhances the individual's survival during that period.

While these data argue that sex choice is a typical part of the life history of at least some species, and that 'stress' causes a shift towards maleness, we do not know how widespread sex choice in plants is. We now turn to a discussion of the meaning of these shifts.

14.2.2 The theory

Several hypotheses may account for the data discussed above. One which has been eliminated in at least a few of the studies is the idea that under stress females simply die faster; that is, the shift toward maleness is a non-adaptive outcome of mortality. In many cases the shift is a physiological change within an individual still alive. If such a shift is adaptive (which seems a very good working assumption), two major factors must play a role and sex choice may be due to either (Charnov & Bull 1977).

(1) Sequential hermaphroditism, where an individual changes sex as it grows older or larger. Theory suggests (Charnov 1982a; Freeman, Harper & Charnov 1980) two important aspects which may be involved. (i) Larger (older) individuals are relatively better at being one sex, for example at attracting pollinators and dispensing pollen (*male function*), or producing and dispersing seeds (*female function*).

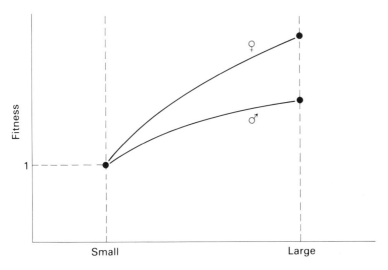

Fig. 14.1 Larger (or older) individuals are better both at being male and at being female. However, the graph illustrates where larger individuals are *relatively better* at being female. For such a situation, most of the small individuals are selected to be males, while females should predominate among the larger. Reproductive success within a sex can be scaled within the sex—the graph assigns small males or females a fitness of one. The large male's fitness is thus measured relative to the small male's; a large female is likewise scaled relative to a small female.

This is illustrated in Fig. 14.1. (ii) Reproducing through one sex is relatively more costly than through the other, so that growth and/or mortality is sex specific (Policansky 1982; Bierzychudek 1981; Freeman, Harper & Charnov 1980). Selection then delays reproduction to a larger size or later age in the sex which pays the higher mortality/ growth cost. The early years are spent as the sex with lower mortality and/or higher growth rates. For example, small individuals may not bear well the cost of being female. Note that this does *not* answer the question as to why the costs differ between the sexes—why, for instance, small individuals do not just allocate much less to being a female. One possibility is that the fixed cost of being (say) female is so high that small individuals cannot afford it (Bierzychudek 1981).

(2) Sex choice may also be a response to a spatially patchy environment. A seemingly uniform habitat might well have microsite variation, so that even assuming discrete generations, some individuals are robust at the time of breeding, while some are small. If size (or condition) translates differentially into male versus female fitness, individuals might be expected to choose the sex appropriate for their size or condition.

These two hypotheses differ in that while each assigns male or female fitness differentially to individuals of various size, the size variation is generated in the first case by age structure, and in the second by spatial variation. In addition, spatial structure (e.g. wet versus dry patches) may affect male versus female fitness opportunities independent of size or age *per se*.

Sequential hermaphroditism and a spatially patchy environment may, of course, interact. For example, one sex may die faster in the dry patches. Freeman, Harper and Ostler (1980) suggested for the US intermountain west that soil moisture is often correlated with sexuality (females in wet) because pollen production takes place in the spring, when moisture is everywhere abundant, while seed and fruit maturation take place later in the year, when patches far from permanent water are very dry. Females would then suffer relatively higher mortality in dry patches than males in those patches.

There are, of course, other possible adaptive explanations for sex choice. For example, if seed dispersal is very local, a female in a good (wet) patch ensures that her offspring are also in that patch. Another possibility is that the relevant environment is the social environment, where sex is chosen according to the identity of one's neighbours (examples in van den Ende 1976 for lower plants).

It is interesting to note that many of these cases require that an individual use 'knowledge' of its immediate breeding opportunities in sex choice. For example, if a 10 g plant is surrounded by plants over 10 g, it is relatively *small*; if it is relatively big (everyone else is <10 g), its breeding options are very different. This is illustrated in Fig. 14.2. A stable environment could select for a fixed size of sex reversal but if this changes much in short time periods, the possible

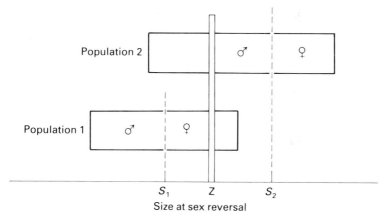

Fig. 14.2 Sex reversal related to size for two populations which differ in their general size distributions. As the general size distribution shifts, so does the size of sex transformation. An individual of size Z should be a male in population 2, but a female in population 1. This figure assumes that the male/female fitnesses are as in Fig. 14.1. It is also possible that the size distributions vary over a small spatial scale—which could make it difficult for individual Z to know the correct sex. Such a situation may favour non-labile sexuality.

inability of a plant to gather and use information about the change may well select against the trait. That is, mistakes of when to alter sex may select for non-labile sexuality. We now consider sex choice in two of the better studied systems.

14.2.3 Some examples

Orchids

Of all orchids, only the three genera of the subtribe Catasetinae have separate male and female flowers. Dodson (1962) noted that in field populations usually less than 5% were females, and suggested sex choice for the group. He picked 15 females (from a species of *Catasetum*) from full sunlight and moved them to dense shade; all then bloomed as males. A control experiment moved 15 females from full sun to full sun—13 of these remained females. By moving plants from shade to sunlight, he also produced the reverse shift. Gregg (1975, 1973) has also studied several species in the genera *Catasetum* and *Cycnoches*. Her data showed that most of the species studied produced significantly more female inflorescences when grown under sun, compared with shade conditions. However, the species were not at all uniform in response. Plants of two species required lower light intensities to produce substantial proportions of female inflorescences, compared with individuals of four other species. Catasetinae lose their leaves just prior to (or just after) flowering, so that all of the resource for reproduction is contained in the storage organ, called a pseudobulb. It appears that plants under good growth conditions typically become females. The picture which emerges from this work is that photosynthate is stored, expended for reproduction, and then replenished. Light, pseudobulb size, and other growth factors influence the resource available for reproduction. These orchids are probably pollinated by large bees, which means that breeding takes place between individuals from many patches (e.g. light and shade areas). The interspecific variation in sexual response to light levels strongly suggests that the species differ in the average light available in their respective environments (see below). There are no data bearing on whether females gain more by being large.

Jack-in-the-pulpit

Arisaema is a large genus (over 100 species) of herbaceous perennials, mostly restricted to forests of Africa and eastern Asia (Wilson 1960). One of the two North American species, *A. triphyllum* (jack-in-the-pulpit), and one Japanese species, *A. japonica*, have been much studied (Schaffner 1922; Maekawa 1924; Policansky 1981; Bierzychudek 1981; Lovett Doust & Cavers 1982) from the viewpoint of sex reversal. *Arisaema triphyllum* overwinters as a corm. The leaf and inflorescence primordia, formed the previous fall, expand in the spring. That is, an individual sets its sexuality (for a given breeding season) the previous fall, after the growing season.

The sexes are usually separate. Small individuals are mostly male, large mostly female. Sex reversal ($\female \rightarrow \male$) may be induced by any of several stress treatments (e.g. mutilation of the corm, leaf removal, etc.). Bierzychudek (1981) and Lovett Doust and Cavers (1982) have documented that reproduction as a female is considerably more costly (energetically) than male reproduction. Policansky (1981) estimated female seed set as a function of size and, as a measure of male reproductive success, estimated number of flowers per inflorescence versus male size. By these measures, males did not gain by being large, while females did. He also calculated an ESS size of sex change and (surprisingly) predicted fairly closely the size when 50% of the population had changed sex. Thus individuals may gain in these two ways by being female only when large.

One final comparison is of interest. Lovett Doust and Cavers (1982) studied *A. triphyllum* near London, Ontario. Data on size versus sex from four locations strongly suggest that if the general size of the individuals is adjusted upwards, so is the size of transformation. In one location the average height of reproductive individuals was *c.* 49 cm, in another location *c.* 37 cm. The average height of females in the first location was 55 cm, in the second 43 cm. The same comparison for males is 41 cm versus 33 cm. Thus females in one population were almost as small as males in the other. Interestingly, these two populations also represent extremes in available sunlight. Of the four populations studied, the location with small plants had the lowest light intensity, the location with large plants the highest.

I can summarize this section on plant sex choice with the suggestion that we have just begun to explore the possibilities plants realize with their differential use of male/female reproductive function. In the next section we ignore this sex choice aspect, and concentrate on the broad-scale male/female reproductive patterns in higher plants.

14.3 BREEDING SYSTEMS IN HIGHER PLANTS

14.3.1 General distribution

The distribution of gender in higher plants is complex because even at a simple morphological level we can talk of the sex of a flower (stamens and pistils, or only one), the sex of an individual (combinations of flower types), or the sexual distribution of a population (various flower and individual types). Darwin (1877) classified flowering plants with such a typology. In Darwin's scheme, monoecy, for example, would refer to a population consisting of individuals each of which produces both male and female flowers. Androdioecy would refer to a population with male individuals and hermaphroditic individuals (probably with perfect flowers). The typology refers to the presence of a morphological type.

Using this typology, Yampolsky and Yampolsky (1922) surveyed published data for about 10^5 angiosperm species. A summary of their data, along with a definition of terms, is shown in Table 14.2. Perfect-

Table 14.2. Sex type in angiosperms (based on flowering morphology).

Sex type	Proportion of species*
Hermaphrodite (perfect-flowered)	0.72
Monoecious	0.05
Dioecious	0.04
Andromonoecious	0.017
Gynomonoecious	0.028
Hermaphrodite + one other sex type	0.07
Monoecious + males and females	0.036
Other mixtures	~ 0.039

* Survey of 121 492 angiosperm species. Data from Yampolsky & Yampolsky 1922.

Definitions:
Hermaphrodite—male and female elements in the same flower
Monoecious—male and female flowers distinct but on the same plant
Dioecious—male and female flowers on separate plants
Andromonoecious—plant with hermaphrodite and male flowers
Gynomonoecious—plant with hermaphrodite and female flowers

flowered species make up almost three-quarters of the flowering plants. In contrast, dioecy and monoecy are rather uncommon (c. 5% each). However, populations consisting of mixtures of two or more sex types (including dioecy) make up almost a quarter of the species. Bawa (1980) suggested that this typology may not be as revealing as it first seems, even though these results are widely cited. It does not specify anything particular about the *functional gender* (Lloyd 1979b, 1980), the extent to which an individual passes genes to the next generation via pollen or seeds. Bawa (1980) noted that recent research with previously little studied tropical floras have turned up many more instances of dioecy, often functional dioecy with perfect flowers (or in taxonomic groups previously described as having perfect flowers). Most plants are *cosexual* (an individual is both male and female in some way—Table 14.2).

In contrast to the large diversity of sexual types in angiosperms, gymnosperms are fairly simple. As reviewed by Givnish (1980), all are wind pollinated and almost all species are either dioecious or monoecious. In monoecious species the flower types are usually spatially segregated. This suggests that, combined with the simplicity of pollen donation and reception (no animal vectors), an individual may itself be patchy with respect to opportunities for male and female function, with some parts relatively better placed for dispersing pollen. The simplicity of flower types correlates well with the absence of animal pollination.

14.3.2   Why all this diversity in breeding systems?

Breeding systems in higher plants have classically been viewed in terms of the regulation of (or avoidance of) inbreeding, primarily selfing (Lewis 1979). The classical view is one of how the population or species as a whole benefits from such, although more recent authors have emphasized individual or gene level benefits (Lloyd 1979a; Charlesworth & Charlesworth 1979). However, several recent authors have challenged this general postulate; they do not deny a role for avoidance or regulation of inbreeding but suggest that we expand our view to recognize that many things plants do in respect to breeding systems probably represent selection operating on various aspects of male and female reproductive success (Charnov *et al.* 1976; Charnov & Bull 1977; Charnov 1979, 1982a; Lloyd 1979b; Willson 1979; Bawa & Beach 1981). Male reproductive function consists of attraction of pollinators (or pollen donation) and the success of that pollen in contributing genes to seeds (e.g. getting pollen transferred to another individual, having that pollen outcompete other pollen for fertilization of ovules, having those ovules incorporated into seeds and fruits). Pollen success may depend partly upon the number of pollen grains, their size, and their resource for growth once on the stigma. Female reproductive success depends upon receiving pollen, allocating resource to seeds and fruit, and having the fruit disperse and/or escape predators. This view is simplistic, yet it begins to suggest how the degree of success through male or female function depends upon various characters and trade-offs. It also seems reasonable to suggest that different forms of sexuality have meaning with respect to the opportunities for an individual facultatively to alter male against female allocation (sex choice).

The suggestion that we view cosexual plants in terms of ecological opportunities for male versus female reproduction has generated a host of new and interesting questions (papers cited above; see also Chapter 8), but as yet few hard data. In this brief discussion, I will simply touch on two of these questions.

(1) *Does availability of pollen limit seed set, or is seed set related rather to the resources allocated to seeds?* Under what conditions does pollen limit seed set? Sexual selection theory in dioecious species is based on the general assumption that females are not limited in egg production by the availability of sperm but by the female's opportunities to garner resources to make eggs (see section 9.2). Male reproductive success is thus limited by access to females and their eggs, although male control over critical resources (e.g. oviposition sites) may be the form of male competition. This is termed *Bateman's principle* and has been discussed and extended by several recent authors. Bateman (1948) suggested that it should also apply to plants.

To rephrase the question—in what ways do floral and other reproductive characters of cosexual plants evolve with the major selective force being male reproductive success, fitness gains through the donation of pollen (Charnov 1979)? If pollen does not in general limit seed set, opportunities arise for reproductive characters to be shaped primarily by male function (although I note here that seed dispersal, a female function, may well select for aspects of floral display). This sexual selection view has been discussed in some specific cases by Bawa (1980), Bawa and Beach (1981), and Willson (1979). A part of the question involves the realization that some forms of the male gain relation (see Fig. 14.4) result in the instability of hermaphroditism itself and selection for dioecy.

(2) *What does natural selection favour in the allocation of resources to male versus female function?* A useful way to think about this question for outcrossed species is illustrated in Fig. 14.3. The problem illustrated is to consider the reproductive consequences (through both male (m) and female (f) function) for the shifting of resources from all male, through various degrees of hermaphroditism, to all female. The curve is the boundary of all possible types. It is easy to show two things (Charnov 1982a).

(i) A *convex* curve implies that outcrossed hermaphroditism (cosexual) is stable; a *concave* curve favours dioecy. The boundary between a convex and a concave trade-off is the straight line $m + f = 1$ (Fig. 14.3) (see also Chapter 8). The intuition of 'convexity favours hermaphroditism' is that here a pure female (or male) giving resource to male (female) function gets more reproductive success back than it

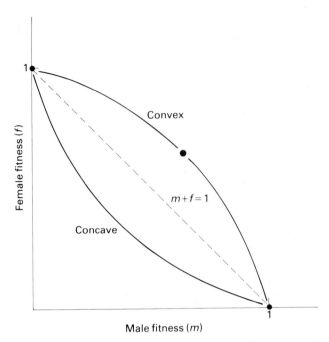

Fig. 14.3 Three possible trade-offs between male (pollen $= m$) and female (seed $= f$) fitness for a cosexual plant. Here, pure males and females (intercepts) are each assigned a fitness of one; the hermaphrodite's fitness through each gender function is scaled relative to the pure sex. For outcrossed hermaphroditism to be stable, the trade-off must be *convex*. For such a situation, the ESS allocation of resources to male versus female reproduction is the point which maximizes the $m \cdot f$ product (dot in the figure). See texts for further discussion of convex versus concave trade-offs.

forfeits. In the most extreme possible case, the hermaphrodite would be almost as good a male as the pure male, and would produce almost as many eggs as a female. To visualize this, imagine pushing the convex curve of Fig. 14.3 outward (intercepts still at $m = 1$, $f = 1$) until the dot is at the coordinates 1, 1.

(ii) The ESS allocation to male versus female function in a cosexual species is that which maximizes the product of the gain through male function (m) times that through female function (f) (dot on the convex curve).

To illustrate these ideas, suppose that $r =$ proportion of reproductive resource given to male function ($1 - r$ to female). Let the seed production be simply proportional to the resource given to seeds (Bateman's principle), or $f = 1 - r$. Let the male reproductive success (m) be represented by $m = r^n$. Thus, pure males and females are assigned a fitness of one, with the hermaphrodite's fitness through each sex function scaled accordingly. Figure 14.4 shows how m alters with r for various assumptions about n. If $n < 1$, male reproductive success shows a *law of diminishing returns*; the first resources put into male function give better reproductive returns than later resources. I invite the reader to plot the relations $m = r^n$ and $f = 1 - r$ in the form of Fig. 14.3. Note that the m, f relation is linear only if $n = 1$ (i.e.

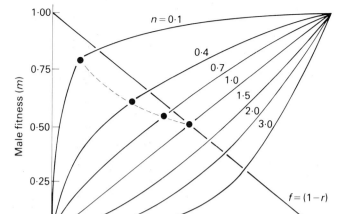

Fig. 14.4 If r is the proportion of resources given to male function, male fitness is given by $m = r^n$. This graph shows the male fitness (the *male gain curve*) for various n. Female fitness is assumed to be proportional to resource input into seeds, or $f = 1 - r$. Hermaphroditism is stable relative to dioecy only if $n < 1$ . . . only if male reproductive success *saturates* or *shows a law of diminishing returns* with resource input. The solid circles show the ESS allocation to male function (r) for four values of n. The faster the saturation of male fitness (i.e. the smaller n), the smaller is the ESS r.

$m + f = 1$); it is convex if $n < 1$ and concave if $n > 1$. For these assumptions, dioecy is favoured if $n > 1$, hermaphroditism if $n < 1$.

The ESS r maximizes $m \cdot f$ or maximizes

$$r^n(1 - r) \qquad (14.1)$$

By elementary calculus, we find the ESS to be $r/(1 - r) = n$. That is, the rate of saturation of male fitness (indexed by n) is the key to the ESS allocation ratio. If m saturates fast ($n \ll 1$), most resource should be given to female function. Note that in general these assumptions imply that greater than half of the resources should be given to female function.

While there are no data bearing on the outcrossed sex allocation predictions of the last section, data do exist to test one such hypothesis. The model just discussed looked at male and female fitness as a hypothetical resource was shifted from all male to all female. Charlesworth and Charlesworth (1981) have extended this model to allow for some degree of self-fertilization. While their derivation is beyond the scope of this book, their final result (provided selfed offspring are anywhere near half as fit as outcrossed offspring) is very simple. If $f = 1 - r$, $m = r^n$ (as before) and if $S =$ fraction of the seeds which are selfed, the ESS $r/(1 - r)$ is

$$\frac{r}{1 - r} \simeq n(1 - S). \qquad (14.2)$$

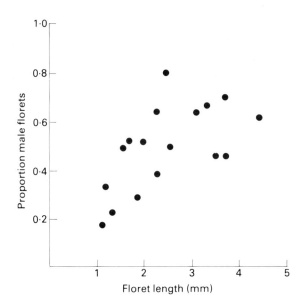

Fig. 14.5 Local mate competition and sex ratio (flower ratio) in monoecious composites of the genus *Cotula*. Each floral head has both male and female florets. Floret length is a rough indicator of the degree of selfing; longer florets meaning more outcrossing (= less selfing). The relative allocation to male function increases with level of outcrossing, for the 17 species studied. (Data from Lloyd 1972b.)

Or, as many previous authors have suggested, selfing downplays opportunities for male fitness gain and selects for more resource to be allocated to female function. Several data sets provide support for this general prediction (review in Charnov 1982a). Fig. 14.5 shows data from Lloyd (1972a, b) for monoecious composites of the genus *Cotula*. The floral heads consist of male and female florets, but the proportion of the florets which are male is positively correlated with the out-crossing rates.

14.3.3 Heterostyly—some ESS calculations

Heterostyly is a reproductive polymorphism whereby a plant population consists of two (distyly) or three (tristyly) morphs, characterized by different style and stamen lengths (Darwin 1877; Fig. 14.6). In distyly the morph with a long style and short stamens is called a pin and the reciprocal a thrum. Botanists have classically viewed this polymorphism as an outcrossing mechanism since in most situations pins and thrums can only cross fertilize.

   In distyly, the pin and thrum morphological characters and incompatibility system are closely linked in a supergene, so that (for example) ss individuals are pin while Ss individuals are thrum. Obligatory cross mating (ss × Ss) will produce equal numbers of each morph. This is not at all necessary, however, since the morph ratio

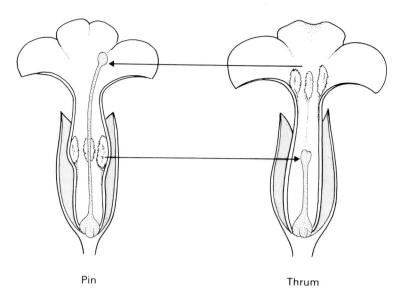

Pin   Thrum

Fig. 14.6 Reciprocal anther and stigma positions in the pin and thrum forms of a distylous plant. The arrows indicate the directions of compatible pollinations. (Modified from Ganders 1979.)

would seem to be easily altered through discrimination either for or against certain pollen (for example, *ss* seed parents accepting mostly *S* pollen) or selective abortion of ovules. The adult morph ratio is often near equality in nature (Ganders 1979). Casper and Charnov (1982) suggested that this is not due to the *ss, Ss* mating system (although this may be a proximate cause) but because equality among the zygotes is the ESS favoured by selection acting on autosomal genes (genes not linked to the *S–s* locus). Suppose that the pin morph produces F_p seeds and M_p pollen grains, while the thrum morph produces F_t and M_t. Consider a rare mutant at an autosomal locus which produces $\hat{q}$ proportion pins among its progeny, while the wild type produces q pins. The fitness (W_t) for the mutant individual in a large population of size N is

$$\text{fitness} = \begin{pmatrix} \text{reproduction} \\ \text{through seeds} \\ \text{of pin progeny} \end{pmatrix} + \begin{pmatrix} \text{reproduction} \\ \text{through seeds} \\ \text{of thrum progeny} \end{pmatrix}$$

$$+ \begin{pmatrix} \text{reproduction} \\ \text{through pollen} \\ \text{of thrum progeny} \end{pmatrix} + \begin{pmatrix} \text{reproduction} \\ \text{through pollen} \\ \text{of pin progeny} \end{pmatrix}$$

or

$$W_t = \hat{q} \cdot F_p + (1 - \hat{q})F_t + \left(\frac{(1 - \hat{q})M_t}{(1 - q)N \cdot M_t} \right)(q \cdot N \cdot F_p)$$

$$+ \left(\frac{\hat{q}M_p}{q \cdot N \cdot M_p} \right)[(1 - q)N \cdot F_t] \quad (14.3)$$

By setting $\partial W_t / \partial \hat{q} = 0$ and solving for q, we find the ESS to be $q^* = 0.5$. Note that this ESS is independent of the values of F and M—thus $q^* = 0.5$ even if pins and thrums suffer differential mortality to adulthood or produce different quantities of seeds or pollen.

While equation 14.3 considers the ESS proportion of pins versus thrums, there are two additional sex allocation problems here, namely the allocation to pollen versus seeds within each of the two morphs. Consider a mutant autosomal gene which when present in a pin alters its resource allocation from (F_p, M_p) to $(\hat{F}_p, \hat{M}_p)$. In a large population with random mating, such a mutant individual will have fitness (W_p):

$$W_p = \hat{F}_p + \left(\frac{\hat{M}_p}{0.5 \cdot N \cdot M_p} \right) (0.5 \cdot N \cdot F_t)$$

or

$$W_p = \hat{F}_p + F_t \cdot \left(\frac{\hat{M}_p}{M_p} \right) \tag{14.4}$$

The analogous fitness for a rare thrum mutant $(\hat{F}_t, \hat{M}_t)$ is, of course,

$$W_t = \hat{F}_t + F_p \cdot \left(\frac{\hat{M}_t}{M_t} \right). \tag{14.5}$$

The ESS (F_t, F_p, M_t, M_p) are values such that the mutants cannot increase W_p, W_t by altering their values ('$\hat{}$' variables) away from the population values. An example is illustrated in Fig. 14.7. I note here

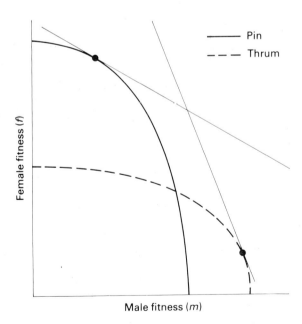

Fig. 14.7 Hypothetical male–female trade-off relations for pins and thrums in a distylous population. I have made the thrums better at being male, as suggested by Beach and Bawa (1980). Also illustrated are the ESS allocations to male versus female function within each morph, as per equations 14.4 and 14.5 (Redrawn from Casper & Charnov 1982.)

two things. First, the formalism is presented in terms of numbers of pollen and seeds; it is easily generalized to the relative abilities to reproduce through pollen and seeds, of which number is only one component. It generalizes, for example, to the abilities of each morph to dispense pollen. Secondly, some shapes (e.g. linear) for the pollen–seed trade-off relations (Fig. 14.7) select for one morph to be male (no F), the other to be female (no M). Dioecy is favoured. Beach and Bawa (1980) reviewed instances where dioecy has apparently evolved from heterostyly. In every case, the thrum (short style) has become the male. They suggest that the initial step towards dioecy is an alteration in the pollination system which results in greater pollen flow from thrums to pins. In terms of Fig. 14.7, the suggestion is that the male/female trade-off within thrums is altered by the change in pollination system.

In summary, if some of the papers being published just now can be used as indicators, the application of sexual selection thinking to plant breeding systems promises to bear much new fruit; as in the case of sex choice, most of the harvest is yet to be.

14.4   CONFLICT OF INTEREST IN PLANT REPRODUCTION

This section briefly discusses some conflict situations for reproduction in higher plants. In specific, I discuss male–female conflict in terms of mate choice (Charnov 1979) and cytoplasmic–nuclear gene conflict over pollen production.

Male reproductive success consists of production and donation of pollen; however, getting one's pollen to conspecific stigmas is not equivalent to male fitness, because fertilization followed by seed production is still to come. The pollen recipient may discriminate for or against certain types of pollen. This choice of pollen is well documented in higher plants, at the level of the stigma or style and (if fertilization occurs) seed development. This again provides the basis for male–female conflict (which pollen is used and which ovules mature?). I suggest that two facets of angiosperm evolution are most easily explicable in terms of this conflict. These are hybridization and double fertilization.

Plants face one situation probably absent in most animals. A male *Drosophila* who displays interest in copulation to a female who turns out to be of the wrong species will almost always be rejected. He is not selected to pursue the attempted copulation, because he probably has other mating options. However, a pollen grain (a male), once on a stigma, has no other options for reproduction. Selection would thus favour the pollen growing, even if the resulting hybrids would be of low vigour. However, the recipient individual would be selected to stop the foreign pollen (except in special situations), particularly if the resulting hybrids would be inviable. Imprecise pollen distribution combined with this male–female conflict may have played a large role in plant evolution. A large percentage of higher plant species are polyploids, and most of these are alloploids—derived from a doubling of chromosome numbers after interspecific (or varietal) hybridization. While botanists have long recognized that mating behaviour of animals limits interspecific crosses and have also stated that mistakes in pollen dispersal favour such crosses, I have found no statement that the crucial difference is the 'no option' situation with regards to male reproductive success.

Suppose that a pollen grain is able to effect fertilization. In angiosperms that will mean double fertilization (Maheshwari 1950). The pollen nucleus divides by mitosis into identical copies; one fertilizes the egg, the other fuses with various numbers of haploid nuclei (usually two, both identical copies of the egg nucleus) to form the endosperm nucleus. The endosperm is a tissue that provides nutrition for its developing zygote (but it may or may not be the food reserve in the mature seed). The nutrients for the endosperm come from the tissues of the maternal plant. The adaptive significance of double fertilization has remained a puzzle, although statements about the advantages of heterozygosity of endosperm and/or the need for endosperm to be an 'aggressive' tissue (competing in some way for nutrients with the surrounding tissue) are common in the literature (Brinck & Cooper 1947). I can suggest two possibilities, both involved with a conflict situation.

(i) The developing zygotes are at best full sibs. The endosperm is in most cases genetically identical (except for a double dose of female-derived genes) to its zygote. Thus, there may well be competition among various endosperms for resources (to be given to their zygotes). The second fertilization clearly raises the relatedness of the endosperm to its own zygote relative to any other zygote and endosperm (except where pollen grains are themselves full sibs). In species in which the female-derived nuclei (destined to be endosperm) are not identical to the egg nucleus, this effect is most marked, thus making it more likely that the endosperm will compete for food for its zygote. This argument derives from the theory of kin selection and treats the endosperm as an individual killing itself to aid a relative (see Chapter 3). In other words, by means of double fertilization the pollen nucleus in the zygote provides itself with a helper.

(ii) There is one other possibility. If abortion of fertilized ovules is common in higher plants, then many pollen grains that fertilize never realize reproductive success. There is thus a conflict between pollen and the maternal plant: it may be advantageous for the mother to sacrifice some fertilized ovules (perhaps on the basis of some assessment of their father's genes). It may even be advantageous for ovules to allow themselves to be killed, because they are sibs to other ovules. However, it is clearly disadvantageous for the pollen nucleus to allow itself to be killed. It may well be that double fertilization is a pollen's way of competing for resources (i.e. becoming part of the food-garnering structure: endosperm) in a situation in which conflict often exists between its interests and those of its mate. Other aspects of mate choice in plants are discussed by Willson and Burley (1983).

Consider one final example of a conflict situation. Consider a nuclear gene in a cosexual species, which causes its bearer to be a female (i.e. to give up pollen production). Such a gene will be selected for (in an outcrossed, cosexual species) if the individual *more than doubles its seed production*. This follows from the simple fact that in a cosexual species each individual produces an average of one seed set and fertilizes one other individual's seed set. To give up male function is to forfeit the seed set the individual would have begotten via pollen. Thus, the female's seed set must at least double to compensate for this loss. However, consider now the same problem from the viewpoint of a gene passed via the maternal cytoplasm (for example mitochondrial DNA). For such a gene, the male (= pollen) pathway does not exist. Here, selection would favour male sterility (= become a female) provided the female simply increased her seed set over the cosexual's. Clearly, there is conflict of interest between nuclear and maternally transmitted DNA over the prospect of being a female. The nuclear DNA requires at least a doubling of seed set, while cytoplasmic DNA requires only a small increase in seed set (Lewis 1941). Indeed, cytoplasmic inheritance of male sterility is not uncommon in plants, and is usually found associated with nuclear genes (called 're-storer genes') which 'combat' the cytoplasmic genes to reinstate pollen production. Such systems are well studied in corn and sunflowers (e.g. Beard 1981), where the cytoplasmic genes are managed for economic gain by the growers.

14.5 CONCLUSION

This brief chapter only hints at the possibilities of (and present work on) evolutionary ecology of plants. While a conclusion is not really possible, I cannot help but suggest that the recurring theme is one of allocation of resources (water, CO_2, light, nutrients) to fitness-enhancing activities. While many interesting questions derive from application of (for example) sexual selection and sex allocation theory to plants, much of the nuts and bolts of the answers lie in the realm of physiological ecology . . . where photosynthesis measurements and measurements of resource flow patterns within a plant are stock-in-trade measurements. The blending of these techniques with behavioural ecology questions is still for the future. However, the simple application of natural selection thinking to photosynthesis itself is quite literally revolutionizing our understanding of this process. Indeed, the means by which plants balance the resource trade-offs necessary for efficient photosynthesis is a case study in adaptation.

Chapter 15
Animal Signals: Mind-Reading and Manipulation

JOHN R. KREBS and RICHARD DAWKINS

Many of the externally visible features of animals, many of their behaviour patterns, many chemical substances and most of the sounds given off by them, are best interpreted as being adapted—'designed by natural selection'—to influence the behaviour of other animals, and are often referred to as 'signals' (some authors reserve 'signals' for morphological features like crests, and use 'displays' for behaviour patterns). Just as a wing performs its normal function by working on the air, so a signal performs its normal function by working on another animal, via its sense organs. With the exception of echo-location sounds, all the sounds produced by special sound-producing organs are signals to other animals. They may attract the other animal, as when a male cricket calls females to his burrow; they may repel the other animal, as in a male robin's territorial song; they may exert some long-term influence on the other animal's physiology, as when the song of a male canary causes his mate's ovaries to ripen over a period of days (Hinde 1970). They may be 'aimed' at the individual's own species, as in the above examples, or at other species, for example the snake-like hiss made by nestlings of hole-nesting birds to scare off predators (Sibley 1955; Krebs 1970). They may, of course, have effects other than those for which natural selection 'designed' them. For example, the song of the cricket *Gryllus integer* is 'designed' to call females, but it also has the (eventually fatal) effect of calling parasitic flies (Cade 1979). The flies are probably 'designed' by natural selection to respond specifically to the cricket song, but most people would not wish to say that the crickets were signalling to the flies. Though definitions vary (Hinde 1972), most authors agree in wanting to exclude such incidental consequences. We have begun with acoustic examples, but signals make use of all sensory modalities.

A dictionary offers two alternative definitions of the word signal. The first is 'any sign, gesture, token etc. that serves to communicate information'; the second is 'anything that acts as an incitement to action' (*Collins English Dictionary*). In the previous edition of this book (Dawkins & Krebs 1978) we gave reasons for, in effect, preferring the second definition over the first (given the proviso that signals are 'designed' for incitement). We went further in the direction that we called 'cynical', and defined a signal as a means by which one animal (the 'actor') exploits another animal's (the 'reactor's') muscle power. In the first part of this chapter we will try to clarify our previous position and extend our discussion from the actor to the reactor. The counterpart to 'manipulation' by the actor is 'mind-reading' by the reactor. The evolution of many animal signals is best seen as an interplay between mind-reading and manipulation. But first we briefly review current understanding of the evolutionary history of animal signals.

15.1 DERIVED ACTIVITIES AND RITUALIZATION

Konrad Lorenz probably did more than anyone to establish the idea that behaviour patterns can be treated like morphological organs, with an evolutionary history that can be traced like the history of morphological organs. Signalling movements, especially in ducks, were among his favourite examples, and indeed, of all behaviour patterns, those concerned with signalling have been most studied from the point of view of evolutionary origins. Some morphological organs, too, have probably evolved solely as a result of selection for signalling function, for instance crests and voice boxes. Sometimes such signalling devices fossilize—the bony resonating chamber in a howler monkey's throat is presumably a good candidate—but generally they do not and their evolution therefore has to be inferred by more indirect means, for instance by using comparative evidence.

One of the main principles of signal evolution could be guessed from common sense even if there were no supporting evidence. This is that signalling movements usually evolve originally from other movements that formerly had no signalling function—the principle of *derived activities* (Tinbergen 1952). Many signalling movements in birds, for instance, can easily be traced back to preening, feather-settling, and temperature-regulating movements (Morris 1956). Others are clearly derived from movements that normally prepare for flight or for drinking. Although signals are presumably derived ultimately from non-signal movements, there are cases in which the ancestral behaviour pattern from which a signal is more immediately derived is another signal, used in a different context. A classic example is the crouching movement used by females of several species of songbird in courtship, which appears to be derived from the food-begging crouch of juveniles.

The reason the principle of derived activities should be expected by common sense is really one of parsimony. It is a special case of a general rule in evolution. The alternative to it is that genetic mutants produced entirely new movements, unrelated to existing movements, and that these were then favoured by natural selection as signals. This is in principle possible, for instance the 'waltzing' mutant behaviour in mice could theoretically evolve into a signal. But behavioural mutations must act by modifying the existing nervous system, and they are most likely to exert some quantitative effect on existing behaviour patterns. More interestingly, there is no particular reason to expect would-be reactors to respond to 'waltzing' or random twitches by actors, whereas they are quite likely already to be in the habit of responding to existing behaviour patterns like preening movements. As we shall see, this is important because unless would-be reactors respond initially to a non-signal movement, there is no reason why it should evolve into a signal.

The process of evolution of signals from non-signal movements is called *ritualization*. Before ritualization, the organ or behaviour pattern concerned performs its normal function, cleaning the feathers, regulating the body temperature, or whatever it is, and it is presumably well designed to do so. It is at some kind of 'utilitarian optimum' for its non-signal function. Before ritualization it is, by definition, not a signal. Although, for reasons to be discussed, it probably has effects on other individuals, these are incidental. After ritualization, the signal has been modified from its old 'utilitarian' optimum, presumably towards a new 'signalling optimum' (see section 15.4). It may, for instance, have become highly repetitive, exaggerated in amplitude, stereotyped in pattern.

Although the word ritualization was originally intended to be applied to behaviour patterns, it is easy to generalize it to structures. For instance, the scent glands that play so large a part in the social life of mammals are clearly 'ritualized' sweat glands and sebaceous glands. Their structure exhibits obvious homology with 'unritualized' glands in the same individuals. In some cases, ritualized organs and behaviour patterns completely lose their original function. For example, in the ritual courtship preening movements of some ducks the bill does not come into contact with the feathers but merely points at them, and so cannot have any cleaning efficacy. In other cases the new function is superimposed on the old. It will usually be hard to decide the issue: who can say, for instance, whether the secretions of a scent gland fulfil some cooling function as they evaporate? The concept of function raises difficult questions of decidability, which we shall not go into here (Hinde 1975). We turn, instead, to a consideration of how natural selection acts on both senders and receivers of signals, and to a justification of our own definition of a signal as a means by which one animal makes use of another animal's muscle power. We shall reconsider the phenomenon of derived activities and ritualization, briefly reviewed above, from the cynical point of view of individuals exploiting other individuals as 'tools'.

15.2 EXPLOITING OTHER ANIMALS AS TOOLS

The world is full of animals whose ancestors succeeded in reproducing. This is trivially true, but non-trivial in its consequences.

Given, in addition, the facts of heredity, we can expect existing animals to have inherited the attributes that made their ancestors successful in becoming ancestors. This is why we feel entitled to regard animals as machines that have been well designed to pass on those attributes—'survival machines'.

Such a well designed machine will tend to use objects in its environment to its own best advantage. These objects will include inanimate ones such as the stone used by a song thrush or by a sea otter to smash mollusc shells. But many of the most important objects in an animal's world will themselves be living bodies with their own nerves and muscles, programmed to work for their own genetic advantage. The possibility of exploitation is there for another living being as it is for a stone. A bolas spider propels its prey towards its mouth by lassoing it and then hauling it in, using the spider's own muscle power. An angler fish propels its prey towards its mouth, not by its own direct muscular force but by waving a tempting lure. Finally, it is the prey fish's own muscles that do the work of propulsion. The angler fish causes this to happen in much the same way as an electrophysiologist might by means of stimulating electrodes. The angler fish has no electrodes, but exerts a similar effect on the prey fish's muscles, via the prey fish's own sense organs. The victims of exploitation do not have to be members of another species as in the case of the angler fish. They might be intended mates, or rivals for a territory. They might be parents exploited by offspring or offspring exploited by parents; hosts exploited by parasites or prey exploited by predators. Whether the manipulated organism is of the same or a different species, or whether it is an inanimate object, makes no difference to the principle.

There are two things that an animal may do if it is to exploit another animal as a tool, and in the previous edition we concentrated on only one of them, manipulation. The other may be called 'mind-reading'. Essentially the difference is that mind-reading involves exploiting the victim's behaviour as it spontaneously emerges from the victim, while manipulation involves actively changing the victim's behaviour. The two may go together or in opposition to each other. One may pave the way for the other in evolutionary time. And both are highly relevant to the study of animal communication.

15.2.1 Manipulation

We are arguing that there is no fundamental difference between the way a living organism might exploit a stone and the way it might exploit another living organism, between the way a male sea otter, say, uses a stone to smash a shellfish, and the way he uses a female sea otter to rear his young. Both are examples of what we are calling manipulation. There is a practical difference between them, resulting from the fact that the female otter is herself a complex machine, while the stone is a simple object. To manipulate the stone an otter need only apply simple muscular forces to it, and the stone will respond according to the ordinary laws of motion. The female otter needs more subtle handling. If pushed, she is apt to depart from the simple laws of motion. She may turn round and bite the male, or she may accelerate away. She may remember the incident, with the result that it changes her behaviour towards the male on subsequent days. Nevertheless, however complex it may be, her behaviour is governed by laws which are ascertainable. This being so, it is in principle possible to exploit those laws in such a way as to manipulate her behaviour. A male sea otter does not have to be consciously aware of these laws, nor does he have to design his techniques to manipulate the female. Natural selection itself will favour male sea otters whose behaviour happens to take advantage of the lawfulness of female behaviour. The effect is that the male manipulates the female in much the same way as he manipulates a stone.

Of course, in any particular relationship between two individuals, it would not be correct to label one the manipulator and the other the victim in any permanent sense. Both individuals will have inherited the manipulating tendencies of their successful ancestors—most of their ancestors may, indeed, be shared—and each will be attempting to manipulate the other. Nevertheless, it is convenient for us to continue to refer to 'the manipulator' and 'the victim', using the words to refer to *roles* which any individual may assume at different times. It should also be said that animals may sometimes benefit by being manipulated, in which case words like 'victim' will seem inappropriate. We shall return to this topic later.

The theoretical possibility of manipulating a living organism is confirmed by the fact that ethologists can do it. By studying the normal triggers of an animal's behaviour, the ethologist can manipulate the behaviour. A direct and potentially powerful way to manipulate behaviour is to interfere with the nervous system itself, or with other aspects of the victim's physiology, either by injecting chemicals, or by making lesions, or by using stimulating electrodes implanted in nervous tissue or in single nerve cells.

Ethologists can achieve equally powerful control over an animal's behaviour by presenting appropriate external stimuli to the animal's sense organs. The classic dummy-presenting studies of Tinbergen and others are the best known examples of this. One of the main conclusions drawn from these studies is that animals respond in mechanical, robot-like fashion to key stimuli. They can usually be 'tricked' into responding to crude dummies that resemble the true, natural stimulus situation only partially, or in superficial respects. To a human observer, for instance, it is surprising that a black-headed gull will show its normal aggressive response to a stuffed gull's head mounted on a stick, with no body (Stout & Brass 1969). A well-known anecdotal example is the aggressive response of Tinbergen's male sticklebacks to the red mail van passing the window. A similar, less well documented anecdote is of the red-coloured *Anolis* lizard that leaped aggressively into a camper's bowl of tomato soup!

Humans are apt to feel superior to sticklebacks aroused to anger by mail vans or to sexual activity by pear-shaped dummies. We think them 'stupid' to be 'fooled' by such crude approximations, since we assume that they, in some sense, 'think' that the mail van really *is* a male stickleback, just because it is red. But a little reflection on our own species helps us to sympathize. A man may be sexually aroused by a picture of a naked woman. A Martian ethologist, observing this, might regard the picture as 'mimicking' the real thing, and assume that the man was 'fooled' into thinking it was a real woman. But nobody who is aroused by such a picture is actually fooled into thinking it *is* the real thing. He knows very well that it is a pattern of printer's ink on paper; it may even be a rather unrealistic caricature; yet it has enough visual stimuli in common with the real thing to have a similar effect on his physiology. We should not ask whether the stickleback 'thinks' the mail van really is a rival, nor whether he is so 'stupid' as to be incapable of distinguishing a mail van from a stickleback. Very probably he can distinguish them very well, but both make him see red! His nervous system is aroused by them to the same emotion, even though it is perfectly capable of seeing the difference between them.

Just as an ethologist can manipulate the behaviour of an animal by stimulating it appropriately, so can another animal. Direct interference with brain tissue is rare, though not unknown: the 'brain worm' *Dicrocoelium dendriticum*, a trematode parasite of sheep and ants, makes a lesion in the ant's brain. This changes the ant's behaviour in such a way that it, and therefore the worm, is more likely to be eaten by the worm's definitive host, a sheep (Wickler 1968). More usually, animals manipulate the behaviour of their victims by stimulating the victim's sense organs. The very fact that animals are susceptible to being 'tricked' by the crude dummies of ethologists, especially supernormal dummies, makes it likely that natural selection will favour similar exploitation by other animals. The most striking and best studied examples of this involve interspecific communication. Bee orchids, for example, present male bees with a supernormal stimulus of the female, and the male collects or transmits pollen while trying to copulate with the flower. 'They have, so to speak, "discovered" the releasing stimuli normally provided by females.' (Proctor & Yeo 1973). These stimuli probably include scent as well as visual cues. So effective is the flowers' stimulus that bees of the genus *Andrena* presented with a choice of real females and flowers of *Ophrys litea* prefer to copulate with the latter!

In our previous article we argued that the evolutionary ritualisation of derived activities can be better understood in terms of selection for effective manipulation than in terms of selection for effective information transfer, a view which for want of a better term we called the 'classical ethological view'. We suggested that ritualized signals are analogous to human advertising signals. In advertising, transfer of information (at least semantic information—see section 15.4.2) is less pertinent than persuasion of the 'victim'. The features that are found to lead to effective advertising include redundancy, rhythmic repetition, bright packaging and supernormal stimuli, features which could be used to characterize a great many signals in the animal world.

It is easy to see that manipulation is a good thing from the manipulator's point of view, provided it can get away with it. But how do manipulation techniques evolve? Why are they victims initially susceptible to manipulation? And why, even if they are initially susceptible, do they remain so in evolutionary time? These are questions which can best be answered after we have considered the other main way in which animals may exploit other animals—'mind-reading'.

15.2.2 Mind-reading

Any animal could benefit if it could behave as if predicting the future behaviour of other animals in its world. At any moment an animal is faced with choosing which of its repertoire of behaviour patterns to perform next: feed, mate, drink, attack, flee, approach, withdraw, etc. (McCleery 1978). The optimal choice will depend on the probable consequences that v ould follow from each choice. For an animal that has any kind of social life, or that is a predator or is preyed upon, these probable consequences will depend crucially on the internal motivational state and probable future behaviour of other animals— rivals, mates, parents, offspring, prey, predators, parasites, hosts. A dog, faced with the choice of approaching or retreating from a rival dog, would do well to take account of any information he can glean as to the mood or motivational state of the rival, and hence, in effect to predict the probable future behaviour of the rival.

Animals can, in principle, forecast the behaviour of other animals, because sequences of animal behaviour follow statistical rules. Ethologists discover the rules systematically by recording long sequences of behaviour and analysing them statistically, for example by transition matrices (e.g. Nelson 1964; Delius 1969), and in the same way an animal can behave as if it is predicting another individual's future behaviour. Without committing ourselves to a view over the philosophical problems of animal mind in the subjective sense (Griffin 1981, 1982), we may use the word 'mind-reading' as a catch-word to describe what we are doing when we use statistical laws to predict what an animal will do next. For an animal, the equivalent of the data-collection and statistical analysis is performed either by natural selection acting on the mind-reader's ancestors over a long period, or by some process of learning during its own lifetime (Lorenz 1966). In both cases, 'experience' of the lawfulness of the behaviour of victims becomes internalized in the brain of the mind-reader. In both cases its mind-reading ability enables it to exploit its victim's behaviour by being 'one jump ahead' of it. The mind-reader is able to optimize its own behavioural choices in the light of the probable future responses of its victim. A dog with its teeth bared is statistically more likely to bite than a dog with its teeth covered. This being a fact, natural selection or learning will shape the behaviour of other dogs in such a way as to take advantage of future probabilities, for example by fleeing from rivals with bared teeth. As with manipulation, mind-reading refers to a *role* that an individual may assume.

Animals will come to be sensitive, then, to the fine clues by which other animals' behaviour may be predicted. The clues that a mind-reader may employ are varied and numerous, and are much discussed in the ethological literature (e.g. Tinbergen 1964, 1952; Morris 1956; Marler 1959; Cullen 1972), albeit often from a less cynical point of view than is implied by our 'mind-reading'. As we shall see, the whole theory of derived activities, discussed above, can be best interpreted in terms of mind-reading as well as manipulation, and the literature on the evolutionary origins from which signals are derived is full of good examples. Any movement of a limb, twitch of a facial muscle, or involuntary catch of the breath is potentially a give-away. Humans use them all the time, and such give-aways as heart rate, breathing patterns, and galvanic skin response are systematically exploited in 'lie-detector' machines. In principle, recording electrodes implanted in the brain should enable even more insidiously accurate mind-reading.

Animals cannot insert electrodes into each other's brains, and, with interesting possible exceptions like electric fish, we know of no evidence that they measure each other's galvanic skin response. But there are plenty of other give-aways. Natural selection will tend to favour animals that become sensitive to available tell-tale clues, however discrete and subtle they may be. A notorious case is Clever Hans, the mathematical horse who 'appeared to be able to do sums in his head and deliver the answer by striking his hoof on the ground the right number of times. It all looked very impressive until it was shown that the horse had learnt a relatively simple trick. He kept on pawing the ground until he received a very small sign from his master that he had got to the right answer whereupon he stopped' (M. Dawkins 1980).

Humphrey (1976) develops the idea of animals as 'Nature's psychologists', and goes so far as to suggest that the whole faculty of subjective consciousness and self-awareness evolved as a device to facilitate reading the minds of others. Whether or not we buy the whole of Humphrey's elegantly argued case on the origins of consciousness, it is entirely reasonable to presume that Clever Hans and all his colleagues among circus animals and household pets are using, in a human context, faculties which their ancestors were selected to use in the wild. Whether it is done consciously or not, 'mind-reading' by means of subtle give-away clues perceived by all the sense organs is an ancient, widespread, and highly developed skill among animals.

15.2.3 Responses to mind-reading

What might an animal whose mind is being read do about it? It is an evolutionary question we are asking, so we should rephrase it. What mechanisms for responding to mind-reading might natural selection build into lineages that are susceptible to being mind-read?

The first thing we need to know is whether the victim suffers or benefits from having its mind accurately read, in other words whether it is really a 'victim' at all, or a willing participant in the process. There may be many occasions on which an animal benefits from having its mood accurately read, and its behaviour accurately predicted. Males of many species are quite likely to attack even females who enter their territory, depending on whether aggression or sexuality dominates their mood. Therefore a female who is willing to mate with a particular male may nevertheless be afraid to approach him (Bastock 1967). The female benefits by reading the male's mind accurately to see whether he is in aggressive mood, and the male benefits by making it easy for her to do so. If the male is a 'victim' of mind-reading here, he is a willing one.

Even in a relationship of seemingly unmitigated enmity, such as the predator–prey relationship, victims of mind-reading may be willing victims. Many predators rely upon surprising their prey, for they cannot outrun them. Once a cat has been seen or otherwise detected by a particular bird, unless it is very close the cat has little chance of catching the bird, which simply takes to the air. To stalk a bird and get close enough to strike is a time-consuming business for a cat, only worthwhile if there is a reasonable chance of success at the end of the stalk. The interesting point is that the bird, too, benefits from making the mind-reading easy for the cat. If he can make the cat give up and slink off, he can continue to feed uninterruptedly on the ground, rather than having to waste time flying off or keeping himself prepared to fly off. It is possible that a number of signals that were once supposed to serve as warnings to conspecifics are in fact aimed at predators (Smythe 1970; Zahavi, quoted in Dawkins 1976; Baker & Parker 1979).

Some of the earlier literature betrays a tacit assumption that co-operation is the norm within species and the exception between species. The theory of natural selection at the genic level gives no obvious grounds for this assumption. In rejecting it in the previous edition of this book, we perhaps gave the misleading impression that cooperation, or an active 'willingness' to be mind-read, was a rarity. We would prefer to say simply that there are no grounds here for distinguishing intraspecific from interspecific relationships. Depending on circumstances, both can be cooperative or the reverse.

What if an animal is an unwilling victim of mind-reading? What countermeasures might it take? Like any victim of spying, it can resort to counter-espionage. Counter-espionage in human warfare or industrial rivalry takes two main forms, concealment and active deception. Concealment consists in making it difficult for the enemy to gain any information at all as to the nation's or the company's true intentions; the equivalent at the individual level is the 'poker face'. Active deception consists in feeding the enemy deliberately misleading information; the equivalent at the individual level is simulating a mood or intention that one does not really have. It is probable that animals do something corresponding to both these forms of counter-espionage (section 15.5.2; Dawkins & Krebs 1978). But there is another way of looking at the countermeasures that a victim of mind-reading might adopt, whether it is a willing or an unwilling victim, and it leads us right back to manipulation and our questions about the origins and evolution of manipulation techniques. The victim of mind-reading might exploit the fact that its mind is being read, in order to manipulate the behaviour of the mind-reader.

We have reached an interim climax in our discussion. Mind-reading and manipulation are not isolated phenomena. They are intimately locked together in evolutionary arms races and feedback loops. Mind-reading is a prerequisite for the evolution of manipulation. Manipulation evolves as an evolutionary response to mind-reading. Mind-reading and manipulation coevolve, and signals are the result of this coevolution. We can use the dog example again, to illustrate how this coevolution might proceed.

A dog, as we have seen, would benefit if he could forecast the probability that a rival will bite him if provoked. It happens to be the case, for fairly obvious practical reasons here, that a dog usually gets its lips out of the way before biting: it bares its teeth. Although initially in evolution this tooth-baring might have been a slight, almost imperceptible movement, it was just detectable by the senses of rivals and so could be used for mind-reading. Now we come to the evolutionary response of 'victims' of the mind-reading. The fact that baring of their teeth has a predictable effect on rivals presents 'victims' of mind-reading with an opportunity to manipulate mind-readers' behaviour. (Again, remember that we are speaking of *roles* not individuals. The same individual may be both mind-reader and manipulator at different times and with different opponents.) Where the mind-reader might be thought to be saying: 'He bares his teeth and, therefore, I prophecy that he will attack', the manipulator can be thought of as saying: 'I bare my teeth and I will make him retreat'. So 'victims' of mind-reading become manipulators.

Mind-reading is not a necessary prerequisite for the evolution of manipulation, but it is probably a common one. When we ask what it is that predisposes an animal to be manipulated, the answer is quite likely to be that its senses are tuned into mind-reading the would-be manipulator, which gives the would-be manipulator a key with which he can unlock the other's nervous system.

15.2.4 Responses to manipulation

We have seen that victims of mind-reading are unlikely to submit passively but will, over evolutionary time, tend to turn the situation to their advantage. The same is surely true of victims of manipulation. What responses or retaliations are open to victims of manipulation?

As in the case of mind-reading, the answer will depend on whether the 'victim' is unwilling, as the word implies, or whether it is, in some sense, a willing victim. Much of the earlier literature tacitly implied the latter, in assuming that signals mediated mutually beneficial cooperation. In the previous edition of this book we perhaps went too far in attempting to redress the balance. There are, of course, many occasions on which both actor and reactor stand to gain from the same outcome. In such cases, even in an obviously cooperative endeavour as the foraging facilitated by the honeybee dance, it is still technically correct to speak, as we did, of the actor using the muscle power of the reactor. But in such cooperative cases it is equally correct to speak of the reactor using the sense organs of the actor. If the reactor's muscles are being used by the actor, they are benefiting the reactor at the same time, and selection would not favour resistance to 'manipulation'.

In other cases, however, it is undoubtedly true that the reactor would benefit from not performing the behaviour which is being urged upon it by the actor. This is obvious for the victims of cuckoos and angler fish, and it is now widely accepted that similar resistance to manipulation is to be expected in some within-species interactions: '. . . selection can act in opposition on the two sexes. Commonly, for a given type of encounter, males will be favoured if they do mate and females if they don't' (Parker 1979). Much the same is true of interactions between parents and offspring (Trivers 1974). In all such cases selection will act simultaneously to increase the power of manipulators *and* to increase resistance to it. '. . . genic selection will foster a skilled salesmanship among the males and an equally well-developed sales-resistance and discrimination among the females' (Williams 1966). Depending upon whether or not the victim of manipulation is a 'willing' victim we can expect to see two kinds of coevolution (see also Markl, in press).

15.2.5 Two kinds of coevolution

Our argument so far may be summarized as follows: the conspicuous, ritualized signals familiar to ethologists are the product of a co-evolutionary race between what we have termed the manipulator and mind-reader roles. This kind of coevolution is to be expected whenever communication is a matter of mutual exploitation rather than cooperation. In our previous article we concluded by suggesting that all signals are products of coevolution between manipulation and sales-resistance, a view which we modified in section 15.2.2 by pointing out that mind-reading by reactors is the other side of the picture. We are now going on to suggest that arms-race coevolution is only part of the story of the evolution of signals, although it is the part which accounts for most of the familiar signals described by ethologists.

The other component of the story is a different kind of coevolution arising from mutual cooperation. In communication, as in any kind of social behaviour, most interactions between individuals are not cooperative and mutually beneficial. It is generally recognized, however, that under certain conditions cooperation may be favoured by selection, the two most important conditions being kinship and reciprocity (see Chapters 3, 12 and 13). Cooperative communication might be favoured because of kinship in hymenopteran colonies, in cooperative breeding groups of birds and mammals, among siblings in a brood, within family flocks of birds, and so on, while reciprocity might play a role in long-term groups such as monkey troops, between the members of a pair, between established territorial neighbours, and whenever there are repeated interactions between individuals.

Can signals arising from cooperation and mutual exploitation be distinguished? We suggest that the two kinds of evolution will lead to different kinds of signals, for the following reason. If the reactor *benefits* from receiving the signal and responding in accord with the actor's interests, instead of heightened sales-resistance leading to exaggeration of the signal during evolution we would expect to see heightened sensitivity to the signal leading to a *reduction* in the amplitude and conspicuousness of the signal. This is because every signal has a cost: it may attract predators, use up time and energy, or reduce the actor's efficiency at doing other things. In the absence of any other consideration, selection on actors should favour a reduction in cost. When signals are cooperative, and reactors are selected to strain their senses to pick up the signal, selection is free (but not entirely free, as we explain in the next section) to favour a reduction in the cost of signalling. In short, the evolution of cooperative signalling should lead not to loud, exaggerated, repetitive, conspicuous signals, but to cost-minimizing conspiratorial whispers. The distinction we are making can be illustrated by an analogy with human communication: contrast the Bible-thumping oratory of a revivalist preacher with the subtle signals, undetected by the rest of the company, between a couple at a dinner party indicating to one another that it is time to go home. The former bears the hallmark of signalling designed for persuasion, the latter of a conspiratorial, cooperative whisper. As we commented earlier, the signals well known to ethologists are probably largely the products of arms-race coevolution: many of the conspiratorial whispers of cooperative signalling may even have not yet been detected.

There may, however, be a constraint on the evolution of conspiratorial whispers, namely the problem of signal detectability. Even a cooperative signal may have to be conspicuous in order for the receiver to detect it.

15.3 SIGNAL DESIGN: DETECTABILITY AND ECONOMICS

In this section we consider the signal design features that might evolve irrespective of whether communication is exploitative or cooperative. Even the conspiratorial whispers of cooperation may be loud, repetitive whispers simply to ensure *detection* by the receiver. However, increasing the detectability of a signal is likely to incur additional costs (energy, risk and so on), so that cooperative signals might evolve to an optimal compromise between detectability and economy. Signals that are a product of arms-race coevolution will not evolve to the same 'engineering' optimum and might be expected instead to be much more costly than detectability considerations alone would lead one to predict.

15.3.1 Signal detection

Wiley (1983) points out that many of the characteristics of ritualized signals can be interpreted in terms of signal detection theory (Green & Swets 1966). In particular he points out that *redundancy, conspicuousness or contrast, small signal repertoires,* and *alerting components* are four common features of ritualized signals which might have evolved to enhance detectability. Detailed studies of bird song have been particularly illuminating in illustrating the role of detectability in signal design. Variation between habitats, both within and between species, in frequency and timing structure of songs has been shown to be correlated with habitat variation in attenuation or degradation of sound (Morton 1975; Nottebohm 1975; Bowman 1979; Hunter & Krebs 1979; Shy 1983; Richards & Wiley 1980). Similarly, there is evidence for variation between habitats in redundancy and repertoire size (Richards & Wiley 1980; Kroodsma 1977) associated with variations in noise level. In noisier habitats, songs tend to contain more repeated elements (redundancy) and in one comparison at least had smaller repertoires. Alerting components have also been identified in bird song (Richards 1981).

15.3.2 Signal economics

A ten-page letter and a two-word telegram ('paper rejected') from a journal editor may convey the same information about the fate of one's latest brainchild, but the telegram is in some sense a more economical way of reducing one's uncertainty. Economy in this case might be measured as actor's time or energy required to generate the signal. The notion of economy seems appropriate for animal signals as well as communications engineers, as it has proved to be in optimization studies of foraging (Chapter 4) and territoriality (Chapter 6).

Pheromonal communication in social insects has already been discussed in economic terms by Wilson (1971). Insect pheromones are usually organic molecules with between 5 and 20 carbon atoms. Wilson suggests that the design reason for this is as follows: with fewer than five atoms, the variety that can be synthesized is too small. Above about 20 the number of distinct molecules increases astronomically to no good purpose yet the energy costs of synthesis go up too. Further, large molecules tend to be less volatile and so travel less far. Thus the observed range 5–20 carbon atoms represents an optimal compromise between variety of distinct signals required and energetic costs of manufacture.

The notion of signal economics may also be put to use in explaining differences in the size of molecule used for different signals. Relatively large molecules tend to be used for sexual attraction and smaller ones as alarm substances. While it is possible that the reason for this is that species specificity (and hence a greater range of possible molecules) is more important for sexual than alarm signals, another interpretation is an economic one, namely that frequently used signals should be the cheapest ones. To understand why, imagine designing a human language with maximal economy for writing. An obvious starting point would be to use the shortest possible codes for the most commonly used words, and then to proceed to longer codes, as the short ones are used up, for less frequently used words. In the same way, if the optimization criterion was to minimize costs of manufacture, frequently used pheromonal signals such as alarm scents should be small molecules, while the less frequently used sexual signals should be allocated the larger, left over, molecules. The relatively simple nature of chemical signals makes them particularly suitable for this kind of analysis, since costs of production can be judged fairly directly from molecule size. Perhaps an example parallel to that of the insect pheromones is the difference in length between the alarm and sexual vocal signals of birds: the former are usually short, the latter often long and complex.

Economy of energetic expenditure is only one form of cost-saving in signal design. Another frequently discussed cost of signalling is the risk of attracting predators (see our Introduction). Among the best known discussions of signal design with respect to predation is Marler's (1955) analysis of the hawk alarm calls of small passerines. The design features of these calls (narrow frequency range, no sharp onset or end) make them hard to locate for human ears and perhaps for avian predators as well (but see Lewis & Coles 1980), although the experimental evidence for this is still equivocal (Shalter 1978; Brown 1982).

15.3.3 Variations in signal design

Signals vary enormously in stereotypy, conspicuousness, and redundancy. The incessant stridulation of a grasshopper is at one end of the spectrum, the subtly variable facial expressions and inter-troop vocal signals of monkeys at the other. Three (mutually compatible) hypotheses can be proposed to account for this variation: (i) There is variation in selection for detectability: signals used over long distances or in noisy channels will have an engineering optimum of greater conspicuousness and redundancy than those used in close encounters or in noise-free environments (Wiley 1983). Variations in repetition frequency and repertoire size of bird sounds referred to earlier are consistent with this line of reasoning. (ii) The benefit of signal transmission varies, and therefore the cost incurred by the actor varies: a male grasshopper calling to attract a female has more at stake than a monkey in a troop squabbling over access to a morsel of food, so the former pays more in signal costs than the latter. (iii) Variations in signal design are related to whether or not signals evolve through a coevolutionary race. As we have already emphasized (section 15.2.5), arms-race signals should evolve greater conspicuousness, repertoire, and redundancy than those used in cooperative communication.

Often the 'coevolution' and 'detectability' accounts of signal design will make similar predictions. Because long-distance (e.g. territorial) signals often are associated with arms-race coevolution while short-distance signals such as those used within an ant colony or a monkey troop will tend to be conspiratorial whispers, both hypotheses predict more extreme development of the four 'signal detection' traits (section 15.3.1) in the former than in the latter. However, it is possible to think of examples where the two accounts differ in their predictions. An example is nestling begging calls. Engineering considerations alone would not predict loud, repetitive nestling begging since the calls are most frequently given when the parent has already arrived at the nest. The problems of detectability are therefore negligible, and, what is more, begging calls are known to attract nest predators (Perrins 1979). Loud, repetitive calls hardly seem to be at an engineering optimum for receiver detection, traded off against the costs of signalling. However, as Trivers (1974) first pointed out, nestling–parent interactions are likely to be characterized by coevolution between persuasion and sales-resistance, involving positive feedback. Loud, incessant begging at the parent's face makes sense within this framework.

As pointed out by Wiley (1983) and Dennet (1983), discussions of whether or not signals transmit information often confuse two uses of the term. Haldane and Spurway (1954), and Wilson (1962) pioneered the use of *information theory* (Shannon & Weaver 1949) to describe animal communication. Information in the 'Shannon' sense means *reduction in uncertainty* of an *observer* about the actor's (broadcast information) or reactor's (transmitted information) behaviour contingent upon a signal. In contrast, discussions such as those of Dawkins and Krebs (1978), Maynard Smith (1982a), Caryl (1979) and Hinde (1981) refer to *semantic* information (Dennet 1983). This is roughly equivalent to the more colloquial meaning of 'information about' something. Animal communication may be about the motivation, age, status, strength and so on of the actor. Measurements of Shannon information do not necessarily reveal anything about semantic information, although they often do. Suppose, for example, that an *observer*'s certainty in predicting the reactor's behaviour goes up from 20% to 80% after the actor has performed a display: one can be sure that Shannon information has been transmitted but one cannot tell whether or not the *reactor* acquired any information about the size, age, etc., of the actor. The recent theoretical discussions about whether or not signals convey information and whether or not this is important in their evolution (Dawkins & Krebs 1978; Hinde 1981) concern only semantic information: by definition signals must transmit Shannon information.

15.4.1   Shannon information

If your newspaper headlines consisted of: 'The sun rose this morning'; 'England is in the northern hemisphere'; 'Yesterday lasted 24 hours'; and similar unsurprising facts, you would probably demand your money back. The reason is that you know it all already: it is not news. The facts are all perfectly true but you do not feel informed by them. This idea that a message, in order to be informative, must be at least somewhat surprising to the receiver, has been used by mathematicians to define information as a precisely measurable commodity (Shannon & Weaver 1949). Although this technical usage of the word information was originally coined for telephone and other engineers, it has been applied on a number of occasions to animal communication.

Mathematically, the information content of a message is measured in terms of the reduction in prior uncertainty caused by the message. Prior uncertainty is measured in terms of probabilities. If the message allows the receiver to decide between two alternatives which had previously been equiprobable, say 'heads' rather than 'tails', or 'boy' rather than 'girl', then one 'bit' of information has been conveyed.

If you pick a card from an ordinary pack, and announce the suit of the card, say 'clubs', the message contains two bits of information. At first sight this is surprising. Since there were four equiprobable alternatives, and the message narrowed uncertainty from four to one why were not four bits of information conveyed? The answer is that it is crucial to the definition of information that it refers to messages which have been recoded in the most economical way possible. The most parsimonious encoding in the card example is first to specify colour (black not red), requiring one bit of information, and then to specify suit (clubs not spades), the second bit of information. The information content is $\log_2 4 = 2$. If the prior probabilities of different behaviours are not equal, a weighted sum of the alternatives is calculated according to the formula:

$$H = \sum p_i \log \left(\frac{1}{p_i} \right)$$

where p_i is the probability of the ith category and H is uncertainty in bits.

Returning to animal communication, we can formally define *transmitted information* as the observer's estimate of H for the receiver before the signal minus H after the signal. Similarly, *broadcast information* is H for the actor before minus H after the signal (Wiley 1983).

In the field of animal communication, information theory was first applied to the bee dance (Haldane & Spurway 1954), and has subsequently been used to describe communication in a range of species and contexts (e.g. Hazlett & Bossert 1965; Wilson 1962). Although it is relatively easy to calculate transmitted information from transition matrices, its quantitative value depends on how the actor's and reactor's behaviours are classified: if, for example, the animals themselves divide up behaviour into more categories than does the observer, H may be underestimated. For this sort of reason it is not straightforward to make interspecific comparisons of Shannon information. In fact, we suggest that the 'economics' side of information theory (section 15.3.2) may be a more useful application to animal signals than its use to quantify the number of 'bits' of information transmitted.

15.4.2 Semantic information

Signals originate because reactors gain some information about the reactor from the signal (p. 386). Some authors have taken the view that the subsequent evolution of signal design is primarily directed by selection pressure on the actor to increase semantic information available to the reactor (Smith 1977; Marler 1959) or to reduce ambiguity of the information (Cullen 1966). In other words, the effectiveness of ritualized signals should be judged by the extent to which they transmit information. Because this view was influential (but not universal

(Hinde 1981)) amongst ethologists of the 1950s and 1960s we referred to it in our earlier article as the 'classical ethological approach', and contrasted it with the viewpoint that was consolidated in the 1970s. Game theoretic (Maynard Smith 1972, 1979; Caryl 1979) and gene selection (Dawkins & Krebs 1978; see Chapter 2) analyses lead to the question of whether actors would ever be selected to increase the efficiency of information transfer in their signals, and we suggested that it might be better to abandon the concept of semantic information altogether in discussions of the ritualization of signals.

In trying to assess whether or not signals do or do not transmit information, and whether or not they ought to on theoretical grounds, it is important to distinguish between three kinds of semantic information: information about *intentions* (what the actor will do next), about *strength, status, size,* or *age* of the actors (*strige* for short), and about the *environment* (see also section 2.5.3). It is also useful to maintain a distinction between whether or not signals in present day populations actually transmit information and whether it is plausible to suppose that they have become ritualized to increase their effectiveness in transmitting information. It will be apparent that most of the evidence discussed below refers primarily to the former problem and only indirectly to the latter.

Intention

It is information about what the actor will do next that poses theoretical problems, for two reasons which refer especially to ritualized contests over resources. First, there is nothing to prevent animals 'lying' about what they will do next, and secondly, for an animal to declare its intention early on in a contest is equivalent to a card player showing his hand to an opponent at the start of the game. It is hard to imagine how selection could favour such behaviour; instead one would expect animals to conceal their eventual intentions until the last possible moment.

The literature on displays performed during contests shows that there are correlations between particular displays and the future behaviour of both actor and reactor, in other words that Shannon information is broadcast and transmitted. This does not necessarily mean, however, that information about intentions is transmitted. When Caryl (1979) re-analysed that data of Stokes (1962) on blue tits Dunham (1966) on grosbeaks, and Andersson (1976) on skuas, he found that displays in these species were in fact rather poor predictors of attack. In blue tits (*Parus caeruleus*), for example, the highest probability of attack following a particular display was only 0.48. Furthermore, the display giving the highest probability at one time of year did not do so at another time, and reactors did not tend to retreat after 'aggressive' displays. In contrast, some displays were good predictors of *retreat*. These do not pose a problem for selection theorists since they presumably save the losing animal from attack once it has decided to surrender. There are two studies of fish displays which also showed that differences between individuals in their displays early in a contest are not good predictors of the eventual outcome (Simpson 1968; Jakobsson *et al.* 1979): in short, the evidence from these studies of birds and fish is largely consistent with the theoretical prediction that signals should not convey the long-term intentions of animals in contests.

Hinde (1981) has criticized Caryl's analysis, arguing that the crucial feature of threat displays is their reflection of motivation conflict and therefore moments of indecision in the actor. 'Threat displays were useful only in moments of indecision: if what an individual would do depended in part on the probable behaviour of the other, threatening by the former might elicit a response from the latter which would precipitate a decision by the initial actor' (Hinde 1981). In other words, displays would not be expected to predict just one activity, say attack, but either attack or something else, say staying put. Hinde goes on to show that Stokes' blue tits do indeed perform 'either a or b' following particular displays, although one has to bear in mind that the greater the number of outcomes included in the analysis, the better the outcomes will correlate with the display, just by chance (Caryl 1982).

While there is no doubt that the 'interactional' view advocated by Hinde is essential for understanding the dynamics of contests (for example, the use of graded threat displays—Dawkins & Krebs 1978), it does not, in Caryl's view, face up to the question of ritualization; he points out that if signals are simply used in moments of indecision to elicit a response from the opponent, there is no reason for them to become ritualized in evolution.

As we have already mentioned, a problem for signals indicating high attack probability is that they are subject to bluff. Andersson (1980) has used this as an evolutionary argument to explain why many species have a variety of different threat signals. He assumes that for each display there is a certain fixed frequency of occasions on which it is followed by attack. If the display is used more often than this, its value as a predictor of attack starts to diminish, so reactors pay less attention to it. Because of this, its frequency of use drops again, and its reliability as a predictor of attack increases. Thus there is a frequency-dependent oscillation of the effectiveness of threat signals, and several different signals could be maintained in equilibrium.

Strige and assessment

Contests often involve assessment. Parker (1974b) coined the term *resource holding potential* (RHP) for the constellation of factors that influence fighting ability; much assessment in contests is assessment

Species	Display/Cue	Reference
Hermit crab (*Clibariarus vitatus*)	Size	Hazlett (1968)
African buffalo (*Syncerus caffer*)	Head-on charge	Sinclair (1977)
Red deer (*Cervus elaphus*)	Roaring tempo	Clutton-Brock & Albon (1979)
Toad (*Bufo bufo*)	Pitch of croak	Davies & Halliday (1978)
Cichlid fish (*Nannacara anomala*)	Mouth wrestling	Jakobsson *et al.* (1979)

of RHP. It seems inevitable that assessment should be based on reliable indicators of RHP, since others could easily be faked (Zahavi 1977b, 1979). Reliable cues are those which are too costly to fake, or which are direct and indirect measures of the factors influencing RHP (size and strength and so on). Some examples are listed in Table 15.1.

Davies & Halliday (1978), for instance, showed that the size of a toad (*Bufo bufo*) is well predicted by the pitch of its croak. They suggest that this is an unfakeable cue—only big toads are physically capable of giving deep croaks—and they showed experimentally that deep croaks are, indeed, more intimidating to toads than high-pitched croaks. It is certainly plausible that selection would favour toads that are intimidated only by genuinely unfakeable advertisements of large size, and that selection would favour the use of such unfakeable advertisements by genuinely large toads. But why do small toads croak at all, since they are, in effect, advertising their small size? Would they not do better to keep silent?

Our answer to this question makes use of the logic, though not the precise mathematics, of ESS theory (Chapter 2). Suppose, in accordance with the last sentence of the previous paragraph, that all toads followed the conditional strategy: 'If larger than a criterion size *s*, croak; if smaller than *s*, keep silent' (the exact value of *s* will, itself, be subject to natural selection). Would this strategy be evolutionarily stable? No, it would not, for the following reason. If a toad croaks, he advertises his exact size; if he keeps silent, he in effect announces that he is smaller than *s*, leaving other toads uncertain exactly how much smaller than *s* he is: in the absence of other good information, they will probably assume that he is close to the average of the set of toads smaller than *s*. It follows that a toad who is only *just* smaller than *s* can improve others' estimate of his size by croaking. Selection will therefore favour a slight reduction in the criterion size *s*. Recursive application of this argument leads to the conclusion that *s* will rapidly decrease under selection until it reaches the size of the smallest toads. In other words, 'always croak, regardless of size' will be the evolutionarily stable strategy.

The argument of the previous paragraph was expressed in terms of the particular example of toads, but it is, of course, general. Something like it may be implicit in Zahavi's (1979) argument that all signals must be 'honest'. He goes further and suggests that the repetitive stereotypy of many displays arises from selection by reactors for a standard performance on the part of actors. The standardization, he argues, allows the reactors better to judge small variations in RHP, in much the same way that a judge of differences in athletic performance depends on all the athletes doing the same task under the same conditions. A prediction of this idea is that the most variable components of displays (those with the largest coefficient of variation) should be the best predictors of RHP.

Zahavi's view may at first sight seem to be quite incompatible with our earlier article in which we emphasized actor manipulation. But as we have already stressed, in actor–reactor coevolution both sides may gain the upper hand. Whether signals are manipulative or reliable, cues may vary from one case to another.

Badges of status

A striking example of an apparently fakeable signal of RHP, for which we coined the term 'badge of status' in our previous article, was described by Rohwer (1977) (see also Chapter 2). He observed that in winter flocks of Harris's sparrow (*Zonotrichia querula*) dominance status at feeding stations is correlated with size of the black bib of feathers below the beak. Although some of the variation in bib size is related to age and sex, even within an age class there is apparently continuous variation related to status. When Rohwer dyed the chins of subordinate birds to enlarge their bibs he observed that they were attacked more often than before by dominant individuals and did not rise in status. However, painting the bib and injecting with testosterone caused subordinates to rise in rank, while hormone treatment alone did not, showing that both a large bib and aggressive behaviour are necessary for a bird to be dominant. The badge alone is not sufficient. These observations still leave open the question of why subordinates do not increase their status by altering both bib size and behaviour, since neither would appear to be very costly. One possibility is that there is simply phenotypic or genotypic variation in ability, but the view favoured by Rohwer and Ewald (1981; see also Rohwer 1982) is that being subordinate in a flock may not after all be a disadvantage. They point out that subordinates are more readily tolerated at feeding sites by dominants and that there may be a frequency-dependent advantage for dominant and subordinate behaviour. It is not yet clear, however, whether this would produce continuous variation in plumage as observed in the Harris's sparrow.

While Rohwer's observations pose a still unsolved problem for
the evolutionary explanation of signals, most badges, such as those
identifying sex, age or species, do not present a comparable difficulty.

401

Animal Signals

Information about the environment

The classical example of communication about the environment is the
dance language of bees mentioned already (von Frisch 1967; Gould
1976). A more recently discussed example is that described by Seyfarth *et al.* (1980b) in vervet monkeys (*Cercopithecus aethiops*). Like
many birds (Marler 1955) and mammals (Sherman 1977) these animals
give alarm calls at the approach or sighting of a predator. What is
intriguing about the vervet monkey, however, is that there are three
different calls for different predators, leopards, eagles and snakes.
Playback of the three calls elicits an appropriate response from other
monkeys in the group: leaping into a tree, scanning the skies and
looking on the ground, in response to the leopard, eagle and snake
calls respectively. The calls apparently transmit information about
particular kinds of predator (see also Dennet 1983).

15.5 SUMMARY

The main points of our argument may be summarized as follows:
(1) The evolution of ritualized signal movements or structures from
their precursors is the product of coevolution between the *roles*. We
have termed these roles 'manipulator' and 'mind-reader'. The manipulator role is selected to alter the behaviour of others to its advantage,
the mind-reader role to anticipate the future behaviour of others.
(2) The consequences of this coevolution depend on whether or not
the signals in question are mutually beneficial. Cooperative communication, in which manipulator and mind-reader roles share a common
interest, should lead to cost-minimizing, muted signals, while non-
cooperative signalling should give rise to conspicuous, repetitive (in
other words 'typical ritualized') signals.
(3) For both types of coevolution, the form of signals is also influenced by environmental constraints on detectability and discrimina-
bility. These may set a lower limit to the degree of muting of
cooperative signals, but in general cooperative signals should evolve
towards an optimal compromise between economy and detectability
while non-cooperative signals should not.
(4) Signals, by definition, transfer information in the technical sense,
reducing the observer's uncertainty about the actor's or reactor's
future behaviour. The extent to which they transmit semantic information about the actor or the environment is less clear cut. Game
theoretic evolutionary arguments suggest that information about long-
term intentions should rarely be transmitted, and that information

about individual quality should be transmitted by uncheatable signals. The literature on these subjects has primarily discussed the dynamics of interactions, for example between two contestants, rather than the evolutionary ritualization of signals.

References

. J exp. Analysis Behav. ⌐

⌐nctional response. Amer.

⌐and guarding by the com-
⌐optera, Halictidae). Insect.

⌐family structured models
theor. Biol. **88**, 743–754.

⌐d, S. (1981) The conflict
between male polygamy and female monogamy: the case of the pied flycatcher Ficedula hypoleuca. Amer. Natur. **117**, 738–753.
Introduction to Part 3

Alatalo R.V., Lundberg A. & Stählbrandt K. (1982) Why do pied flycatcher
females mate with already-mated males? Anim. Behav. **30**, 585–593.
Introduction to Part 3

Alcock J., Jones C.E. & Buchmann S.L. (1977) Male nesting strategies in the
bee Centris pallida Fox (Anthophoridae: Hymenoptera). Amer. Natur.
111, 145–155.
2.3, 2.7.2, 2.8

Alexander R.D. (1974) The evolution of social behavior. A. Rev. Ecol. Syst. **5**,
325–383.
5.1, 5.2.1, 12.3.1, 12.5, 13.6

Alexander R.D. & Sherman P.W. (1977) Local mate competition and parental
investment in social insects. Science **196**, 494–500.
9.3.2, 13.5

Alexander R.D., Hoogland J.L., Howard R., Noonan K.M. & Sherman P.W.
(1979) Sexual dimorphism and breeding systems in pinnipeds, ungulates,
primates and humans. In: Evolutionary Biology and Human Social Behaviour (ed. N.A. Chagnon & W.D. Irons), pp. 402–435. Duxbury Press,
North Scituate, Mass.
1.1

Alexander R. McN. (1982) Optima for Animals. Edward Arnold, London.
4.1, 4.1.1

Allison, A.C. (1954) Notes on sickle-cell polymorphism. Ann. hum. Genet. **19**,
39–57.
3.2.2

403

Altmann S.A. (1974) Baboons, space, time and energy. *Amer. Zool.*, **14**, 221–248.

5.1.1, 5.1.2, 5.2, 5.2.1, 5.3.2

Altmann S.A. & Altmann J. (1970) *Baboon Ecology, African Field Research.* University of Chicago Press, Chicago.

5.1.1

Altmann S.A., Wagner S.S. & Lenington S. (1977) Two models for the evolution of polygyny. *Behav. Ecol. Sociobiol.* **2**, 397–410.

10.3.1

Andersson M. (1976) Social behaviour and communication in the great skua. *Behaviour* **58**, 40–77.

15.4.2

Andersson M. (1978) Optimal foraging area: size and allocation of search effort. *Theor. Popul. Biol.* **13**, 397–409.

6.3.1

Andersson M. (1980) Why are there so many threat displays? *J. theor. Biol.* **86**, 773–781.

15.4.2

Andersson M. (1982a) Sexual selection, natural selection and quality advertisement. *Biol. J. Linn. Soc.* **17**, 375–393.

2.3, 9.2.3, 9.2.4, 9.2.5

Andersson M. (1982b) Female choice selects for extreme tail length in a widowbird. *Nature, Lond.* **299**, 818–820.

6.4, 9.2.4

Andersson M. & Krebs J.R. (1978) On the evolution of hoarding behaviour. *Anim. Behav.* **26**, 707–711.

2.3, 7.3.4

Andersson M. & Wicklund C.G. (1978) Clumping versus spacing out: experiments on nest predation in fieldfares (*Turdus pilaris*). *Anim. Behav.* **26**, 1207–1212.

5.3.2

Anxolabehere D., Goux J.M. & Periquet G. (1982) A bias in estimation of viabilities from competition experiments. *Heredity* **48**, 271–282.

9.5

Appleby M.C. (1982) The consequences and causes of high social rank in red deer stags. *Behaviour* **80**, 259–273.

9.2.3

Arak A. (1982) Male–male competition and mate choice in frogs and toads. Unpublished PhD thesis, University of Cambridge.

9.2.4

Arak P.A. (1983) Sexual selection by male–male competition in natterjack toad choruses. *Nature, Lond.* **306**, 261–262.

6.5.1, 9.2.4

Arnold S.J. (1978) The evolution of a special class of modifiable behaviors in relation to environmental pattern. *Amer. Natur.* **112**, 415–427.

7.2.2

Arnold S.J. (1983) Sexual selection: the interface of theory and empiricism. In: *Mate Choice* (ed. P.P.G. Bateson), pp. 67–107. Cambridge University Press, Cambridge.

9.1, 9.2.2, 10.1

Assem J. van den (1967) Territory in the three-spined stickleback (*Gasterosteus aculeatus*). *Behaviour*, Suppl. **16**, 1–164.

10.3.1

Averhoff W.W. & Richardson R.H. (1974) Pheromonal control of mating patterns in *Drosophila melanogaster*. *Behav. Genet.* **4**, 207–225.
9.5

Axelrod R. & Hamilton W.D. (1981) The evolution of cooperation. *Science* **211**, 1390–1396.
6.5.4, 12.4, 13.7

Bachmann C. & Kummer H. (1980) Male assessment of female choice in hamadryas baboons. *Behav. Ecol. Sociobiol.* **6**, 315–321.
2.5.2

Baker H.G. (1955) Self-compatibility and establishment after 'long-distance' dispersal. *Evolution* **9**, 347–348.
8.3

Baker H.G. (1959) Reproductive methods as factors in speciation in flowering plants. *Cold Spring Harbor Symp. Quant. Biol.* **24**, 177–191.
8.3

Baker M.C. (1982) Vocal dialect recognition and population genetic consequences. *Amer. Natur.* **22**, 561–569.
9.4

Baker M.C., Belcher C.S., Deutsch L.C., Sherman G.L. & Thompson D.B. (1981) Foraging success in junco flocks and the effects of social hierarchy. *Anim. Behav.* **29**, 137–142.
5.2.1

Baker R.R. & Parker G.A. (1979) The evolution of bird colouration. *Phil. Trans. R. Soc. B.* **287**, 63–130.
1.2.2, 15.2.3

Balda R.R. & Bateman G.C. (1971) Flocking and the annual cycle in the pinon jay, *Gymnorhinus cyanocephalus*. *Condor* **73**, 287–302.
12.4

Baldwin J. & Krebs H.A. (1981) The evolution of metabolic cycles. *Nature, Lond.* **291**, 381–382.
4.1

Barash D.P. (1974) An adaptive advantage to winter flocking in the blackcapped chickadee, *Parus atricapillus*. *Ecology* **55**, 674–676.
5.2.2

Barash D.P. (1980) Predictive sociobiology: mate selection in damselfishes and brood defense in white-crowned sparrows. In: *Sociobiology: Beyond Nature/Nurture?* (ed. G.W. Barlow & J. Silverberg), pp. 209–226. Westview Press, Boulder, Colorado.
3.3.2

Barker P., Freeman D.C. & Harper K.T. (1982) Sexual flexibility in *Acer grandidentatum*. *Forest Sci.*
14.2.1

Barnard C.J. (1980) Equilibrium flock size and factors affecting arrival and departure in feeding house sparrows. *Anim. Behav.* **28**, 503–511.
5.1.1

Barnard C.J. & Brown C.A.J. (1981) Prey size selection and competition in the common shrew. *Behav. Ecol. Sociobiol.* **8**, 239–243.
7.4.3

Barnard C.J. & Sibly R.M. (1981) Producers and scroungers: a general model and its applications to captive flocks of house sparrows. *Anim. Behav.* **29**, 543–550.
2.7.2, 5.1.2

Bartz, S.H. & Hölldobler B. (1982) Colony founding in *Myrmecocystus mimicus* Wheeler (Hymenoptera: Formicidae) and the evolution of foundress associations. *Behav. Ecol. Sociobiol.* **10**, 137–147.
5.2.2, 5.3.3, 13.6, 13.7

Bastock M. (1967) *Courtship: a zoological study.* London, Heinemann.
15.2.3

Bateman A.J. (1948) Intra-sexual selection in *Drosophila. Heredity* **2**, 349–368.
14.3.2

Bateson P.P.G. (1979) How do sensitive periods arise and what are they for?
Anim. Behav. **27**, 470–486.
7.3.1, 7.3.2

Bateson P.P.G. (1982) Preferences for cousins in Japanese quail. *Nature, Lond.* **295**, 236–237.
7.3.2, 9.1, 9.3.1

Batra S.W.T. (1964) Behavior of the social bee, *Lasioglossum zephyrum,* within the nest. *Insect. Soc.* **11**, 159–186.
13.7

Baum W.M. (1981) Optimization and the matching law as accounts of instrumental behaviour. *J. exp. analysis behav.* **36**, 386–403.
4.5

Bawa K.S. (1980) Evolution of dioecy in flowering plants. *A. Rev. Ecol. Syst.* **11**, 15–39.
14.3.1, 14.3.2

Bawa K.S. & Beach J.H. (1981) Evolution of sexual systems in flowering plants. *Ann. Mo. Bot. Gard.* **68**, 259–275.
14.3.2

Bawa K.S. & Opler P.A. (1975) Dioecism in tropical forest trees. *Evolution* **29**, 167–179.
8.3.

Bayer R.D. (1982). How important are bird colonies as information centers?
Auk **99**, 31–40.
5.3.1

Baylis J.R. (1981) The evolution of parental care in fishes, with reference to Darwin's rule of male sexual selection. *Env. Biol. Fish.* **6**(2), 223–251.
10.3.2

Beach J.H. & Bawa K.S. (1980) Role of pollinators in the evolution of dioecy from distyly. *Evolution* **34**, 1138–1143.
14.3.3

Beard B.H. (1981) The sunflower crop. *Scient. Am.* **244**, 150–162.
14.4

Bekoff M. & Wells M.C. (1980) The social ecology of coyotes. *Scient. Am.* **242**, 130–151.
12.1

Bell G. (1980) The costs of reproduction and their consequences. *Amer. Natur.* **116**, 45–76
11.2, 11.5

Bell G. (1982) *The Masterpiece of Nature: The Evolution and Genetics of Sexuality.* University of California Press, Berkeley.
14.1

Bell R.H.V. (1969) The use of the herb layer by grazing ungulates in the Serengeti. In: *Animal Populations in Relation to their Food Resources* (ed. A. Watson), pp. 111–128. Blackwell Scientific Publications, Oxford.
1.1

Bellman R.E. (1957) *Dynamic Programming*. Princeton University Press, Princeton, N.J.
4.4

Belovsky G.E. (1978) Diet optimization in a generalist hervibore; the moose. *Theor. Popul. Biol.* **14**, 105–134.
4.2.2

Bendegem J.P. van, Gibo D.L. & Alloway T.M. (1981) Effects of colony division on foundress associations in *Polistes fuscatus* (Hymenoptera: Vespidae) *Can. Entomol.* **113**, 551–556.
13.7

Bengtsson B.O. (1978) Avoid in-breeding: at what cost? *J. theor. Biol.* **73**, 439–444.
8.3

Berry J.F. & Shine R. (1980) Sexual size dimorphism and sexual selection in turtles (Order Testudines). *Oecologia* **44**, 185–191.
1.1

Bertness M.D. (1981) Pattern and plasticity in tropical hermit crab growth and reproduction. *Amer. Natur.* **117**, 754–773.
11.2

Bertram B.C.R. (1975) Social factors influencing reproduction in wild lions. *J. Zool.* **177**, 462–482.
5.2.2, 12.1, 12.4

Bertram B.C.R. (1976) Kin selection in lions and in evolution. In: *Growing Points in Ethology* (ed. P.P.G. Bateson & R.A. Hinde), pp. 281–301. Cambridge University Press, Cambridge.
3.3.3, 3.4.3

Bertram B.C.R. (1978) Living in groups: predators and prey. In: *Behavioural Ecology: An Evolutionary Approach*, 1st edn. (ed. J.R. Krebs & N.B. Davies), pp. 64–96. Blackwell Scientific Publications, Oxford.
5.1, 5.2.1, 5.4, 5.4.3, 12.3.1

Bertram B.C.R. (1980a) Vigilance and group size in ostriches. *Anim. Behav.* **28**, 278–286.
5.3.2

Bertram B.C.R. (1980b) Breeding system and strategies of ostriches. *Proc. XVII Int. Ornith Congr.*, pp. 890–894. Berlin, Germany.
11.3

Berven K.A. (1981) Mate choice in the wood frog, *Rana sylvatica*. *Evolution* **35**, 707–722.
9.2.3

Bibby C.J. & Green R.E. (1980) Foraging behaviour of migrant pied flycatchers (*Ficedula hypoleuca*) on temporary territories. *J. Anim. Ecol.* **49**, 507–521.
6.3.3

Bierzychudek P. (1981) The demography of jack-in-the-pulpit, a forest perennial that changes sex. PhD dissertation, Cornell University, Ithaca, NY.
14.2.2, 14.2.3

Birkhead T.R. & Clarkson K. (1980) Mate selection and precopulatory guarding in *Gammarus pulex*. *Z. Tierpsychol.* **52**, 365–380.
9.4

Bishop D.T. & Cannings C. (1978) A generalised war of attrition. *J. theor. Biol.* **70**, 85–124.
2.5.1

Bishop D.T., Cannings C. & Maynard Smith J. (1978) The war of attrition

with random rewards. *J. theor. Biol.* **74**, 377–388.
2.5.3, 2.8

Bitterman M.E. (1975) The comparative analysis of learning. *Science* **188**, 699–709.
7.3.2

Black R. (1971) Hatching success in the three-spined stickleback, *Gasterosteus aculeatus*, in relation to changes in behaviour during the parental phase. *Anim. Behav.* **19**, 532–541.
6.3.1

Blaustein A.R. & O'Hara R.K. (1981) Genetic control for sibling recognition. *Nature, Lond.* **290**, 246–248.
9.3.1

Blaxter K.L. (1971) The comparative biology of lactation. In: *Lactation* (ed. I.R. Falconer), pp. 51–69. Butterworths, London.
1.2.2, 1.2.4

Blumer L.S. (1979) Male parental care in the bony fishes. *Q. Rev. Biol.* **54**, 149–161.
10.3.2

Bobisud L.I. & Portratz C.J. (1976) One-trial versus multi-trial learning for a predator encountering a model–mimic system. *Amer. Natur.* **110**, 121–128.
7.2.2

Boggs C.L. (1981) Selection pressures affecting male nutrient investment at mating in Heliconine butterflies. *Evolution* **35**, 931–940.
9.2.4

Boggs C.L. & Gilbert L.E. (1979) Male contribution to egg production in butterflies: evidence for transfer of nutrients at mating. *Science* **206**, 83–84.
9.2.4

Boggs C.L. & Watt W.B. (1981) Population structure of pierid butterflies IV. Genetic and physiological investment in offspring by male *Colias. Oecologia* **50**, 320–324.
9.2.4

Bolles R.C. (1970) Species-specific defense reactions and avoidance learning. *Psychol. Rev.* **77**, 32–48.
7.2.1

Bolles R.C. (1980) Some functionalistic thoughts about regulation. In: *Analysis of Motivational Processes* (ed. F.M. Toates & T.R. Halliday), pp. 63–75. Academic Press, London.
4.5.1

Bonner J.T. (1965) *Size and Cycle: An Essay on the Structure of Biology.* Princeton University Press, Princeton.
1.2.3

Borgia G. (1979) Sexual selection and the evolution of mating systems. In: *Sexual Selection and Reproductive Competition in Insects* (ed. M.S. Blum & N.A. Blum), pp. 19–80. Academic Press, New York.
10.3.3

Bossema I. (1979) Jays and oaks: An ecoethological study of a symbiosis. *Behaviour* **70**, 1–117.
7.3.4

Boswell M.T., Ord J.K. & Patil G.P. (1979) Chance mechanisms underlying univariate distributions. In: *Statistical Distributions in Ecological Work*

(ed. J.K. Ord, G.P. Patil & C. Taillie). International Co-operative Publishing House, Fairland, Maryland.
5.1.1

Bowen W.D. (1981) Variation in coyote social organisation: the influence of prey size. *Can. J. Zool.* **59**, 639–652.
12.1

Bowman R.I. (1979) Adaptive morphology of song dialects in Darwin's finches. *J. Orn. Lpz.* **120**, 353–390.
15.3.1

Bradbury J.W. (1977) Lek mating behavior in the hammer-headed bat. *Z. Tierpsychol.* **45**, 225–255.
10.3.3

Bradbury J.W. (1981) The evolution of leks. In: *Natural Selection and Social Behavior: Recent Research and New Theory* (ed. R.D. Alexander & D.W. Tinkle), pp. 138–169. Chiron Press, New York.
10.3.3

Bradbury J.W. & Gibson R. (1980) Leks and mate choice. In: *Mate Choice* (ed. P.P.G. Bateson), pp. 109–138. Cambridge University Press, Cambridge.
10.3.3

Breder C.N. & Rosen D.E. (1966) *Modes of Reproduction in Fishes*. Natural History Press, New York.
8.4

Breed M.D. & Gamboa G.J. (1977) Behavioral control of workers by queens in primitively eusocial bees. *Science* **195**, 694–696.
13.7

Brian M.V. (1979) Caste differentiation and division of labor. In: *Social Insects* (ed. H.R. Hermann), Vol. I, pp.121–222. Academic Press, New York.
13.2

Brian M.V. (1980) Social control over sex and caste in bees, wasps and ants. *Biol. Rev.* **55**, 379–415.
13.6, 13.7

Brian M.V., Jones R.M. & Wardlaw J.C. (1981) Quantitative aspects of queen control over reproduction in the ant *Myrmica*. *Insect. Soc.* **28**, 191–207.
13.6, 13.7

Brinck R.A. & Cooper D.C. (1947) The endosperm in seed development. *Bot. Rev.* **13**, 423–541.
14.4

Brockmann H.J. & Dawkins R. (1979) Joint nesting in a digger wasp as an evolutionarily stable preadaptation to social life. *Behaviour* **71**, 203–245.
2.8, 13.4

Brockmann H.J., Grafen A. & Dawkins R. (1979) Evolutionarily stable nesting strategy in a digger wasp. *J. theor. Biol.* **77**, 473–496.
2.3, 2.8, 3.2.1, 11.3

Brothers D.J. (1974) Phylogeny and classification of the aculeate Hymenoptera, with special reference to Mutillidae. *Univ. Kansas. Sci. Bull.* **50**, 483–648.
13.3

Brothers D.J. & Michener C.D. (1974) Interactions in colonies of primitively social bees. III. Ethometry of division of labor in *Lasioglossum zephyrum* (Hymenoptera: Halictidae). *J. Comp. Physiol.* **90**, 129–168.
13.7

Brower J. van Z. (1960) Experimental studies of mimicry. IV. The reactions of starlings to different proportions of models and mimics. *Amer. Natur.* **94**, 271–282.
7.2.2

Brown C.H. (1982) Ventriloquial and locatable vocalizations in birds. *Z. Tierpsychol.* **59**, 338–350.
15.3.2

Brown J.L. (1963) Social organization and behavior of the Mexican jay. *Condor* **65**, 126–153.
12.1

Brown J.L. (1964) The evolution of diversity in avian territorial systems. *Wilson Bull.* **76**, 160–169.
5.2.1, 6.1, 10.0, 10.3.1

Brown J.L. (1969) Territorial behavior and population regulation in birds: A review and re-evaluation. *Wilson Bull.* **81**, 293–329.
5.2.1, 5.3.3, 5.4.3, 12.3.3

Brown J.L. (1974) Alternate routes to sociality in jays with a theory for the evolution of altruism and communal breeding. *Amer. Zool.* **14**, 63–80.
5.1.1, 5.3.3, 12.3.2, 12.3.3, 12.4

Brown J.L. (1975) *The Evolution of Behavior.* W.W. Norton and Co. Inc., New York.
3.3.3, 5.1

Brown J.L. (1978) Avian communal breeding systems. *A. Rev. Ecol. Syst.* **9**, 123–155.
5.3.1, 5.3.3, 12.3.2, 12.4

Brown J.L. (1982) Optimal group size in territorial animals. *J. theor. Biol.* **95**, 793–810.
5.2.1, 5.4, 5.4.2, 5.4.3, 6.5.1, 6.5.3, 6.5.4

Brown J.L. & Brown E.R. (1980) Reciprocal aid-giving in a communal bird. *Z. Tierpsychol.* **53**, 313–324.
12.4

Brown J.L. & Brown E.R. (1981a) Kin selection and individual fitness in babblers. In: *Natural Selection and Social Behavior: Recent Results and New Theory* (ed. R.D. Alexander & D.W. Tinkle), pp. 244–256. Chiron Press, New York.
5.1.1, 5.3.1, 5.3.3, 12.4

Brown J.L. & Brown E.R. (1981b) Extended family system in a communal bird. *Science* **211**, 959–960.
12.1

Brown J.L. & Orians G.H. (1970) Spacing patterns in mobile animals. *A. Rev. Ecol. Syst.* **1**, 239–262.
5.2.1

Brown J.L., Dow D.D., Brown E.R. & Brown S.D. (1978) Effects of helpers on feeding of nestlings in the grey-crowned babbler (*Pomatostomus temporalis*). *Behav. Ecol. Sociobiol.* **4**, 43–59.
3.3.3, 5.3.1, 12.2

Brown J.L., Brown E.R., Brown S.D. & Dow D.D. (1982) Helpers: effects of experimental removal on reproductive success. *Science* **215**, 421–422.
3.3.3, 12.1, 12.2.

Bruning D.F. (1974) Social structure and reproductive behaviour in the greater rhea. *Living Bird* **13**, 251–294.
10.3.2

Bryant D.M. (1979) Reproductive costs in the house martin (*Delichon urbica*). *J. Anim. Ecol.* **48**, 655–676.
11.5

Bryant E.H. (1979) Inbreeding and heterogametic mating: an alternative to Aveshoff and Richardson. *Behav. Genet.* **9**, 249–256.
9.2.4

Bryant E.H., Kence A. & Kimball K.T. (1980) A rare-male advantage in the housefly induced by wing clipping and some general considerations for *Drosophila*. *Genetics* **96**, 975–993.
9.5

Bucher T.L., Ryan M.J. & Bartholomew G.A. (1981) Oxygen consumption during resting, calling, and nest building in the frog, *Physalaemus pustulosus*. *Physiol. Zool.* **55**, 10–22.
10.3.3

Buckle G.R. & Greenberg L. (1981) Nestmate recognition in sweat bees (*Lasioglossum zephyrum*): does an individual recognize its own odour or only odours of its nestmates? *Anim. Behav.* **29**, 802–809.
13.5

Buckley P.A. & Buckley F.G. (1972) Individual egg and chick recognition by adult royal terns (*Sterna maxima maxima*). *Anim. Behav.* **20**, 457–462.
7.3.3

Buechner H.K. & Roth D.H. (1974) The lek system in the Uganda kob antelope. *Amer. Zool.* **14**, 145–162.
9.2.4

Buechner H.K. & Schloeth R. (1965) Ceremonial mating behaviour in Uganda kob (*Adenota kob thomasi* Neumann). *Z. Tierpsychol.* **22**, 209–225.
9.2.4, 10.3.3

Bull J.J. (1980) Sex determination in reptiles. *Q. Rev. Biol.* **55**, 3–21.
8.2

Bull J.J. (1981) Coevolution of haplo-diploidy and sex determination in the Hymenoptera. *Evolution* **35**, 568–580.
13.5

Bulmer M.G. & Taylor P.D. (1980) Sex ratio under the haystack model. *J. theor. Biol.* **86**, 83–89.
3.4.3

Burger J. (1981) Super territories: a comment. *Amer. Natur.* **118**, 578–580.
6.4

Buschinger A. (1978) Genetisch bedingte Entstehung geflugelter Weibchen bei der sklavenhaltenden Ameise *Harpagoxenus sublaevis* (Nyl.) (Hym., Form.) *Insect. Soc.* **25**, 163–172.
13.2

Bush G.L. (1974) The mechanism of sympatric host race formation in the true fruit flies (Tephritidae). In: *Genetic Mechanisms of Speciation in Insects* (ed. M.J.D. White), pp. 3–23. Australian and New Zealand Book Company, Sydney.
9.6

Buskirk R.E. (1981) Sociality in the Arachnida. In: *Social Insects* (ed. H.R. Hermann), Vol. II, pp. 281–367. Academic Press, New York.
13.2, 13.6

Bygott J.D., Bertram B.C.R. & Hanby J.P. (1979) Male lions in large coalitions gain reproductive advantages. *Nature, Lond.* **282**, 839–841.
12.1, 12.4

Cade W. (1979) The evolution of alternative male reproductive strategies in field crickets. In: *Sexual Selection and Reproductive Competition in Insects* (ed. M.S. Blum & N.A. Blum), pp. 343–379. Academic Press, New York.
9.2.5, 15.0

Cade W. (1980) Alternative male reproductive behaviours. *Fla. Ent.* **63**, 30–45.
9.2.5

Camenzind F.J. (1978) Behavioral ecology of coyotes (*Canis latrans*) on the National Elk Refuge, Jackson, Wyoming. Unpublished PhD dissertation.
12.1, 12.3

Campanella P.J. & Wolf L.L. (1974) Temporal leks as a mating system in a temperate zone dragonfly (Odonata: Anisoptera) I. *Plathemis lydia.* (Drury). *Behaviour* **51**, 49–87.
5.3.3, 6.5.1

Caraco T. (1979a) Time budgeting and group size: A test of theory. *Ecology* **60**, 618–627.
5.1.1, 5.2.1, 5.3.2, 5.4.2

Caraco T. (1979b) Time budgeting and group size: A theory. *Ecology* **60**, 611–617.
5.1.2, 5.2.2, 5.4, 5.4.2

Caraco T. (1980) Stochastic dynamics of avian foraging flocks. *Amer. Natur.* **115**, 262–275.
5.1.1, 10.0

Caraco T. (1981a) Energy budgets, risk and foraging preferences in dark-eyed juncos (*Junco hymelais*). *Behav. Ecol. Sociobiol.* **8**, 213–217.
4.2.3

Caraco T. (1981b) Risk-sensitivity and foraging groups. *Ecology* **62**, 527–531.
5.2.1, 5.3.1

Caraco T. (1982) Aspects of risk-aversion in foraging white-crowned sparrows. *Anim. Behav.* **30**, 719–727.
5.3.1

Caraco T. & Pulliam H.R. (1980) Time budgets and flocking dynamics. *Proc. XVII Int. Ornith. Congr.* Berlin, Germany.
5.1.1, 5.4.2

Caraco T. & Wolf L.L. (1975) Ecological determinants of group sizes of foraging lions. *Amer. Natur.* **109**, 343–352.
5.3, 5.3.1, 5.4.1

Caraco T., Martindale S. & Whitham T.S. (1980a) An empirical demonstration of risk-sensitive foraging preferences. *Anim. Behav.* **28**, 820–830.
4.2.3, 5.3.2, 5.4.2

Caraco T., Martindale S. & Pulliam H.R. (1980b) Avian flocking in the presence of a predator. *Nature, Lond.* **285**, 400–401.
4.3, 5.3.1, 5.3.2

Caraco T., Martindale S. & Pulliam H.R. (1980c) Avian time budgets and distance to cover. *Auk* **97**, 872–875.
5.3.2

Carey M.D. & Nolan V. (1975) Polygyny in indigo buntings: a hypothesis tested. *Science* **190**, 1296–1297.
10.3.1

Carlson A. & Moreno J. (1982) The loading effect in central place foraging wheatears *Oenanthe oenanthe* L. *Behav. Ecol. Sociobiol.* **11**, 173–184.
4.2.1

Carpenter F.L. & MacMillen R.E. (1976) Threshold model of feeding territoriality and test with a Hawaiian honeycreeper. *Science* **194**, 639–642.
6.3.1

Carson H.L. (1982) Evolution of *Drosophila* on the newer Hawaiian volcanoes. *Heredity* **48**, 3–25.
9.6

Caryl P.G. (1979) Communication by agonistic displays: what can games theory contribute to ethology? *Behaviour* **68**, 136–169.
2.5.3, 15.4, 15.4.2

Caryl P.G. (1982) Animal signals: a reply to Hinde. *Anim. Behav.* **30**, 240–244.
15.4.2

Casper B.B. & Charnov E.L. (1982) Sex allocation in heterostylous plants. *J. theor. Biol.* **96**, 143–149.
14.3.3

Caswell H. (1981) The evolution of 'mixed' life histories in marine invertebrates and elsewhere. *Amer. Natur.* **117**, 529–536.
11.5

Chapin F.S. (1980) The mineral nutrition of wild plants. *A. Rev. Ecol. Syst.* **11**, 233–260.
14.1

Charlesworth B. (1978) Some models of the evolution of altruistic behaviour between siblings. *J. theor. Biol.* **72**, 297–319.
3.3.4, 13.6

Charlesworth B. (1980) Models of kin selection. In: *Evolution of Social Behaviour: Hypotheses and Empirical Tests* (ed. H. Markl). Verlag Chemie, Weinheim.
3.3.3, 3.3.4

Charlesworth B. & Charnov E.L. (1980) Kin selection in age-structured populations. *J. theor. Biol.* **88**, 103–119.
11.5

Charlesworth D. & Charlesworth B. (1979) The evolutionary genetics of sexual systems in flowering plants. *Proc. R. Soc. Lond. B.* **205**, 513–530.
14.3.2

Charlesworth D. & Charlesworth B. (1981) Allocation of resources to male and female functions in hermaphrodites. *Biol. J. Linn. Soc.* **15**, 57–74.
14.3.2

Charnov E.L. (1976) Optimal foraging: the marginal value theorem. *Theor. Popul. Biol.* **9**, 129–136.
4.2.1

Charnov E.L. (1976) Optimal foraging: attack strategy of a mantid. *Amer. Natur.* **110**, 141–151.
7.4.3

Charnov E.L. (1977) An elementary treatment of the genetical theory of kin selection. *J. theor. Biol.* **66**, 541–550.
3.3.1

Charnov E.L. (1978) Evolution of eusocial behavior: offspring choice or parental parasitism? *J. theor. Biol.* **75**, 451–456.
13.5, 13.6

Charnov E.L. (1979) Simultaneous hermaphroditism and sexual selection.

Proc. natl Acad. Sci., USA **76**, 2480–2484.
14.3.2, 14.4

Charnov E.L. (1981) Kin selection and helpers at the nest: effects of paternity and biparental care. *Anim. Behav.* **29**, 631–632.
12.2, 12.4

Charnov E.L. (1982a) *The Theory of Sex Allocation.* Princeton University Press, Princeton.
8.2, 8.3, 11.3, 13.5, 14.2.2, 14.3.2

Charnov E.L. (1982b) Parent–offspring conflict over reproductive effort. *Amer. Natur.* **119**, 736–737.
11.5

Charnov E.L. & Bull J.J. (1977) When is sex environmentally determined? *Nature, Lond.* **266**, 828–830.
8.2, 14.2.2, 14.3.2

Charnov E.L. & Krebs J.R. (1975) The evolution of alarm calls: altruism or manipulation? *Amer. Natur.* **109**, 107–112.
5.3.2

Charnov E.L. & Schaffer W.M. (1973) Life history consequences of natural selection: Cole's result revisited. *Amer. Natur.* **107**, 791–793.
11.2

Charnov E.L., Maynard Smith J. & Bull J.J. (1976) Why be a hermaphrodite? *Nature, Lond.* **263**, 125–126.
8.3, 14.3.2

Charnov E.L., Orians G.H. & Hyatt K. (1976) The ecological implications of resource depression. *Amer. Natur.* **110**, 247–259.
4.2.1

Cheng K.M., Shoffner R.N., Phillips R.E. & Lee F.B. (1978) Mate preference in wild and domesticated (game-farm) mallards (*Anas platyrhynchos*). I. Initial preference. *Anim. Behav.* **26**, 996–1003.
9.4

Cheverton J.R. (1983) *Which flower next? Foraging decisions of bumblebees.* DPhil thesis, Oxford University.
7.4.2

Clark A.B. (1978) Sex ratio and local resource competition in a prosimian primate. *Science* **201**, 163–165.
8.2

Clark C.J. (1976) *Mathematical Bioeconomics.* John Wiley & Sons, New York.
11.5

Clarke B. & O'Donald P. (1964) Frequency-dependent selection. *Heredity* **19**, 201–206.
2.1.2

Cleveland L.R., Hall S.R., Sanders E.P. & Collier J. (1934) The wood-feeding roach *Cryptocercus*, its Protozoa, and the symbiosis between Protozoa and roach. *Mem. Am. Acad. Arts Sci.* **17**, 185–342.
13.4

Clutton-Brock T.H. (1974) Primate social organisation and ecology. *Nature, Lond.* **250**, 539–542.
1.1, 1.2.1

Clutton-Brock T.H. (1975) Feeding behaviour of red colobus and black and white colobus in East Africa. *Folia primatol.* **23**, 165–207.
5.2.1

Clutton-Brock T.H. (1983) Selection in relation to sex. In: *From Molecules to*

Men (ed. D.S. Dendall on behalf of Darwin College, Cambridge), pp. 457–481. Cambridge University Press, Cambridge.
9.2.1

Clutton-Brock T.H. & Albon S.D. (1979) The roaring of red deer and the evolution of honest advertisement. *Behaviour* **69**, 145–170.
2.5.2, 15.4.2

Clutton-Brock T.H. & Albon S.D. (1982) Parental investment in male and female offspring in mammals. In: *Current Problems in Sociobiology* (ed. King's College Sociobiology Group), pp. 223–248. Cambridge University Press, Cambridge.
1.1

Clutton-Brock T.H. & Harvey P.H. (1976) Evolutionary rules and primate societies. In: *Growing Points in Ethology* (ed. P.P.G. Bateson & R.A. Hinde), pp. 195–237. Cambridge University Press, Cambridge.
1.1, 9.2.5

Clutton-Brock T.H. & Harvey P.H. (1977) Primate ecology and social organisation. *J. Zool.* **183**, 1–39.
1.1, 1.2.2, 1.2.5, 1.2.6, 5.1.2, 5.2

Clutton-Brock T.H. & Harvey P.H. (1979a) Comparison and adaptation. *Proc. R. Soc. Lond. B.* **205**, 547–565.
1.2.3, 1.2.5

Clutton-Brock T.H. & Harvey P.H. (1979b) Home range, population density and phylogeny in primates. In: *Primate Ecology and Human Origins* (ed. I.S. Bernstein & E.O. Smith), pp. 201–214. Garland STPM Press, New York.
1.1, 1.2.3, 1.3

Clutton-Brock T.H., Harvey P.H. & Rudder B. (1977) Sexual dimorphism, socionomic sex ratio and body weight in primates. *Nature, Lond.* **269**, 797–800.
1.1

Clutton-Brock T.H., Albon S.D., Gibson R.M. & Guinness F.E. (1979) The logical stag: adaptive aspects of fighting in red deer (*Cervus elaphus* L.). *Anim. Behav.* **27**, 211–225.
2.8, 9.2.5

Clutton-Brock T.H., Albon S.D. & Harvey P.H. (1980) Antlers, body size and breeding group size in the Cervidae. *Nature, Lond.* **285**, 565–567.
1.1

Clutton-Brock T.H., Guinness F.E. & Albon S.D. (1982) *Red Deer: the Behavior and Ecology of Two Sexes*. Chicago University Press, Chicago.
1.1, 4.1.1

Cody M.L. & Cody C.B.J. (1972) Territory size, clutch size and food in populations of wrens. *Condor* **74**, 473–477.
6.3.4

Cohen D. & Eshel I. (1976) On the founder effect and the evolution of altruistic traits. *Theor. Popul. Biol.* **10**, 276–302.
3.4.3

Cohen J.E. (1969) Natural primate troops and a stochastic population model. *Amer. Natur.* **103**, 455–477.
5.1.1

Cohen J.E. (1971) *Casual Groups of Monkeys and Men: Stochastic Models of Elemental Social Systems*. Oxford University Press, London.
5.1.1

Cohen J.E. (1972) Markov population processes as models of primate social and population dynamics. *Theor. Popul. Biol.* **3**, 119–134.
5.1.1

Cole B.J. (1981) Dominance hierarchies in *Leptothorax* ants. *Science* **212**, 83–84.
13.7

Cole L.C. (1954) The population consequences of life history phenomena. *Q. Rev. Biol.* **29**, 103–137.
11.2

Colgan P. (1979) Is a super-territory strategy stable? *Amer. Natur.* **114**, 604–605.
6.4

Collier G.H. & Rovee-Collier C.K. (1981) A comparative analysis of optimal foraging behavior: Laboratory simulations. In: *Foraging Behavior: Ecological, Ethological, and Psychological Approaches* (ed. A.C. Kamil & T.D. Sargent), pp. 39–76. Garland STPM Press, New York.
7.4.3

Colwell R.K. (1974) Predictability, constancy, and contingency of periodic phenomena. *Ecology* **55**, 1148–1153.
5.2.1

Colwell R.K. (1981) Group selection is implicated in the evolution of female-biased sex ratios. *Nature, Lond.* **290**, 401–404.
3.4.3

Comins H.N., Hamilton W.D. & May R.M. (1980) Evolutionarily stable dispersal strategies. *J. theor. Biol.* **82**, 205–230.
11.4

Constantz, G.D. (1975) Behavioural ecology of mating in the male Gila topminnow, *Poeciliopsis occidentalis* (Cyprinodentiformes: Poeciliidae). *Ecology* **56**, 966–973.
6.5.1

Cook R.M. & Cockrell B.J. (1978) Predator ingestion rate and its bearing on feeding time and the theory of optimal diets. *J. Anim. Ecol.* **47**, 529–547.
4.2.1

Cook R.M. & Hubbard S.F. (1977) Adaptive strategies in insect parasites. *J. Anim. Ecol.* **46**, 115–125.
4.5

Cooke F. (1978) Early learning and its effect on population structure. Studies of a wild population of snow geese. *Z. Tierpsychol.* **46**, 344–358.
9.4

Cooke F., Finney G.H. & Rockwell R.F. (1976) Assortative mating in lesser snow geese (*Anser caerulescens*). *Behav. Genet.* **6**, 127–140.
9.4

Councilman J.J. (1977) A comparison of two populations of the grey-crowned babbler. *Bird Behavior* **1**, 43–82.
12.1, 12.2, 12.3

Cowan D.P. (1979) Sibling matings in a hunting wasp: adaptive inbreeding? *Science* **205**, 1403–1405.
9.3.2

Cowie R.J. (1977) Optimal foraging in great tits *Parus major*. *Nature, Lond.* **268**, 137–139.
4.2.1, 7.4.4, 7.5.2

Cowie R.J. & Krebs J.R. (1979) Optimal foraging in patchy environments. In:

Population Dynamics (ed. R.M. Anderson, B.D. Turner & R.L. Taylor), pp. 183–205. Blackwell Scientific Publications, Oxford.
7.4.4

Cowie R.J., Krebs J.R. & Sherry D.F. (1981) Food storing by marsh tits. *Anim. Behav.* **29**, 1252–1259.
7.3.4

Cox C.R. (1981) Agonistic encounters among male elephant seals: frequency, context and the role of female preference. *Amer. Zool.* **21**, 197–209.
9.2.4

Cox C.R. & Le Boeuf B.J. (1977) Female incitation of male competition: a mechanism of mate selection. *Amer. Natur.* **111**, 317–335.
9.2.3, 9.2.4

Craig J.L. (1979) Habitat variation in the social organization of a communal gallinule, the pukeko, *Porphyrio p. melanotis*. *Behav. Ecol. Sociobiol.* **5**, 331–358.
12.1, 12.3.3

Craig J.L. (1980) Pair and group breeding behaviour of a communal gallinule, the pukeko, *Porphyrio porphyrio melanotus*. *Anim. Behav.* **28**, 593–603.
12.1, 12.5

Craig J.L. (in press) Are communal pukeko caught in the prisoner's dilemma? *Behav. Ecol. Sociobiol.*
6.5.4

Craig R. (1979) Parental manipulation, kin selection, and the evolution of altruism. *Evolution* **33**, 319–334.
13.6

Craig R. & Crozier R.H. (1979) Relatedness in the polygynous ant *Myrmecia pilosula*. *Evolution* **33**, 335–341.
13.6

Craig R.B., DeAngelis D.L. & Dixon K.R. (1979) Long and short-term dynamic optimization models with application to the feeding strategy of the loggerhead shrike. *Amer. Natur.* **113**, 31–51.
4.4.3

Crawford R.D. (1977) Polygynous breeding of short-billed marsh wren. *Auk* **94**, 359–362.
10.3.1

Crewe R.M. & Velthuis H.H.W. (1980) False queens: A consequence of mandibular gland signals in worker honeybees. *Naturwissenschaften* **67**, 467–469.
13.7

Crook J.H. (1964) The evolution of social organization and visual communication in the weaver birds (Ploceinae). *Behaviour, Suppl.* **10**, 1–178.
1.2.1, 10.0

Crook J.H. (1965) The adaptive significance of avian social organisations. *Symp. Zool. Soc. Lond.* **14**, 181–218.
1.1, 1.2.1, 5.3.1, 10.0

Crook J.H. (1970) The socio-ecology of primates. In: *Social Behaviour in Birds and Mammals* (ed. J.H. Crook). Academic Press, London.
5.1, 5.2

Crook J.H. (1972) Sexual selection, dimorphism, and social organization in the primates. In: *Sexual Selection and the Descent of Man, 1871–1971* (ed. B. Campbell). Aldine, Chicago.
5.1.2, 5.2, 5.2.1, 5.3.2

Crook J.H. & Gartlan J.S. (1966) Evolution of primate societies. *Nature, Lond.* **210**, 1200–1203.
1.2.1, 5.2, 5.2.1

Crossley S. (1975) Changes in mating behavior produced by selection for ethological isolation between *ebony* and *vestigial* mutants of *Drosophila melanogaster*. *Evolution* **28**, 631–647.
9.6

Crow J.F. & Kimura M. (1965) Evolution in sexual and asexual populations. *Amer. Natur.* **99**, 439–450.
8.1

Crozier R.H. (1970) Coefficients of relationship and the identity of genes by descent in the Hymenoptera. *Amer. Natur.* **104**, 216–217.
13.5

Crozier R.H. (1977) Evolutionary genetics of the Hymenoptera. *A. Rev. Entomol.* **22**, 263–288.
13.2, 13.3, 13.5, 13.6

Crozier R.H. (1979) Genetics of sociality. In: *Social Insects* (ed. H.R. Hermann), Vol. I, pp. 223–286. Academic Press, New York.
13.2, 13.5, 13.6

Crozier R.H. & Dix M.W. (1979) Analysis of two genetic models for the innate components of colony odor in social Hymenoptera. *Behav. Ecol. Sociobiol.* **4**, 217–224.
13.5

Cruz Y.P. (1981) A sterile defender morph in a polyembryonic hymenopterous parasite. *Nature, Lond.* **294**, 446–447.
13.3

Cuellar O. (1976) Intraclonal histocompatibility in a parthenogenetic lizard, evidence of genetic homogeneity. *Science* **193**, 150–153.
8.1.1

Cullen E. (1957) Adaptations in the kittiwake to cliff nesting. *Ibis* **99**, 275–302.
1.1, 7.3.3

Cullen J.M. (1960) Some adaptations in the nesting behaviour of terns. *Proc. XII Int. Ornith. Congr.* pp. 153–157. Helsinki, Finland.
1.1

Cullen J.M. (1966) Reduction of ambiguity through ritualization. *Phil. Trans. R. Soc. Lond. B.* **251**, 363–374.
15.4.2

Cullen J.M. (1972) Some principles of animal communication. In: *Non-verbal Communication* (ed. R.A. Hinde), pp. 101–122. Cambridge University Press, Cambridge.
15.2.2

Cullen J.M. & Ashmole N.P. (1963) The black noddy *Arious tenuirostris* in Ascension Island. *Ibis* **103**, 423–443.
1.1

Curio E.B. (1973) Towards a methodology of teleonomy. *Experientia* **29**, 1045–1058.
1.3

Curio E.B. (1978) The adaptive significance of avian mobbing. *Z. Tierpsychol.* **48**, 175–183.
5.3.2

Daly M., Rauschenberger, J. & Behrends P. (1982) Food-aversion learning in

kangaroo rats: a specialist–generalist comparison. *Anim. Learning Behav.* **10**, 314–320.
    7.3.4

Damuth J. (1981) Population density and body size in mammals. *Nature, Lond.* **290**, 699–700.
    1.1

Darwin C. (1859) *On the Origin of Species by Means of Natural Selection, or, the Preservation of Favoured Races in the Struggle for Life.* John Murray, London.
    2.1.3, 13.1

Darwin C. (1877) *The Different Forms of Flowers on Plants of the Same Species.* John Murray, London.
    8.3, 14.3.1, 14.3.3

Davies N.B. (1977) Prey selection and social behaviour in wagtails. (Aves: Motacillidae). *J. Anim. Ecol.* **46**, 37–57.
    4.2.1

Davies N. B. (1978a) Ecological questions about territorial behaviour. In: *Behavioural Ecology. An Evolutionary Approach* (ed. J.R. Krebs & N.B. Davies), pp. 317–350. Blackwell Scientific Publications, Oxford.
    9.2.4

Davies N.B. (1978b) Territorial defence in the speckled wood butterfly (*Pararge aegeria*), the resident always wins. *Anim. Behav.* **26**, 138–147.
    6.1, 9.2.5

Davies N.B. (1980) The economics of territorial behaviour in birds. *Ardea* **68**, 63–74.
    5.2.1

Davies N.B. (1981) Calling as an ownership convention on pied wagtail territories. *Anim. Behav.* **29**, 529–534.
    6.3.5

Davies N.B. (1982) Behaviour and competition for scarce resources. In: *Current Problems in Sociobiology* (ed. King's College Sociobiology Group), pp. 363–380. Cambridge University Press, Cambridge.
    2.9.2, 9.2.5

Davies N.B. & Halliday T.R. (1978). Deep croaks and fighting assessment in toads *Bufo bufo*. *Nature, Lond.* **274**, 683–685.
    2.5.2, 15.4.2

Davies N.B. & Halliday T.R. (1979) Competitive mate searching in common toads, *Bufo bufo. Anim Behav.* **27**, 1253–1267.
    2.3, 2.7.1, 9.2.3

Davies N.B. & Houston A.I. (1981) Owners and satellites: the economics of territory defence in the pied wagtail, *Motacilla alba. J. Anim. Ecol.* **50**, 157–180.
    5.3, 6.3.5, 6.5.1, 7.4.2, 7.4.4

Davies N.B. & Houston A.I. (1983) Time allocation between territories and flocks and owner–satellite conflict in foraging pied wagtails, *Motacilla alba. J. Anim. Ecol.* **52**, 621–634.
    6.3.1, 6.3.5, 6.5.3

Davis J.W.F. & O'Donald P. (1976) Territory size, breeding time and mating preference in the Arctic skua. *Nature, Lond.* **260**, 774–775.
    6.3.1

Davison G.W.H. (1981) Sexual selection and the mating system of *Argusianus argus* (Aves: Phasianidae). *Biol. J. Linn. Soc.* **15**, 91–104.
    9.2.4

Dawkins M. (1980) *Animal Suffering: the Science of Animal Welfare.* Chapman & Hall, London.
15.2.2

Dawkins R. (1976) *The Selfish Gene.* Oxford University Press, Oxford.
3.4.1, 12.4, 15.2.3

Dawkins R. (1980) Good strategy or evolutionarily stable strategy? In: *Sociobiology: Beyond Nature/Nuture?* (ed. G.W. Barlow & S. Silverberg), pp. 331–367. Westview Press, Boulder, Colorado.
2.1.2, 2.2, 2.9.2, 3.2.1

Dawkins R. (1982a) *The Extended Phenotype.* W.H. Freeman, Oxford.
3.2, 3.2.1, 3.4.1

Dawkins R. (1982b) Replicators and vehicles. In: *Current Problems in Sociobiology* (ed. King's College Sociobiology Group). Cambridge University Press, Cambridge.
3.4.1

Dawkins R. & Carlisle T.R. (1976) Parental investment, mate desertion, and a fallacy. *Nature, Lond.* **262**, 131–133.
8.4, 10.3.2

Dawkins R. & Krebs J.R. (1978) Animal signals: information or manipulation. In: *Behavioural Ecology: An Evolutionary Approach* (ed. J.R. Krebs & N.B. Davies), pp. 282–309. Blackwell Scientific Publications, Oxford.
2.5.2, 15.0, 15.2.3, 15.4, 15.4.2

Dawkins R. & Krebs J.R. (1979) Arms races between and within species. *Proc. R. Soc. Lond. B.* **205**, 489–511.
9.2.2

De Benedictis P.A., Gill F.B., Hainsworth F.R., Pyke G.H. & Wolf L.L. (1978) Optimal meal size in hummingbirds. *Amer. Natur.* **112**, 301–316.
4.3

DeGroot P. (1980) Information transfer in a socially roosting weaver bird (*Quelea quelea*: Ploceinae): an experimental study. *Anim. Behav.* **28**, 1249–1254.
5.3.1

Delius J.D. (1969) A stochastic analysis of the maintenance behaviour of skylarks. *Behaviour* **33**, 137–178.
15.2.2

Dennet D.C. (1983) Intentional systems in cognitive ethology: The 'Panglossian Paradigm' defended. *Brain Behav. Sci.* (in press).
4.1.1, 15.4, 15.4.2

De Steven D. (1980) Clutch size, breeding success, and parental survival in the tree swallow (*Iridoprocne bicolor*). *Evolution* **34**, 278–291.
11.5

deVries T. (1979) The breeding biology of the Galapagos hawk, *Buteo galapoensis. Gerfaut* **65**, 29–57.
12.5

Dewsbury D.A. (1982) Ejaculate cost and male choice. *Amer. Natur.* **119**, 601–610.
9.2.1

Dickinson A. (1980) *Contemporary Animal Learning Theory.* Cambridge University Press, Cambridge.
7.2.1

Dill L.M. (1978) An energy-based model of optimal feeding territory size. *Theor. Popul. Biol.* **14**, 396–429.
6.3.1

Dittus W.P.J. (1977) The social regulation of population density and age–sex distribution in the toque monkey. *Behaviour* **63**, 281–322.
5.1.1, 5.1.2

Dodson C.H. (1962) Pollination and variation in the subtribe Catasetinae (Orchidaceae). *Ann. Missouri Bot. Gard.* **49**, 35–56.
14.2.3

Domjan M. (1980) Ingestional aversion learning: Unique and general processes. *Adv. Study Behav.* **11**, 275–336.
7.2.1

Domjan M. (in press, a) Selective associations in aversion learning. In: *Quantitative Analyses of Behavior, Vol. 3* (ed. M.L. Commoms, R.J. Herrnstein & A.R. Wagner). Ballinger, Cambridge, Mass.
7.2.1

Domjan M. (in press, b) Biological constraints on instrumental and classical conditioning: Implications for general process theory. *Psychology Learning Motiv.* **17**.
7.2.1

Domjan M. & Wilson N.E. (1972) Specificity of cue to consequence in aversion learning in the rat. *Psychonom. Science* **26**, 143–145.
7.2.1

Dow D.D. (1977) Reproductive behavior of the noisy miner, a communally breeding honeyeater. *Living Bird* **16**, 163–185.
12.1, 12.3.3, 12.5

Downhower J.F. & Armitage K.B. (1971) The yellow-bellied marmot and the evolution of polygyny. *Amer. Natur.* **105**, 355–370.
10.3.1

Downhower J.F. & Brown L. (1981) The timing of reproduction and its behavioural consequences for mottled sculpins, *Cottus bairdi.* In: *Natural Selection and Social Behaviour: Recent Research and New Theory* (ed. R.D. Alexander & D.W. Tinkle), pp. 78–95. Chiron Press, New York.
9.2.4

Drent R.H. & Sweirstra P. (1977) Goose flocks and food: field experiments with barnacle geese (*Branta leucopsis*) in winter. *Wildfowl* **28**, 15–20.
5.2.1

Drent R.H. & van Eerden M. (1980) Goose flocks and food exploitation: how to have your cake and eat it. *Proc. XVII Int. Ornith. Congr.* Berlin, Germany.
5.2.1

Drickamer L.C. (1974) A ten year summary of reproductive data for free-ranging *Macaca mulatta. Folia primatol.* **21**, 61–80.
5.1.2

Dunbar R.I.M. (1982) Intraspecific variations in mating strategy. In: *Perspectives in Ethology, Vol. 5* (ed. P.P.G. Bateson & P.H. Klopfer), pp. 385–431. Plenum Press, New York.
2.3, 9.2.5, 11.3

Dunbar R. & Dunbar E. (1977) Dominance and reproductive success among female gelada baboons. *Nature, Lond.* **266**, 351–352.
5.1.2

Dunham D.W. (1966) Agonistic behaviour in captive rose-breasted grosbeaks, *Pheucticus ludovicianus* (L.). *Behaviour* **27**, 160–173.
15.4.2

Dunn E. (1977) Predation by weasels, *Mustela nivalis*, on breeding tits, *Parus* spp., in relation to the density of tits and rodents. *J. Anim. Ecol.* **46**, 634–652.

6.3.1

Dyrcz A. (1977) Polygyny and breeding success among great reed warblers, *Acrocephalus arundinaceus*, at Milicz, Poland. *Ibis* **119**, 73–77.

10.3.1

Dyson-Hudson R. & Smith E.A. (1978) Human territoriality: an ecological reassessment. *Am. Anthrop.* **80**, 21–41.

5.2.1

Eberhard W.G. (1972) Altruistic behavior in a sphecid wasp: support for kin-selection theory. *Science* **175**, 1390–1391.

13.7

Eberhard W.G. (1974) The natural history and behaviour of the wasp *Trigonopsis cameronii* Kohl (Sphecidae). *Trans. R. ent. Soc. Lond.* **125**, 295–328.

13.7

Edwards A.W.F. (1970) The search for genetic variability of the sex ratio. *J. biosoc. Sci.* Suppl. 2, 55–60.

8.2

Edwards W. & Tversky A. (eds.) (1967) *Decision Making*. Penguin Books, London.

4.1.2

Ehleringer J. & Forseth I. (1980) Solar tracking by plants. *Science* **210**, 1094–1098.

14.1

Eickwort G.C. (1975) Gregarious nesting of the mason bee *Hoplitis anthocopoides* and the evolution of parasitism and sociality among megachilid bees. *Evolution* **29**, 142–150.

13.7

Eickwort G.C. (1981) Presocial insects. In: *Social Insects* (ed. H.R. Hermann), Vol. II, pp. 199–280. Academic Press, New York.

13.3, 13.4, 13.5

Eisenberg J.F. (1966) The social organization of mammals. *Handbk. Zool.* **8**, 1–92.

10.0

Eisenberg J.F. (1981) *The Mammalian Radiations*. Chicago University Press, Chicago.

1.1, 1.2.5

Eisenberg J.F. & Wilson D.E. (1978) Relative brain size and feeding strategies in the Chiroptera. *Evolution* **32**, 740–775.

1.1

Eisenberg J.F., Muckenhirn N.A. & Rudran R. (1972) The relation between ecology and social structure in primates. *Science* **176**, 863–874.

1.1, 5.1.2, 5.2

Eiserer L.A. (1980) Development of filial attachment to static visual features of an imprinting object. *Anim. Learning Behav.* **8**, 159–166.

7.3.1

Elgar M.A. & Catterall C.P. (1981) Flocking and predator surveillance in house sparrows: test of an hypothesis. *Anim. Behav.* **29**, 868–872.

5.3.2

Elliott N.B. & Shlotzhauer T. (1980) Presocial behavior in *Cerceris wat-*

lingensis Elliott & Salbert (Hymenoptera: Sphecidae). *Jl N.Y. ent. Soc.* **88**, 45–46.
13.6

Elliot P.F. (1975) Longevity and the evolution of polygamy. *Amer. Natur.* **109**, 281–287.
10.3.1

Elner R.W. & Hughes R.N. (1978) Energy maximization in the diet of the shore crab, *Carcinus maenas*. *J. Anim. Ecol.* **47**, 103–116.
4.2.1, 7.4.3

Emlen S.T. (1970) Celestial rotation: its importance in the development of migratory orientation. *Science* **170**, 1198–1201.
7.3.1

Emlen S.T. (1978) The evolution of cooperative breeding in birds. In: *Behavioural Ecology: An Evolutionary Approach* (ed. J.R. Krebs & N.B. Davies), pp. 245–281. Blackwell Scientific Publications, Oxford.
3.3.3, 5.3.3, 12.3.2

Emlen S.T. (1981) Altruism, kinship, and reciprocity in the white-fronted bee-eater. In: *Natural Selection and Social Behavior: Recent Research and New Theory* (ed. R.D. Alexander & D.W. Tinkle), pp. 245–281. Chiron Press, New York.
12.1, 12.1.1, 12.2, 12.3.3, 12.4

Emlen S.T. (1982a) The evolution of helping. I. An ecological constraints model. *Amer. Natur.* **119**, 29–39.
12.1.1, 12.3.2, 12.3.3, 12.3.4

Emlen S.T. (1982b) The evolution of helping. II. The role of behavioral conflict. *Amer. Natur.* **119**, 40–53.
11.4, 12.5

Emlen S.T. & Oring L.W. (1977) Ecology, sexual selection, and the evolution of mating systems. *Science* **197**, 215–223.
1.1, 5.3.3, 10.3.2, 10.3.3

Emlen S.T. & Vehrencamp S.L. (1983) Cooperative breeding strategies among birds. In: *Perspectives in Ornithology* (ed. A.H. Brush & G.A. Clark, Jr.), pp. 93–120. Cambridge University Press, Cambridge.
12.1

Ende H. van den (1976) *Sexual Interactions in Plants. The Role of Specific Substances in Sexual Reproduction.* Academic Press, New York.
14.2.2

Erkhardt R.B. (1975) The relative body weights of Bornean and Sumatran orangutans. *Am. J. of phys. Anthrop.* **42**, 349–350.
1.2.2

Eshel I. & Cavalli-Sforza L.L. (1982) Assortment of encounters and evolution of cooperativeness. *Proc. natl Acad. Sci., USA* **79**, 1331–1335.
13.7

Estes R.D. (1974) Social organization of the African Bovidae. In: *The Behavior of Ungulates and its Relation to Management* (ed. V. Geist & F. Walther). I.U.C.N., Switzerland.
5.1.1

Evans H.E. (1958) The evolution of social life in wasps. *Proc. 10th Int. Congr. Entomol. 1956.* **2**, 449–457.
13.4

Evans H.E. (1973) Burrow sharing and nest transfer in the digger wasp *Philanthus gibbosus* (Fabricius). *Anim. Behav.* **21**, 302–308.
13.6

Evans H.E. (1977) Extrinsic versus intrinsic factors in the evolution of insect sociality. *BioScience* **27**, 613–617.
13.3, 13.5

Evans R.M. (1982) Foraging-flock recruitment at a black-billed gull colony: implications for the information center hypothesis. *Auk* **99**, 24–30.
5.3.1

Ewald P.W. & Rohwer S. (1980) Age, coloration and dominance in non-breeding hummingbirds: a test of the asymmetry hypothesis. *Behav. Ecol. Sociobiol.* **7**, 273–279.
6.5.1

Ewald P.W., Hunt G.L. & Warner M. (1980) Territory size in western gulls: importance of intrusion pressure, defense investments, and vegetation structure. *Ecology* **61**, 80–87.
6.3.1

Ewing A.W. (1978) An investigation into selective mechanisms capable of maintaining balanced polymorphisms. Unpublished PhD thesis, Portsmouth Polytechnic.
9.5

Ewing A.W. & Bennet-Clark H.C. (1968) The courtship songs of *Drosophila*. *Behaviour* **31**, 288–301.
9.2.4

Faaborg J.F., deVries T., Patterson C.B. & Griffin C.R. (1980) Preliminary observations on the occurrence and evolution of polyandry in the Galapagos hawk (*Buteo galapagoensis*). *Auk* **97**, 581–590.
10.3.2, 12.5

Falconer D.S. (1981) *Introduction to Quantitative Genetics*, 2nd edn. Longman, London.
9.2.3, 9.3.1

Farr J.A. (1976) Social facilitation of male sexual behavior, intrasexual competition, and sexual selection in the guppy, *Poecilia reticulata* (Pisces: Poeciliidae). *Evolution* **30**, 707–717.
9.2.5

Farr J.A. (1977) Male rarity or novelty, female choice behavior and sexual selection in the guppy *Poecilia reticulata* Peters (Pisces: Poeciliidae). *Evolution* **31**, 162–168.
9.5

Felsenstein J. (1974) The evolutionary advantage of recombination. *Genetics* **78**, 737–756.
8.1

Felsenstein J. (1976) The theoretical population genetics of variable selection and migration. *A. Rev. Genet.* **10**, 253–280.
9.2.3

Fentress J.C. & C.J. Ryan (1982) A long-term study of distributed pup feeding and associated behavior in captive wolves. In: *Wolves of the World* (ed. F. Harrington & P. Paquet), pp. 238–259. Noyes Press, Park Ridge, New Jersey.
12.1

Fernald R.D. & Hirata N.R. (1977) Field study of *Haplochromis burtoni*: quantitative behavioural observations. *Anim. Behav.* **25**, 964–975.
6.5.1

Fisher R.A. (1930) *The Genetical Theory of Natural Selection*. Clarendon Press, Oxford.
2.1.3, 8.1, 8.2, 9.2.4, 9.2.5, 11.5

Fisher R.A. (1954) *Statistical Methods for Research Workers*. 12th edn. Oliver & Boyd, Edinburgh.
1.2.4

Fisher R.A. (1958) *The Genetical Theory of Natural Selection*, 2nd edn. Dover, New York.
3.4.3

Forsyth A.B. (1975) Usurpation and dominance behavior in the polygynous social wasp *Metapolybia cingulata* (Hymenoptera: Vespidae: Polybiini). *Psyche* 299–303.
13.7

Forsyth A.B. (1980) Worker control of queen density in hymenopteran societies. *Amer. Natur.* **116**, 895–898.
13.7

Frame L.H. & Frame G.W. (1977) Female African wild dogs emigrate. *Nature, Lond.* **263**, 227–229.
12.1, 12.4

Frame L.H., Malcolm J.R., Frame G.W. & Lawick H. van (1979) Social organization of African wild dogs (*Lycaon pictus*) on the Serengeti Plains, Tanzania. 1967–1978. *Z. Tierpsychol.* **50**, 225–249.
12.1, 12.3.3, 12.4

Freeland W.J. & Janzen D.H. (1974) Strategies in herbivory by mammals: the role of plant secondary compounds. *Amer. Natur.* **108**, 269–284.
4.2.2

Freeman D.C., Harper K.T. & Charnov E.L. (1980) Sex change in plants: Old and new observations and new hypotheses. *Oecologia.* **47**, 222–232.
14.2.1, 14.2.2

Freeman D.C., Harper K.T. & Ostler W.K. (1980) Ecology of plant dioecy in the intermountain region of western North America and California. *Oecologia* **44**, 410–417.
14.2.2

Freeman S. & McFarland D.J. (1982) The Darwinian objective function and adaptive behaviour. In: *Functional Ontogeny* (ed. D.J. McFarland), pp. 24–59. Pitman, London.
4.1.2, 4.4.3

Fretwell S.D. (1972) *Populations in a Seasonal Environment*. Princeton University Press, Princeton.
5.1.2, 5.4.3, 10.3.1

Fretwell S.D. & Lucas H.L. (1970) On territorial behaviour and other factors influencing habitat distribution in birds. *Acta biotheor.* **19**, 16–36.
2.1.3, 2.3, 2.7.1

Frisch K. von (1967) *The Dance Language and Orientation of Bees*. Belknap Press, Cambridge, Mass.
15.4.2

Frith H.J. (1962) *The Mallee-Fowl*. Angus & Robertson, Sydney.
8.4

Fry C.H. (1972) The social organization of bee-eaters (Meropidae) and co-operative breeding in hot-climate birds. *Ibis* **114**, 1–14.
12.1, 12.3.3

Futuyma D.J. (1979) *Evolutionary Biology*. Sinauer, New York.
1.3

Futuyma D.J. & Mayer G.C. (1980) Non-allopatric speciation in animals. *Syst. Zool.* **29**, 254–271.
9.6

Gadgil M. (1971) Dispersal: population consequences and evolution. *Ecology* **52**, 253–261.

11.4

Gadgil M. (1972) Male dimorphism as a consequence of sexual selection. *Amer. Natur.* **106**, 574–580.

2.1.3, 9.2.1

Gadgil M. (1982) Changes with age in the strategy of social behaviour. *Persp. Ethol.* **5**, 489–501.

11.5

Gamboa G.J. (1978) Intraspecific defense: advantage of social cooperation among paper wasp foundresses. *Science* **199**, 1463–1465.

13.7

Gamboa G.J. (1980) Comparative timing of brood development between multiple- and single-foundress colonies of the paper wasp, *Polistes metricus. Ecol. Entomol.* **5**, 221–225.

13.5

Gamboa G.J. & Dropkin J.A. (1979) Comparisons of behaviors in early vs. late foundress associations of the paper wasp, *Polistes metricus* (Hymenoptera: Vespidae). *Can. Entomol.* **111**, 919–926.

13.7

Gamboa G.J., Heacock B.D. & Wiltjer S.L. (1978) Division of labor and subordinate longevity in foundress associations of the paper wasp, *Polistes metricus* (Hymenoptera: Vespidae). *J. Kans. Entomol. Soc.* **51**, 343–352.

13.7

Ganders F.R. (1979) The biology of heterostyly. *N.Z. J. Bot.* **17**, 607–635.

14.3.3

Garcia J.F. & Koelling R.A. (1966) Relation of cue to consequence in avoidance learning. *Psychonomic Science* **4**, 123–124.

7.2.1

Garcia J.F., McGowan B.K. & Green K.F. (1972) Biological constraints on conditioning. In: *Classical Conditioning II. Current Research and Theory* (ed. A.H. Black & W.F. Prokasy), pp. 3–27. Appleton-Century-Crofts, New York.

7.2.1

Gass C.L. (1979) Territory regulation, tenure, and migration in rufous hummingbirds. *Can. J. Zool.* **57**, 914–923.

6.3.2

Gass C.L., Angehr G. & Centa J. (1976) Regulation of food supply by feeding territoriality in the rufous hummingbird. *Can. J. Zool.* **54**, 2046–2054.

6.3.2

Gaston A.J. (1973) The ecology and behaviour of the long-tailed tit. *Ibis* **115**, 330–351.

12.4

Gaston A.J. (1978a) Demography of the jungle babbler *Turdoides striatus. J. Anim. Ecol.* **47**, 845–879.

12.1, 12.3.2, 12.3.3, 12.4

Gaston A.J. (1978b) The evolution of group territorial behavior and cooperative breeding. *Amer. Natur.* **112**, 1091–1100.

6.5.1, 12.5

Gaston A.J. (1978c) Ecology of the common babbler, *Turdoides caudatus. Ibis* **120**, 415–432.

12.1, 12.2

Geist V. (1971) *Mountain Sheep*. Chicago University Press, Chicago.
  2.5.2

Gerling D. & Hermann H.R. (1978) Biology and mating behavior of *Xylocopa virginica* L. (Hymenoptera, Anthophoridae). *Behav. Ecol. Sociobiol.* **3**, 99–111.
  13.6

Getty T. (1981) Competitive collusion: the preemption of competition during the sequential establishment of territories. *Amer. Natur.* **118**, 426–431.
  6.4

Ghiselin M.T. (1969) The evolution of hermaphroditism among animals. *Q. Rev. Biol.* **44**, 189–208.
  8.3

Gibbon J. & Church R.M. (1981) Time left: linear vs. logarithmic subjective time. *J. exp. Psychol. Anim. Behav. Proc.* **7**, 87–107.
  4.2.1, 7.4.3

Gibbons J.R.H. (1979) A model for sympatric speciation in *Megarhyssa* (Hymenoptera: Ichneumonidae): competitive speciation. *Amer. Natur.* **114**, 719–741.
  9.6

Gibo D.L. (1978) The selective advantage of foundress associations in *Polistes fuscatus* (Hymenoptera: Vespidae). A field study of the effects of predation on productivity. *Can. Entomol.* **110**, 519–540.
  13.5

Gibo D.L. & Metcalf R.A. (1978) Early survival of *Polistes apachus* (Hymenoptera: Vespidae) colonies in California: a field study of an introduced species. *Can. Entomol.* **110**, 1339–1343.
  13.5

Gilebe B.D. & Leggett W.C. (1981) Latitudinal differences in energy allocation and use during the freshwater migrations of American shad (*Alosa sapidissima*) and their life history consequences. *Can. J. Fish. Aquat. Sci.* **38**, 806–820.
  11.2

Gill F.B. & Wolf L.L. (1975) Economics of feeding territoriality in the golden-winged sunbird. *Ecology* **56**, 333–345.
  5.2.1, 6.3.1, 6.3.3

Gill F.B. & Wolf L.L. (1977) Non-random foraging by sunbirds in a patchy environment. *Ecology* **58**, 1284–1296.
  5.2.1, 6.3.3

Giraldeau L.A. & Kramer D.L. (1982) The marginal value theorem: a quantitative test using load size variation in a central place forager, the eastern chipmunk *Tamias striatus. Anim. Behav.* **30**, 1036–1042.
  4.2.1

Gittleman J.L. (1983) The behavioural ecology of carnivores. Unpublished PhD thesis, University of Sussex.
  1.1, 1.2.4

Gittleman J.L. & Harvey P.H. (1980) Why are distasteful prey not cryptic? *Nature, Lond.* **286**, 149–150.
  7.2.2

Gittleman J.L. & Harvey P.H. (1982) Carnivore home-range size, metabolic needs and ecology. *Behav. Ecol. Sociobiol.* **10**, 57–64.
  1.1, 1.2.2

Givnish T.J. (1980) Ecological constraints on the evolution of breeding systems in seed plants: dioecy and dispersal in gymnosperms. *Evolution* **34**, 959–972.

14.1, 14.3.1

Glesener R.R. & Tilman D. (1978) Sexuality and the components of environmental uncertainty. *Amer. Natur.* **112**, 659–673.

8.1.2

Goodman D. (1982) Optimal life histories, optimal notation, and the value of reproductive value. *Amer. Natur.* **119**, 803–823.

11.5

Gosling L.M. (1982) A reassessment of the function of scent marking in territories. *Z. Tierpsychol.* **60**, 89–118.

6.1

Goss-Custard J.D. (1976) Variation in the dispersion of redshank (*Tringa totanus*) on their winter feeding grounds. *Ibis* **118**, 257–263.

5.3.1

Goss-Custard J.D. (1977) Predator responses and prey mortality in the redshank *Tringa totanus* (L.) and a preferred prey *Corophium volutator* (Pallas). *J. Anim. Ecol.* **46**, 21–36.

4.2.2

Gottlieb G. (1975) Development of species identification in ducklings: I. nature of perceptual deficit caused by embryonic auditory deprivation. *J. Comp. Physiol. Psychol.* **89**, 387–399.

7.5.1

Gould J.L. (1976) The dance language controversy. *Q. Rev. Biol.* **57**, 211–244.

15.4.2

Gould S.J. (1975) Allometry in primates with emphasis on scaling and the evolution of the brain. *Contr. Primatol.* **5**, 244–292.

1.1, 1.2.3, 1.2.5

Gould S.J. (1981) *The Mismeasure of Man.* W.W. Norton, New York.

1.2.5

Gould S.J. & Lewontin R.C. (1979) The spandrels of San Marco and the Panglossian Paradigm: a critique of the adaptationist programme. *Proc. R. Soc. Lond. B.* **205**, 581–598.

4.1.1, 4.6

Gould S.J. & Vrba E.S. (1982) Exaptation—a missing term in the science of form. *Paleobiology* **8**, 4–15.

4.6

Goux J.M. & Anxolabehere D. (1980) The measurement of sexual isolation and selection: a critique. *Heredity* **45**, 255–262.

9.5

Grafen A. (1979) The hawk–dove game played between relatives. *Anim. Behav.* **27**, 905–907.

2.6.2

Grafen A. (1982) How not to measure inclusive fitness. *Nature, Lond.* **298**, 425–426.

3.3, 3.3.2, 12.4

Grafen A. & Sibly R. (1978) A model of mate desertion. *Anim. Behav.* **26**, 645–652.

10.3.2

Grant B., Burton S., Contoreggi C. & Rothstein M. (1980) Outbreeding via frequency-dependent mate selection in the parasitoid wasp, *Nasonia* (*Mormoniella*) *vitripennis* Walker. *Evolution* **34**, 983–992.

9.5

References

Grant B., Snyder G.A. & Glesner S.F. (1974) Frequency-dependent mate selection in *Mormoniella vitripennis. Evolution* **28**, 259–264.
9.5

Grant V. (1958) The regulation of recombination in plants. *Cold Spring Harbor Symp. Quant. Biol.* **23**, 337–363.
8.3

Grant V. (1971) *Plant Speciation.* Columbia University Press, New York.
14.1

Grau H.J. (1982) Kin recognition in white-footed deermice (*Peromyscus leucopus*). *Anim. Behav.* **30**, 497–505.
9.3.1

Graul W.D., Derrickson S.R. & Mock D.W. (1977) The evolution of avian polyandry. *Amer. Natur.* **111**, 812–816.
10.3.2

Green D.M. & Swets J.A. (1966) *Signal Detection Theory and Psychophysics.* John Wiley & Sons, New York.
15.3.1

Green L. & Snyderman M. (1980) Choice between rewards differing in amount and delay: toward a choice model of self-control. *J. exp Analysis Behav.* **34**, 135–147.
7.4.4

Green R.F. (1980) Bayesian birds: A simple example of Oaten's stochastic model of optimal foraging. *Theor. Popul. Biol.* **18**, 244–256.
4.2.3, 7.4.4

Greenberg L. (1979) Genetic component of bee odor in kin recognition. *Science* **206**, 1095–1097.
9.3.1, 13.5, 13.6

Greenwood P.J. (1980) Mating systems, philopatry and dispersal in birds and mammals. *Anim. Behav.* **28**, 1140–1162.
1.1, 9.3.1, 11.4

Greenwood P.J., Harvey P.H. & Perrins C.M. (1978) Inbreeding and dispersal in the great tit. *Nature, Lond.* **271**, 52–54.
9.3.1

Gregg K.B. (1973) Studies on the control of sex expression in the genera *Cycnoches* and *Catasetum*, subtribe Catasetinae, Orchidaceae. PhD dissertation, University of Miami, Coral Gables, Florida.
14.2.3

Gregg K.B. (1975) The effect of light intensity on sex expression in species of *Cycnoches* and *Catasetum* (Orchidaceae). *Selbyana* **1**, 101–113.
14.2.3

Griffin D.R. (1981) *The Question of Animal Awareness*, 2nd edn. Rockefeller University Press, New York.
15.2.2

Griffin D.R. (ed.) (1982) *Animal Mind – Human Mind.* Springer-Verlag, Berlin.
15.2.2

Grime J.P. (1979) *Plant Strategies and Vegetation Processes.* John Wiley & Sons, New York.
14.1

Grime J.P. & Mowforth M.A. (1982) Variation in genome size—an ecological interpretation. *Nature, Lond.* **299**, 151–153.
1.1

Grimes L.G. (1976) The occurrence of cooperative breeding behavior in

African birds. *Ostrich* **47**, 1–15.
12.3.3

Gross M.R. & Charnov E.L. (1980) Alternative male life histories in bluegill sunfish. *Proc. natl Acad. Sci., USA* **77**, 6937–6940.
2.8, 9.2.5

Gross M.R. & Shine R. (1981) Parental care and mode of fertilization in ectothermic vertebrates. *Evolution* **35**, 775–793.
10.3.2

Gwynne D.T. (1981) Sexual difference theory: mormon crickets show role reversal in mate choice. *Science* **213**, 779–780.
9.2.4

Hailman J.P. (1965) Cliff nesting adaptations of the Galapagos swallow-tailed gull. *Wilson Bull.* **77**, 346–362.
1.1

Haldane J.B.S. & Spurway H. (1954) A statistical analysis of communication in 'Apis mellifera' and a comparison with communication in other animals. *Insect. Soc.* **1**, 247–283.
15.4, 15.4.1

Halliday T.R. (1978) Sexual selection and mate choice. In: *Behavioural Ecology: An Evolutionary Approach* (ed. J.R. Krebs & N.B. Davies), pp. 180–213. Blackwell Scientific Publications, Oxford.
9.2.2, 9.2.4, 9.6

Halliday T.R. (1983) The study of mate choice. In: *Mate Choice* (ed. P.P.G. Bateson), pp. 3–32. Cambridge University Press, Cambridge.
9.2.2

Halliday T.R. & Houston A.I. (1978) The newt as an honest salesman. *Anim. Behav.* **26**, 1273–1274.
9.2.4

Hamilton W.D. (1964) The genetical evolution of social behaviour. *J. theor. Biol.* **7**, 1–52
3.3.1, 3.3.2, 3.3.3, 3.4.3, 5.3.3, 12.3.2, 12.4, 13.5, 13.6, 13.7

Hamilton W.D. (1967) Extraordinary sex ratios. *Science* **156**, 477–488.
2.1.3, 3.4.3, 8.2, 9.3.2, 11.4

Hamilton W.D. (1971) Geometry for the selfish herd. *J. theor. Biol.* **31**, 295–311.
1.3, 5.3.2

Hamilton W.D. (1972) Altruism and related phenomena, mainly in social insects. *A. Rev. Ecol. Syst.* **3**, 193–232.
3.4.3, 13.5, 13.7

Hamilton W.D. (1975) Innate social aptitudes of man: an approach from evolutionary genetics. In: *Biosocial Anthropology* (ed. R. Fox), pp. 133–155. John Wiley & Sons, New York.
3.3.3, 3.4.2, 3.4.3

Hamilton W.D. (1979) Wingless and fighting males in fig wasps and other insects. In: *Sexual Selection and Reproductive Competition in Insects* (ed. M.S. Blum & N.A. Blum), pp. 167–220. Academic Press, London.
5.1.2, 11.4

Hamilton W.D. (1980) Sex versus non-sex parasite. *Oikos* **35**, 282–290.
8.1.2

Hamilton W.D. & May R.M. (1977) Dispersal in stable habitats. *Nature, Lond.* **269**, 578–581.
11.4

Hamilton W.D. & Zuk M. (1982) Heritable true fitness and bright birds: a role for parasites? *Science* **218**, 384–387.
1.2.2, 9.2.3

Hammerstein P. (1981) The role of asymmetries in animal contests. *Anim. Behav.* **29**, 193–205.
2.5.2

Hammerstein P. & Parker G.A. (1982) The asymmetric war of attrition. *J. theor. Biol.* **96**, 647–682.
2.5.2

Harcourt A.H., Harvey P.H., Larson S.G. & Short R.V. (1981) Testis weight, body weight and breeding system in primates. *Nature, Lond.* **293**, 55–57.
1.1, 1.2.6, 9.2.1

Hardy J.W. (1961) Studies in behavior and phylogeny of certain New World jays (Garrulinae). *Univ. Kansas Sci. Bull.* **42**, 13–149.
12.3.3

Hardy J.W., Webber T.A. & Raitt R.J. (1981) Communal social biology of the Southern San Blas jay. *Bull. Fla St. Mus.* **26**, 203–264.
12.3.3

Harestad A.S. & Bunnell F.L. (1979) Home range and body weight—a re-evaluation. *Ecology* **60**, 389–402.
1.1

Harley C.B. (1981) Learning the evolutionarily stable strategy. *J. theor. Biol.* **89**, 611–633.
4.2.3, 7.4.4

Harmeson J.P. (1974) Breeding ecology of the dickcissel. *Auk* **91**, 348–359.
10.3.1

Harpending H. & Davis H. (1977). Some implications for hunter–gatherer ecology derived from the spacial structure of resources. *World Arch.* **8**, 275–286.
5.2.1

Harper D.G.C. (1982) Competitive foraging in mallards: 'ideal free' ducks. *Anim. Behav.* **30**, 575–584.
2.3, 2.7.1, 5.4.3

Harper J.L. (1977) *Population Biology of Plants.* Academic Press, New York.
14.1, 14.2

Harper J.L. (1982) After description. In: *The Plant Community as a Working Mechanism* (ed. E.I. Norman). Blackwell Scientific Publications, Oxford.
4.6

Harris R.N. (1979) Aggression, superterritories and reproductive success in tree swallows. *Can. J. Zool.* **57**, 2072–2078.
6.4

Harrison C.J.O. (1969) Helpers at the nest in Australian passerine birds. *Emu* **69**, 30–40.
12.3.3

Hartzler J.E. (1972) An analysis of sage grouse lek behavior. PhD dissertation thesis, University of Montana, Missoula, Montana.
10.3.3

Harvey P.H. (1982) On rethinking allometry. *J. theor. Biol.* **95**, 37–41.
1.2.4

Harvey P.H. & Arnold S.J. (1982) Female mate choice and runaway sexual selection. *Nature, Lond.* **297**, 533–534.
9.2.4

Harvey P.H. & Clutton-Brock T.H. (1981) Primate home-range size and meta-

bolic needs. *Behav. Ecol. Sociobiol.* **8**, 151–155.
1.1, 1.2.2

Harvey P.H. & Greene P.J. (1981) Group composition: an evolutionary perspective. In: *Group Cohesion: Theoretical and Clinical Perspectives* (ed. H. Kellerman), pp. 148–169. Grune & Stratton, New York.
1.3

Harvey P.H. & Harcourt A.H. (1982) Sperm competition, testes size and breeding system in primates. In: *Sperm Competition and the Evolution of Animal Mating Systems* (ed. R.C. Smith). Academic Press, London.
1.1

Harvey P.H. & Mace G.M. (1982) Comparisons between taxa and adaptive trends: problems of methodology. In: *Current Problems in Sociobiology* (ed. King's College Sociobiology Group), pp. 343–362. Cambridge University Press, Cambridge.
1.2.1, 1.2.3, 1.2.4

Harvey P.H., Clutton-Brock T.H. & Mace G.M. (1980) Brain size and ecology in small mammals and primates. *Proc. natl Acad. Sci, USA* **77**, 4387–4389.
1.1, 1.2.3, 1.2.5

Harvey P.H., Kavanagh M. & Clutton-Brock T.H. (1978) Sexual dimorphism in primate teeth. *J. Zool.* **186**, 475–486.
1.1

Harvey P.H., Bull J.J., Pemberton M. & Paxton R.J. (1982) The evolution of aposematic coloration in distasteful prey: A family model. *Amer. Natur.* **119**, 710–719.
7.2.2

Hassell M.P. (1980) Foraging strategies, population models, and biological control: a case study. *J. Anim. Ecol.* **49**, 603–628.
4.2.4

Hausfater G. (1975) Dominance and reproduction in baboons (*Papio cynocephalus*). *Contr. Primatol.* **7**.
5.1.2

Hay D.E. & McPhail J.D. (1975) Mate selection in three-spined sticklebacks. *Can. J. Zool.* **53**, 441–450.
9.4

Hazlett B.A. (1968) Size relations and aggressive behaviour in the hermit crab *Clibanarius vitatus*. *Z. Tierpsychol.* **25**, 608–614.
15.4.2

Hazlett B.A. & Bossert W.H. (1965) A statistical analysis of the aggressive communications systems of some hermit crabs. *Anim. Behav.* **13**, 357–373.
15.4.1

Hazlett B.A., Rubenstein D.I. & Rittschoff D. (1975) Starvation, aggression and energy reserves in the crayfish *Orconectes virilis*. *Crustaceana.* **28**, 11–16.
2.5.2

Hegner R.E., Emlen S.T. & Demong N.J. (1982) Spatial organization of the white fronted bee-eater. *Nature, Lond.* **298**, 264–266.
12.1, 12.1.1, 12.4

Heisler I.L. (1981) Offspring quality and the polygyny threshold: a new model for the 'sexy son' hypothesis. *Amer. Natur.* **117**, 316–328.
10.3.1

Heller R. & Milinski M. (1979) Optimal foraging of sticklebacks *Gasterosteus aculeatus* on swarming prey. *Anim. Behav.* **27**, 1127–1141.
4.4.2

Hermann H.R. (1979) Insect sociality—an introduction. In: *Social Insects* (ed. H.R. Hermann), Vol. I, pp. 1–33. Academic Press, New York.
13.7

Herreid C.F. II & Schlenker E.H. (1980) Energetics of mice in stable and unstable social conditions: evidence of an air-borne factor affecting metabolism. *Anim. Behav.* **28**, 20–28.
5.3.1

Herrnstein R.J. (1970) On the law of Effect. *J. exp. Analysis Behav.* **13**, 243–266.
4.5

Herrnstein R.J. (1974) Formal properties of the matching law. *J. exp. Analysis Behav.* **21**, 159–164.
5.4.3

Herrnstein R.J. & Vaughan W. (1980) Melioration and behavioral allocation. In: *Limits to Action: The Allocation of Individual Behavior* (ed. J.E.R. Staddon), pp. 143–176. Academic Press, New York.
5.4.3, 7.4.4

Heslop-Harrison J. (1957) The experimental modification of sex expression in flowering plants. *Biol. Rev.* **32**, 38–90.
14.2.1

Heslop-Harrison J. (1972) Sexuality of angiosperms. In: *Plant Physiology*, Vol. 6C (ed. F.C. Stewart), pp. 133–290. Academic Press, New York.
14.2.1

Hespenheide H.A. (1971) Food preference and the extent of overlap in some insectivorous birds, with special reference to the Tyrannidae. *Ibis* **113**, 59–72.
1.1

Heyman G.M. & Luce R.D. (1981) Matching is not a logical consequence of maximizing reinforcement rate. *Anim. Learning Behav.* **7**, 133–140.
4.5

Hilden O. & Vuolanto S. (1972) Breeding biology of the red-necked phalarope, *Phalaropus lobatus*, in Finland. *Ornis Fenn.* **49**, 57–85.
10.3.2

Hill J.L. (1974) *Peromyscus*: effect of early pairing on reproduction. *Science* **186**, 1042–1044.
8.3, 9.3.1

Hinde R.A. (1970) *Animal Behaviour*. 2nd edn. McGraw Hill, New York.
15.0

Hinde R.A. (ed.) (1972) *Non-verbal Communication*. Cambridge University Press, Cambridge.
15.0

Hinde R.A. (1975) The concept of function. In: *Function and Evolution in Behaviour* (ed. G.P. Baerends, C. Beer & A. Manning), p. 415. Clarendon Press, Oxford.
4.6, 15.1, 15.4.2

Hinde R.A. (1981) Animal signals: ethological and games-theory approaches are not incompatible. *Anim. Behav.* **29**, 535–542.
15.4, 15.4.2

Hinton H.E. (1961) Sperm transfer in insects and the evolution of haemo-coelic insemination. In: *Insect Reproduction* (ed. K.C. Highnam), pp. 95–107. Royal Entomological Society, London.
9.2.3

Hirshfield M.F. (1980) An experimental analysis of reproductive effort and

cost in the Japanese medaka *Oryzios tatipes. Ecology* **61**, 282–292.
11.2

Hirshfield M.F. & Tinkle D. (1975) Natural selection and the evolution of reproductive effort. *Proc. natl Acad. Sci., USA* **72**, 2227–2231.
11.2

Hodos W. (1982) Some perspectives on the evolution of intelligence and the brain. In: *Animal Mind – Human Mind* (ed. D.R. Griffin), pp. 33–55. Springer-Verlag, New York.
7.3.2

Hoffman H.S. & Ratner A.M. (1973) A reinforcement model of imprinting: implications for socialization in monkeys and men. *Psychol. Rev.* **80**, 527–544.
7.3.1

Hogan-Warburg A.J. (1966) Social behaviour of the ruff, *Philomachus pugnax* L. *Ardea* **54**, 109–229.
6.5.1, 9.2.5

Högstedt G. (1980) Evolution of clutch size in birds: adaptive variation in relation to territory quality. *Science* **210**, 1148–1150.
11.5

Hoikkala A., Lakovaara S. & Romppainen E. (1982) Mating behavior and male courtship sounds in the *Drosophila virilis* group. In: *Advances in Genetics, Development and Evolution of Drosophila* (ed. S. Lakovaara), pp. 407–421. Plenum Press, New York.
9.6

Hölldobler B. (1976) The behavioral ecology of mating in harvester ants (Hymenoptera: Formicidae: *Pogonomyrmex*). *Behav. Ecol. Sociobiol.* **1**, 405–423.
13.6

Hölldobler B. & Haskins C.P. (1977) Sexual calling behavior in primitive ants. *Science* **195**, 793–794.
13.7

Hölldobler B. & Lumsden C.J. (1980) Territorial strategies in ants. *Science* **210**, 732–739.
6.2, 13.7

Hölldobler B. & Michener C.D. (1980) Mechanisms of identification and discrimination in social Hymenoptera. In: *Evolution of Social Behavior: Hypotheses and Empirical Tests* (ed. H. Markl), pp. 35–58. Verlag Chemie, Deerfield Beach, Florida.
13.5

Hölldobler B. & Wilson E.O. (1977) The number of queens: An important trait in ant evolution. *Naturwissenschaften* **64**, 8–15.
13.2, 13.7

Hölldobler B. & Wilson E.O. (1983) Queen control in colonies of weaver ants (Hymenoptera: Formicidae: *Oecophylla*). *Ann. ent. Soc. Am.* **76**, 235–238.
13.7

Holling C.S. (1959) Some characteristics of simple types of predation and parasitism. *Can. Entomol.* **91**, 385–398.
4.2.4

Holling C.S. (1965) The functional response of predators to prey density and its role in mimicry and population regulation. *Mem. Ent. Soc. Can.* **45**, 1–60.
4.2.4

Holm C.H. (1973) Breeding sex ratios, territoriality, and reproductive success

in the red-winged blackbird (*Agelaius phoeniceus*). *Ecology* **54**, 356–365.
10.3.1

Honk C.G.J. van, Röseler R.F., Velthuis H.H.W. & Hoogeveen J.C. (1981) Factors influencing the egg laying of workers in a captive *Bombus terrestris* colony. *Behav. Ecol. Sociobiol.* **9**, 9–14.
13.7

Hoogland J.L. (1979a) The effects of colony size on individual alertness of prairie dogs (Sciuridae: *Cynomys* spp.). *Anim. Behav.* **27**, 394–407.
5.3.2

Hoogland J.L. (1979b) Aggression, ectoparasitism and other possible costs of prairie dog (Sciuridae: *Cynomys* spp.) coloniality. *Behaviour* **69**, 1–35.
5.3.2

Hoogland J.L. & Sherman P.W. (1976) Advantages and disadvantages of bank swallow (*Riparia riparia*) coloniality. *Ecol. Monogr.* **46**, 33–58.
5.2.1, 5.3.2, 5.3.3, 12.3.1

Horn H.S. (1968) The adaptive significance of colonial nesting in the Brewer's blackbird *Euphagus cyanocephalus*. *Ecology* **49**, 682–694.
49, 682–694.
5.2.1, 5.3.1, 5.3.2, 6.2

Horn H.S. (1978) Optimal tactics of reproduction and life-history. In: *Behavioural Ecology: An Evolutionary Approach* (ed. J.R. Krebs & N.B. Davies), pp. 411–429. Blackwell Scientific Publications, Oxford
11.2

Horn H.S. (1981) Sociobiology. In: *Theoretical Ecology: Principles and Applications* (ed. R.M. May), pp. 272–294. Blackwell Scientific Publications, Oxford.
11.4

Horn H.S. (1983) Some theories about dispersal. In: *The Ecology of Animal Movement* (ed. I.R. Swingland & P.J. Greenwood). Oxford University Press, Oxford.
11.4

Houston A.I. (1980) Godzilla v. the creature from the Black Lagoon. In: *The Analysis of Motivational Processes* (ed. F.M. Toates & T.R. Halliday), pp. 297–318. Academic Press, London.
4.4, 7.4.1

Houston A.I. (1983) Optimality theory and matching. *Behav. Analysis Lett.* **3**, 1–5.
4.5

Houston A.I. & McNamara J. (1981) How to maximize reward rate on two variable-interval paradigms. *J. exp. Analysis Behav.* **35**, 367–396.
6.2, 7.4.4

Houston A.I. & McNamara J. (1982) A sequential approach to risk-taking. *Anim. Behav.* **30**, 1260–1261.
4.2.3

Houston A.I., Kacelnik A. & McNamara J. (1982) Some learning rules for acquiring information. In: *Functional Ontogeny* (ed. D.J. McFarland), pp. 140–191. Pitman, London.
4.2.3

Howard R.D. (1978) The evolution of mating strategies in bullfrogs, *Rana catesbiana*. *Evolution* **32**, 850–871.
5.1.2, 6.5.1

Howard R.D. (1979) Big bullfrogs in a little pond. *Nat. Hist. Mag.* **88**, 30–36.
9.2.5

Howe H.F. (1977) Nestling sex ratio adjustment among common grackles. *Science* **198**, 744–746.
8.2

Howe M.A. (1975a) Behavioral aspects of the pair bond in Wilson's phalarope. *Wilson Bull.* **87**, 248–270.
10.3.2

Howe M.A. (1975b) Social interactions in flocks of courting Wilson's phalaropes (*Phalaropus tricolor*). *Condor* **77**, 24–33.
10.3.2

Hubbard S.F. & Cook R.M. (1978) Optimal foraging by parasitoid wasps. *J. Anim. Ecol.* **47**, 593–604.
4.5

Hughes R.N. (1979) Optimal diets under the energy maximization premise: the effects of recognition time and learning. *Amer. Natur.* **113**, 209–221.
7.4.2

Humphrey N.K. (1976) The social function of intellect. In: *Growing Points in Ethology* (ed. P.P.G. Bateson & R.A. Hinde), pp. 303–317. Cambridge University Press, Cambridge.
15.2.2

Hunt G.L. & Hunt M.W. (1976) Gull chick survival: the significance of growth rates, timing of breeding and territory size. *Ecology* **57**, 62–75.
6.3.1

Hunt J.H. (1982) Trophallaxis and the evolution of eusocial Hymenoptera. In: *The Biology of Social Insects: Proceedings of the Ninth Congress of the International Union for the Study of Social Insects, Boulder, Colorado, August 1982* (ed. M.D. Breed, C.D. Michener & H.E. Evans), pp. 201–205. Westview Press, Boulder, Colorado.
13.6

Hunter M.L. & Krebs J.R. (1979) Geographical variation in the song of the great tit (*Parus major*) in relation to ecological factors. *J. Anim. Ecol.* **48**, 759–786.
15.3.1

Inglis I.R. & Lazarus J. (1981) Vigilance and flock size in brent geese: the edge effect. *Z. Tierpsychol.* **57**, 193–200.
5.3.2

Iwasa Y., Higashi M. & Yamamura N. (1981) Prey distribution as a factor determining the choice of optimal foraging strategy. *Amer. Natur.* **117**, 710–723.
4.2.3, 7.4.4

Jakobsson S., Radesäter T. & Järvi T. (1979) On the fighting behaviour of *Nannacara anomala* (Pisces: Cichlidae) ♂♂. *Z. Tierpsychol.* **49**, 210–220.
15.4.2

Janetos A.C. & Cole B.J. (1981) Imperfectly optimal animals. *Behav. Ecol. Sociobiol.* **9**, 203–210.
4.5

Jarman M. (1975) The quintessential antelope: a study of the behaviour of the impala. *Afr. Wild. Leadership Fed. News* **8**, 2–7.
2.8

Jarman P.J. (1974) The social organisation of antelope in relation to their ecology. *Behaviour* **48**, 215–267.
1.1, 5.1.1, 5.1.2, 5.2.1, 5.3.2, 6.2

Jarman P.J. (1982) Prospects for interspecific comparisons in sociobiology. In: *Current Problems in Sociobiology* (ed. King's College Sociobiology Group), pp. 323–342. Cambridge University Press, Cambridge.
1.2.2

Järvi T., Sillén-Tullberg B. & Wiklund C. (1981) The cost of being aposematic. An experimental study of predation of larvae of *Papilio machaon* by the great tit, *Parus major*. *Oikos* **36**, 267–272.
7.2.2

Jarvinen O. (1976) Migration, extinction and alary morphism in water striders. *Ann. Acad. Sci. Fenn. Ser. A IV Biol.* **206**, 107.
11.4

Jenni D.A. (1974) Evolution of polyandry in birds. *Amer. Zool.* **14**, 129–144.
8.4, 9.2.1, 10.3.2

Jenni D.A. & Betts B.J. (1978) Sex differences in nest construction, incubation, and parental behaviour in the polyandrous American jacana (*Jacana spinosa*). *Anim. Behav.* **26**, 207–218.
10.3.2

Jensen B. (1973) Movements of the red fox (*Vulpes vulpes*) in Denmark investigated by marking and recovery. *Danish Rev. Game Biol.* **8**, 1–20.
12.3.3

Jerison H.J. (1973) *Evolution of the Brain and Intelligence.* Academic Press, New York.
1.2.5

Johnson L.K. (1982) Sexual selection in a brentid weevil. *Evolution* **36**, 251–262.
9.2.3, 9.4

Johnston T.D. (1981) Contrasting approaches to a theory of learning. *Behav. Brain Sci.* **4**, 125–139.
7.1, 7.2.1, 7.3.1, 7.5.1

Johnston T.D. (1982) Selective costs and benefits in the evolution of learning. *Adv. Study Behav.* **12**, 65–106.
7.3.2, 7.3.3

Johnston T.D. & Turvey M.T. (1980) An ecological metatheory for theories of learning. In: *The Psychology of Learning and Motivation: Advances in Research and Theory Vol 14* (ed. G.H. Bower), pp. 147–205. Academic Press, New York.
7.1, 7.5.1

Jones J.S. & Partridge L. (1983) Tissue rejection: the price of sexual acceptance? *Nature, Lond.* **304**, 484–485.
9.4

Jorde L.B. & Spuhler J.N. (1974) A statistical analysis of selected aspects of primate demography, ecology and social behavior. *J. Anthropol. Res.* **30**, 199–224.
1.2.5

Kacelnik A. (in press) Central place foraging in starlings (*Sturnus vulgaris*) I: patch residence time. *J. Anim. Ecol.*
4.2.1

Kacelnik A., Houston A.I. & Krebs J.R. (1981) Optimal foraging and territorial defence in the great tit *Parus major*. *Behav. Ecol. Sociobiol.* **8**, 35–40.
4.2.1, 4.3

Kalat J.W. & Rozin P. (1971) Role of interference in taste-aversion learning. *J. Comp. Physiol. Psychol.* **77**, 53–58.
7.2.1

Kamil A.C. (1978) Systematic foraging for nectar by amakihi, *Loxops virens. J. Comp. Physiol. Psychol.* **92**, 388–396.
6.3.3, 7.4.2

Kamil A.C. & Sargent T.D. (ed.) (1981) *Foraging Behaviour: Ecological, Ethological, and Psychological Approaches.* Garland STPM Press, New York.
7.4.1

Kamil A.C. & Yoerg A.C. (1982) Learning and foraging behavior. *Persp. Ethol.* **5**, 325–346.
7.1, 7.4.1

Kamin L.J. (1969) Predictability, surprise, attention, and conditioning. In: *Punishment and Abusive Behavior* (ed. B.A. Campbell & R.M. Church), pp. 279–296. Appleton-Century-Crofts, New York.
7.2.2, 7.3.3

Kareem A.M. & Barnard C.J. (1982) The importance of kinship and familiarity in social interactions between mice. *Anim. Behav.* **30**, 594–601.
9.3.1

Kasuya E. (1982) Central place water collection in a Japanese paper wasp, *Polistes chinensis antennalis. Anim. Behav.* **30**, 1010–1014.
4.2.1

Keeton W.T. (1981) The ontogeny of bird orientation. In: *Behavioral Development* (ed. G.W. Barlow, L. Petrinovich & M. Main), pp. 509–517. Cambridge University Press, New York.
7.3.1

Kelly F.P. (1979) *Reversibility and Stochastic Networks.* John Wiley & Sons, New York.
5.1.1

Kence A. (1981) A rare-male advantage in *Drosophila*: a possible source of bias in experimental design. *Amer. Natur.* **117**, 1027–1028.
9.5

Kendeigh S.C. (1952) Parental care and its evolution in birds. *Ill. Biol. Monogr.* **22**, 1–356.
10.3.2

Kenward R.E. (1978) Hawks and doves: attack success and selection in goshawk flights at wood-pigeons. *J. Anim. Ecol.* **47**, 449–460.
1.3, 5.3.2

Kermack K.A. & Haldane J.B.S. (1950) Organic correlation and allometry. *Biometrika* **37**, 30–41.
1.2.4

Killeen P. (1981) Averaging theory. In: *Quantification of Steady-State Operant Behavior* (ed. C.M. Bradshaw, E. Szabadi & C.F. Lowe), pp. 21–34. Elsevier, New York.
4.2.3, 7.4.4

Kinnaird M.F. & Grant P.R. (1982) Cooperative breeding by the Galapagos mockingbird, *Nesomimus parvulus. Behav. Ecol. Sociobiol.* **10**, 65–73.
5.3.1

Kirkpatrick M. (1982) Sexual selection and the evolution of female choice. *Evolution* **36**, 1–12.
9.1, 9.2.4, 10.3.3

Klahn J.E. (1979) Philopatric and nonphilopatric foundress associations in the social wasp *Polistes fuscatus. Behav. Ecol. Sociobiol.* **5**, 417–424.
13.5, 13.6

Kleiber M. (1961) *The Fire of Life: An Introduction to Animal Energetics.* John

Wiley & Sons, New York.

1.1

Kleiman D.G. & Eisenberg J.F. (1973) Comparisons of canid and felid social systems from an evolutionary perspective. *Anim. Behav.* **21**, 637–659.

5.1, 12.3.2

Knerer G. & Schwarz M. (1976) Halictine social evolution: the Australian enigma. *Science* **194**, 445–448.

13.3

Knowlton N. (1982) Parental care and sex role reversal. In: *Current Problems in Sociobiology* (ed. King's College Sociobiology Group, Cambridge), pp. 203–222. Cambridge University Press, Cambridge.

9.2.1, 10.3.2

Knowlton N. & Parker G.A. (1979) An evolutionarily stable strategy approach to indiscriminate spite. *Nature, Lond.* **279**, 419–421.

2.3, 2.6.1, 6.4

Knox P.B. (1967) Apoximis: seasonal and population differences in a grass. *Science* **157**, 325–326.

8.1.2

Kobel H.R., Du Pasquier L. & Tinsley R.C. (1981) Natural hybridisation and gene introgression between *Xenopus gilli* and *Xenopus laevis laevis* (Anura: Pipidae). *J. Zool. Lond.* **194**, 317–322.

9.6

Kodric-Brown A. & Brown J.H. (1978) Influence of economics, interspecific competition, and sexual dimorphism on territoriality of migrant rufous hummingbirds. *Ecology* **59**, 285–296.

6.3.2, 6.5.1

Koenig W.D. (1981a) Reproductive success, group size, and the evolution of cooperative breeding in the acorn woodpecker. *Amer. Natur.* **117**, 421–443.

5.3.3, 12.2, 12.3.2

Koenig W.D. (1981b) Space competition in the acorn woodpecker: Power struggles in a cooperative breeder. *Anim. Behav.* **29**, 396–409.

12.3.3, 12.4

Koenig W.D. & Pitelka F.A. (1979) Relatedness and inbreeding avoidance: Counterploys in the communally nesting acorn woodpecker. *Science* **206**, 1103–1105.

12.1

Koenig W.D. & Pitelka F.A. (1981) Ecological factors and kin selection in the evolution of cooperative breeding in birds. In: *Natural Selection and Social Behavior: Recent Research and New Theory* (ed. R.D. Alexander & D.W. Tinkle), pp. 261–280. Chiron Press, New York.

5.3.3, 12.1, 12.3.2, 12.3.3, 12.5

Konishi M. & Nottebohm F. (1969) Experimental studies in the ontogeny of avian vocalizations. In: *Bird Vocalizations* (ed. R.A. Hinde), pp. 29–48. Cambridge University Press, London.

7.3.2

Krebs J.R. (1970) A study of territorial behaviour in the great tit *Parus major* L. Unpublished DPhil thesis, Oxford University.

15.0

Krebs J.R. (1971) Territory and breeding density in the great tit, *Parus major* L. *Ecology* **52**, 2–22.

5.3.2, 5.4.3, 6.3.1, 10.3.1

Krebs J.R. (1974) Colonial nesting and social feeding as strategies for exploiting food resources in the great blue heron (*Ardea herodias*). *Behaviour* **51**, 99–134.
5.1.1, 5.2.1, 5.3.1

Krebs J.R. (1978) Optimal foraging: decision rules for predators. In: *Behavioural Ecology: An Evolutionary Approach* (ed. J.R. Krebs & N.B. Davies), pp. 23–63. Blackwell Scientific Publications, Oxford.
4.2

Krebs J.R. (1980) Optimal foraging, predation risk and territorial defence. *Ardea* **60**.
4.1.1

Krebs J.R. & Davies N.B. (ed.) (1978) *Behavioural Ecology: An Evolutionary Approach*. Blackwell Scientific Publications, Oxford.
1.3

Krebs J.R. & Davies N.B. (1981) *An Introduction to Behavioural Ecology*. Blackwell Scientific Publications, Oxford.
5.4.2

Krebs J., MacRoberts M. & Cullen J. (1972) Flocking and feeding in the great tit *Parus major*: an experimental study. *Ibis* **114**, 507–530.
5.2.1, 5.3.1

Krebs J.R., Ryan J.C. & Charnov E.L. (1974) Hunting by expectation or optimal foraging? A study of patch use by chickadees. *Anim. Behav.* **22**, 953–964.
7.4.4

Krebs J.R., Ashcroft R. & Webber M.I. (1978) Song repertoires and territory defence in the great tit (*Parus major*). *Nature, Lond.* **271**, 539–542.
6.1

Krebs J.R., Kacelnik A. & Taylor P. (1978) Test of optimal sampling by foraging great tits. *Nature, Lond.* **275**, 27–31.
4.2.3, 7.4.1, 7.4.4

Krebs J.R., Stephens D.W. & Sutherland W.J. (1983) Perspectives in optimal foraging. In: *Perspectives in Ornithology* (ed. G.A. Clark & A.H. Brush). Cambridge University Press, New York.
4.2, 4.2.1, 4.2.4, 4.6, 7.4.1, 7.4.4

Krebs J.R., Erichsen J.T., Webber M.I. & Charnov E.L. (1977) Optimal prey selection in the great tit (*Parus major*). *Anim. Behav.* **25**, 30–38.
4.2.1, 7.4.3

Krombein K.V., Hurd P.D., Smith D.R. & Burks B.D. (1979) *Catalog of Hymenoptera in America North of Mexico*. Smithsonian Institution Press, Washington D.C.
13.3

Kroodsma D.E. (1977) Correlates of song organization among North American wrens. *Amer. Natur.* **111**, 995–1008.
15.3.1

Kroodsma D.E. (1981) Ontogeny of bird song. In: *Behavioral Development* (ed. K. Immelmann, G.W. Barlow, L. Petrinovich & M. Main), pp. 520–532. Cambridge University Press, New York.
7.3.2

Kruijt J.P. & Hogan J.A. (1967) Social behavior on the lek in black grouse. *Ardea* **55**, 203–240.
10.3.3

Kruijt J.P., Vos G.J. de & Bossema I. (1972) The arena system of black

grouse. *Proc. XV Int. Ornith. Congr.*, pp. 399–423. E.J. Brill, Leiden.
10.3.3

Kruuk H. (1964) Predators and anti-predator behaviour of the black-headed gull (*Larus ridibundus*). *Behaviour* Suppl. **11**, 1–129.
5.3.2

Kruuk H. (1972) *The Spotted Hyena*. University of Chicago Press, Chicago and London.
5.1.1, 5.2.2, 5.3.1

Kugler J., Motro M. & Ishay J.S. (1979) Comb building abilities of *Vespa orientalis* L. queenless workers. *Insect. Soc.* **26**, 147–153.
13.7

Kukuk P.F., Breed M.D., Sobti A. & Bell W.J. (1977) The contributions of kinship and conditioning to nest recognition and colony member recognition in a primitively social bee, *Lasioglossum zephyrum* (Hymenoptera: Halictidae). *Behav. Ecol. Sociobiol.* **2**, 319–327.
13.5

Kullmann E.J. (1972) Evolution of social behavior in spiders (Araneae: Eresidae and Theridiidae). *Amer. Zool.* **12**, 419–426.
13.2

Kummer H. (1968) *Social Organization of Hamadryas Baboons: A Field Study*. University of Chicago Press, Chicago.
5.1.1, 5.2.1

Kummer H., Gotz W. & Angst W. (1974) Triadic differentiation: an inhibitory process protecting pair bonds in baboons. *Behaviour* **49**, 62–87.
2.5.2

Lack D. (1968) *Ecological Adaptations for Breeding in Birds*. Methuen, London.
1.1, 1.2.1, 1.2.3, 8.4, 10.3.2

Lacy R.C. (1979) Adaptiveness of a rare male mating advantage under heterosis. *Behav. Genet.* **9**, 51–55.
9.5

Lamprecht J. (1979) Field observations on the behaviour and social system of the bat-eared fox, *Otocyon megalotis* Desmarest. *Z. Tierpsychol.* **49**, 260–284.
12.1

Lande R. (1976) The maintenance of genetic variability by mutation in a polygenic character with linked loci. *Genet. Res.* **26**, 221–235.
9.2.3

Lande R. (1981) Models of speciation by sexual selection on polygenic traits. *Proc. natl Acad. Sci., USA* **78**, 3721–3725.
9.1, 9.2.4, 10.3.3

Lande R. (1982) Rapid origin of sexual isolation and character divergence within a cline. *Evolution* **36**, 213–223.
9.6

Lande R. & Arnold S.J. (in press) The measurement of selection on correlated characters. *Evolution*.
10.2

Larkin S.B.C. (1981) Time and energy in decision making. Unpublished DPhil thesis, Oxford University.
4.4.1

Lawick H. van (1973) *Solo*. Collins, London.
12.1

Lawton M.F. & Guindon C.F. (1981) Flock composition, breeding success and learning in the brown jay. *Condor* **83**, 27–33.
12.4

Lazarus J. (1979) The early warning function of flocking in birds: an experimental study with captive quelea. *Anim. Behav.* **27**, 855–865.
5.3.2

Lea S.E.G. (1979) Foraging and reinforcement schedules in the pigeon: Optimal and non-optimal aspects of choice. *Anim. Behav.* **27**, 875–886.
7.4.3

Lea S.E.G. (1981) Correlation and contiguity in foraging behavior. In: *Advances in Analysis of Behavior 2* (ed. P. Harzem & M. Zeiler), pp. 355–406. John Wiley & Sons, New York.
7.4.1, 7.4.3, 7.4.4

Lea S.E.G. (1982) The mechanism of optimality in foraging. In: *Quantitative Analyses of Behavior. 2: Matching and Maximizing Accounts* (ed. M.L. Commons, R.J. Herrnstein & H. Rachlin). Ballinger, Cambridge, Mass.
7.4.1

Le Boeuf B.J. (1974) Male–male competition and reproductive success in elephant seals. *Amer. Zool.* **14**, 163–176.
2.8, 5.1.2

Le Croy M. (1981) The genus *Paradisea*—display and evolution. *American Museum Novitates*, No. 2714.
9.2.4

Le Croy M., Kulupi A. & Peckover W.S. (1980) Goldie's bird of paradise: display, natural history and traditional relationships of people to the bird. *Wilson Bull.* **92**, 289–301.
9.2.4

Lee R.B. (1969) !Kung bushman subsistence: an input–output analysis. In: *Environment and Cultural Behavior* (ed. A.P. Vayda). The Natural History Press, New York.
5.2.1

Lehrman D.S. (1970) Semantic and conceptual issues in the nature–nurture problem. In: *Development and Evolution of Behavior* (ed. L.R. Aronson, E. Tobach, D.S. Lehrman & J.S. Rosenblatt), pp. 17–52. W.H. Freeman, San Francisco.
7.1

Lendrem D. (1983) Predation risk and vigilance in the blue tit. *Behav. Ecol. Sociobiol.* **13**.
4.3

Lenington S.G. (1980) Female choice and polygyny in red-winged blackbirds. *Anim. Behav.* **28**, 347–361.
10.3.1

Lerner I.M. (1954) *Genetic Homeostasis*. Oliver & Boyd, Edinburgh.
8.3

Lester L.J. & Selander R.K. (1981) Genetic relatedness and the social organization of *Polistes* colonies. *Amer. Natur.* **117**, 147–166.
13.6

Lester N.P. (in press) The 'feed–feed' decision: How goldfish solve the patch depletion problem. *Behaviour*.
7.4.4

Levin D.A. (1975) Pest pressure and recombination systems in plants. *Amer.*

8.1.2

Lewis D. (1941) Male sterility in natural populations of hermaphroditic plants. *New Phytol.* **40**, 56–63.
14.4

Lewis D. (1979) *Sexual Incompatibility in Plants.* Edward Arnold, London.
14.3.2

Lewis D.B. & Coles R.B. (1980) Sound localization in birds. *Trends Neurosci.* May 1980, 102–105.
15.3.2

Lewontin R.C. (1978) Adaptation. *Scient. Am.* **239**, 212–230.
1.1, 1.2.1

Lewontin R.C. (1979) Fitness, survival and optimality. In: *Analysis of Ecological Systems* (ed. D.J. Horn, R.D. Mitchell & G.R. Stairs), pp. 3–21. Ohio State University Press, Columbus.
4.1.1, 5.2, 9.1

Ligon J.D. (1981) Demographic patterns and communal breeding in the green woodhoopoe, *Phoeniculus purpureus.* In: *Natural Selection and Social Behavior: Recent Research and New Theory* (ed. R.D. Alexander & D.W. Tinkle), pp. 231–243. Blackwell Scientific Publications, Oxford.
12.1, 12.4

Ligon J.D. & S.H. Ligon (1978a) The communal social system of the green woodhoopoe in Kenya. *Living Bird* **17**, 159–198.
12.1, 12.2, 12.3.3, 12.4

Ligon J.D. & Ligon S.H. (1978b) Communal breeding in green woodhoopoes as a case for reciprocity. *Nature, Lond.* **276**, 496–498.
12.4

Lill A. (1968a) An analysis of sexual isolation in the domestic fowl: I. The basis of homogamy in males. *Behaviour* **30**, 107–126.
9.4

Lill A. (1968b) An analysis of sexual isolation in the domestic fowl: II. The basis of homogamy in females. *Behaviour* **30**, 127–145.
9.4

Lill A. (1974) Social organisation and space utilisation in the lek-forming white-bearded manakin, *M. manacus trinitatis* Hortert. *Z. Tierpsychol.* **36**, 513–530.
9.2.4, 10.3.3

Lill A. (1976) Lek behavior in the golden-headed manakin (*Pipra erythrocephala*) in Trinidad (West Indies). In: *Forts Verhaltensforch.*, Heft 18, pp. 1–84. Verlag Paul Parey, Berlin/Hamburg.
10.3.3

Lill A. & Wood-Gush D.G.M. (1965) Potential ethological isolating mechanisms and assortative mating in the domestic fowl. *Behaviour* **25**, 16–44.
9.4

Lima S.L. (1983) Downy woodpecker foraging behaviour: efficient sampling in simple stochastic environments. *Ecology* (in press).
4.2.3

Lin N. (1964) Increased parasitic pressure as a major factor in the evolution of social behavior in halictine bees. *Insect. Soc.* **11**, 187–192.
13.5

Lin N. & Michener C.D. (1972) Evolution of sociality in insects. *Q. Rev. Biol.* **47**, 131–159.
13.4, 13.5, 13.6, 13.7

Lindauer M. (1961) *Communication Among Social Bees.* Harvard University Press, Cambridge, Mass.
13.2

Linsenmair K.E. & Linsenmair C. (1972) Die Biedutung familienspezifischer 'Abzeichen' für den Familienzusammenhalt bei der sozialen Wüstenassel *Hemilepistus reaumuri* Audouin u. Savigny (Crustacea, Isopoda, Oniscoidea). *Z. Tierpsychol.* **31**, 131–162.
9.3.1

Litte M. (1979) *Mischocyttarus flavitarsis* in Arizona: social and nesting biology of a polistine wasp. *Z. Tierpsychol.* **50**, 282–312.
13.7

Littlejohn M.J. (1965) Premating isolation in the *Hyla ewingi* complex (Anura: Hylidae). *Evolution* **19**, 234–243.
9.6

Littlejohn M.J. & Loftus-Hills J.J. (1968) An experimental evaluation of premating isolation in the *Hyla ewingi* complex (Anura: Hylidae). *Evolution* **22**, 659–663.
9.6

Lloyd D.G. (1972a) Breeding systems in *Cotula* L. (Compositae, Anthemideae). I. The array of monoclinous and diclinous systems. *New Phytol.* **71**, 1181–1194.
14.3.2

Lloyd D.G. (1972b) Breeding systems in *Cotula* L. (Compositae, Anthemideae). II. Monoecious populations. *New Phytol.* **71**, 1195–1202.
14.3.2

Lloyd D.G. (1977) Genetic and phenotypic models of natural selection. *J. theor. Biol.* **69**, 543–560.
3.2

Lloyd D.G. (1979a) Some reproductive factors affecting the selection of self-fertilization in plants. *Amer. Natur.* **113**, 67–79.
14.3.2

Lloyd D.G. (1979b) Parental strategies of angiosperms. *N.Z. J. Bot.* **17**, 595–606.
14.3.1, 14.3.2

Lloyd D.G. (1980) Sexual strategies in plants III. A quantitative method for describing the gender of plants. *N.Z. J. Bot.* **18**, 103–108.
14.3.1

Lloyd J.E. (1979) Mating behavior and natural selection. *Fla. Entomol.* **62**, 17–34.
9.2.2, 9.2.3

Lorenz K. (1966) *On Aggression.* Methuen, London.
15.2.2

Lovett Doust J. & Cavers P.B. (1982) Sex and gender dynamics in jack-in-the-pulpit *Arisaema triphyllum* (L.) Schott (Araceae). *Ecology* **63**, 797–808.
14.2.3

Lubin Y.D. & Robinson M.H. (1982) Dispersal by swarming in a social spider. *Science* **216**, 319–321.
13.2

Lundberg A. (1981) Population ecology of the Ural owl *Strix uralensis* in Central Sweden. *Ornis Scand.* **12**, 111–119.
6.3.1

McArthur E.D. (1977) Environmentally induced changes of sex expression in

Atriplex canescens. Heredity **38**, 97–103.
14.2.1

McArthur E.D. & Freeman D.C. (1982) Sex reversal in *Atriplex canescens. Bot. Gaz.*
14.2.1

MacArthur R.H. (1968) Selection for life tables in periodic environments. *Amer. Natur.* **102**, 381–383.
11.6

MacArthur R.H. & Pianka E.R. (1966) On the optimal use of a patchy environment. *Amer. Natur.* **100**, 603–609.
4.2.1

MacArthur R.H. & Wilson E.O. (1967) *The Theory of Island Biogeography.* Princeton University Press, Princeton.
11.3

McCann T.S. (1982) Aggressive and maternal activities of female southern elephant seals (*Mirounga leonina*). *Anim. Behav.* **30**, 268–276.
5.1.2

McClintock M.K., Anisko J.J. & Adler N.T. (1982) Group mating among Norway rats II. The social dynamics of copulation: competition, cooperation, and mate choice. *Anim. Behav.* **30**, 410–425.
5.1.2

McDiarmid R.W. (1978) Evolution of parental care in frogs. In: *The Development of Behavior: Comparative and Evolutionary Aspects* (ed. G.M. Burghardt & M. Beckoff), pp. 127–147. Garland STPM Press, New York.
10.3.2

Macdonald D.W. (1979) Helpers in fox society. *Nature, Lond.* **282**, 69–71.
12.1, 12.3.3, 12.5

Macdonald D.W. (1980) Social factors affecting reproduction in the red fox, *Vulpes vulpes.* In: The Red Fox. Symp. on Behavior and Ecology, *Biogeographica* **18** (ed. E. Zimen). W. Junk, The Hague.
12.1, 12.3.3, 12.4, 12.5

Macdonald D.W. & Apps P.J. (1978) The social behavior of a group of semi-dependent farm cats. *Felis catus. Carnivore Genetics Newsletter* **3**, 256–268
12.1

Macdonald D.W. & Moehlman P.D. (1982) Cooperation, altruism, and restraint in the reproduction of carnivores. In: *Perspectives in Ethology,* Vol. 5 (ed. P. Klopfer & P. Bateson), pp. 433–466. Plenum Press, New York.
12.1, 12.3.2, 12.4, 12.5

Mace G.M. (1979) The evolutionary ecology of small mammals. Unpublished DPhil thesis, University of Sussex.
1.2.2, 1.2.3

Mace G.M. & Harvey P.H. (1983) Energetic constraints on home-range size. *Amer. Natur.* (in press).
1.1

Mace G.M., Harvey P.H. & Clutton-Brock T.H. (1981) Brain size and ecology in small mammals. *J. Zool.* **193**, 333–354.
1.2.5

Mace G.M., Harvey P.H. & Clutton-Brock T.H. (1983) Vertebrate home-range sizes and metabolic requirements. In: *The Ecology of Animal Movement* (ed. I. Swingland & P.J. Greenwood), pp. 32–53. Oxford University Press, Oxford.
1.1

Macevicz S. & Oster G. (1976) Modelling social insect populations II. Optimal

reproductive strategies in annual eusocial insect colonies. *Behav. Ecol. Sociobiol.* **1**, 265–282.
4.4.1

McCleery R.H. (1977) On satiation curves. *Anim. Behav.* **25**, 1005–1015.
4.4.1

McCleery R.H. (1978) Optimal behaviour sequences and decision making. In: *Behavioural Ecology: An Evolutionary Approach* (ed. J.R. Krebs & N.B. Davies), pp. 377–410. Blackwell Scientific Publications, Oxford.
15.2.2

McCleery R.H. (1983) Interactions between activities. In: *Causes and Effects (Animal Behaviour, Vol. I)* (ed. T.R. Halliday & P.J.B. Slater), pp. 132–165. Blackwell Scientific Publications, Oxford.
4.6

McFarland D.J. & Houston A.I. (1981) *Quantitative Ethology. The State Space Approach*. Pitman, London.
4.1.2, 4.2.2, 4.4.2, 7.3.1

McGregor P.K. & Krebs J.R. (1982a) Mating and song types in the great tit. *Nature, Lond.* **297**, 60–61.
9.3.1

McGregor P.K. & Krebs J.R. (1982b) Song types in a population of great tits (*Parus major*): Their distribution, abundance, and acquisition by individuals. *Behaviour* **79**, 126–152.
7.3.2

McGregor P.K., Krebs J.R. & Perrins C.M. (1981) Song repertoires and lifetime reproductive success in the great tit (*Parus major*). *Amer. Natur.* **118**, 149–159.
3.2.1

McLain D.K. (1982) Behavioural and morphological correlates of male dominance and courtship in the blister beetle *Epicauta pennsylvanica* (Coleoptera: Meloidae). *Amer. Midl. Nat.* **107**, 396–403.
9.4

MacLean S.F. & Seastedt T.R. (1979) Avian territoriality: sufficient resources or interference competition. *Amer. Natur.* **114**, 308–312.
6.4

McNair J.N. (1981) A stochastic foraging model with predator training effects. II. Optimal diets. *Theor. Popul. Biol.* **19**, 147–162.
7.4.2

McNair J.N. (1982) Optimal giving up times and the marginal value theorem. *Amer. Natur.* **119**, 511–529.
4.5, 7.4.4

McNair M.R. & Parker G.A. Models of parent–offspring conflict. III. Intrabrood conflict. *Anim. Behav.* **27**, 1202–1209.
4.2.1

McNamara J.M. (1982) Optimal patch use in a stochastic environment. *Theor. Popul. Biol.* **21**, 269–288.
4.2.3

McNamara J.M. & Houston A.I. (1982) Short term behaviour and lifetime fitness. In: *Functional Ontogeny* (ed. D.J. McFarland), pp. 60–87. Pitman, London.
4.4.3, 4.5.1, 6.3.5, 7.4.4

McNamara J.M. & Houston A.I. (1983) Optimal responding on a variable interval schedule. *Behav. Analysis Lett.* **3**, 157–170.
7.4.4

MacRoberts M.H. & MacRoberts B.R. (1976) Social organization and behavior

of the acorn woodpecker in Central Coastal California. *Ornithol. Monogr.* **21**, 1–115.
12.1, 12.3.3, 12.3.4, 12.4

Mader W.J. (1975) Biology of the Harris' hawk in Southern Arizona. *Living Bird* **14**, 59–85.
12.5

Mader W.J. (1979) Breeding behavior of a polyandrous trio of Harris' hawks in Southern Arizona. *Auk* **96**, 776–788.
12.5

Maekawa T. (1924) On the phenomena of sex transition in *Arisaema japonica*. *Jour. Coll. Agr. Hokkaido Imp. Univ.* **13**, 217–305.
14.2.3

Maheshwari P. (1950) *An Introduction to the Embryology of Angiosperms*. McGraw-Hill, New York.
14.4

Majerus M.E.N., O'Donald P. & Weir J. (1982) Female mating preference is genetic. *Nature, Lond.* **300**, 521–523.
9.1

Major P. (1978) Predator–prey interactions in two schooling fishes, *Caranx ignobilis* and *Stolephorus purpureus*. *Anim. Behav.* **26**, 760–777.
5.3.1

Malyshev S.I. (1968) *Genesis of the Hymenoptera and the Phases of their Evolution*. Methuen, London.
13.3, 13.4

Manning A. (1976) Animal learning: ethological approaches. In: *Neural Mechanisms of Learning and Memory* (ed. M.R. Rosenzweig & E.L. Bennett), pp. 147–158. MIT Press, Cambridge, Mass.
7.2.1, 7.3.3, 7.5.2

Manning J.T. (1975) Male discrimination and investment in *Asellus aquaticus* (L.) and *A. meridianus* Racovitsza (Crustacea: Isopoda). *Behaviour* **55**, 1–14.
9.2.4, 9.4

Markl H. (1984) Modulating signals in animal communication. *Fortschr. Zool.* (in press).
15.2.4

Marler P. (1955) Characteristics of some alarm calls. *Nature, Lond.* **176**, 6–8.
15.3.2, 15.4.2

Marler P. (1959) Developments in the study of animal communication. In: *Darwin's Biological Work. Some Aspects Reconsidered* (ed. P.R. Bell), pp. 150–206. John Wiley & Sons, New York.
15.2.2, 15.4.2

Marler P. & Peters S. (1982) Long-term storage of learned birdsongs prior to production. *Anim. Behav.* **30**, 479–482.
7.3.1

Marshall L.D. (1982) Male nutrient investment in the Lepidoptera: what nutrients should males invest? *Amer. Natur.* **120**, 273–279.
9.2.4

Martin R.D. (1981a) Relative brain size and basal metabolic rate in terrestrial vertebrates. *Nature, Lond.* **293**, 57–60.
1.1, 1.2.2, 1.2.5

Martin R.D. (1981b) Field studies of primate behaviour. *Symp. Zool. Soc. Lond.* **46**, 287–336.
1.2.1, 1.2.4

Martin S.G. (1974) Adaptations for polygynous breeding in the bobolink,

Dolichonyx oryzivorous. Amer. Zool. **14**, 109–119.
10.3.1

Martindale S. (1982) Nest defence and central place foraging: a model and experiment. *Behav. Ecol. Sociobiol.* **10**, 85–90.
4.3

Matessi C. & Jayakar S.D. (1976) Conditions for the evolution of altruism under Darwinian selection. *Theor. Popul. Biol.* **9**, 360–387.
3.4.3

Matthews R.W. (1968a) *Microstigmus comes*: sociality in a sphecid wasp. *Science* **160**, 787–788.
13.3

Matthews R.W. (1968b) Nesting biology of the social wasp *Microstigmus comes* (Hymenoptera: Sphecidae, Pemphredoninae). *Psyche* **75**, 23–45.
13.3

Matthews R.W. (1982) Social parasitism in yellowjackets (*Vespula*). In: *Social Insects in the Tropics* (ed. P. Jaisson), pp. 194–202. Université Paris-Nord.
13.7

Matthews R.W. & Starr C.K. (in press) *Microstigmus comes* wasps have a method of nest construction unique among social insects. *Biotropica*.
13.3

May R.M. & Rubenstein D.I. (1982) Reproductive strategies. In: *Reproductive Fitness (Reproduction in Mammals, Volume 4)* (ed. C.R. Austin & R.V. Short). Cambridge University Press, Cambridge.
1.2.3, 11.3

Maynard Smith J. (1964) Group selection and kin selection. *Nature, Lond.* **201**, 1145–1147.
8.1

Maynard Smith J. (1966) Sympatric speciation. *Amer. Natur.* **100**, 637–650.
9.6

Maynard Smith J. (1968) Evolution in sexual and asexual populations. *Amer. Natur.* **102**, 469–473.
8.1

Maynard Smith J. (1972) *On Evolution.* Edinburgh University Press, Edinburgh.
2.1, 2.1.2, 2.1.3, 2.5, 15.4.2

Maynard Smith J. (1974a) The theory of games and the evolution of animal conflicts. *J. theor Biol.* **47**, 209–221.
2.1.3, 2.5, 2.5.1, 5.4.3

Maynard Smith J. (1974b) *Models in Ecology.* Cambridge University Press, Cambridge.
10.3.1

Maynard Smith J. (1976a) Evolution and the theory of games. *Am. Scient.* **64**, 41–45.
2.5.2, 5.1.2, 11.3

Maynard Smith J. (1976b) Group selection. *Q. Rev. Biol.* **51**, 277–283.
3.4.2, 3.4.3, 8.1

Maynard Smith J. (1976c) A short-term advantage for sex and recombination through sib-competition. *J. theor. Biol.* **63**, 245–258.
8.1.2

Maynard Smith J. (1977) Parental investment—a prospective analysis. *Anim. Behav.* **25**, 1–9.
8.4, 10.3.2

Maynard Smith J. (1978a) Optimization theory in evolution. *A. Rev. Ecol.*

Syst. **9**, 31–56.

3.2.1, 4.1, 9.1

Maynard Smith J. (1978b) *The Evolution of Sex*. Cambridge University Press, Cambridge.

8.1, 8.1.2, 9.3.2

Maynard Smith J. (1979) Game theory and the evolution of behaviour. *Proc. R. Soc. Lond. B* **205**, 475–488.

2.2, 2.7.1, 15.4.2

Maynard Smith J. (1980a) Selection for recombination in a polygenic model. *Genet. Res.* **35**, 269–277.

8.1.2

Maynard Smith J. (1980b) A new theory of sexual investment. *Behav. Ecol. Sociobiol.* **7**, 247–251.

8.2

Maynard Smith J. (1981) Will a sexual population evolve to an ESS? *Amer. Natur.* **117**, 1015–1018.

2.9.1, 9.1

Maynard Smith J. (1982a) *Evolution and the Theory of Games*. Cambridge University Press, Cambridge.

2.1, 2.2, 2.3, 2.5.3, 2.6.2, 2.7.1, 2.9.1, 3.2.2, 6.5.4, 9.1, 10.0, 10.3.1, 10.3.2, 15.4

Maynard Smith J. (1982b) The evolution of social behaviour—a classification of models. In: *Current Problems in Sociobiology* (ed. King's College Sociobiology Group). Cambridge University Press, Cambridge.

3.3.1

Maynard Smith J. & Parker G.A. (1976) The logic of asymmetric contests. *Anim. Behav.* **24**, 159–175.

2.2, 2.5, 2.5.1, 2.5.2, 2.5.3, 5.1.2

Maynard Smith J. & Price G.R. (1973) The logic of animal conflict. *Nature, Lond.* **246**, 15–18.

2.1.3, 2.5, 3.2.1

Maynard Smith J. & Ridpath M.G. (1972) Wife sharing in the Tasmanian native hen, *Tribonyx mortierrii*: a case of kin selection? *Amer. Natur.* **106**, 447–452.

10.3.2

Mayr E. (1963) *Animal Species and their Evolution*. Belknap Press, Cambridge, Mass.

Mazur J.E. (1981) Optimization theory fails to predict performance of pigeons in a two-response situation. *Science* **214**, 823–825.

4.5, 7.4.4

Mech L.D. (1970) *The Wolf: The Ecology and Behaviour of an Endangered Species*. Natural History Press, New York.

5.3.1, 12.1, 12.4, 12.5

Mech L.D. (1975) Disproportionate sex ratios of wild dogs. *J. Wildl. Manag.* **39**, 737–740.

12.3.3

Metcalf R.A. & Whitt G.S. (1977a) Intra-nest relatedness in the social wasp *Polistes metricus*. A genetic analysis. *Behav. Ecol. Sociobiol.* **2**, 339–351.

13.6

Metcalf R.A. & Whitt G.S. (1977b) Relative inclusive fitness in the social wasp *Polistes metricus*. *Behav. Ecol. Sociobiol.* **2**, 353–360.

3.3.3, 5.3.3, 13.6

Meyer P.L. (1970) *Introductory Probability and Statistical Applications*.

Addison-Wesley, London.
1.2.4, 10.0, 10.3.1

Michener C.D. (1958) The evolution of social behavior in bees. *Proc. X Int. Congr. Entomol.* 1956 **2**, 442–447.
13.5

Michener C.D. (1969) Comparative social behavior of bees. *A. Rev. Ent.* **14**, 299–342.
13.3, 13.4, 13.5

Michener C.D. (1974) *The Social Behavior of the Bees: A Comparative Study.* Harvard University Press, Cambridge, Mass.
13.2, 13.3, 13.4, 13.5, 13.7

Michener C.D. & Brothers D.J. (1974) Were workers of eusocial Hymenoptera initially altruistic or oppressed? *Proc. natl Acad. Sci., USA* **71**, 671–674.
13.6

Michener C.D. & Lange R.B. (1958a) Distinctive type of primitive social behavior among bees. *Science* **127**, 1046–1047.
13.6

Michener C.D. & Lange R.B. (1958b) Observations on the behavior of Brasilian halictid bees (Hym., Apoidea), V. *Chloralictus. Insect. Soc.* **5**, 379–407.
13.7

Michod R.E. (1982) The theory of kin selection. *A. Rev. Ecol. Syst.* **13**, 23–55.
3.3.4

Michod R.E. & Hamilton W.D. (1980) Coefficients of relatedness in sociobiology. *Nature, Lond.* **288**, 694–697.
3.3.4

Milinski M. (1978) Kin selection and reproductive value. *Z. Tierpsychol.* **47**, 328–329.
11.5

Milinski M. (1979a) Can an experienced predator overcome the confusion of swarming prey more easily? *Anim. Behav.* **27**, 1122–1126.
5.3.2

Milinski M. (1979b) An evolutionarily stable feeding strategy in sticklebacks. *Z. Tierpsychol.* **51**, 36–40.
2.3, 2.7.1, 5.4.3

Milinski M. & Heller R. (1978) Influence of a predator on the optimal foraging behaviour of sticklebacks *Gasterosteus aculeatus. Nature, Lond.* **275**, 642–644.
4.4.2

Millar J.S. (1977) Adaptive features of mammalian reproduction. *Evolution* **31**, 370–386.
1.1, 1.2.2, 1.2.3, 1.2.4

Millar J.S. (1981) Pre-partum reproductive characteristics of eutherian mammals. *Evolution* **36**, 1149–1163.
1.1

Mills M.G.L. (1982) Mating system of the Brown hyaena. *Behav. Ecol. Sociobiol.* **10**, 131–136.
12.1

Milton K. (1979) Factors influencing leaf choice by howler monkeys: a test of some hypotheses of food selection by generalist herbivores. *Amer. Natur.* **114**, 362–368.
4.2.2

Milton K. & May M.L. (1976) Body weight, diet and home range area in primates. *Nature, Lond.* **259**, 459–462.
1.1, 1.2.2

Mockford E.L. (1971) Parthenogenesis in psocids (Insecta: Psocoptera). *Amer. Zool.* **11**, 327–339.
8.1.1

Moehlman P.D. (1979) Jackal helpers and pup survival. *Nature, Lond.* **277**, 382–383.
12.1, 12.1.1, 12.2, 12.4

Moehlman P.D. (1983) Socioecology of silverbacked and golden jackals, *Canis mesomelas* and *C. aureus*. In: *Recent Advances in the Study of Mammalian Behavior* (ed. J.F. Eisenberg & D.G. Kleiman). Special publication number 7 of the American Society of Mammalogists.
12.1, 12.1.1, 12.2, 12.4

Mooney H.A. & Gulmon (1982) Constraints on leaf structure and function in reference to herbivory. *BioScience* **32**, 198–206.
14.1

Moore N.J. (1972). The ethology of the Mexican junco (*Junco phaeonotus palliatus*). PhD dissertation, University of Arizona, Tucson.
5.1.1, 5.1.2

Moore W.S. (1976) Components of fitness in the unisexual fish *Poeciliopsis monacha-occidentalis*. *Evolution* **30**, 564–578.
8.1.1

Morris D.J. (1956) The feather postures of birds and the problem of the origin of social signals. *Behaviour* **9**, 75–114.
15.1', 15.2.2

Morse D.H. (1970) Ecological aspects of some mixed species foraging flocks of birds. *Ecol. Monogr.* **40**, 119–168.
5.3.1

Morton E.S. (1975) Ecological sources of selection on avian sounds. *Amer. Natur.* **109**, 17–34.
15.3.1

Muggleton J. (1979) Non-random mating in wild populations of polymorphic *Adalia bipunctata*. *Heredity* **42**, 57–65.
9.5

Muller H.J. (1932). Some genetic aspects of sex. *Amer. Natur.* **66**, 118–138.
8.1

Muller H.J. (1964) The relation of recombination to mutational advance. *Mutation Res.* **1**, 2–9.
8.1

Myers J.P., Connors P.G. & Pitelka F.A. (1979) Territory size in wintering sanderlings: the effects of prey abundance and intruder density. *Auk* **96**, 551–561.
6.3.4

Myers J.P., Connors P.G. & Pitelka F.A. (1981) Optimal territory size and the sanderling: compromise in a variable environment. In: *Foraging Behaviour: Ecological, Ethological and Psychological Approaches* (ed. A.C. Kamil & T.D. Sargent), pp. 135–158. Garland STPM Press, New York.
6.3.1

Nakatsuru K. & Kramer D.L. (1982) Is sperm cheap? Limited male fertility and female choice in the lemon tetra (Pisces, Characidae). *Science* **216**, 753–755.
9.2.1, 9.2.4

Nalepa C.A. (1982) Colony composition of the woodroach *Cryptocercus punctulatus*. In: *The Biology of Social Insects: Proceedings of the Ninth Congress of the International Union for The Study of Social Insects, Boulder, Colorado,*

August 1982 (ed. M.D. Breed, C.D. Michener & H.E. Evans), p. 181. West-view Press, Boulder, Colorado.
13.4

Nalepa C. (1984) Colony composition, protozoan transfer and some life history characteristics of the woodroach *Cryptocercus punctulatus* Scudder (Dictyoptera: Cryptocercidae). *Behav. Ecol. Sociobiol.* (in press).
13.4, 13.6

Neal E. (1970) The banded mongoose, *Mungos mungo. East Afr. Wildl. J.* **8**, 63–71.
12.1, 12.1.2

Neal E.G. (1977) *Badgers.* Blandford Press, Poole, Dorset, England.
12.1, 12.3

Neill S.R. St. J. & Cullen J.M. (1974) Experiments on whether schooling by their prey affects the hunting behaviour of cephalopods and fish predators. *J. Zool.* **172**, 549–569.
4.4.2

Nelson K. (1964) The temporal patterning of courtship behaviour in the glandulocaudine fishes (*Ostariophysi, Characidae*). *Behaviour* **24**, 90–146.
15.2.2

Newton I. & Marquiss M. (1979) Sex ratio among nestlings of the European sparrowhawk. *Amer. Natur.* **113**, 309–315.
8.2

Noonan K.M. (1978) Sex ratio of parental investment in colonies of the social wasp *Polistes fuscatus. Science* **199**, 1354–1356.
13.7

Noonan K.M. (1981) Individual strategies of inclusive-fitness-maximizing in *Polistes fuscatus* foundresses. In: *Natural Selection and Social Behavior: Recent Research and New Theory* (ed. R.D. Alexander & D.W. Tinkle), pp. 18–44. Chiron Press, New York.
3.3.3, 13.5, 13.6, 13.7

Noordwijk A.J. van & Scharloo W. (1981) Inbreeding in an island population of the great tit. *Evolution* **35**, 674–688.
9.3.1

Norman R.F., Taylor P.D. & Robertson R.J. (1977) Stable equilibrium strategies and penalty functions in a game of attrition. *J. theor. Biol.* **65**, 571–578.
2.5.1

Nottebohm F. (1975) Continental patterns of song variability in *Zonotrichia capensis*: some possible ecological correlates. *Amer. Natur.* **109**, 605–624.
15.3.1

Nudds T. (1978) Convergence of group size strategies by mammalian social carnivores. *Amer. Natur.* **112**, 957–960.
5.3.1

Oaten A. (1977) Optimal foraging in patches: case for stochasticity. *Theor. Popul. Biol.* **12**, 263–285.
4.2.3

O'Donald P. (1967) A general model of sexual and natural selection. *Heredity* **22**, 499–518.
9.2.4

O'Donald P. (1980) *Genetic Models of Sexual Selection.* Cambridge University Press, Cambridge.
9.1, 9.2.4

O'Donald P. (1982) The concept of fitness in population genetics and socio-biology. In: *Current Problems in Sociobiology* (ed. King's College Sociobiology Group), pp. 65–86. Cambridge University Press, Cambridge.
2.1.2

O'Donald P. & Muggleton J. (1979) Melanic polymorphism in ladybirds maintained by sexual selection. *Heredity* **43**, 143–148.
9.5

Ollason J.G. (1980) Learning to forage—optimally? *Theor. Popul. Biol.* **18**, 44–56.
4.2.3, 7.4.4

Olton D.S. & Samuelson R.J. (1976) Remembrance of places passed: spatial memory in rats. *J. Exp. Psychol.: Anim. Behav. Processes* **2**, 97–116.
7.4.2

Ord J.K. (1979) Time-series and spatial patterns in ecology. In: *Spatial and Temporal Analysis in Ecology* (ed. R.M. Cormack & J.K. Ord). International Cooperative Publishing House, Fairland, Maryland.
5.2.1

Orians G.H. (1961) The ecology of blackbirds (*Agelaius*) social systems. *Ecol. Monogr.* **31**, 285–312.
1.1, 1.2.1, 10.0

Orians G.H. (1969) On the evolution of mating systems in birds and mammals. *Amer. Natur.* **103**, 589–603.
9.2.1, 10.3.1, 11.3

Orians G.H. (1980) *Some Adaptations of Marsh-Nesting Blackbirds.* Princeton University Press, Princeton, N.J.
1.2.1

Orians G.H. (1981) Foraging behavior and the evolution of discriminatory abilities. In: *Foraging Behavior: Ecological, Ethological, and Psychological Approaches* (ed. A.C. Kamil & T.D. Sargent), pp. 389–405. Garland STPM Press, New York.
7.4.3, 7.4.4, 7.5.2

Orians G.H. & Collier G. (1963) Competition and blackbird social systems. *Evolution* **17**, 449–459.
5.2.2

Orians G.H. & Pearson N.E. (1979) On the theory of central place foraging. In: *Analysis of Ecological Systems* (ed. D.J. Horn, R. Mitchell & G.R. Stair), pp. 155–177. Ohio State University Press, Columbus.
4.2.1

Oring L.W. (1982) Avian mating systems. In: *Avian Biology, Vol. VI* (ed. D. Farner, J. King & K. Parkes), pp. 1–92. Academic Press, New York.
10.3.2, 10.3.3

Oring L.W. & Knudson M.L. (1972) Monogamy and polyandry in the spotted sandpiper. *Living Bird* **11**, 59–73.
10.3.2

Oring L.W. & Maxson S.S. (1978) Instances of simultaneous polyandry by spotted sandpipers, *Actitis mascularia*. *Ibis* **120**, 349–353.
10.3.2

Oster G.F. & Alberch P. (1982) Evolution and bifurcation in developmental programmes. *Evolution* **36**, 444–459.
4.1.1

Oster G.F. & Wilson E.O. (1978) *Caste and Ecology in the Social Insects.* Princeton University Press, Princeton.
4.1, 4.4.1, 5.3.1, 5.4.3, 13.2

Owens D. & Owens M. (1979a) Communal denning and clan association in brown hyena (*Hyaena brunnea*) of the Central Kalahari Desert. *Afr. J. Ecol.* **17**, 35–44.
12.1

Owens D. & Owens M. (1979b) Notes on the social organization and behavior in brown hyena. *J. Mammal.* **60**, 405–408.
12.1

Owen Smith N. & Novellie P. (1982) What should a clever ungulate eat? *Amer. Natur.* **119**, 151–178.
4.2.2

Packer C. (1975) Male transfer in olive baboons. *Nature, Lond.* **255**, 219–220.
8.3

Packer C. (1979) Inter-troop transfer and inbreeding avoidance in *Papio anubis*. *Anim. Behav.* **27**, 1–36.
9.3.1, 9.3.2

Pamilo P. (1982) Genetic population structure in polygynous *Formica* ants. *Heredity* **48**, 95–106.
13.6

Pardi L. (1948) Dominance order in *Polistes* wasps. *Physiol. Zool.* **21**, 1–13.
13.7

Parker E.D. & Selander R.K. (1976) The organization of genetic diversity in the parthenogenetic lizard *Cnemidophorus tesselatus*. *Genetics* **84**, 791–805.
8.1.1

Parker G.A. (1970a) The reproductive behaviour and the nature of sexual selection in *Scatophaga stercoraria* L. (Diptera: Scatophagidae) II. The fertilization rate and the spatial and temporal relationships of each sex around the site of mating and oviposition. *J. Anim. Ecol.* **39**, 205–228.
2.1.3, 2.3

Parker G.A. (1970b) Sperm competition and its evolutionary effect on copula duration in the fly *Scatophaga stercoraria*. *J. Insect Physiol.* **16**, 1301–1328.
1.1, 2.1.3

Parker G.A. (1970c) The reproductive behavior and the nature of sexual selection in *Scatophaga stercoraria* L. (Diptera: Scatophagidae). VII. The origin and evolution of the passive phase. *Evolution* **24**, 774–788.
2.1.3

Parker G.A. (1974a) The reproductive behavior and the nature of sexual selection in *Scatophaga stercoraria* L. (Diptera: Scatophagidae). IX. Spatial distribution of fertilization rates and evolution of male search strategy within the reproductive area. *Evolution* **28**, 93–108.
2.3, 2.7.1

Parker G.A. (1974b) Assessment strategy and the evolution of fighting behaviour. *J. theor. Biol.* **47**, 223–243.
2.5, 2.5.2, 9.2.5, 15.4.2

Parker G.A. (1978a) Evolution of competitive mate searching. *A. Rev. Ent.* **23**, 173–196.
2.2, 2.7.1

Parker G.A. (1978b) Searching for mates. In: *Behavioural Ecology: An Evolutionary Approach* (ed. J.R. Krebs & N.B. Davies), pp. 214–244. Blackwell Scientific Publications, Oxford.
2.3, 2.7.1, 4.2.1, 11.3

Parker G.A. (1979) Sexual selection and sexual conflict. In: *Sexual Selection and Reproductive Competition in Insects* (ed. M.S. Blum & N.A. Blum), pp. 123–166. Academic Press, New York.
9.3.2, 15.2.4

Parker G.A. (1982) Phenotype-limited evolutionarily stable strategies. In: *Current Problems in Sociobiology* (ed. King's College Sociobiology Group), pp. 173–201. Cambridge University Press, Cambridge.
2.2, 2.3, 2.8, 9.2.3, 9.2.4, 9.2.5

Parker G.A. (1983a) Arms races in evolution—an ESS to the opponent—independent costs game. *J. theor. Biol.* (in press).
2.3

Parker G.A. (1983b) Mate quality and mating decisions. In: *Mate Choice* (ed. P.P.G. Bateson), pp. 141–164. Cambridge University Press, Cambridge.
9.2.4

Parker G.A. & Knowlton N. (1980) The evolution of territory size—some ESS models. *J. theor. Biol.* **84**, 445–476.
2.3, 2.6.1, 2.8, 6.4

Parker G.A. & MacNair M.R. (1978) Models of parent–offspring conflict. I. Monogamy. *Anim. Behav.* **26**, 97–110.
2.6.2

Parker G.A. & MacNair M.R. (1979) Models of parent–offspring conflict. II. Suppression: evolutionary retaliation of the parent. *Anim. Behav.* **27**, 1210–1235.
2.3, 2.6.2

Parker G.A. & Rubenstein D.I. (1981) Role assessment reserve strategy, and acquisition of information in asymmetric animal contests. *Anim. Behav.* **29**, 221–240.
2.5.2, 2.5.3, 2.9.1

Parker G.A., Baker R.R. & Smith V.G.F. (1972) The origin and evolution of gamete dimorphism and the male–female phenomenon. *J. theor. Biol.* **36**, 529–553.
2.1.3

Parry V. (1973) The auxiliary social system and its effect on territory and breeding in kookaburras. *Emu* **73**, 81–100.
12.1

Partridge L. (1978) Habitat selection. In: *Behavioural Ecology: An Evolutionary Approach* (ed. J.R. Krebs & N.B. Davies), pp. 351–376. Blackwell Scientific Publications, Oxford.
9.3.1

Partridge L. (1980) Mate choice increases a component of offspring fitness in fruit flies. *Nature, Lond.* **283**, 290–291.
9.2.3, 10.3.3

Partridge L. (1983) Non-random mating and offspring fitness. In: *Mate Choice* (ed. P.P.G. Bateson). Cambridge University Press, Cambridge.
9.1, 9.5

Partridge L. & Farquhar M. (1981) Sexual activity reduces lifespan of male fruitflies. *Nature, Lond.* **294**, 580–582.
9.2.1, 9.2.3

Partridge L. & Farquhar M. (1983) Lifetime mating success of male fruitflies (*Drosophila melanogaster*) is related to their size. *Anim. Behav.* **31**, 871–877.
9.2.4

Payne R. & Payne K. (1977) Social organization and mating success in local song populations of village indigo birds, *Vidua chalybeata. Z. Tierpsychol.* **45**, 113–173.
10.3.3

Pennycuick C.J. (1979) Energy costs of locomotion and a concept of 'foraging radius'. In: *Serengeti: Dynamics of an Ecosystem* (ed. A.R.E. Sinclair & M. Norton-Griffiths), pp. 164–184. University of Chicago Press, Chicago.
1.3

Perrill S.A., Gerhardt H.C. & Daniel R.E. (1978) Mating strategy shifts in male green treefrogs (*Hyla cinerea*): an experimental study. *Anim. Behav.* **30**, 43–48.
6.5.1, 9.2.5

Perrins C.M. (1979) *British Tits.* Collins, London.
6.3.1, 15.3.3

Perrone M. & Zaret T.M. (1979) Parental care patterns of fishes. *Amer. Natur.* **113**, 351–361.
10.3.2

Petit C. (1958) Le déterminisme génétique et psychophysiologique de la compétition sexuelle chez *Drosophila melanogaster. Bull. biol. Fr. Belg.* **92**, 248–329.
9.5

Petrie M. (1983) Mate choice in role-reversed species. In: *Mate Choice* (ed. P.P.G. Bateson). Cambridge University Press, Cambridge.
9.2.1

Petrinovich L., Patterson T. & Baptista L.F. (1981) Song dialects as barriers to dispersal: a re-evaluation. *Evolution* **35**, 180–188.
9.4

Pianka E.R. & Parker W.S. (1975) Age-specific reproductive tactics. *Amer. Natur.* **109**, 453–464.
11.5

Pitcher T.J., Magurran A.E. & Winfield I.J. (1982) Fish in larger shoals find food faster. *Behav. Ecol. Sociobiol.* **10**, 149–151.
5.3.1

Pleszczynska W.K. (1978) Microgeographic prediction of polygyny in the lark bunting. *Science* **201**, 935–937.
10.3.1, 11.3

Pleszczynska W. & Hansell R. (1980) Polygyny and decision theory: testing of a model in lark buntings (*Calamospiza melanocorys*). *Amer. Natur.* **116**, 821–830.
10.3.1

Plotkin H.C. & Odling Smee F.J. (1979) Learning, change, and evolution: an enquiry into the teleonomy of learning. *Adv. Stud. Behav.* **10**, 1–41.
7.1

Policansky D. (1981) Sex choice and the size advantage model in jack-in-the-pulpit (*Arisaema triphyllum*). *Proc. natl Acad. Sci., USA* **78**, 1306–1308.
14.2.3

Policansky D. Sex change, differential sex-specific mortality and a puzzle: Why don't more organisms change sex? (in prep.)
14.2.2

Popp J.L. & Devore I. (1979) Aggressive competition and social dominance theory: synopsis. In: *The Great Apes* (ed. D.A. Hamburg & E.R. McCown), pp. 316–338. Benjamin/Cummings, California.
2.5.2

Post D., Hausfater G. & McCluskey S. (1980) Feeding behavior of yellow baboons (*Papio cynocephalus*): relationship to age, gender, and dominance rank. *Folia primatol.* **34**, 170–195.
5.1.2

Powell G.V.N. (1974) Experimental analysis of the social value of flocking by starlings (*Sturnus vulgaris*) in relation to predation and foraging. *Anim. Behav.* **22**, 501–505.
5.3.2

Price F.E. & Bock C.E. (1973) Polygyny in the dipper. *Condor* **75**, 457–459.
10.3.1

Price G.R. (1970) Selection and covariance. *Nature, Lond.* **227**, 520–521.
3.3.3, 3.4.2

Price G.R. (1972) Extension of covariance selection mathematics. *Ann. hum. Genet.* **35**, 485–490.
3.3.3, 3.4.2

Proctor M. & Yeo P. (1973) *The Pollination of Flowers*. Collins, London.
15.2.1

Pulliam H.R. (1973a) On the advantages of flocking. *J. theor. Biol.* **38**, 419–422.
5.2.1, 5.3.2

Pulliam H.R. (1973b) Comparative feeding ecology of a tropical grassland finch (*Tiaris olivacea*). *Ecology* **54**, 284–299.
5.2.2

Pulliam H.R. (1976) The principle of optimal behavior and the theory of communities. In: *Perspectives in Ethology*, Vol. 2 (ed. P.P.G. Bateson & P.H. Klopfer), pp. 311–332. Plenum Press, New York.
5.1.2, 5.3, 5.4, 5.4.2

Pulliam H.R. (1981) Learning to forage optimally. In: *Foraging Behaviour: Ecological, Ethological, and Psychological Approaches* (ed. A.C. Kamil & T.D. Sargent), pp. 379–388. Garland STPM Press, New York.
7.1

Pulliam H.R. & Dunford C. (1980) *Programmed to Learn: An Essay on the Evolution of Culture*. Columbia University Press, New York.
7.1, 7.5.2

Pulliam H.R. & Millikan G.C. (1982) Social organization in the non-reproductive season. In: *Avian Biology* (ed. D.S. Farner & J.R. King). Academic Press, New York.
5.2.1, 5.3.1

Pulliam H.R. & Mills G.S. (1977) The use of space by wintering sparrows. *Ecology* **58**, 1393–1399.
5.3.2

Pulliam H.R., Pyke G.H. & Caraco T. (1982) The scanning behavior of juncos: a game-theoretical approach. *J. theor. Biol.* **95**, 89–103.
5.3.2, 5.4.2

Pyke G.H. (1979) Optimal foraging in bumblebees: Rule of movement between flowers within inflorescences. *Anim. Behav.* **27**, 1167–1181.
7.4.2

Pyke G.H. (1979) The economics of territory size and time budget in the golden-winged sunbird. *Amer. Natur.* **114**, 131–145.
5.2.1, 6.3.3

Pyke G.H. (1981) Optimal foraging in nectar-feeding animals and coevolution with their plants. In: *Foraging Behavior: Ecological, Ethological, and Psychological Approaches* (ed. A.C. Kamil & T.D. Sargent), pp. 19–38.

Garland STPM Press, New York.
7.4.2

Pyke G.H., Pulliam H.R. & Charnov E.L. (1977) Optimal foraging: a selective review of theory and tests. *Q. Rev. Biol.* **52**, 137–154.
4.2, 7.4.1, 7.4.3, 7.4.4

Rachlin H., Battalio R., Kagel J. & Green L. (1981) Maximization theory in behavioral psychology. *Behav. Brain Sci.* **4**, 371–417.
4.1.2

Rapport D.J. (1980) Optimal foraging for complementary resources. *Amer. Natur.* **116**, 324–326.
4.2.2

Real L.A. (1980) Fitness, uncertainty and the role of diversification in evolution and behavior. *Amer. Natur.* **115**, 623–638.
5.3.1

Rechten C., Avery M.I. & Stevens T.A. (1983) Optimal prey selection: why do great tits show partial preferences? *Anim. Behav.* **31**, 576–584.
4.2.1, 7.4.3

Reighord J. (1920) The breeding behavior of the suckers and minnows. *Biol. Bull.* **38**, 1–32.
10.3.2

Reiss M.J. (1982) *Functional Aspects of Reproduction: Some Theoretical Considerations.* Unpublished DPhil thesis, University of Cambridge.
1.2.2

Rescorla R.A. & Holland P.C. (1976) Some behavioral approaches to the study of learning. In: *Neural Mechanisms of Learning and Memory* (ed. M.R. Rosenzweig & E.L. Bennett), pp. 165–192. MIT Press, Cambridge, Mass.
7.1, 7.5.1

Rettenmeyer C.W. (1970) Insect mimicry. *A. Rev. Ent.* **15**, 43–74.
7.2.2

Revusky S. (1977) Learning as a general process with an emphasis on data from feeding experiments. In: *Food Aversion Learning* (ed. N.W. Milgran, L. Krames & T.M. Alloway), pp. 1–47. Plenum Press, New York.
7.2.1

Reyer H.-U. (1980) Flexible helper structure as an ecological adaptation in the pied kingfisher (*Ceryle rudis*). *Behav. Ecol. Sociobiol.* **6**, 219–227.
12.1, 12.3.3

Reznick D. & Endler J.A. (1982) The impact of predation on life history evolution in Trinidadian guppies (*Poecilia reticulata*). *Evolution* **36**, 160–177.
11.2

Rhijn J.G. van (1973) Behavioural dimorphism in male ruffs *Philomachus pugnax* L. *Behaviour* **47**, 153–229.
6.5.1, 9.2.5

Richards A.J. (1973) The origin of *Taraxacum* agamo-species. *Bot. J. Linn. Soc.* **66**, 189–211.
8.1.1

Richards D.G. (1981) Alerting and message components in songs of rufous-sided towhees. *Behaviour* **76**, 223–249.
15.3.1

Richards D.G. & Wiley R.H. (1980) Reverberations and amplitude fluctuations in the propagation of sound in forests: implications for animal com-

munication. *Amer. Natur.* **115**, 381–399.
15.3.1

Richards O.W. & Richards M.J. (1951) Observations on the social wasps of South America (Hymenoptera: Vespidae). *Proc. R. ent. Soc. Lond.* **102**, 1–170.
13.5, 13.6

Ricklefs R.E. (1975) The evolution of cooperative breeding in birds. *Ibis* **117**, 531–534.
12.3.2

Ricklefs R.E. (1981) Fitness, reproductive value, age structure, and the optimization of life-history patterns. *Amer. Natur.* **117**, 819–825.
11.5

Ridley M. (1978) Paternal care. *Anim. Behav.* **26**, 904–932.
10.3.2

Ridley M. (1981) How the peacock got his tail. *New Scient.* **91**, 398–401.
9.2.4

Ridley M. (1983) *The Explanation of Organic Diversity.* Clarendon Press, Oxford.
1.2.6

Ridley M. & Thompson D.J. (1979) Size and mating in *Asellus aquaticus* (Crustacea: Isopoda). *Z. Tierpsychol.* **51**, 380–397.
9.4

Ridpath M.G. (1972) The Tasmanian native hen, *Tribonyx mortierrii. CSIRO Wildl. Res.* **17**, 1–118.
10.3.2, 12.1, 12.3.3, 12.4, 12.5

Riechert S.E. (1978) Games spiders play: behavioral variability in territorial disputes. *Behav. Ecol. Sociobiol.* **3**, 135–162.
2.5.2

Roberts W.A. (1981) Retroactive inhibition in rat spatial memory. *Anim. Learn. & Behav.* **9**, 566–574.
7.4.2

Robertson D.R., Sweatman, H.P.A., Fletcher E.A. & Cleland M.G. (1976) Schooling as a mechanism for circumventing the territoriality of competitors. *Ecology* **57**, 1208–1230.
5.2.2

Robinson A. (1956) The annual reproductive cycle of the magpie, *Gymnorhina dorsalis* Campbell, in South-Western Australia. *Emu* **56**, 233–336.
12.5

Rodman P.S. (1981) Inclusive fitness and group size with a reconsideration of group sizes of lions and wolves. *Amer. Natur.* **118**, 275–283.
12.5

Roell A. (1978) Social behaviour of the jackdaw, *Corvus monedula*, in relation to its niche. *Behaviour* **64**, 1–124.
5.2.2

Rohwer S. (1977) Status signalling in Harris' sparrows: some experiments in deception. *Behaviour* **61**, 107–129.
15.4.2

Rohwer S. (1982) The evolution of reliable and unreliable badges of fighting ability. *Amer. Zool.* **22**, 531–546.
2.5.3, 15.4.2

Rohwer S. & Rohwer F.C. (1978) Status signalling in Harris' sparrows: Experimental deceptions achieved. *Anim. Behav.* **26**, 1012–1022.
5.1.2

Rohwer S. & Ewald P.W. (1981) The cost of dominance and advantage of subordination in a badge signalling system. *Evolution* **35**, 441–454.
15.4.2

Rood, J.P. (1974) Banded mongoose males guard young. *Nature, Lond.* **248**, 176.
12.1, 12.1.2

Rood J.P. (1978) Dwarf mongoose helpers at the den. *Z. Tierpsychol.* **48**, 277–287.
12.1, 12.1.2

Rood J.P. (1980) Mating relationships and breeding suppression in dwarf mongoose. *Anim. Behav.* **28**, 143–150.
12.1, 12.1.2, 12.5

Roper T.J. (1983) Learning as a biological phenomenon. In: *Genes, Development and Learning* (*Animal Behaviour, Vol. 3*) (ed. T.R. Halliday & P.J.B. Slater), pp. 178–212. Blackwell Scientific Publications, Oxford.
7.2.1

Rose M.R. (1982) Antagonistic pleiotropy, dominance and genetic variation. *Heredity* **48**, 63–78.
9.2.3

Rose M.R. & Charlesworth B. (1981a) Genetics of life history in *Drosophila melanogaster*. 1. Sib analysis of adult females. *Genetics* **97**, 173–186.
9.2.3

Rose M.R. & Charlesworth B. (1981b) Genetics of life history in *Drosophila melanogaster*. 2. Exploratory selection experiments. *Genetics* **97**, 187–196.
9.2.3

Rosenzweig M.L. (1968) The strategy of body size in mammalian carnivores. *Amer. Midl. Nat.* **80**, 299–315.
1.1

Ross N.M. & Gamboa G.J. (1981) Nestmate discrimination in social wasps (*Polistes metricus*, Hymenoptera: Vespidae). *Behav. Ecol. Sociobiol.* **9**, 163–165.
13.5

Rothstein S.I. (1979) Gene frequencies and selection for inhibiting traits, with special emphasis on the adaptiveness of territoriality. *Amer. Natur.* **113**, 317–331.
6.4

Rowley I. (1965) The life history of the superb blue wren, *Malurus cyaneus*. *Emu* **64**, 251–297.
3.3.3, 12.1, 12.3.3, 12.3.4, 12.4

Rowley I. (1968) Communal species of Australian birds. *Bonn Zool. Beitr.* **19**, 362–370.
12.3.3

Rowley I. (1976) Cooperative breeding in Australian birds. *Proc. XVI Int. Ornithol. Congr*, pp. 657–666. Canberra, Australia.
12.3.3

Rowley I. (1978) Communal activities among white-winged choughs, *Corcorax melanorhamphus*. *Ibis* **120**, 178–197.
12.1

Rowley I. (1981) The communal way of life in the splendid wren, *Malurus splendens*. *Z. Tierpsychol.* **55**, 228–267.
12.1, 12.3.4, 12.4

Royama T. (1970) Factors governing the hunting behaviour and selection of food by the great tit. (*Parus major*, L.). *J. Anim. Ecol.* **39**, 619–668.
4.2.2

Rozin P. (1976) The evolution of intelligence and access to the cognitive unconscious. *Prog. Psychobiol. Physiol. Psychol.* **6**, 245–280.
7.3.2

Rozin P. (1977) The significance of learning mechanisms in food selection: Some biology, psychology, and sociology of science. In: *Learning Mechanisms in Food Selection* (ed. L.M. Barker, M.R. Best & M. Domjan), pp. 557–583. Baylor University Press, Waco, Texas.
7.2.1

Rozin P. & Kalat J.W. (1971) Specific hungers and poison avoidance as adaptive specializations of learning. *Psychol. Rev.* **78**, 459–486.
7.1, 7.2.1, 7.3.2

Rubenstein D.I. (1977) Population density, resource patterning, and mechanisms of competition in the Everglades pygmy sunfish. PhD dissertation, Duke University. University Microfilms, Ann Arbor, MI.
2.5.2

Rubenstein D.I. (1978) On predation competition, and the advantages of group living. *Persp. Ethol.* **3**, 205–231.
5.2.1

Rubenstein D.I. (1980) On the evolution of alternative mating strategies. In: *Limits to Action* (ed. J.R. Staddon), pp. 65–100. Academic Press, New York.
2.3, 11.3

Rubenstein D.I. (1981a) Individual variation and competition in the Everglades pygmy sunfish. *J. Anim. Ecol.* **50**, 337–350.
11.2

Rubenstein D.I. (1981b) Population density, resource patterning and territoriality in the Everglades pygmy sunfish. *Anim. Behav.* **29**, 155–172.
6.2

Rubenstein D.I. (1982a) Reproductive value and behavioral strategies: coming of age in monkeys and horses. *Persp. Ethol.* **5**, 467–485.
11.5

Rubenstein D.I. (1982b) Risk, uncertainty and evolutionary strategies. In: *Current Problems in Sociobiology* (ed. King's College Sociobiology Group). Cambridge University Press, Cambridge.
10.0, 10.3, 11.6

Rudder B.C.C. (1979) The allometry of primate reproductive parameters. Unpublished PhD thesis, University of London.
1.2.4

Russell E.M. (1982) Patterns of parental care and parental investment in marsupials. *Biol. Rev.* **57**, 1423–1486.
1.2.2, 1.2.4

Rutowski R.L. (1979) The butterfly as an honest salesman. *Anim. Behav.* **27**, 1269–1270.
9.2.4

Rutowski R.L. (1980) Courtship solicitation by females of the checkered white butterfly, *Pieris protodice*. *Behav. Ecol. Sociobiol.* **7**, 113–117.
9.2.4

Ruttner F. (1977) The problem of the cape bee (*Apis mellifera capensis* Escholtz): parthenogenesis—size of population—evolution. *Apidologie* **8**, 281–294.
13.7

Ryan M.J. (1980) Female mate choice in a neotropical frog. *Science* **209**, 523–525.
10.3.3

Sacher G.A. (1959) Relationship of lifespan to brain weight and body weight in mammals. In: *C.I.B.A. Foundation Symposium on the Lifespans of Animals* (ed. G.E.W. Wolstenholme & M. O'Connor), pp. 115–133. Little, Brown & Co., Boston, Mass.
1.2.2

Sade D., Cushing K., Cushing P., Dunaif J., Figuerola A., Kaplan J., Lauer C., Rhodes D. & Schneider J. (1977) Population dynamics related to social structure on Cayo Santiago. *Yearb. Phys. Anthrop.* **20**, 253–262.
5.1.2

Sadleir R.M.F.S. (1980a) Milk yield of black tailed deer. *J. Wildl. Manag.* **44**, 472–478.
1.2.2

Sadleir R.M.F.S. (1980b) Energy and protein intake in relation to growth of suckling black tailed deer fawns. *Can. J. Zool.* **58**, 1347–1354.
1.2.2

Sadleir R.M.F.S. (in press) Reproduction of female cervids. *Biology and Management of the Cervidae.* Smithsonian, Washington.
1.2.2

Sakagami S.F. & Maeta Y. (1977) Some presumably presocial habits of Japanese *Ceratina* bees, with notes on various social types in Hymenoptera. *Insect. Soc.* **24**, 319–343.
13.3

Salthe S.N. & Mecham J.S. (1974) Reproductive and courtship patterns. In: *Physiology of the Amphibia*, Vol. 2 (ed. B. Lofts), pp. 309–521. Academic Press, New York.
10.3.2

Schaffer W.M. (1974) Selection for optimal life histories: the effects of age structure. *Ecology* **55**, 291–303.
11.5, 11.6

Schaffer W.M. & Elson P.F. (1975) The adaptive significance of variations in life history among local populations of Atlantic salmon in N. America. *Ecology* **56**, 577–590.
11.2

Schaffner J.H. (1922) Control of sexual state in *Arisaema triphyllum* and *A. dracontium. Am. J. Bot.* **9**, 72–78.
14.2.3

Schaller G.B. (1972) *The Serengeti Lion.* University of Chicago Press, Chicago.
5.1.1, 5.2.2, 5.3.1, 5.4.1, 12.1, 12.4

Schamel D. & Tracy D. (1977) Polyandry, replacement clutches, and site tenacity in the red phalarope (*Phalaropus fulicarius*) at Barrow, Alaska. *Bird-Banding* **48**, 314–324.
10.3.2

Schantz T. von (1981) Female cooperation, male competition, and dispersal in the red fox, *Vulpes vulpes. Oikos* **37**, 63–68.
12.3.3, 12.4, 12.5

Schoener T.W. (1968) Sizes of feeding territories among birds. *Ecology* **49**, 123–141.
1.1

Schulman S.R. & Chapais B. (1980). Reproductive value and rank relations among macaque sisters. *Amer. Natur.* **115**, 580–593.
11.5

Searcy W.A. (1979a) Female choice of mates: a general model for birds and its application to red-winged blackbirds (*Agelaius phoeniceus*). *Amer.*

Natur. **114**, 77–100.
10.3.1

Searcy W.A. (1979b) Male characteristics and pairing success in red-winged blackbirds. *Auk* **96**, 353–363.
10.3.1

Searcy W.A. & Yasukawa K. (1981) Does the 'sexy son' hypothesis apply to mate choice in red-winged blackbirds? *Amer. Natur.* **117**, 343–348.
10.3.1

Seger J. (1981) Kinship and covariance. *J. theor. Biol.* **91**, 191–213.
3.3.3, 3.3.4

Seger J. (1982) Conditional relatedness, recombination, and the chromosome number of insects. In: *Essays in Honor of E.E. Williams* (ed. Special Publications of the Museum of Comparative Zoology, Harvard University).
1.1

Seger J. (1983) Partial bivoltinism may cause alternating sex-ratio biases that favour eusociality. *Nature, Lond.* **301**, 59–62.
13.5

Selander R.K. (1964) Speciation in wrens of the genus *Campylorhynchus*. *Univ. Calif. Pub. Zool.* **74**, 1–224.
12.3.3

Selander R.K. (1970) Behavior and genetic variation in natural populations. *Amer. Zool.* **10**, 53–66.
5.1.1

Selander R.K. (1972) Sexual selection and dimorphism in birds. In: *Sexual Selection and the Descent of Man* (ed. B. Campbell), pp. 180–230. Heinemann, London.
9.2.2

Seligman M.E.P. (1970) On the generality of the laws of learning. *Psychol. Rev.* **77**, 406–448.
7.2.1

Selten R. (1980) A note on evolutionarily stable strategies in asymmetric animal conflicts. *J. theor. Biol.* **84**, 93–101.
2.5.2

Seyfarth R.M., Cheney D.L. & Marler P. (1980a) Vervet monkey alarm calls: semantic communication in a free-ranging primate. *Anim. Behav.* **28**, 1070–1094.
5.3.2

Seyfarth R.M, Cheney D.L. & Marler P. (1980b) Monkey responses to three different alarm calls: evidence of predator classification and semantic communication. *Science* **210**, 801–803.
15.4.2

Shalter M.D. (1978) Localisation of passerine seet and mobbing calls by goshawks and pygmy owls. *Z. Tierpsychol.* **46**, 260–267.
15.3.2

Shannon C.E. & Weaver W. (1949) *The Mathematical Theory of Communication*. University of Illinois Press, Urbana.
15.4, 15.4.1

Shellman J.S. & Gamboa G.J. (1982) Nestmate discrimination in a social wasp: the role of exposure to nest and nestmates (*Polistes fuscatus*, Hymenoptera: Vespidae). *Behav. Ecol. Sociobiol.* **11**, 51–53.
13.5

Shepher J. (1971) Mate selection among second generation kibbutz adolescents and adults: incest avoidance and negative imprinting. *Arch. sex.*

Behav. **1**, 293–307.

8.3

Sherman P.W. (1977) Nepotism and the evolution of alarm calls. *Science* **197**, 1246–1253.

15.4.2

Sherman P.W. (1979) Insect chromosome number and eusociality. *Amer. Natur.* **113**, 925–935.

1.1, 1.2.6

Sherman P.W. (1981) Reproductive competition and infanticide in Belding's ground squirrels and other animals. In: *Natural Selection and Social Behavior: Recent Research and New Theory* (ed. R.D. Alexander & D.W. Tinkle), pp. 311–331. Chiron Press, New York.

6.3.1

Sherry D.F., Krebs J.R. & Cowie R.J. (1981) Memory for the location of stored food in marsh tits. *Anim. Behav.* **24**, 1260–1266.

7.3.4

Shettleworth S.J. (1972a) Constraints on learning. *Adv. Study Behav.* **4**, 1–68.

7.2.1, 7.3.3

Shettleworth S.J. (1972b) Stimulus relevance in the control of drinking and conditioned fear responses in domestic chicks, *Gallus gallus. J. Comp. Physiol. Psychol.* **80**, 175–198.

7.2.2

Shettleworth S.J. (1972c) The role of novelty in learned aversion to unpalatable 'prey' by domestic chicks. *Anim. Behav.* **20**, 29–35.

7.2.2

Shettleworth S.J. (1983a) Memory in food-hoarding birds. *Scient. Am.* **248**(3), 102–110.

7.3.4

Shettleworth S.J. (1983b) Function and mechanism in learning. In: *Advances in Analysis of Behavior*, Vol.3 (ed. M. Leilu & C. Harzem), pp. 1–37. John Wiley & Sons, Chichester.

7.1, 7.3.1

Shettleworth S.J. & Krebs J.R. (1982) How marsh tits find their hoards: the roles of site preference and spatial memory. *J. Exp. Psychol: Anim. Behav. Processes* **8**, 354–375.

7.4.2, 7.5.2

Shields W.M. (1982) *Philopatry, Inbreeding and the Evolution of Sex*. State University of New York Press, Albany.

9.3.1

Shine R. (1978) Sexual size dimorphism and male combat in snakes. *Oecologia* **33**, 269–277.

1.1

Shine R. (1979) Sexual selection and sexual dimorphism in the amphibia. *Copeia* **2**, 297–306.

1.1

Short R.V. (1979) Sexual selection and its component parts, somatic and genital selection as illustrated by man and the great apes. *Adv. Study Behav.* **5**, 131–158.

1.1

Shy E. (1983) The relation of geographical variation in song to habitat characteristics and body size in North American tanagers (Thraupinae: *Piranga*). *Behav. Ecol. Sociobiol.* (in press).

15.3.1

Sibley G.C. (1955) Behavioral mimicry in the titmice (*Paridae*) and certain other birds. *Wilson Bull.* **67**, 128–132.
15.0

Sibly R.M. (1983) Optimal group size is unstable. *Anim. Behav.* **31**, 947–948.
5.4.3

Sibly R.M. & McFarland D.J. (1976) On the fitness of behavior sequences. *Amer. Natur.* **110**, 601–617.
4.4.1, 4.4.2

Siegfried W.R. & Underhill L.G. (1975) Flocking as an anti-predator strategy in doves. *Anim. Behav.* **23**, 504–508.
5.3.2

Sih A. (1980) Optimal behaviour: can foragers balance two conflicting demands? *Science* **210**, 1041–1043.
4.3

Silk J.B., Clark-Wheatley C.B., Rodman P. & Samuels A. (1981) Differential reproductive success and facultative adjustment of sex ratios among captive female bonnet macaques (*Macaca radiata*). *Anim. Behav.* **29**, 1106–1120.
5.1.2

Simmons M.J. & Crow J.F. (1977) Mutations affecting fitness in *Drosophila* populations. *A. Rev. Genet.* **11**, 49–78.
9.3.1

Simon C.A. (1975) The influence of food abundance on territory size in the iguanid lizard, *Scleropus jarrovi*. *Ecology* **56**, 993–998.
6.3.4

Simon H.A. (1956) Rational choice and the structure of the environment. *Psychol. Rev.* **63**, 129–138.
4.5.1

Simpson M.J.A. (1968) The display of Siamese fighting fish, *Betta splendens*. *Anim. Behav. Monogr.* **1**, 1–73.
15.4.2

Sinclair A.R.E. (1977) *The African Buffalo. A study of resource limitation of populations*. University of Chicago Press, Chicago.
15.4.2

Skutch A.F. (1957) The incubation patterns of birds. *Ibis* **99**, 69–93.
10.3.2

Slaney P.A. & Northcote T.G. (1974) Effects of prey abundance on density and territorial behavior of young rainbow trout *Salmo gairdneri* in laboratory stream channels. *J. Fish. Res. Board Can.* **31**, 1201–1209.
6.3.1

Slobodkin L.B. & Rapoport A. (1974) An optimal strategy of evolution. *Q. Rev. Biol.* **49**, 181–200.
7.1

Smeeton L. (1981) The source of males in *Myrmica rubra* L. (Hym. Formicidae). *Insect Soc.* **28**, 263–278.
13.7

Smith C.C. & Fretwell S.D. (1974) The optimal balance between size and number of offspring. *Amer. Natur.* **108**, 499–506.
4.2.1

Smith E.A. (1980) The application of optimal foraging theory to the analysis of hunter–gatherer group size. In: *Hunter–Gatherer Foraging Strategies* (ed. B. Winterhalder & E.A. Smith). University of Chicago Press, Chicago.
5.4.1, 5.4.2

Smith J.N.M. (1981) Does high fecundity reduce survival in song sparrows? *Evolution* **35**, 1142–1148.
11.5

Smith R.H. (1979) On selection for inbreeding in polygynous animals. *Heredity* **43**, 205–211.
9.3.2

Smith R.J. (1980) Rethinking allometry. *J. theor. Biol.* **87**, 97–111.
1.2.4, 1.2.5

Smith R.L. (1980) Evolution of exclusive post-copulatory parental care in the insect. *Fla. Entomol.* **63**, 65–78.
10.3.2

Smith W.J. (1977) *The Behaviour of Communicating: an ethological approach.* Harvard University Press, Cambridge, Mass.
15.4.2

Smythe N. (1970) On the existence of 'pursuit invitation' signals in mammals. *Amer. Natur.* **104**, 491–494.
15.2.3

Snelling R.R. (1981) Systematics of social Hymenoptera. In: *Social Insects* (ed. H.R. Hermann), Vol. II, pp. 369–453. Academic Press, New York.
13.3

Snyderman M. (1983) Optimal prey selection: Partial selection, delay of reinforcement, and self control. *Behav. Analysis Lett.* **3**, 131–147.
4.2.1, 7.4.3

Sokal R.R. & Rohlf F.J. (1969) *Biometry.* W.H. Freeman, San Francisco.
1.2.4, 1.2.6

Solbrig O.T., Jain S., Johnson G.B. & Raven P.H. (ed.) (1979) *Topics in Plant Population Biology.* Columbia University Press, New York.
14.1

Southwood T.R.E. (1977) Habitat, the templet for ecological strategies? *J. Anim. Ecol.* **46**, 337–365.
11.3

Southwood T.R.E. (1981) Bionomic strategies and population parameters. In: *Theoretical Ecology: Principles and Applications* (ed. R.M. May). Blackwell Scientific Publications, Oxford.
1.2.3, 11.2

Southwood T.R.E., May R.M., Hassell M.P. & Conway G.R. (1974) Ecological strategies and population parameters. *Amer. Natur.* **108**, 791–804.
11.3, 11.4

Spiess E.B. & Ehrman L. (1978) Rare male mating advantage. *Nature, Lond.* **272**, 188–189.
9.5

Spradbery J.P. (1973) *Wasps: An Account of the Biology and Natural History of Solitary and Social Wasps.* University of Washington Press, Seattle.
13.2, 13.4, 13.5

Spuhler J.N. & Jorde L.B. (1975) Primate phylogeny, ecology and social behavior. *J. Anthropol. Res.* **31**, 376–405.
1.2.5

Stacey P.B. (1979a) Habitat saturation and communal breeding in the acorn woodpecker. *Anim. Behav.* **27**, 1153–1166.
12.1, 12.3.3, 12.3.4

Stacey P.B. (1979b) Kinship, promiscuity, and communal breeding in the acorn woodpecker. *Behav. Ecol. Sociobiol.* **6**, 53–66.
12.1, 12.5

Stacey P.B. & Bock C.E. (1978) Social plasticity in the acorn woodpecker.

Science **202**, 1298–1300.
12.3.4

Staddon J.E.R. (1980) Optimality analyses of operant behavior and their relation to optimal foraging. In: *Limits to Action: The Allocation of Individual Behavior* (ed. J.E.R. Staddon), pp. 101–141. Academic Press, New York.
4.5, 7.4.1, 7.4.4

Stallcup J.A. & Woolfenden G.E. (1978) Family status and contributions to breeding by Florida scrub jays. *Anim. Behav.* **26**, 1144–1156.
12.2

Stamps J.A. & Metcalf R.A. (1978) Parent–offspring conflict. In: *Sociobiology: Beyond Nature/Nurture?* (ed. G.W. Barlow & J. Silverberg), pp. 589–618. Westview Press, Boulder, Colorado.
13.5

Starr C. (1979) Origin and evolution of insect sociality: A review of modern theory. In: *Social Insects* (ed. H.R. Hermann), pp. 35–79. Academic Press, New York.
13.7

Stearns S.C. (1976) Life-history tactics: a review of the ideas. *Q. Rev. Biol.* **51**, 3–47.
11.2

Steinmann E. (1976) Über die Nahorientierung solitärer Hymenopteren: Individuelle Markierung der Nesteingänge. *Mitt. Schweizer. Ent. Gesellsch.* **49**, 253–258.
13.5

Stenger J. (1958) Food habits and available food of ovenbirds in relation to territory size. *Auk* **75**, 335–346.
6.3.4

Stephen W.P., Bohart G.E. & Torchio P.F. (1969) The biology and external morphology of bees. Agricultural Experiment Station, Oregon State University, Corvalis.
13.5

Stephens D.W. (1981) The logic of risk-sensitive foraging preferences. *Anim. Behav.* **29**, 628–629.
4.2.3

Stephens D.W. (1982) Stochasticity in foraging theory: risk and information. Unpublished DPhil thesis, Oxford University.
4.2.3

Stephens D.W. & Charnov E.L. (1982) Optimal foraging: some simple stochastic models. *Behav. Ecol. Sociobiol.* **10**, 251–263.
4.2.3

Stiles F.G. & Wolf L.L. (1979) Ecology and evolution of lek mating in the long-tailed hermit hummingbird. *AOU Ornithol. Monogr.* **27**, 1–78.
10.3.3

Stokes A.W. (1962) Agonistic behaviour among blue tits at a winter feeding station. *Behaviour* **19**, 118–138.
15.4.2

Stout J.F. & Brass M.E. (1969) Aggressive communication by *Larus glaucescens*. II: Visual communication. *Behaviour* **34**, 42–54.
15.2.1

Strassmann J.E. (1981a) Evolutionary implications of early male and satellite nest production in *Polistes exclamans* colony cycles. *Behav. Ecol. Sociobiol.* **8**, 55–64.
13.5

Strassmann J.E. (1981b) Kin selection and satellite nests in *Polistes exclamans*.

In: *Natural Selection and Social Behavior: Recent Research and New Theory* (ed. R.D. Alexander & D.W. Tinkle), pp. 45–58. Chiron Press, New York. 13.6, 13.7

Strassmann J.E. (1981c) Parasitoids, predators, and group size in the paper wasp, *Polistes exclamans. Ecology* **62**, 1225–1233. 13.5

Strassmann J.E. (1981d) Wasp reproduction and kin selection: reproductive competition and dominance hierarchies among *Polistes annularis* foundresses. *Fla. Entomol.* **64**, 74–88. 13.7

Struhsaker T.T. (1967) Ecology of vervet monkeys (*Cercopithecus aethiops*) in the Masai Amboseli game reserve, Kenya. *Ecology* **48**, 891–904. 1.3

Struhsaker T.T. (1975) *The Red Colobus Monkey.* University of Chicago Press, Chicago. 1.1

Struhsaker T.T. & Leland L. (1979) Socioecology of five sympatric monkey species in the Kibale Forest. *Adv. Study Behav.* **9**, 159–227. 1.1

Stubbs M. (1977) Density dependence in the life-cycles of animals and its importance in *K*- and *r*-strategies. *J. Anim. Ecol.* **46**, 677–688. 11.3

Swingland I.R. & Greenwood P.J. (ed.) (1983) *The Ecology of Animal Movement.* Oxford University Press, Oxford. 11.4

Tauber C.A. & Tauber M.J. (1977) A genetic model for sympatric speciation through habitat diversification and seasonal isolation. *Nature, Lond.* **268**, 702–705. 9.6

Taylor H.M., Gourley R.S., Lawrence C.E. & Kaplan R.S. (1974) Natural selection of life history attributes: an analytical approach. *Theor. Popul Biol.* **5**, 104–122. 11.5

Taylor R.J. (1979) The value of clumping to prey when detectability increases with group size. *Amer. Natur.* **113**, 299–301. 5.3.2

Tenaza R. (1971) Behavior and nesting success relative to nest location in Adelie penguins (*Pygoscelis adeliae*). *Condor* **73**, 81–92. 5.3.2

Thompson W.D., Vertinsky I. & Krebs J.R. (1974) The survival value of flocking in birds: a simulation model. *J. Anim. Ecol.* **43**, 785–820. 5.3.1

Thorne B. (1982a) Multiple primary queens in termites: phyletic distribution, ecological context, and a comparison to polygyny in Hymenoptera. In: *The Biology of Social Insects: Proceedings of the Ninth Congress of the International Union for the Study of Social Insects, Boulder, Colorado, August 1982* (ed. M.D. Breed, C.D. Michener & H.E. Evans), pp. 206–211. Westview Press, Boulder, Colorado. 13.2, 13.4

Thorne B.L. (1982b) Polygyny in termites: multiple primary queens in colonies of *Nasutitermes corniger* (Motschulsky) (Isoptera: Termitidae). *Insect Soc.* **29**, 102–117. 13.2

Thornhill R. (1979a) Adaptive female-mimicking behavior in a scorpionfly. *Science* **205**, 412–414.
2.3, 2.8, 9.2.5

Thornhill R. (1979b) Male and female sexual selection and the evolution of mating strategies in insects. In: *Sexual Selection and Reproductive Competition* (ed. M.S. Blum & N.H. Blum), pp. 81–121. Academic Press, New York.
9.2.5

Thornhill R. (1980) Competitive, charming males and choosy females: was Darwin correct? *Fla. Entomol.* **63**, 5–30.
9.2.2

Thornhill R. (1981) *Panorpa* (Mecoptera: Panorpidae) scorpionflies: systems for understanding resource-defence polygyny and alternative male reproductive efforts. *A. Rev. Ecol. Syst.* **12**, 355–386.
9.2.2, 9.2.4, 9.2.5

Thornhill R. (1983) Alternative hypotheses for traits believed to have evolved by sperm competition. In: *Sperm Competition and the Evolution of Mating Systems* (ed. R.L. Smith). Academic Press, New York.
2.8

Tinbergen J. (1981) Foraging decisions in starlings (*Sturnus vulgaris*). *Ardea* **69**, 1–67.
4.2.1, 4.2.2

Tinbergen N. (1951) *The Study of Instinct*. Clarendon Press, Oxford.
5.3.2

Tinbergen N. (1952) Derived activities: their causation, biological significance, origin and emancipation during evolution. *Q. Rev. Biol.* **27**, 1–32.
15.1, 15.2.2

Tinbergen N. (1953) *The Herring Gull's World*. Collins, London.
7.3.3

Tinbergen N. (1963) On aims and methods of ethology. *Z. Tierpsychol.* **20**, 410–433.
1.1, 1.3

Tinbergen N. (1964) The evolution of signalling devices. In: *Social Behavior and Organisation Among Vertebrates* (ed. W. Etkin), pp. 206–230. University of Chicago Press, Chicago.
15.2.2

Tinbergen N., Kruuk H. & Paulette M. (1962) Eggshell removal by the black-headed gull (*Larus r. ridibundus*) II. The effects of experience on the response to colour. *Bird Study* **9**, 123–131.
1.1

Tinbergen N., Brockhuysen G.J., Feekes F., Houghton J.C.W., Kruuk H. & Szulc E. (1963) Egg shell removal by the black headed gull, *Larus ridibundus* L.: a behaviour component of camouflage. *Behaviour* **19**, 74–117.
1.1

Tinbergen N., Impekoven M. & Franck D. (1967) An experiment on spacing out as a defence against predators. *Behaviour* **28**, 307–321.
5.3.2

Tomlinson J. (1966) The advantage of hermaphroditism and parthenogenesis. *J. theor. Biol.* **11**, 54–58.
8.3

Toro M.A. & Charlesworth B. (1982) An attempt to detect genetic variation in sex ratio in *Drosophila melanogaster*. *Heredity* **49**, 199–209.
8.2

Townsend C.R. & Calow P. (ed.) (1981) *Physiological Ecology*. Blackwell Scien-

tific Publications, Oxford.
14.1

Trail P.W. (1980) Ecological correlates of social organization in a communally breeding bird, the acorn woodpecker, *Melanerpes formicivorous*. *Behav. Ecol. Sociobiol*. **7**, 83–92.
12.3.3

Trail P.W., Strahl S.D. & Brown J.L. (1981) Infanticide in relation to individual and flock histories in a communally breeding bird, the Mexican jay (*Aphelocoma ultramarina*). *Amer. Natur*. **118**, 72–82.
12.5

Treisman M. (1975). Predation and the evolution of gregariousness. I. Models for concealment and evasion. *Anim. Behav*. **23**, 779–800.
5.3.2

Trivers R.L. (1971) The evolution of reciprocal altruism. *Q. Rev. Biol*. **46**, 35–57.
11.4, 12.4, 13.7

Trivers R.L. (1972) Parental investment and sexual selection. In: *Sexual Selection and the Descent of Man* (ed. B. Campbell). Aldine Press, Chicago.
2.6.2, 8.4, 9.2.1, 9.2.2, 10.3.2, 10.3.3

Trivers R.L. (1974) Parent–offspring conflict. *Amer. Zool*. **14**, 249–265.
2.6.2, 11.5, 12.5, 15.2.4, 15.3.3

Trivers R.L. & Hare H. (1976) Haplodiploidy and the evolution of the social insects. *Science* **191**, 249–263.
13.5

Trivers R.L. & Willard D.E. (1973) Natural selection of parental ability to vary the sex ratio of offspring. *Science* **179**, 90–92.
13.5

Trune D.R. & Slobodchikoff C.N. (1976) Social effects of roosting on the metabolism of the pallid bat (*Antrozous pallidus*). *J. Mammal*. **57**, 656–663.
5.3.1

Tschinkel W.R. & Howard D.F. (1978) Queen replacement in orphaned colonies of the fire ant, *Solenopsis invicta*. *Behav. Ecol. Sociobiol*. **3**, 297–310.
13.7

Tullock G. (1979) On the adaptive significance of territoriality: comment. *Amer. Natur*. **113**, 772–775.
6.4

Turillazzi S. & Pardi L. (1977) Body size and hierarchy in polygynic nests of *Polistes gallicus* (L) (Hymenoptera: Vespidae). *Monit. Zool. Ital*. **11**, 101–112.
13.7

Turner F.B., Jeinrich R.I. & Weintraub J.D. (1969) Home ranges and body size of lizards. *Ecology* **50**, 1076–1081.
1.1

Turner J.R. (1967) Why does the genome not congeal? *Evolution* **21**, 645–656.
8.1.2

Tuttle M. & Ryan M.J. (1981) Bat predation and the evolution of frog vocalizations in the neotropics. *Science* **214**, 677–678.
10.3.3

d'Udine B. & Partridge L. (1981) Olfactory preferences of inbred mice (*Mus musculus*) for their own strain and for siblings: effects of strain, sex and

cross-fostering. *Behaviour* **78**, 314–324.
9.4

Van Valen L. (1973) A new evolutionary law. *Evol. Theory* **1**, 1–30.
8.1.2

Vasek F. (1966) The distribution and taxonomy of three western junipers. *Brittonia* **18**, 350–372.
14.2.1

Vehrencamp S.L. (1977) Relative fecundity and parental effort in communally nesting anis, (*Crotophaga sulcirostris*). *Science* **197**, 403–405.
12.1, 12.1.2

Vehrencamp S.L. (1978) The adaptive significance of communal nesting in groove-billed anis (*Crotophaga sulcirostris*). *Behav. Ecol. Sociobiol.* **4**, 1–33.
5.3.3, 12.1, 12.1.2, 12.5

Vehrencamp S.L. (1979) The roles of individual, kin, and group selection in the evolution of sociality. In: *Handbook of Behavioral Neurobiology*, Vol. 3 (ed. P. Marler & J.G. Vandenbergh), pp. 351–394. Plenum Press, New York.
12.4, 12.5

Vehrencamp S.L. (1982) Body temperatures of incubating versus non-incubating roadrunners. *Condor* **84**, 203–207.
10.3.2

Vehrencamp S.L. (1983) A model for the evolution of despotic versus egalitarian societies. *Anim. Behav.* **31**, 667–682.
12.5

Velthuis H.H.W. (1976) Environmental, genetic and endocrine influences in stingless bee caste determination. In: *Phase and Caste Determination in Insects* (ed. M. Lüscher), pp. 35–53. Pergamon Press, Oxford.
13.2

Velthuis H.H.W. (1977) Egg laying, aggression and dominance in bees. *Proc. XV Int. Congr. Entomol.* 1977, 436–449.
13.7

Verner J. (1964) Evolution of polygamy in the long-billed marsh wren. *Evolution* **18**, 252–261.
6.3.1, 10.0, 10.3.1, 11.3

Verner J. (1977) On the adaptive significance of territoriality. *Amer. Natur.* **111**, 769–775.
6.4

Verner J. & Engelsen G.H. (1970) Territories, multiple nest building, and polygyny in the long-billed marsh wren. *Auk* **87**, 557–567.
10.3.1

Verner J. & Willson M.F. (1966) The influence of habitats on mating systems of North American passerine birds. *Ecology* **47**, 143–147.
1.1, 10.0, 10.3.1

Verrell P.A. (1982a) Male newts prefer large females as mates. *Anim. Behav.* **30**, 1254–1255.
9.2.1

Verrell P.A. (1982b) The sexual behaviour of the red-spotted newt, *Notophthalmus viridescens* (Amphibia: Urodela: Salamandridae). *Anim. Behav.* **30**, 1224–1236.
9.2.5

Verrell P.A. (1983) The influence of ambient sex ratio and intermale competition on the sexual behaviour of the red-spotted newt, *Notophthalmus*

viridescens. Behav. Ecol. Sociobiol., **13**, 307–313.
9.2.5

Vine I. (1973) Detection of prey flocks by predators. *J. theor. Biol.* **40**, 207–210.
5.3.2

Vos G.J. de (1983) Social behavior of black grouse; an observational and experimental field study. *Ardea* **71**, 1–103.
10.3.3

Waage J.K. (1979a) Dual function of the damsel fly penis: sperm removal and transfer. *Science* **203**, 916–918.
9.2.3

Waage J.K. (1979b) Foraging for patchily-distributed hosts by the parasitoid *Nemeritis canescens. J. Anim. Ecol.* **48**, 353–371.
4.5, 7.4.3

Waage J.K. (1982) Sib-mating and sex-ratio strategies in scelionid wasps. *Ecol. Entomol.* **7**, 103–112.
9.3.2

Waddington C.H. (1977) *Tools for Thought.* Jonathan Cape, London.
4.6

Waddington K.D. (1982) Optimal diet theory: sequential simultaneous encounter models. *Oikos* **39**, 278–280.
4.2.1

Waddington K.D. & Heinrich B. (1979) The foraging movements of bumblebees on vertical 'inflorescences': an experimental analysis. *J. Comp. Physiol.* **134**, 113–117.
4.5

Waddington K.D. & Holden R. (1979) Optimal foraging: on flower selection by bees. *Amer. Natur.* **114**, 179–196.
4.2.1

Wade M.J. (1982) The effect of multipie inseminations on the evolution of social behaviors in diploid and haplo-dipoid organisms. *J. theor. Biol.* **95**, 351–368.
13.5

Wade M.J. & Arnold S.J. (1980) The intensity of sexual selection in relation to male sexual behaviour, female choice and sperm precedence. *Anim. Behav.* **28**, 446–461.
9.2.2, 10.1

Waldbauer G.P. & Sheldon J.K. (1971) Phenological relationships of some aculeate hymenoptera, their dipteran mimics, and insectivorous birds. *Evolution* **25**, 371–381.
7.2.2

Wall S.B. vander (1982) An experimental analysis of cache recovery in Clark's nutcracker. *Anim. Behav.* **30**, 84–94.
7.3.4

Wall S.B. vander & Balda R.P. (1981) Ecology and evolution of food-storage behavior in conifer-seed-caching corvids. *Z. Tierpsychol.* **56**, 217–242.
7.3.4

Waller D.M. & Green D. (1981) Implications of sex for the analysis of life histories. *Amer. Natur.* **117**, 810–813.
11.2

Ward P. (1965) Feeding ecology of the black-faced dioch *Quelea quelea* in Nigeria. *Ibis* **107**, 173–214.
5.2.1, 5.3.1

Ward P. & Zahavi A. (1973) The importance of certain assemblages of birds as 'information centres' for food finding. *Ibis* **115**, 517–534.
5.3.1

Warner R.R., Robertson D.R. & Leigh E.G. (1975) Sex change and sexual selection. *Science* **190**, 633–638.
11.3

Waser P.M. (1976) *Cercocebus albigena*: site attachment, avoidance and inter-group spacing. *Amer. Natur.* **110**, 911–935.
1.3

Weatherhead P.J. & Robertson R.J. (1977a) Harem size, territory quality, and reproductive success in the redwinged blackbird (*Agelaius phoeniceus*). *Can. J. Zool.* **55**, 1261–1267.
10.3.1

Weatherhead P.J. & Robertson R.J. (1977b) Male behaviour and female recruitment in the red-winged blackbird. *Wilson Bull.* **89**, 583–592.
10.3.1

Weatherhead P.J. & Robertson R.J. (1979) Offspring quality and the poly-gyny threshold: 'the sexy son hypothesis'. *Amer. Natur.* **113**, 201–208.
10.3.1

Weatherhead P.J. & Robertson R.H. (1981) In defense of the 'sexy son' hypothesis. *Amer. Natur.* **117**, 349–356.
10.3.1

Werner E.E. & Hall D.J. (1974) Optimal foraging and the size selection of prey by the bluegill sunfish *Lepomis macrochirus*. *Ecology* **55**, 1042–1052.
4.2.1

Werren J.H. (1980) Sex ratio adaptation to local mate competition in a parasitic wasp. *Science* **208**, 1157–1159.
8.2

Werren J.H. & Charnov E.L. (1978) Facultative sex ratios and population dynamics. *Nature, Lond.* **272**, 349–350.
13.5

West M.J. (1967) Foundress associations in polistine wasps: dominance hierarchies and the evolution of social behavior. *Science* **157**, 1584–1585.
13.7

West Eberhard M.J. (1969) The social biology of polistine wasps. *Misc. Publ. Mus. Zool. Univ. Michigan* **140**, 1–101.
13.7

West Eberhard M.J. (1975) The evolution of social behaviour by kin-selection. *Q. Rev. Biol.* **50**, 1–33.
12.4, 13.6

West Eberhard M.J. (1977) The establishment of reproductive dominance in social wasp colonies. *Proc. Eighth Int. Congress Intl. Union Study of Social Insects*, pp. 223–227. Centre for Agricultural Publ. and Documentation, Wageningen, Netherlands.
13.6, 13.7

West-Eberhard M.J. (1978a) Polygyny and the evolution of social behavior in wasps. *J. Kans. ent. Soc.* **51**, 832–856.
13.5, 13.7

West-Eberhard M.J. (1978b) Temporary queens in *Metapolybia* wasps: non-reproductive helpers without altruism? *Science* **200**, 441–443.
13.6, 13.7

West-Eberhard M.J. (1979) Sexual selection, social competition, and evolution. *Proc. Am. phil. Soc.* **123**, 222–234.
9.2.2, 12.3.3

West-Eberhard M.J. (1981) Intragroup selection and the evolution of insect societies. In: *Natural Selection and Social Behavior: Recent Research and New Theory* (ed. R.D. Alexander & D.W. Tinkle), pp. 3–17. Chiron Press, New York.
13.5, 13.7

West-Eberhard M.J. (1982) Introduction to the evolution and ontogeny of eusociality symposium. In: *The Biology of Social Insects: Proceedings of the Ninth Congress of the International Union for the Study of Social Insects, Boulder, Colorado, August 1982* (ed. M.D. Breed, C.D. Michener & H.E. Evans), pp. 185–186. Westview Press, Boulder, Colorado.
13.7

Western D. (1979) Size, life history and ecology in mammals. *Afr. J. Ecol.* **17**, 185–205.
1.1, 1.2.3, 11.3

Wheeler W.M. (1928) *The Social Insects: Their Origin and Evolution.* Kegan, Paul, Trench, Trubner and Co., Ltd, London.
13.4

Whitham T.G. (1980) The theory of habitat selection examined and extended using *Pemphigus* aphids. *Amer. Natur.* **115**, 449–466.
5.4.3

Wickler W. (1968) *Mimicry in Plants and Animals.* McGraw-Hill, New York.
15.2.1

Wiens J.A. (1976) Population responses to patchy environments. *A. Rev. Ecol. Syst.* **7**, 81–120.
5.2.1

Wiley R.H. (1973) Territoriality and non-random mating in the sage grouse *Centrocercus urophasianus. Anim. Behav. Monogr.* **6**, 87–169.
9.2.4, 10.3.3

Wiley R.H. (1974) Evolution of social organization and life history patterns among grouse (*Aves: Tetraonidae*). *Q. Rev. Biol.* **49**, 201–227.
10.3.3

Wiley R.H. (1983) The evolution of communication: Information and manipulation. In: *Communication* (*Animal Behaviour, Vol. 2*) (ed. T.R. Halliday & P.J.B. Slater), pp. 156–189. Blackwell Scientific Publications, Oxford.
15.3.1, 15.3.3, 15.4, 15.4.1

Wiley R.H. & K.N. Rabenold (1981) Kin selection and delayed reciprocity across generations in the evolution of cooperative breeding by striped-backed wrens. Paper presented at the XVII Int. Eth. Congr., Oxford.
12.3.3, 12.4

Williams G.C. (1966) *Adaptation and Natural Selection.* Princeton University Press, Princeton, N.J.
1.1, 3.4.1, 4.6, 8.1, 10.3.3, 15.2.4

Williams G.C. (1975) *Sex and Evolution.* Princeton University Press, Princeton, N.J.
8.1, 8.1.2

Willis E.O. (1972) Do birds flock in Hawaii, a land without predators? *Cal. Birds* **3**, 1–8.
5.2.2

Willson M.F. (1966) The breeding ecology of the yellow-headed blackbird. *Ecol. Monogr.* **36**, 51–77.
10.0, 10.3.1

Willson M.F. (1979) Sexual selection in plants. *Amer. Natur.* **113**, 777–790.
14.3.2

Willson M.F. (1981) On the evolution of complex life cycles in plants: a review and an ecological perspective. *Ann. Mo. Bot. Garden.* **68**, 275–300.
14.1

Willson M.F. & Burley N. (1983) *Mate Choice in Plants: Tactics, Mechanisms, and Consequences.* Princeton University Press, Princeton, N.J.
14.4

Wilson D.S. (1975) A theory of group selection. *Proc. natl Acad. Sci., USA* **72**, 143–146.
3.4.3

Wilson D.S. (1977) Structured demes and the evolution of group-advantageous traits. *Amer. Natur.* **111**, 157–185.
12.3.2

Wilson D.S. (1980) *The Natural Selection of Populations and Communities.* Benjamin/Cummings, Menlo Park, California.
3.4.3

Wilson D.S. & Colwell R.K. (1981) Evolution of sex ratio in structured demes. *Evolution* **35**, 882–897.
3.4.3

Wilson E.O. (1953) The origin and evolution of polymorphism in ants. *Q. Rev. Biol.* **28**, 136–156.
13.2

Wilson E.O. (1962) Chemical communication among workers of the fire ant *Solenopsis saevissima* (Fr. Smith): 1, the organization of mass-foraging: 2, an information analysis of the odour trail: 3, the experimental induction of social responses. *Anim. Behav.* **10**, 134–164.
15.4, 15.4.1

Wilson E.O. (1971) *The Insect Societies.* Belknap Press, Cambridge, Mass.
13.2, 13.3, 13.4, 13.7, 15.3.2

Wilson E.O. (1975) *Sociobiology, the Modern Synthesis.* Harvard University Press, Cambridge, Mass.
1.3, 3.3.2, 5.1, 5.3, 5.4, 5.4.2

Wilson K.A. (1960) The genera of the Arales in the southeastern United States. *Jour. Arnold Arb.* **41**, 47–71.
14.2.3

Wilson M.E., Gordon T.P. & Bernstein I.S. (1978) Timing of births and reproductive success in rhesus monkey social groups. *J. med. Primat.* **7**, 202–212.
5.1.2

Winston M.L. & Michener C.D. (1977) Dual origin of highly social behavior among bees. *Proc. natl Acad. Sci., USA* **74**, 1135–1137.
13.3

Wirtz P. (1978) The behaviour of the Mediterranean *Tripterygion* species (Pisces, Blennioidei). *Z. Tierpsychol.* **48**, 142–174.
6.5.1

Wirtz P. (1981) Territorial defence and territory take-over by satellite males in the waterbuck, *Kobus ellipsiprymnus* (Bovidae). *Behav. Ecol. Sociobiol.* **8**, 161–162.
6.5.1

Wirtz P. (1982) Territory holders, satellite males and bachelor males in a high density population of waterbuck (*Kobus ellipsiprymnus*) and their associations with conspecifics. *Z. Tierpsychol.* **58**, 277–300.
6.5.1

Wittenberger J.F. (1978a) The breeding biology of an isolated bobolink

population in Oregon. *Condor* **80**, 355–371.
10.3.1

Wittenberger J.F. (1979) The evolution of mating systems in birds and mammals. In: *Handbook of Behavioral Neurobiology* (ed. P. Marler & J. Vandenbergh), Vol 3 (Social Behavior and Communication), pp. 271–349. Plenum Press, New York.
10.3.1

Wittenberger J.F. (1980) Group size and polygamy in social mammals. *Amer. Natur.* **115**, 197–222.
5.3

Wittenberger J.F. (1981a) Time; a hidden dimension in the polygyny threshold model. *Amer. Natur.* **118**, 803–822.
10.3.1

Wittenberger J.F. (1981b) Male quality and polygyny: the 'sexy son' hypothesis revisited. *Amer. Natur.* **117**, 329–342.
10.3.1

Wolf L.L. (1975) 'Prostitution' behavior in a tropical hummingbird. *Condor* **77**, 140–144.
10.3.1

Wolf L.L. (1978) Aggressive social organisation in nectarivorous birds. *Amer. Zool.* **18**, 765–778.
6.3.1

Wolf L.L. & Stiles F.G. (1970) Evolution of pair cooperation in a tropical hummingbird. *Evolution* **24**, 759–773.
10.3.1

Woolfenden G.E. (1973) Nesting and survival in a population of Florida scrub jays. *Living Bird* **12**, 25–49.
12.1.1

Woolfenden G.E. (1975). Florida scrub jay helpers at the nest. *Auk* **92**, 1–15.
3.3.3, 12.1, 12.1.1, 12.3.3

Woolfenden G.E. (1978) Growth and survival of young Florida scrub jays. *Wilson Bull.* **90**(1), 1–58.
12.2

Woolfenden G.E. (1981) Selfish behavior by Florida scrub jay helpers. In: *Natural Selection and Social Behavior: Recent Research and New Theory* (ed. R.D. Alexander & D.W. Tinkle), pp. 257–260. Chiron Press, New York.
12.2

Woolfenden G.E. & Fitzpatrick J.W. (1978) The inheritance of territory in group-breeding birds. *BioScience* **28**, 104–108.
5.3.3, 11.4, 12.1, 12.1.1, 12.3.2, 12.3.3

Woolfenden G.E. & Fitzpatrick J.W. (in press) The Florida scrub jay: Demographic attributes of a cooperative breeder.
12.1, 12.1.1

Wrangham R.W. (1981) An ecological model of female-bonded primate groups. *Behaviour* **75**, 262–300.
1.1, 5.1.2

Wrangham R.W. (1982) Mutualism, kinship and social evolution. In: *Current Problems in Sociobiology* (ed. King's College Sociobiology Group), pp. 269–289. Cambridge University Press, Cambridge.
13.7

Wu H.M.H., Holmes W.G., Medina S.R. & Sackett G.P. (1980) Kin preference in infant *Macaca nemestrina*. *Nature, Lond.* **285**, 225–227.
9.3.1

Wynne-Edwards V.C. (1962) *Animal Dispersion in Relation to Social Behaviour*. Oliver & Boyd, Edinburgh and London.
3.4.2

Wynne-Edwards V.C. (1978) Intrinsic population control: an introduction. In: *Population Control by Social Behaviour* (ed. F.J. Ebling & D.M. Stoddart), pp. 1–22. Institute of Biology, London.
3.4.2

Yamazaki K., Boyse E.A., Mike V., Thaler H.T., Mathieson B.J., Abbott J., Boyse J., Zayas Z. & Thomas L. (1976) Control of mating preferences in mice by genes in the major histocompatibility complex. *J. exp. Med.* **144**, 1324–1335.
9.4

Yamazaki K., Yamaguchi M., Andrews P.W., Peake B. & Boyse E.A. (1978) Mating preferences of F2 segregants of crosses between MHC-congenic mouse strains. *Immunogenetics* **6**, 253–259.
9.4

Yamazaki K., Beauchamp G.K., Bard J., Thomas L. & Boyse E.A. (1982) Chemosensory recognition of phenotypes determined by the T1a and H-2K regions of chromosome 17 of the mouse. *Proc. natl Acad. Sci., USA* **79**, 7828–7831.
9.4

Yampolsky E. & Yampolsky H. (1922) Distribution of sex forms in phanerogamic flora. *Bibl. Genet.* **3**, 1–62.
14.3.1

Yanai U. & McClearn G.E. (1972) Assortative mating in mice and the incest taboo. *Nature, Lond.* **238**, 281–282.
9.4

Yanai U. & McClearn G.E. (1973) Assortative mating in mice. II. Strain differences in female mating preference, and the question of possible sexual selection. *Behav. Genet.* **3**, 65–74.
9.4

Yasukawa K. (1979) Territory establishment in red-winged blackbirds: importance of aggressive behavior and experience. *Condor* **81**, 258–264.
10.3.1

Ydenberg R.C. (1982) Territorial vigilance and foraging, a study of tradeoffs. Unpublished DPhil thesis, Oxford University.
4.4.2

Yeaton R.I. & Cody M.L. (1974) Competitive release in island song sparrow populations. *Theor. Popul. Biol.* **5**, 42–58.
1.3

Zahavi A. (1971) The social behaviour of the white wagtail *Motacilla alba alba* wintering in Israel. *Ibis* **113**, 203–211.
5.2.1, 6.2

Zahavi A. (1974) Communal nesting by the Arabian babbler: A case of individual selection. *Ibis* **116**, 84–87.
12.1, 12.3.3, 12.4, 12.5

Zahavi A. (1975) Mate selection—a selection for a handicap. *J. theor. Biol.* **53**, 205–214.
9.2.4, 10.3.3

Zahavi A. (1976) Cooperative nesting in Eurasian birds. *Proc. XVI Int. Ornith. Congr.*, pp. 685–693. Canberra, Australia.
12.5

Zahavi A. (1977a) The cost of honesty (further remarks on the handicap principle). *J. theor. Biol.* **67**, 603–605.

9.2.4

Zahavi A. (1977b) Reliability in communication systems and the evolution of altruism. In: *Evolutionary Ecology* (ed. B. Stonehouse & C.M. Perrins), pp. 253–259. Macmillan, London.

15.4.2

Zahavi A. (1979) Ritualisation and the evolution of movement signals. *Behaviour* **72**, 77–81.

15.4.2

Zimen E. (1975) Social dynamics of the wolf pack. In: *The Wild Canids* (ed. M.W. Fox), pp. 336–362. Van Nostrand Reinhold Publications, New York.

12.5

Zimen E. (1976) On the regulation of pack size in wolves. *Z. Tierpsychol.* **40**, 300–341.

12.1, 12.5

Zimen E. (1981) *The Wolf. Its place in the natural world.* Souvenir Press, London.

12.5

Zimmerman J.L. (1966) Polygyny in the dickcissel. *Auk* **83**, 534–546.

10.3.1

Subject Index

Organism Index